Reproduction and
Human Welfare

Reproduction and
Human Welfare:
A Challenge to Research

A Review of the Reproductive
Sciences and Contraceptive
Development

Roy O. Greep
Marjorie A. Koblinsky
Frederick S. Jaffe

Sponsored by the Ford Foundation

The MIT Press
Cambridge, Massachusetts, and London, England

Copyright © 1976 by
The Ford Foundation

All rights reserved. No part of this book may be reproduced in any form or by any means, electronic or mechanical, including photocopying, recording, or by any information storage and retrieval system, without permission in writing from the publisher.

This book was set in IBM Composer Century by Techdata Associates, printed on Nashoba Smooth Hi White and bound in Columbia Milbank Cloth by The Colonial Press Inc. in the United States of America.

Library of Congress Cataloging in Publication Data

Greep, Roy Orval, 1905-
  Reproduction and Human Welfare.

  Includes index.
  1. Conception—Prevention—Research. 2. Human reproduction—Research. 3. Contraceptives—Research.
I. Koblinsky, Marjorie A., joint author. II. Jaffe, Frederick S., joint author. III. Ford Foundation. IV. Title. [DNLM: 1. Reproduction. 2. Contraception. 3. Family planning. WQ205 G816r]
RC136.G68      613.9'43      76-47792
ISBN 0-262-07067-7

# Contents

List of Participants
ix

Foreword
by McGeorge Bundy
xvii

Preface
by Roy O. Greep
xix

Summary of Findings and Recommendations
1

I
The Challenge to Reproductive Research
37

1. The Challenge to Reproductive Research
39

2. The Uses and Limits of Current Contraceptive Technology
53

3. The Human Requirements of an Adequate Technology for Fertility Regulation
68

II
Major Advances and New Opportunities
77

Introduction to Part II
79

4. The Female Reproductive System
81

5. The Male Reproductive System
165

6. New Methods for Fertility Regulation: Status Report
278

## III Institutional Setting and Constraints
297

7. The Institutional Base for Research and Training in the Reproductive Sciences
299

8. Contraceptive Development in Industry and the Public Sector
326

9. The New Climate for Research and Development
347

## IV Financing: Past, Present, and Future
365

10. The Historical Background
367

11. The Decade of Growth: 1965-1974
393

12. Toward an Adequate Worldwide Research Effort
427

## Appendixes
439

A. Bibliography of Reports
441

B. Human Reproduction and Prospective Means of Fertility Control
446

C. Country Funding Data
481

D. Agency Funding Data
535

E. Estimates of Funding Needs for Adequate Exploitation of Existing Knowledge
552

List of Tables
585

List of Figures
591

Contents of Companion Volume,
*Frontiers in Reproduction and
Fertility Control: A Review of
the Reproductive Sciences and
Fertility Control*
595

Name Index
597

Subject Index
605

# List of Participants

## Review of Reproductive Sciences and Contraceptive Development

**Project Director**
Roy O. Greep
Harvard Medical School

**Assistant Project Director**
Marjorie A. Koblinsky
Ford Foundation

**Administrative Assistant**
Jacqueline Ans
Rockefeller Foundation

**Staff Support**
Joyann Brody
Rockefeller Foundation

Ardell Wilbur
Harvard Medical School

**Consultants**
Egon R. Diczfalusy
Karolinska Institute, Sweden

Don W. Fawcett
Harvard Medical School

Frederick S. Jaffe
Alan Guttmacher Institute

Kenneth J. Ryan
Harvard Medical School

Sheldon J. Segal
Population Council

**Steering Committee**
George F. Brown
International Development Research Centre, Canada

Elizabeth B. Connell
Rockefeller Foundation

Egon R. Diczfalusy
Karolinska Institute, Sweden

Don W. Fawcett
Harvard Medical School

John Gill
International Development Research Centre, Canada

Roy O. Greep
Harvard Medical School

Oscar Harkavy
Ford Foundation

Frederick S. Jaffe
Alan Guttmacher Institute

Marjorie A. Koblinsky
Ford Foundation

Seymour Lieberman
College of Physicians and Surgeons
Columbia University

Richard T. Mahoney
Ford Foundation

Kenneth J. Ryan
Harvard Medical School

Sheldon J. Segal
Population Council

Anna L. Southam
Ford Foundation

**Advisory Committee**
B. Kwaku Adadevoh
University of Ibadan, Nigeria

# List of Participants

Allan C. Barnes
Rockefeller Foundation

Robert W. Berliner
Yale University School of Medicine

Carlos Beyer
Instituto Mexicano de Seguro
Social, Mexico

George F. Brown
International Development Research
Centre, Canada

Philip A. Corfman
Center for Population Research
National Institute for Child Health
and Human Development

Heinz Gibian
Schering AG, West Germany

Louis M. Hellman
Department of Health, Education and
Welfare

Bryan Hudson
University of Melbourne, Australia

Seymour Lieberman
College of Physicians and Surgeons
Columbia University

Hans R. Lindner
Weizman Institute of Science,
Israel

Yoshimasa Nishikawa
Kyoto University, Japan

Gerhard A. Overbeek
Organon Scientific Group,
Netherlands

Gordon W. Perkin
Ford Foundation, Brazil

M. R. N. Prasad
University of Delhi, India

Neena B. Schwartz
Northwestern University

Rodney P. Shearman
University of Sydney, Australia

Roger V. Short
MRC Unit of Reproductive Biology
Scotland

Anna L. Southam
Ford Foundation

J. Joseph Speidel
Agency for International Development

David I. Weisblat
The Upjohn Company

**Contributors**
B. Kwaku Adadevoh
University of Ibadan, Nigeria

Yoon Ok Ahn
Seoul National University, Korea

Everett Anderson
Harvard Medical School

Ruben Apelo
UP-PGH Medical Center, Philippines

Anna Bing Arnold
Bing Fund

C. R. Austin
Physiological Laboratory, England

## List of Participants

Om P. Bahl
State University of New York

Janice M. Bahr
University of Illinois

D. T. Baird
MRC Unit of Reproductive Physiology
Scotland

R. M. Barakat
Reza Phalavi Medical Center, Iran

Bernard Barber
Barnard College

J. Michael Bedford
Cornell Medical School

Louise Bennett
International Development Research
Centre, Canada

Heinz Berendes
Center for Population Research
National Institute of Child Health
and Human Development

John Biggers
Harvard Medical School

Richard J. Blandau
University of Washington

R. Jeffrey Bogumil
Mt. Sinai School of Medicine

John L. Boling
University of Washington

Benjamin G. Brackett
University of Pennsylvania
School of Medicine

James Brauer
Tufts University

George Braun
Ortho Pharmaceutical Company

Robert M. Brenner
Oregon Regional Primate Research
Center

Stevan Broderson
University of Washington

Frank T. Brooks
United States Atomic Energy
Commission

C. Lalor Burdick
Lalor Foundation

E. R. Casillas
Oregon Regional Primate Research
Center

Lauro F. Cavazos
Tufts University Medical School

John R. G. Challis
Oxford University, England

Min Chueh Chang
Worcester Foundation for
Experimental Biology

Graham Chedd
NOVA-WGBH, Boston

Yves Clermont
McGill University, Canada

Donald Collins
Scaife Family Charitable Trust

## List of Participants

O. A. Dada
University of Ibadan, Nigeria

Katherine Darabi
Columbia University

Phillip Darney
Boston Hospital for Women

Julian Davidson
Stanford University

Egon R. Diczfalusy
Karolinska Institute, Sweden

Donald W. Dierschke
University of Wisconsin

Gordon W. Duncan
Battelle

Nikorn Dusitsin
Chulalonghorn Hospital, Thailand

Jane Dustan
Foundation for Child Development

Martin Dym
Harvard Medical School

Anibal Faundes
Consezo Nacional de Poplacion,
Dominican Republic

C. A. Finn
The Royal Veterinary College,
England

Don W. Fawcett
Harvard Medical School

Paul Franchimont
Hôpital Universitaire de Baviere,
Belgium

Henry G. Friesen
University of Manitoba, Canada

T. Gaudin
Recherche Industrielle et Innovation,
France

Carl A. Gemzell
University Hospital, Sweden

Ian Gibbons
University of Hawaii

T. D. Glover
University of Queensland,
Australia

Roger Guillemin
Salk Institute for Biological
Studies

Erlio Gurpide
Mt. Sinai School of Medicine

Kerstin Hagenfeldt
Karolinska Institute, Sweden

David W. Hamilton
Harvard Medical School

Charles Hamner
University of Virginia Medical
School

Oscar Harkavy
Ford Foundation

Robin A. Harrison
ARC Unit-Reproductive Physiology
and Biochemistry, England

Fouad Hefnawi
Al-Azhar University, Egypt

## List of Participants

Wylie Hembree
College of Physicians and Surgeons
Columbia University

A. F. Holstein
University of Hamburg, Germany

Richard Horton
University of Southern California,
Los Angeles

J. Robin Hoult
Kings College Cambridge, England

R. J. F. Hunter
University of Edinburgh, Scotland

Peter B. Hutt
Food and Drug Administration

H. Jackson
University of Manchester, England

Frederick S. Jaffe
Alan Guttmacher Institute

Elwood V. Jensen
University of Chicago

A. D. Johnson
University of Georgia

Lawrence J. Kane
Human Life Foundation

Alexander Kessler
World Health Organization,
Switzerland

Ernst Knobil
University of Pittsburg

Claude Kordon
National Institute of Health,
France

Dorothy Krieger
Mt. Sinai School of Medicine

Thomas J. Lardner
University of Illinois

Kenneth A. Laurence
Population Council

Allen Lein
University of California,
San Diego

George Lewerenz
National Institute for Child
Health and Human Development

Seymour Lieberman
College of Physicians and Surgeons
Columbia University

Hans R. Lindner
Weizman Institute of Science,
Israel

Mortimer Lipsett
Case Western Reserve University

John Loupos
Plymouth, Massachusetts

Ho Kei Ma
University of Hong Kong,
Hong Kong

Samuel McCann
University of Texas-Southwestern
Medical School

Charles E. McCormack
Chicago Medical School

Anne McLaren
University College London,
England

## List of Participants

Sheila Arvin McLean
Ford Foundation

Anastasia Makris
Harvard Medical School

Thaddeus R. R. Mann
University of Cambridge,
England

A. Mantingh
Academisch Ziekenhuis,
Netherlands

Luciano Martini
University of Milano, Italy

Luigi Mastroianni, Jr.
Hospital of the University of
Pennsylvania

Daniel R. Mishell, Jr.
University of Southern California
Medical School

H. J. van der Molen
Erasmus Universiteit Rotterdam,
Netherlands

John McLean Morris
Yale University School of Medicine

William Moyle
Harvard Medical School

Fredrick Naftolin
Harvard Medical School

Andrew V. Nalbandov
University of Illinois

Harold Nash
Population Council

William B. Neaves
University of Texas-Southwestern
Medical School

Peter Ofner
Tufts University Medical School

D. A. Olatunbosun
University of Ibadan, Nigeria

Bert W. O'Malley
Baylor College of Medicine

Edward Ortiz
Food and Drug Administration

Harold Papkoff
University of California Medical
School

Alvin Paulsen
United States Public Health
Service Hospital, Seattle

David Porter
University of Bristol, England

M. R. N. Prasad
University of Delhi, India

Thomas Pullum
University of California, Davis

H. G. Madhwa Raj
Harvard Medical School

Peter W. Ramwell
Georgetown Medical Center

Ralph M. Richart
College of Physicians and Surgeons
Columbia University

List of Participants

Elijah Romanoff
National Science Foundation

Griff T. Ross
National Institutes of Health

Robert J. Ryan
Mayo Foundation

Nafis Sadik
United Nations Fund for Population
Activities

Hilton Salhanick
Harvard Medical School

Charles Sawyer
University of California,
Los Angeles

John J. Schrogie
Schering Corporation

Neena B. Schwartz
Northwestern University

Sheldon J. Segal
Population Council

Brian P. Setchell
ARC-Institute of Animal Physiology,
England

Roger V. Short
MRC Unit of Reproductive Physiology,
Scotland

S. P. Spain
National Institute of Mental
Health

J. Joseph Speidel
Agency for International
Development

Tabitha Standley
World Health Organization,
Switzerland

Emil Steinberger
University of Texas Medical
School at Houston

Ronald S. Swerdloff
University of California,
Los Angeles

Howard J. Tatum
Population Council

C. Thibault
National Institute for Agronomic
Research, France

Irwin E. Thompson
Peter Bent Brigham Hospital

Michael S. Teitelbaum
Ford Foundation

Christopher Tietze
Population Council

Robert Utiger
University of Pennsylvania School
of Medicine

G. R. Venning
G. D. Searle & Company, Ltd.,
England

Klaus-Dieter Voigt
Universitat Hamburg, Germany

Darrell N. Ward
University of Houston

Helen Walters
National Institute of Child Health
and Human Development

William A. Walters
Monash University, Australia

Paul W. Waltz
Equation Models Associates,
Chicago

I. G. White
University of Sydney, Australia

W. Wollmer
Deutsche Forschungsgemeinschaft,
Germany

Richard J. Wurtman
Massachusetts Institute of
Technology

R. Yanagimachi
University of Hawaii

Samuel S. C. Yen
University of California,
San Diego

# Foreword
## McGeorge Bundy

It is an honor to write a few words of introduction to this extraordinary book. Together with its companion volume of scientific essays, *Frontiers in Reproduction and Fertility Control*, it constitutes a truly comprehensive review of one of the most important of all fields of human understanding and ignorance.

What Dr. Greep and his colleagues have done is to gather and order what we know and do not know about the reproductive process. They have also reviewed and assessed our ways of learning, the financial and institutional constraints upon them, and the levels of achievement that we can reasonably hope to reach if the necessary steps are taken. They make findings and recommendations of high importance on the needed size and shape of basic research, on ethical standards, on professional recognition, and on public reporting and analysis. Taken as a whole, the volume is an uncommonly persuasive summons to action.

That summons is issued to all who can help, but, as the authors conclude, the *financial* help that is needed must of necessity come mainly from governments, and most of all from the government of the United States. If American federal support for the reproductive sciences is not massively increased, there is little hope that the next decade will see effective exploitation of the extraordinary range of scientific opportunities and challenges described in this book. Our federal research dollar cannot—and should not—carry the whole burden, but there is no other possible source of the necessary primary leadership.

But federal funding agencies are seldom unmoved movers. In the biomedical sciences they will act primarily in response to what is wanted by two groups: the biomedical scientists and those who act for the general public on Capitol Hill and in the White House.

Thus this volume must be commended to both the biomedical world and the general public. If the authors are right in their conclusions about what more money can do in the reproductive sciences, then we need a substantial and even dramatic adjustment of our priorities. That will not happen until it is wanted by those who together decide on the levels and directions of federal support for basic health research.

I speak of "health research" advisedly because that is exactly what this subject is. The quality of our understanding of the reproductive sciences is an inescapable determinant of what can be done in the future in a field of human behavior central to the health of the species—and especially to women of childbearing age.

In the largest sense, this is a universal matter. Basic science has no geographic boundaries. What American dollars can help to discover is needed by the whole human family. But sometimes, because the population crisis seems likely to have its most acute and desperate impact in other countries, people make the casual assumption that this kind of knowledge is not a critical part of *American* health care. This assumption is wrong.

What is universal is not unimportant to the people of the United States. The massive reliance of Americans on the present generation of birth-control agents gives us the most pressing and direct interest in learning how they can be improved. They embody remarkable advances over what was available to Americans less than twenty years ago. But as this volume reminds us, they are not risk-free and they can have undesirable side effects. That is the immediate domestic meaning of the authoritative conclusion of this book, that "current technology cannot be regarded as adequate to meet individual *or* societal needs in either industrial *or* developing nations."

A dramatic expansion of federal support for basic research in the reproductive sciences would be amply justified if our universe were no larger than the fifty American states and our concern no wider than the health of the Americans who use fertility control. When we think of the whole human race, and our inescapable membership therein, the case becomes overwhelming.

But this is not a matter to be settled by rhetoric. The argument rests instead on a sober and compelling demonstration of what the reproductive sciences have achieved, where they are heading, and what is needed now to fulfill their promise. That is what this book provides, and its authors have put us all in their debt.

# Preface
## Roy O. Greep

This study was initiated by the Ford Foundation, with the collaboration of the Rockefeller Foundation and the International Development Research Centre of Canada, to assess the gains made in understanding the complex morphological, physiological, and biochemical processes of the male and female reproductive systems and to determine what financial and human resources have been and are being brought to bear on this field of science, a field that has as its social imperative a major objective of more effective control of human fertility. A strong motivating force behind this study was the generally held view that (1) the expansion of the support and the manpower pool during the past decade has greatly intensified progress in fundamental knowledge and research techniques and (2) the opportunity for building on these gains over the next decade appears exceptionally promising in terms of extending to humanity the material and health benefits of safe contraceptive practices. It will be evident from this report that an immense effort has been mounted. It is also fair to state that within the past decade the chronicle of brilliant successes in solving technically difficult problems of the most critical nature has stirred the admiration and imagination of the biomedical world.

This two-year study has been carried out with the help of a large number of participants who are listed on the front pages of this report. In essence, the operational structure has comprised a small staff and a worldwide network of consultants and advisors. Administrative decisions have been made by a steering committee with helpful guidance from an international advisory committee representing the scientific community, the pharmaceutical industry, and the major public and private donor agencies.

The two major aspects of the survey, scientific advances and funding in the reproductive sciences and contraceptive development, have been carried out independently, but the results have been compared for evidence of correlations. The two sets of data have been examined for trends and projections made in respect to (1) problems in urgent need of solution and (2) the financial support that will be required to develop a program of basic and applied research and training commensurate with the magnitude of the human problem it is designed to alleviate.

The survey of scientific advances has been obtained through the medium of forty-one essays, each focusing on specific aspects of reproductive research. Some were prepared by individual specialists and others by small groups of experts. The essays have served as the basis for the summary of major advances and remaining gaps in knowledge set forth in tables. The essays and summary tables will be published in a companion volume.

The funding data have been compiled from many sources and from many individuals in different countries who obtained data specific to their own country or region. Some funding data are simply not obtainable. There can be no doubt, however, that the data presented here represent the watershed of financial support for this field of endeavor.

There is a goodly degree of timeliness to the present survey. Reproductive endocrinology is rounding out its first half century of meaningful experimentation. The total annual support of reproductive research over that period has grown from virtually nothing to somewhere in the neighborhood of $120 million. With these gains in activity it seemed an appropriate time to take an inventory of where the field stands, how it is progressing, what its needs are, and how further progress might be more effectively gained.

Numerous people in research laboratories, pharmaceutical firms, and national and international funding agencies have contributed to the gathering of information for this review. The authors are especially grateful to the contributors listed at the beginning of this volume and to the innumerable scientists who filled in questionnaires, participated in interviews and review sessions, and wrote comments. The authors are indebted to the administrators of pharmaceutical firms and funding agencies located in various countries throughout the world who helped in the preparation and compilation of data for the funding survey.

In addition, Jacqueline Ans, R. Jeffrey Bogumil, Joyann Brody, and Ardell Wilbur deserve special thanks for their invaluable aid in assembling this material for publication.

**Reproduction and
Human Welfare**

# 1                    Summary of Findings and Recommendations

In 1959, Zuckerman summed up the first large-scale international conference on reproductive physiology and the regulation of fertility in these words: "vast areas of the subject are still cloaked in an ignorance which prevents a rational and scientific approach to the problem of population control. . . . The subject . . . is still littered with legends" (p. 1263).

The ensuing decade and a half witnessed a substantial expansion of research in the fundamental reproductive sciences and contraceptive development, stimulated primarily by mounting concern over rapid population growth. Scientists and research institutions in both industrialized and developing nations have contributed to an explosion in basic knowledge that laid to rest some of the traditional legends and clarified many long-standing problems in the field. This worldwide effort makes possible a systematic effort in the last quarter of the twentieth century to develop, on a rational and scientific basis, improved means of regulating human reproduction.

These are the principal conclusions of an intensive review of the reproductive sciences and contraceptive development initiated in 1974 by the Ford Foundation, with the participation of the Rockefeller Foundation and the International Development Research Centre of Canada. More than 160 experts in 26 nations participated in a collaborative effort to evaluate progress and to identify persisting gaps in our knowledge of these vital processes, to identify sources and track levels of support for the field, and to outline the scope of a program of research and development adequate to the urgent needs of individuals and nations throughout the world. The review was both detailed and wide ranging and yielded a diverse set of findings of a substantive scientific nature as well as of relevance to policy formation. This volume presents these findings, with emphasis on those questions bearing on policy determination. The major scientific assessments are presented in a companion volume that contains forty-one invited papers, each by a recognized expert, on the separate subjects into which the overall field logically can be divided. This opening section is a summary of the review's major findings. The social context of this field of inquiry is addressed first, followed by a synoptic view of its substantive development—the major advances since 1960, the principal hiatuses in the knowledge base, and the emerging possibilities for new modalities of fertility regulation. The professional and institutional resources at work in the field are then described. Finally an analysis is presented of trends in the financial support for this branch of research—by country, sector, and purpose. We conclude with the principal recommendations that emerge from the information assembled and the analyses carried out.

**The Challenge to Reproductive Research**

Whether one's perspective is limited to the personal effects of high fertility on individuals and families or its aggregate effects on nations and the world community, improved regulation of fertility and the resulting reduction in fertility levels that could be expected are urgent objectives in both industrial and developing nations. The inability of millions of individuals adequately to regulate pregnancy—to determine whether and when they will bear children—contributes to the current unprecedented rate of world population growth that retards economic development in many nations. It is also a major factor associated with the high incidence of adverse pregnancy outcomes—prematurity, infant mortality, retardation, congenital defects, maternal mortality and morbidity—and with intractable social and economic problems—illegitimacy, early marriage, family disruption, educational disadvantage, and the perpetuation of poverty. The causes are multiple, ranging from our ignorance of many aspects of the basic biological processes of reproduction to the grinding poverty and malnutrition that affect much of the world. Reproductive research aimed at improved methods of fertility regulation thus links the search for solutions of the personal problems of individuals and the social problems of societies with some of the most critical overarching problems facing the world as a whole. Because a complex web of social, psychological, economic, medical, and environmental factors are intertwined in the problems of population change and adverse reproductive outcome, the solutions will necessarily also be complex. The development of improved fertility control methods does not imply a technological panacea. Technological change, how-

## Summary

ever, can have an important impact on human behavior and lead to more rapid fertility decline than would otherwise be the case. Whatever the level of motivation to regulate fertility, more effective, safe, and acceptable methods that require less costly and sophisticated delivery systems would increase the effectiveness of fertility regulation and the magnitude of the reduction in fertility. Moreover, the costs of even an optimum research effort in this field are modest compared to the costs of other development or social programs.

Reproductive research thus aims at improved human welfare through improved fertility control technology. Progress toward that goal is dependent upon fundamental knowledge of the human reproductive process because safer and more efficient regulation is possible only as the mechanisms become better understood. As a result, much of the work in the field necessarily consists of basic investigations, the findings of which are relevant to a wide range of health problems associated with childbearing.

Some degree of fertility decline would be expected as a consequence of social transformation. But in many nations and subgroups within nations, high fertility itself retards the process of social transformation. If fertility could be reduced concomitantly with other development efforts, it would have a multiplier effect on the overall process. Many nations initiated national family planning programs in the last decade, and more vigorous efforts to distribute the current technology through these programs could bring about significant reductions in fertility levels. But the existing technology has many limitations. New methods, particularly those better adapted to mass application in nations lacking sufficient health resources, would make a large difference in the worldwide effort.

The potential importance of technological change is suggested by the experience of the last fifteen years. The emergence of oral contraceptives and intrauterine devices revolutionized contraceptive practice in the United States and other industrial nations and spurred the initiation of large-scale family planning programs in the poorer nations. As a result, millions of men and women in all parts of the world altered their fertility regulation practices. But the current technology has limitations with respect to efficacy, safety, acceptability, and continuity of use that result in unacceptably high failure rates and require large-scale resort to abortion as a back-up measure in order to reduce fertility to the low levels recently experienced in industrial nations. The safety issue is sufficiently problematic to make the most effective methods unacceptable to some women and to induce levels of concern that militate against regular use by others. At the least these factors make more difficult and costly the delivery of family planning services and limit the extent of effective fertility regulation that can be expected.

Thus, current technology cannot be regarded as adequate to meet individual *or* societal needs in either industrial *or* developing nations. A measure of the gap is that of the approximately 500 million women throughout the world risking unwanted pregnancy, an estimated 70 percent were using no contraceptive method at all in 1971. The heterogeneity of personal, cultural, religious, and economic circumstances of human life, as well as the varying needs of individuals at different stages in the life cycle, impose diverse demands upon the technology. It is thus likely that there will never be an "ideal" contraceptive for all circumstances. What is needed is a broad array of contraceptive methods that require less complex distribution systems; that are safer, less discomforting, and more convenient than current methods; that combine high acceptability with high continuity of use; and that are suited to the diverse requirements imposed by the diversity of conditions throughout the world.

Major Advances and New Opportunities

Progress toward new methods of fertility regulation can best be viewed in the light of the recent past. Twenty years ago clinical testing of an oral contraceptive, based on earlier findings of fundamental research, was initiated in the United States, Mexico, and Puerto Rico. By 1960, the original pill was approved and marketed. At about the same time the first clinical evaluations of modern intrauterine devices began and by 1963 sufficient information on their efficacy and safety had been assembled to permit general distribution throughout the world. In the ensuing years, the contraceptive development effort placed high priority on improvements in these two methods. Pharmaceutical companies devoted millions of dollars to the search for competing products that would reduce the pill's side effects without sacrificing efficacy. IUDs with fewer side effects and a wider range of usefulness among women of differing reproductive states were sought.

From these efforts, lower steroid doses were tested and found to be effective and potent new synthetic steroids that could be used at even lower doses were developed. This phase of contraceptive research in the 1960s reduced risk and improved the acceptability of oral contraceptives. The first pills contained 10 milligrams of synthetic steroids. In 1976, some pills in use contain less than 5 percent of that amount. Improved intrauterine devices were not assigned a high priority by major companies, and no dramatic breakthrough occurred in new design for many years. But public sector research efforts made considerable progress in analyzing IUD performance by developing new means of epidemiological study and in understanding the devices' safety and mechanism of action through animal studies.

New methods of fertility control, beyond modification of those already in use, have emerged in just the past few years (for greater detail see chapter 6). The development of pharmacologically active IUDs introduced an entirely new principle of contraception that extends the availability of intrauterine contraception to young nulliparous women for whom previous devices were unsuited. One form of this new method employs copper as the antifertility agent and is a direct result of basic research findings published only seven years ago on the effect of metallic ions on uterine enzymes in laboratory rabbits.

In 1967, chemists synthesized the ubiquitous natural substances termed prostaglandins that were discovered thirty years earlier and erroneously believed to be solely a product of the prostate gland. In the short interval that these synthetic prostaglandins and their derivatives have become available, renewed interest has led to the observation that they can cause the pregnant uterus to contract. Although it has not been possible thus far to use natural prostaglandins as nonsurgical means to terminate pregnancy at the time of the first missed period or within the first twelve weeks of gestation, these compounds have a useful application for midtrimester abortions. Prostaglandin analogs, such as a 15α-methyl prostaglandin $F_{2\alpha}$, appear to be effective earlier. These results suggest the potential use of prostaglandins may eventually prove to be much broader than second trimester abortions.

Surgical techniques for both abortion and sterilization have been improved in the last decade. Rapid and safe vacuum aspiration has virtually replaced surgical curettage for early abortion and surgical sterilizations have been simplified to reduce the period of hospitalization required or even to eliminate it entirely.

Work is now underway on methods for couples for whom current methods seem to be unsuited. The immunological approach is illustrated by the testing of a vaccine against the human pregnancy hormone, chorionic gonadotropin (hCG). The recent unraveling of the chemical structure of gonad stimulating hormones including hCG provided a needed breakthrough. Although there are difficult stages of testing ahead, particularly to provide assurance of safety and reversibility, the goal of many years of research may have been reached with the identification of a highly specific antigen that may serve as the basis for a pregnancy vaccine.

Work continues in the search for a luteolytic agent, an approach that could serve as the basis for a once-a-month pill alleviating the problem of risks associated with the continuous dosage forms of hormonal contraception.

New long-acting forms of steroidal contraception are almost certain to emerge in the near future in the form of subdermal capsules or rods, vaginal rings, injections, contraceptive bracelets, or drug-releasing intracervical or intrauterine devices. Prototypes have been tested and found to be both effective and acceptable. The most advanced, the subdermal implant, is the product of ten years of investigations in fundamental chemistry and basic clinical pharmacology in animals. The first results on animals were published in 1966, the first clinical tests reported in 1969, and the first large-scale field trial could begin as soon as 1977 or 1978.

New methods of fertility regulation for men also appear to be in the offing. The feasibility of a contraceptive drug for men has been established, although research continues in order to identify the optimal drug, dose, and mode of administration. Sperm production can be halted temporarily by using doses of steroid hormones that do not elevate blood levels of testosterone beyond the normal range, but research has yet to alleviate the unacceptable side effects caused by the hormones. Another approach seeks to prevent sperm from achieving the capacity to fertilize even though normal quantities are produced. With the former approach, the transition from animal observations to serious clinical evaluation of the concept took twenty years. With the latter, the observation in animals was reported in 1971 but remains controversial. Crucial studies in men have begun.

Thus, during the past two decades the introduction of the pill and the modern IUD has been followed by significant improvements in these methods and in others. Many findings of basic research have concurrently led to new principles for fertility regulation, and these have evolved into new methods now widely used or in advanced stages of testing. From this learning process have emerged other basic advances that hold promise for future practical applications to fertility regulation and problems of reproductive biomedicine. Many of these are described in detail in chapters 4 and 5; here we briefly summarize some of the highlights.

The isolation and total synthesis of the polypeptide produced in the brain that passes to the pituitary gland and triggers the production and release of gonad stimulating hormones (gonadotropins) stands out as a notable achievement. In the 1950s, indirect evidence mounted for the existence of this substance (called luteinizing hormone releasing factor, LRF). In the hypothalamus in the floor of the brain, certain neurons with the characteristics of secretory cells had been identified and a system of portal veins that provided a connection between this portion of the brain and the pituitary gland had been described. When a barrier was created experimentally between the brain and pituitary, gonadal function was interrupted. In 1960, direct evidence for the existence of LRF was obtained by demonstrating that extracts of hypothalamus would raise the blood level of luteinizing hormone (LH) and induce ovulation. This information created a strong case for the existence of hypothalamic neurosecretory control of the gonadotropin producing cells of the pituitary. Before alternate explanations could be abandoned, however, hypothalamic factors responsible for the release of gonadotropins had to be isolated. It was a difficult search; even the class of chemical compounds to which these factors belonged first had to be determined. Millions of the hypothalamic portions of brains from sheep and hogs were extracted and studied. After a decade of work the active substance, the releasing factor, was found to be a small peptide. The structure of the luteinizing hormone releasing factor was determined in 1971 and shown to be a peptide consisting of ten amino acid residues. Almost immediately thereafter, a total synthesis of this molecule was achieved. Thus, in less than twenty-five years the releasing factor governing the control of the production and release of gonadotropins moved from a theoretical concept to the identification of the hormone and its subsequent synthesis. The releasing factor is now being used to study and control the production of gonadotropic hormones in both men and women.

Chemists had other notable successes. The precise structures of the sex hormones of the reproductive system were established, and the biochemical pathways by which these steroids are biosynthesized in the body were determined. By the end of the 1930s, the structures of these essential sex hormones were known. During the next two decades the enzymatic reactions necessary for their formation were elucidated in some detail and the cellular organelles in which these enzyme systems reside were identified. Even the unique coordination of fetal and maternal cells required for the production of some steroid hormones of pregnancy has been established. The pathways of degradation leading to inactivation and excretion of steroid metabolites into the urine has also been clarified to a very large extent.

For the protein hormones, the gonadotropins, the state of knowledge twenty years ago was more primitive. All that was known was that they were large molecular weight proteins with sizes ranging from 35,000 to over 50,000 daltons and that they were glycoproteins (they had carbohydrate components linked at special positions of the polypeptide chain). Knowledge of the chemistry of the gonadotropins has advanced dramatically since the 1960s. Their precise size is now known as well as their amino acid composition; the sequence of these nearly two hundred residues; the type, number, and location of the carbohydrates moieties; and how the various gonadotropins differ from each other. Large segments of these molecules have been synthesized in the laboratory and some structure-activity relationships have been established. The rapid progress in gonadotropic chemistry has already led to the effort to apply this basic knowledge to the immunological regulation of fertility as noted above.

An entirely new dimension and understanding of the mechanism of hormone action has been forthcoming in recent years. Twenty years ago, much was known about what hormones did but little if anything was known about how they mediate their physiological action. Concerning sex hormones, for example, it was known that there is hormone and target cell specificity (certain cells responded to specific hormones and not others and, conversely, specific hormones act on certain cells and not others). Various theories sought to explain how hormones activate cells, but no single theory fit all the available facts; the theories mainly focused on cell membrane permeability or changes in rate of enzyme reactions. With the advent of radioactively labeled hormones with high specific activity, tracing of the fate of hormones became feasible. Milestone discoveries were made in rapid sequence. First the sequestering of steroids by target cells was explained when it was found that target cells are characterized by specific receptor molecules for binding specific hormones. Next these cytoplasmic receptors were shown to carry the steroid hormones to the nucleus, where the steroid-receptor complex interacts with genetic material DNA to change the pattern of the cell's production of the informational macromolecule called messenger RNA, which in turn alters the cell's program of protein synthesis. In retrospect, the interaction of hormone with the genetic material of cell nuclei as a basic mechanism of steroid action has been one of the major conceptual advances of the past two decades of endocrine research. It carries our understanding of reproductive events to the subcellular level and redirects research into new avenues that may lead to the development of new methods for fertility regulation.

A further triumph of chemical and clinical endocrinology during the past decade has been the development of techniques for the measurement of the hormones' secretion rates and the circulating blood levels of the principal hormones of the reproductive system. Twenty years ago such determinations were not possible. The production of gonadotropic hormones could then only be crudely determined by concentrating the hormone excreted in a twenty-four-hour urine sample and undertaking a bioassay based upon measurable changes in the reproductive organs of various animals, such as rats, mice, frogs, toads, rabbits, or weaver finches. Textbooks of the 1960s described "rat units" of activity, for example. Steroid hormones were measured by complicated and time-consuming procedures involving hydrolysis of urinary conjugates and relatively nonspecific chemical determinations (of total 17-ketosteroids, for example) or by biological assays. Scientific papers referred to the "cockerel's comb growth" as a measure of testosterone, the "latticed endometrium" of the rabbit in response to progesterone, and "vaginal cornification of the castrate female rat" to detect estrogenic activity. In the 1950s, the development of a relatively quick colorimetric method for the measurement of total estrogens had a sensitivity in the level of micrograms of combined estradiol, estrone, and estriol and was considered a major advance. With the advent in the late 1960s, however, of radioimmunoassay procedures for both steroid hormones and gonadotropins, the detection of infinitesimally small amounts of these hormones in biological fluids is no longer a problem. Their quantitative determination in small samples of blood can be carried out with precision at a level of sensitivity of picograms (one-trillionth of a gram). With gonadotropins, a minute droplet of blood from a finger prick can be used to measure the circulating hormones at a level of sensitivity one thousandfold greater than that possible in the early 1960s, when twenty-four-hour urine samples were necessary for bioassays. Because of this vast improvement in the capability of quantitative estimation, the past five years have brought forth the first clear and accurate picture of hormone levels in men, prepubertal children, women of different reproductive statuses, and individuals with various abnormal conditions. These new findings have had broad application in diagnosis and management of various endocrine conditions including infertility and have greatly facilitated the evaluation of methods for regulating fertility.

Our understanding of reproductive phenomena in man has benefited from other methodological advances. Electron microscopic study of the testis has led to new concepts of testicular function. The Sertoli cells, which surround the sperm-producing cells during their development, were formerly thought to have only a supportive function. Now it has been found that special sealing junctions between the supporting cells serve to isolate the germ cells in a compartment distinct from the general extracellular space of the testis. The secretory products of the Sertoli cells create in that compartment a special microenvironment favorable for development of the germ cells into spermatozoa. The existence of a permeability barrier between the blood and the germ cells (blood-testis barrier) comparable to the blood-brain barrier serves to protect the germinal cells and is important in preventing sperm proteins from reaching the blood and inducing an autoimmune reaction against them. Contractile movements of the Sertoli cells also clearly provide a mechanism for moving clones of spermatogenic cells upward from the basal membrane of the seminiferous tubules toward the lumen, ultimately releasing them when they have completed the first phase of their maturation. The process of spermatogenesis itself has now been described in exquisite detail due to the availability of electron microscopes of high resolution and magnification. A population of spermatogonia always held in reserve but capable of dividing and initiating new cycles of spermatogenesis can now be recognized. The process and the schedule of their renewal, which makes possible continuous sperm production throughout the life of the male, have been thoroughly studied by labeling the dividing cells with radioactive compounds. Significant advances have also been made in understanding the hormonal environment required for sperm production and maturation. Local testosterone levels in the testis must be maintained at much higher levels than in the peripheral blood. This maintenance is accomplished in part by a recently discovered androgen binding protein, produced by Sertoli cells, which facilitates uptake of testosterone from the Leydig cells into the spermatogenic tubules and its transport to the epididymis with the spermatozoa. The need for posttesticular maturation of spermatozoa has become evident in the past decade. The epididymis was formerly believed to serve simply as an organ of sperm storage and transport. Now it is known that important maturational changes in the spermatozoa take place during their passage through this organ. These changes include alterations in structure, metabolism, pattern of motility, and fertilizing capacity. These findings have focused attention upon the epididymis as a possible site for interference with the processes by which spermatozoa acquire the ability to fertilize ova.

These are but a few examples of significant advances in the recent past. Not only have there been a large number of significant findings, but the pace of discovery has itself quickened; of the 446 key papers published after 1960 that are cited in chapters 4 and 5, 15 percent were published between 1960 and 1965, 30 percent between 1966 and 1970, and 55 percent between 1971 and 1975. The total number of papers published in all internationally recognized journals has increased from 6,557 in 1970 to 9,150 in 1975. New instruments and novel research methods have allowed investigation to proceed to the subcellular and macromolecular levels, thus greatly augmenting our fundamental understanding of the physiological phenomena of reproduction. These powerful tools continue to open up new avenues for research, providing answers to long-standing gaps in our knowledge and raising challenging new questions. At the same time a variety of new opportunities for fertility control have been identified. From the molecular level (such as elucidation of the structure of gonadotropins) to the physiological level (such as establishment of a role for prolactin in the reproductive process), the persisting gaps in our knowledge must be bridged to bring these new opportunities from the laboratory to clinical application. Research workers consulted in this review have listed more than 230 such gaps in our present understanding of the reproductive processes (see appendix E). The preconditions have thus been established for a variety of approaches to fertility regulation and for thorough understanding of the biological and behavioral aspects of human reproductive activity. These promising leads await adequate exploitation.

**Institutional Setting and Constraints**

Research and training in the reproductive sciences related to contraceptive development is highly decentralized. Fundamental research is primarily a university function, with important work also underway at a few free-standing institutes, while more focused applied studies are typically undertaken by research institutes, public sector programs, and the pharmaceutical industry. The membership lists of the key relevant professional societies suggest that there are perhaps three thousand investigators in the field throughout the world.

Based on the 1973 records of the principal funding agencies, 176 research groups could be identified in the United States; 106 included the full-time equivalent of one senior scientist supported by junior investigators, 49 had at least two full-time senior scientists and their coworkers, and 21 could be considered "institute" level, with 5 to 10 senior investigators and substantial supporting staffs. In the other industrial nations, there are 162 institutions, with clusters in the United Kingdom, Sweden, France, Germany, and Japan. In developing nations, 81 institutions could be identified. Between 1970 and 1975, more than 3,500 papers were published by developing world investigators in internationally recognized scientific journals, about 7 percent of all published papers in the field; about 40 percent of the developing nation papers were prepared by Indian scientists. Outside of India, there are high quality fundamental research groups in Mexico, Brazil, Argentina, and Chile and clinics and laboratories also carry on important clinical studies in other nations.

The last decade has witnessed the emergence of public sector agencies supported by governmental and/or philanthropic funds with extensive programs in contraceptive development, traditionally the pharmaceutical industry's domain. Pharmaceutical firms have not abandoned the search for new contraceptives, but the resources they allocate to this effort have been limited as a result of increasing costs and risks. As a legacy of the development of hormonal contraceptives, a new standard of needed information with respect to pharmacology, metabolism, side effects and acceptability, theoretical and use-effectiveness, reversibility in long-term use, and potential carcinogenic effects of chronic exposure to drugs and/or devices has been adopted.

Programs in public sector agencies, now probably the principal agencies seeking to develop improved methods of fertility regulation, often collaborate with private industry in bringing a product to market. Many of these public sector organizations not only provide grant or contract funds to other investigators, but also carry out some of the functions traditionally performed by industry in the development process. These include the Population Council's International Committee for Contraception Research (ICCR), the World Health Organization's Expanded Programme of Research, Development and Research Training in Human Reproduction, the Center for Population Research of the U.S. National Institute of Child Health and Human Development, the Research Division of the U.S. Agency for International Development's Office of Population, and the Indian Council for Medical Research.

Research in this field is affected by trends influencing scientific inquiry generally. The review commissioned several papers to assess two factors that have particularly affected the reproductive sciences related to fertility regulation: increased regulatory requirements for drugs and devices and ethical questions concerning the conduct of research with human subjects. In the United States and other nations, these generic trends have had considerable impact on this field because of increasing concern over the risks of existing contraceptives used, for the most part, by healthy individuals for long periods of time. Our assessment indicated that regulation of drug research and development is likely to become more rigorous in all nations, a process that in the long run should prove beneficial to the field as a whole. The testing of new drugs and devices is inevitably associated with some risk to the experimental subject; and much effort in recent years, particularly in the United States, has been devoted to defining more explicit guidelines for experimentation with human subjects. Two major means of reconciling the benefits of biomedical research with its risks have emerged and are now employed in the United States and some other nations: the principles of informed voluntary consent by the experimental subjects and of peer reviews of research proposals, for their ethical implications as well as for scientific merit. Both approaches are acknowledged to provide safeguards against abuses. Discussion and debate on ethically appropriate criteria for the conduct of research is likely to increase in the future, which should ultimately lead to a closer perceived community of interest in the research enterprise by all parties.

In the short run, however, these trends make contraceptive research a lengthier and costlier process that is less, rather than more, attractive to the pharmaceutical industry.[1] It therefore does not seem likely that the industry will play the substantial role in the development of new methods originally envisioned by some observers, although policy changes might conceivably create greater incentives for industrial involvement. In any case, the development of new contraceptives will now surely take longer than in the past because more time will necessarily elapse between the conceptualization of a new approach and its emergence as an approved product. The costs of product development will also be higher because both the additional technical requirements of longer studies in animals and humans and the new ethical requirements mean that more time and effort must go into each study. We do not believe these new requirements could, or should, be waived, even though improved fertility regulation methods are urgently needed. We do believe that the principal immediate consequence of these trends is that the search for improved contraceptive methods now depends, even more than in the past, on decisions made and activities undertaken by the noncommercial sector—government and international agencies and the philanthropic community.

---

[1] In addition to these factors, industry's role in the future may be limited because some new types of contraceptives may be inherently less profitable than the current ones. An example is the development of long-acting contraceptives, a formulation desired by many users.

## Financing: Past, Present, and Future

Our review of the historical record demonstrated that the reproductive sciences related to the regulation of fertility were virtually excluded from the rapid expansion of support for biomedical research in the United States and other industrial nations in the two decades following World War II. To sustain serious research in the field has required, for more than half a century, special efforts by interested individuals and private organizations, most of whom were outside the biomedical mainstream. U.S. funds for this branch of research in the fifteen years following World War II constituted a fraction of 1 percent of philanthropic funds for medical research and only a slightly larger proportion of pharmaceutical industry research funds, despite the anticipation that a new contraceptive would be profitable. Although government became the principal source of support for biomedical research generally in the United States in the 1950s, governmental funds were not available for this field of research as a matter of policy.

In the first half of the 1960s, mounting concern over population growth led to the beginnings of policy change in the U.S. National Institutes of Health, the United Nations, and the World Health Organization. Similar changes occurred five years later in the medical research establishments of Western European nations. By the end of the 1960s, important programs had emerged in three private foundations and a governmental medical research agency in the United States, six international development assistance agencies, and the medical research arms of five other nations. As a result, funding levels increased beginning in the mid-1960s, leading to an expansion of work in laboratories in many countries and the improved knowledge of the reproductive process summarized above. But it is questionable if these recent increases have redressed imbalances in the field caused by decades of taboo and underfinancing.

The review assembled for the first time systematic information on financial support for the reproductive sciences related to fertility regulation from government agencies, industry, and private philanthropy in the United States, fifteen other industrial nations, and nine developing nations and regions. These data are summarized in figures S.1 through S.4. In current U.S. dollars, the worldwide level of support increased from $31 million in 1965 to $119 million in 1974; in constant U.S. dollars, however, the increase was less marked—from $38 million to $91 million (1970=100). Moreover, real dollar support for the field peaked in 1973 and declined precipitously in 1974 (figure S.1).[2]

Most of the funds throughout the decade came from U.S. sources, and U.S. laboratories did the bulk of the work. But the decade also witnessed an increasing trend toward a more equitable international division of responsibility for both financing and conduct of reproductive research: The share of funds originating in the industrial nations of Europe doubled and reached 31 percent in 1974 and their share of the work increased from 19 to 27 percent (figure S.2).

The bulk of financial support for the field came from government sources throughout the decade; in fact, the governmental share increased from 44 percent in 1965 to 69 percent in 1974 (figure S.3). Concomitantly, the share coming from private philanthropy declined from 22 percent to 14 percent, while the proportion contributed by the pharmaceutical industry was cut by more than half—from 34 percent to 16 percent—apparently as a result of revised assessments of the potential profitability of new contraceptive methods. The 1965 to 1974 trends make evident that it would be unrealistic to depend on industry for the bulk of funds needed to develop an improved fertility control technology.

Throughout the decade, most of the funds supported fundamental studies of the reproductive sciences, with the share devoted to this purpose increasing slightly from 55 percent in 1965 to 57 percent in 1974 (figure S.4). The proportion allocated to contraceptive development declined from 35 percent to 30 percent, while training's share

---

[2] Expressing funding levels in constant dollars may still overstate the amount of real dollars available to support research activities as a result of changes in accounting practices. In the United States, for example, an increasing proportion of research funds in the last decade has been allocated to "indirect costs" of university administration and overhead, while at the same time many universities have increasingly required tenured professors to obtain a larger proportion of their salaries from grant budgets. Both of these practices absorb a significant share of funds allocated for research.

18  Summary

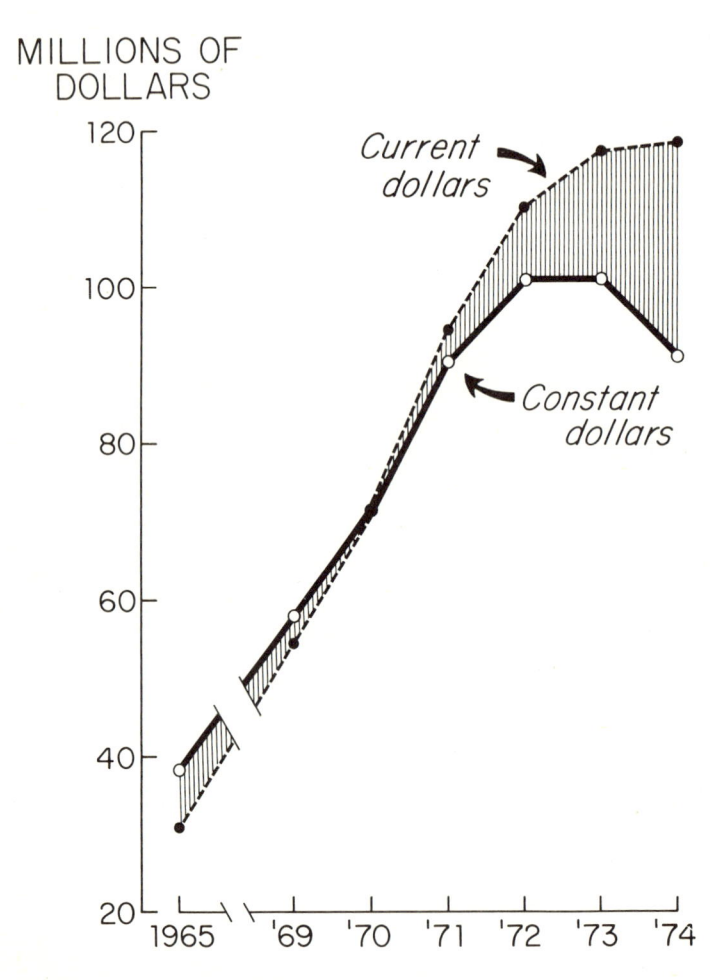

MILLIONS OF
DOLLARS

Figure S.1 Total worldwide expenditures for the reproductive sciences and contraceptive development (current U.S.-constant U.S. dollars, 1970=100).

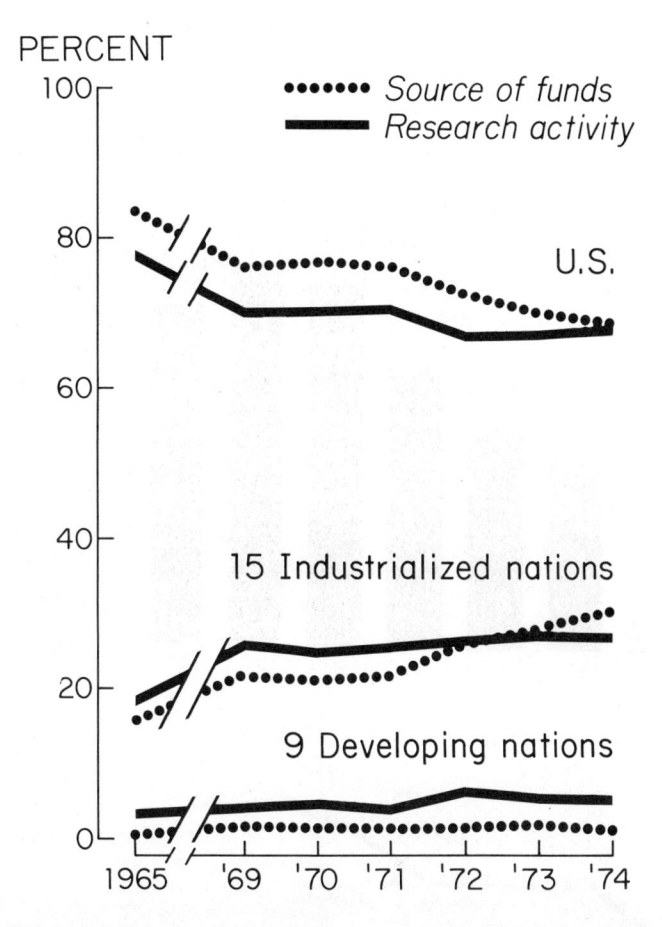

**Figure S.2** Percent of worldwide expenditures for reproductive sciences and contraceptive development by geographic distribution: source of funds and location of research activity (based on constant U.S. dollars, 1970=100).

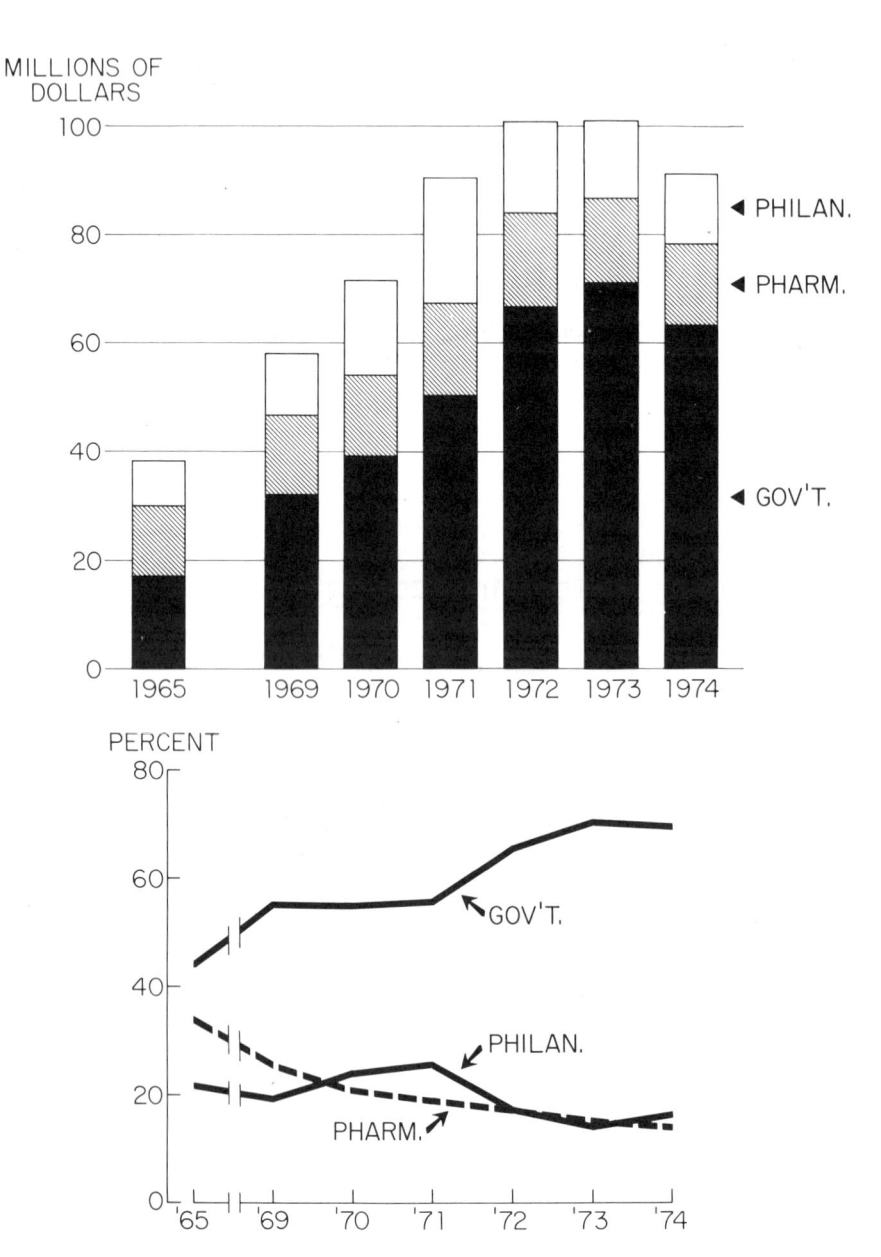

Figure S.3 Total and percent of worldwide expenditures for reproductive sciences and contraceptive development by sector: government, philanthropy, and pharmaceutical firms (based on constant U.S. dollars, 1970=100).

declined from 7 to 5 percent. The proportion devoted to evaluation of the safety of existing methods increased during the same period from 3 to 9 percent.

In chapter 11, we present several ways in which the adequacy of these funding levels can be assessed. In aggregate, the total funds available for this field in 1973 represent only 37 cents per woman of reproductive age in the nineteen countries for which funding and census data are available. Governmental allocations to the field by 1965 constituted 0.9 percent of total governmental medical research expenditures in the nations represented in the study. This proportion increased to a high point of 2.2 percent in 1973 and dropped back to 1.6 percent in 1974, proportions which do not convey the impression that the field is regarded as a priority effort in any nation. The escalation of worldwide concern over rapid population growth in the last decade, symbolized by the convening of the first intergovernmental population conference in Bucharest in 1974, thus has not been reflected in increased priority for the field from governmental medical research agencies, which tend to emphasize disease-oriented research.

Nor has reproductive research been perceived as a priority effort by governmental and international development assistance agencies. The field received approximately 9 percent of funds allocated to population activities by nine international, governmental, and philanthropic agencies that were the principal institutions in the development community supporting the population or family planning work in 1974. Total expenditures on reproductive research by these agencies, however, constituted barely more than 0.2 percent of total official development assistance funds provided by member nations of the Organization for Economic Cooperation and Development's Development Assistance Committee.

Increased investment in the field during the 1960s, accomplished by training a cadre of qualified investigators and attracting others to its scientific and social importance, laid the groundwork for a substantial increase in research activity in the mid-1970s. But survey respondents from sixteen nations reported that they anticipate sharp reductions in available funds from all sources. Unless these trends are

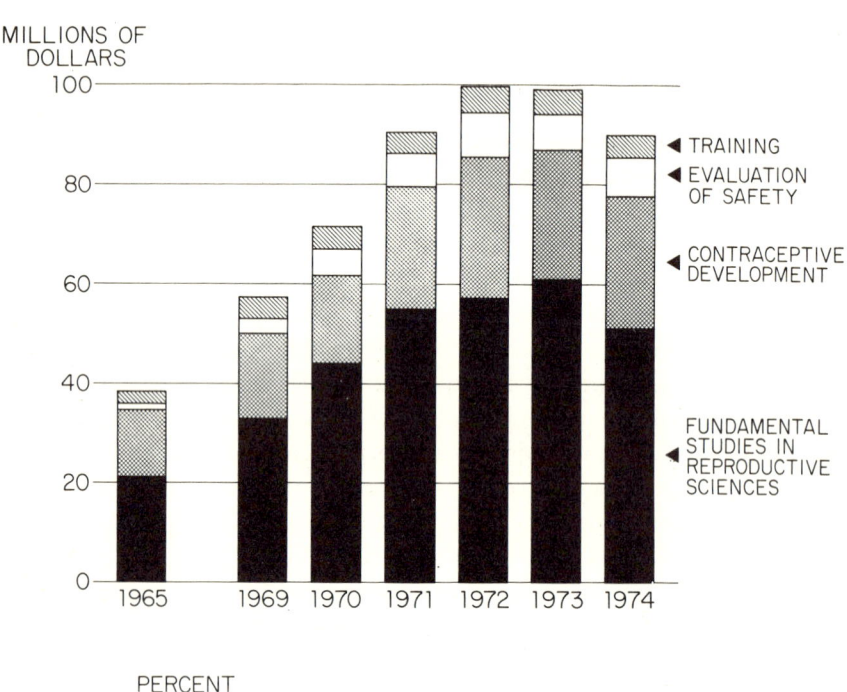

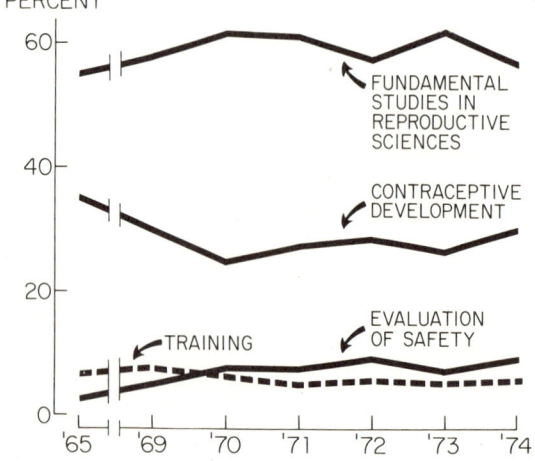

Figure S.4 Total and percent of worldwide expenditures for reproductive sciences and contraceptive development by purpose: fundamental reproductive research, contraceptive development, evaluation of safety of contraceptives, and training to strengthen professional capacity (based on constant U.S. dollars, 1970=100).

reversed and greater resources are found to support this field of research, there will clearly be less activity and less progress toward improved fertility control technology in the next decade.

Finally, we sought to project the field's future funding needs. Three approaches to a 1976-1980 projection are shown in figure S.5. Simply to maintain the 1974 level of activity would require, as a result of anticipated inflation, that the worldwide funding level increase to nearly $200 million in 1980. At the other extreme, a high priority effort receiving 10 percent of medical research expenditures would need $566 million in 1976 and $766 million in 1980. (It should be noted that two disease-oriented research programs—heart and lung, and cancer—receive 16 percent and 33 percent, respectively, of the research funds of the U.S. National Institutes of Health.) Our own estimates encompass a program to exploit adequately existing knowledge and opportunities. The building blocks for these estimates are the experts' assessments of the state of the field—what is known, what remains at issue, what could be resolved with current techniques and instrumentation. In essence, the projected program, built up inductively from detailed analysis of the field's component subjects, is one for which the prerequisite knowledge already exists and a significant professional and institutional capacity is in place. The total required for such a program, $361 million in 1976, is three times the amount currently available.

Research on fundamental reproductive research, contraceptive development, and training could usefully absorb approximately three times their current levels of funding. However, our analysis of important remaining questions regarding the potential risks of fertility regulating methods now in use led to an estimate that safety studies require about $72 million in 1976, or approximately seven times the funding level currently available.

Taking anticipated inflation into account, the costs of the entire projected program would increase to about $500 million in 1980. Given recent trends, neither industry nor the philanthropic community can be expected to be the major source of support for the field in the next decade; the increased funds required will have to come from governmental agencies principally in the United States and the

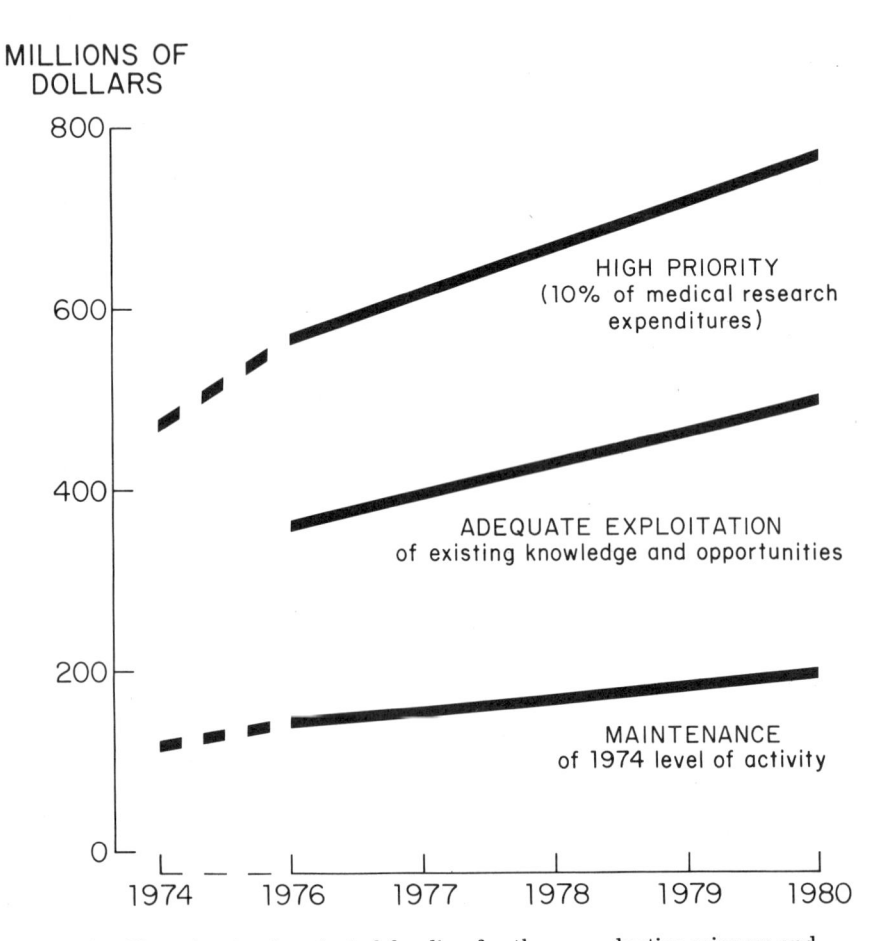

Figure S.5 Three levels of projected funding for the reproductive sciences and contraceptive development, 1976-1980 (current U.S. dollars).

industrial nations of Europe, Canada, Israel, Australia, and Japan and the international assistance agencies.

The projected program would move in the direction of carrying out the mandate of the World Population Plan of Action adapted unanimously in Bucharest in 1974. The plan calls upon the nations of the world to give high priority to a variety of research activities, including "the assessment and improvement of existing and new means of fertility regulation by means of research, including basic biological and applied research . . . and the evaluation of the impact of different methods of family planning on the health conditions of women and members of their families" (United Nations 1974, p. 28).

Recommendations

These principal findings lead us to recommend to governments, philanthropic foundations, the scientific community, and the pharmaceutical industry three overriding objectives, as well as numerous instrumental steps designed to advance the achievement of these goals.

*Recommendation 1. A variety of safe and effective methods of fertility regulation beyond those now available is urgently needed by the world's diverse population living under different conditions and circumstances. This requires increased efforts ranging from fundamental research on reproductive processes to targeted activities in contraceptive development.*

Two decades of experience with oral contraceptives and intrauterine devices have made it evident that these methods have significant limitations. New methods that have fewer health hazards than existing methods, that are more acceptable and convenient for individual couples, and that are simpler to distribute are needed.

The new opportunities for development of improved fertility regulation methods have emerged from the fundamental research of the past fifteen years. A long lead time is needed before knowledge can be translated into usable technologies. Any potential method that has not yet reached the stage of preliminary testing in humans is at least five to ten years away from practical application, given the time needed to establish efficacy and safety. Any new discovery of basic research in 1976 could not be expected to yield a usable product before 1986. The need for a variety of effective, safe, and acceptable methods of fertility regulation will confront men and women throughout the world not only in the next decade, but in 1990, 2000, 2010, and 2050. If new methods are to be available for widespread use by the next generation, the research effort must be expanded now. This is particularly true because no one can guarantee when—or if—the critical basic discoveries will occur. But it can be reasonably asserted that the greater the number of scientists and research institutions devoting attention to problems of the reproductive sciences and contraceptive development, the greater the probability that new findings will close important knowledge gaps and lead to new applications for fertility regulation.

*Recommendation 2. Within this priority effort, greater attention must be given to studies of intermediate and long-term safety of methods of fertility regulation now widely used and new methods yet to be developed. In addition to expanded work in fundamental research, this requires selective and specific procedures to measure parameters of biomedical events, to identify appropriate animal models for evaluation of risk factors, and to carry out sophisticated clinical studies as well as carefully designed prospective and retrospective epidemiological surveys.*

When drugs or medical devices of any type are introduced for widescale use, it is virtually impossible to anticipate all possible safety issues. A primary problem is whether data on safety can be extrapolated from the original study group to all populations. Furthermore, a small increase in rare events can be discerned only after the drug or device has been used by a large number of persons. Side effects that occur after a long period of latency or in subsequent generations can be detected only after many years of use.

These principles apply to fertility regulating drugs as well as to all drugs. For most contraceptive methods now in use, numerous health-related issues remain on the agenda of the scientific community for systematic investigation. A greatly increased effort to study the safety and health-related aspects of presently available fertility regulating methods is therefore necessary to seek answers to questions that frequently arise, causing anxiety and distress to many individuals—particularly women—and to broaden the knowledge base so that all methods may be used by the people of the world under conditions of optimal benefits and minimal risk.

From both individual and societal perspectives, research in the reproductive sciences and contraceptive development thus warrants higher priority by governments, private sector funding sources, and the scientific community. Our third central recommendation, therefore, is designed to address this issue.

*Recommendation 3. By 1980, allocations for research in the reproductive sciences related to contraceptive development and evaluation should comprise substantially higher proportions of total expenditures for medical research and development assistance than is presently the case. For an adequate effort on a worldwide basis $361 million would be required in 1976, increasing to close to $500 million in 1980, taking anticipated inflation into account.*

These funding targets are justified not only because of the urgent problems this field seeks to solve, but in order to exploit adequately the new knowledge of reproductive processes already in hand. They will obviously not be achieved this year, but the 1980 target could be reached if changes in research allocation policies were initiated now by governmental, multilateral, and philanthropic agencies.

That the field currently has considerable unused capacity that could quickly begin to work in an expanded effort is indicated by the fact that in recent years between half and two-thirds of approved grant applications have not been funded. A phased-in increase in total funding over the next five years would not only tap this existing capacity but would also allow time to generate the additional capacity required for a worldwide program approximately triple the size of the current effort.

To achieve these targets, support from U.S. sources would have to increase to $328 million by 1980; $162 million would be needed from other industrial nations and $7 million from developing nations. Medical research and development assistance agencies in the United States and other industrial nations would need to increase their levels of support for this field, while the international assistance agencies such as the United Nations Fund for Population Activities and the World Bank would have to include among their priorities the funding of biomedical research aimed at improved regulation of fertility.

In the course of this review, the authors examined a range of issues concerning how an expanded worldwide effort might best be organized. Our recommendations on these questions are derived from this examination as well as from the diverse viewpoints expressed by researchers with varying backgrounds and long experience in the field.

**Emphasis within the Field**

The prospects for the development of new contraceptive modalities depend on a continuing flow of information derived from fundamental research on the reproductive processes and other aspects of the biomedical sciences. During the past decade this basic principle has been expressed in the allocation of funds within the field to the four categories into which it can logically be divided: fundamental research, directed work aimed at new contraceptive development, studies of the safety of current methods, and the training of new research personnel. As noted previously, the proportion allocated to fundamental studies actually increased somewhat over the decade, despite the mission-oriented nature of the field. At the same time, the proportion devoted to safety studies also increased.

Our overall recommendation for expansion of the total field is based on our considered judgment that effort must be increased substantially in each of these categories. We recommend special emphasis on safety studies, which implies a greater proportional allocation to this category. But we stress that this recommendation is within the framework of considerably increased funding for the entire field. If this objective is achieved, funding of safety studies can be increased more rapidly than the other components; but the allocations for fundamental work, contraceptive development, and training will also increase in absolute terms.

*Recommendation 4. Expansion of the field requires a balanced effort*

## Summary

*rapidly to increase fundamental research in the reproductive sciences, contraceptive development, safety studies, and training. The special emphasis proposed for evaluation of safety should lead to a greater proportional allocation to these studies, which should account for at least 15 percent of total funds by 1980.*

Our review has demonstrated the obvious: More is known about the reproductive process in the female than in the male and the bulk of ongoing work is devoted to studies of the female. There are numerous reasons for this, one of which is a tradition in some Western nations that has in this century assigned to the female responsibility for limitation of family size; a common failure of the laity to distinguish clearly between potency and fertility; and the prevalent impression that pharmacological suppression of fertility would not be acceptable to men. Among scientists, there has been the debatable assumption that interference with a single event, ovulation, occurring once a month would be more easily achieved than interruption of spermiation, a continuous process releasing tens of millions of gametes daily. Whatever the historic reasons, the result is an imbalance in the field that is scientifically inappropriate and clinically self-defeating. Men are coparticipants in the reproductive process. Effective antifertility preparations for male use could markedly improve the ability to regulate childbearing. Moreover, sexual partners could then have the possibility of alternating responsibility for contraception every few months, thus minimizing for both any risk of cumulative undesirable side effects from medication.

Research on male reproductive processes has in fact accelerated during the last decade as a result of the introduction of new instruments and methods such as electron microscopy, radioautography, radioimmunoassay, micropuncture and cannulation techniques, and dissociation of the cellular components of the testis for biochemical analysis. These have made possible significant advances in our understanding of the endocrine environment of spermatogenesis, the mechanisms of sperm release and sperm motility, the nature of the posttesticular maturation of spermatozoa, and the structural and chemical bases for sperm penetration during fertilization. Several leading laboratories devoted to the study of the male have emerged and a Society of Andrology has been established in the United States. The World Health Organization has established a task force on the male, and the Center for Population Research now devotes almost half of its contraceptive development funds to studies of the male. We believe these trends should be encouraged.

*Recommendation 5. More attention needs to be devoted to study of the male reproductive process, with particular emphasis upon the site and mechanism of action of hormones in the endocrine control of spermatogenesis, the identification of vulnerable biochemical events in germ cell differentiation, and the discovery of means for selective interference with those epididymal functions essential for sperm maturation. These and other promising areas now seem ready for intensive investigation.*

## Geographical Distribution of Responsibility

The data presented in figure S.2 show that increasing proportions of the funds available for the field have supported work by investigators in the industrial nations of Western Europe, Canada, Japan, Israel, and Australia as well as in the developing countries. We regard this as a salutary trend and believe it should be further encouraged. The problems of human reproduction that this field addresses are worldwide in scope and their solution will require the active involvement of researchers and laboratories in many nations. Although the capacity of the scientist and institution to pursue the indicated research should be the principal criterion for determining the allocation of funds, there is every reason to believe that appropriate policy changes could help stimulate the participation of qualified researchers in developing as well as industrial nations and thus bring about a more equitable international distribution of responsibility for the research effort.

*Recommendation 6. To ensure the increased participation of scientists in developing nations will require special efforts. While the need for clinical research specific to each nation is obvious, centers of high-quality fundamental research presently exist in developing nations and more could be encouraged. We believe that funding agen-*

cies should develop mechanisms to provide full-time salary support to key research personnel in developing nations.

*Recommendation 7. Based on the trends observed from 1965 to 1974 (table 11.6), it seems feasible to move toward a more balanced division of effort by 1980, so that 40 percent of the work is done in industrialized nations other than the United States and 10 percent is conducted in developing nations.*

*Recommendation 8. To move in this direction, the medical research councils of Western Europe, Canada, Japan, and Australia will need to allocate a greater proportion of their budgets to this field. We further urge that national development assistance agencies fund research in industrialized as well as developing nations, while multilateral agencies, such as the United Nations Fund for Population Activities (UNFPA) and the World Bank, include among their priorities the support of reproductive research and adopt policies to fund studies in both industrial and developing nations.*

As the international base of the research effort broadens, questions will inevitably arise regarding the differing standards of research conduct in different countries. In principle, ethical standards in carrying out research involving human subjects should be applied uniformly with full respect for local circumstances. This will require particular attention to the different risk-to-benefit ratios applicable to conditions in different nations. A related issue concerns freedom in the dissemination of scientific knowledge. Scientists in all countries should have access to scientific information and should be able to participate in scientific exchange with their colleagues throughout the world.

*Recommendation 9. We therefore recommend that in carrying out studies involving human subjects, ethical principles based on informed consent by the subject and continuing peer review of projects should be followed in all countries. Those responsible for approving studies of methods of fertility regulation should take into consideration the balance between potential benefit and risk appropriate to each nation in making these determinations.*

*Recommendation 10. We recommend that appropriate scientific organizations in each country and international agencies undertake responsibility for more systematic efforts to encourage greater international cooperation, exchange of scientific information, and collaborative projects. We realize that WHO has a broad mandate from the World Health Assembly and the proper facilities and knowhow and continuously is involved in activities of this kind but is hampered by the paucity of available funds. We recommend therefore that additional funds be made available to WHO to enable the organization to increase its activities in this field by sponsoring periodic regional and international conferences and by promoting the systematic translation into other languages of scientific articles, including those published in the USSR, the People's Republic of China, the United States, and Europe.*

**Strengthening Institutional and Professional Capacity**

Although our review has demonstrated that growth of the field's capacity during the last decade has been impressive, it is nonetheless apparent that greater institutional and human resources are needed throughout the world. A number of recommendations focus on the objective of increasing the participation of industrial, academic, and public sector institutions. Both the initiation of new approaches and the applied phase of contraceptive development will require major participation of publicly supported institutions and programs for the foreseeable future. In undertaking this responsibility, the public sector should cooperate to the maximum extent possible with industry and assure the full utilization of this important segment of the world's scientific expertise, while making arrangements for a fair distribution of both the benefits and the costs.

During the past twenty years the two major new methods in the field of fertility regulation now widely used were developed largely through the initiative of publicly supported institutions. During the next two decades new approaches to fertility regulation will in all likelihood depend even more on publicly supported programs. Most of the current work on advanced leads toward development of new methods is proceeding under the direction of such programs.

It can be expected that many, if not all, of the new methods developed will eventually be licensed to industry to assure widespread distribution and competent product management. In such arrangements with the pharmaceutical industry, the publicly supported programs should seek to protect the public interest.[3]

*Recommendation 11. We recommend, accordingly, that public sector programs supported by governmental or philanthropic funds should seek to protect patentable inventions, with the right to require royalty payments in return for exclusive or nonexclusive licenses. This would make it possible to negotiate effectively with industry to guarantee low cost and availability of the products for use in the public sector and to assure that profits from the private sector sale of the products will contribute to further research and evaluation.*

At the same time it seems evident that greater initiative by industry itself is not forthcoming in part because of the increased investment necessitated by current regulatory requirements. We are convinced that the field has much to gain from strengthened, rational regulation based upon careful examination of the issues of safety and efficacy inherent in any modalities to be used by large numbers of persons for long periods of time. These requirements, however, will increase the costs of research and lengthen the time between initial discovery and ultimate use. Research policy has to comprehend these realities and compensate for them in appropriate ways.

*Recommendation 12. We recommend, therefore, that serious consideration be given to proposals designed to create additional incentives for industry to participate in the contraceptive development effort. These include proposals that would extend the life of patents in the United States for ten years after the New Drug Application is issued and would have government bear some of the risks of the development process (with reimbursement from royalties on sales of approved products). In particular, consideration should be given to*

---

[3] It is of interest that in its first purchase of the Copper T for use in developing country programs, the savings to UNFPA as a result of ICCR public sector pricing arrangements with the manufacturing company nearly offset the total cost to ICCR of this product research and development.

*government support of expensive toxicology studies, either through direct subsidy or through establishment of public sector institutions to conduct these studies on behalf of industry.*

*Recommendation 13. All nations have an interest in the safety and efficacy of the products used by their citizens, yet current regulatory requirements vary widely in different nations. We believe that these requirements should be reexamined with a view toward making them more valid scientifically and achieving the intended purpose of assuring optimal availability of fertility regulating methods at minimal risk.*

University-based scientists and university laboratories comprise perhaps the principal resource for this field in both industrialized and developing nations. National and international funding agencies will need to develop policies aimed at ensuring the development of an increasing cadre of basic scientists knowledgeable in the field and at supporting a larger number of institutions in which they can learn and work.

*Recommendation 14. We recommend that increased attention be given to mechanisms, such as postdoctoral fellowships, long-term institutional grants for training, career development awards, and established investigatorships, to attract more scientists to devote their careers to the field.*

*Recommendation 15. To enhance the status of the field within the academic community, we recommend the establishment of formal departments of reproductive sciences in medical and graduate schools, encompassing male as well as female reproductive processes. Such a step would help to ensure adequate attention to the field, particularly in those institutions in which it is presently a part-time program.*

*Recommendation 16. Numerous institutes or research centers devoted to the reproductive sciences have the capacity to expand considerably their levels of activity and research policy should facilitate this expansion. For the enlarged effort proposed, however, additional institutes or centers will be needed to work on studies that cannot be carried out by individuals or that require multidisciplinary inputs. We recommend the creation of additional facilities with major commitments to this field of inquiry.*

*Recommendation 17. The emphasis we place on safety studies suggests the need for additional specialized institutes capable of conducting such studies both in the clinical testing stage and after a product is in general use. We recommend the creation of an adequate number of centers expert in carrying out clinical trials and studies of human pharmacology. We also recommend that several institutes for the long-term monitoring and evaluation of contraceptive methods be established. In the entire world, there is not one authoritative institution with a continuing responsibility for such epidemiological studies, despite the widespread concern over the safety of contraceptive modalities.*

Every effort should be made to increase industry's participation, but it would be unwise to expect that industry will play the dominant role in the contraceptive development process. For the reasons we have cited, we believe that the initiative will increasingly have to come from specialized programs supported by governmental or philanthropic funds. The contraceptive development programs that have emerged in the last several years have already begun to make significant contributions.

*Recommendation 18. We urge national and international assistance agencies to increase the levels of support for these public sector programs to develop new and safe contraceptive methods and assure their appropriate introduction and widespread availability.*

## Monitoring the Field's Progress

The results of the reproductive research enterprise are of immense practical importance to hundreds of millions of persons throughout the world. It can be said that the enterprise is too important to be left to the scientists—or to the policy makers. Those who are consumers have a right to know about the field's progress and the dissemination of such knowledge can only have beneficial effects, in the long run, on its development. There are doubtless many reasons for the low priority accorded the field, but one of them is that consumers who would directly benefit from its findings have been given little information and have essentially been left out of the decision-making process.

The collection of information about the field and its analysis to identify trends, problems, and opportunities has been a principal purpose of this review. The importance of such monitoring is indicated by the systematic picture that has emerged, for the first time, of the field as a whole. We believe this effort should be continued on a regular basis.

*Recommendation 19. We recommend that an appropriate mechanism be established to assemble periodically information on the field's funding levels, personnel and institutional resources, research emphases, and allocations throughout the world. The systematic analysis of such information and its publication in appropriate forms would be of immense value to science policy makers in every nation.*

*Recommendation 20. We recommend support for dissemination of authoritative information to the public on safety and efficacy of current methods, research findings that offer the hope of emerging as new methods, funding levels and sources and patterns of support for research, and related subjects of general concern. The importance of such an effort in building public understanding has been amply demonstrated in other fields of science.*

### References

United Nations, Center for Economic and Social Information. 1974. *United Nations World Population Conference—Action Taken at Bucharest.* (CESI/WPY-22).

Zuckerman, S. 1959. Mechanisms involved in conception. *Science* 130: 1260-1264.

# I  The Challenge to Reproductive Research

# 1 The Challenge to Reproductive Research

All adult human beings are involved in the process of reproduction. Whether these processes actually result in the fruition of live birth, how many births occur, whether the newborn are healthy and the mothers survive, and whether these variables are affected by rational and effective regulation are questions that confront each individual at some time in life; and they must also be faced by families, communities, and nations.

As we enter the last quarter of the twentieth century a partial table of accounts of the human reproductive process would look something like table 1.1.

Last year there were approximately 335 million conceptions throughout the world; of these perhaps 160 million[1] terminated in spontaneous miscarriages and stillbirths, about 50 million were terminated by induced abortion, and the remainder (about 125 million) resulted in live births.[2] The evidence suggests that 30 to 55 million of these births were neither planned nor wanted by parents at the time of conception.[3] One-fifth of all births were to women below age twenty or above age thirty-five (Berelson 1971).

Of the 125 million babies born:
- Twenty million were born prematurely, and many of of them were unlikely to survive or grow normally.
- About 14 million died before the end of their first year (UN 1974).
- An additional 18 million would be expected to die before age fifteen.
- At least 6 million had chromosomal abnormalities, hereditary biochemical disorders, or major congenital defects at birth, with handi-

[1] Much depends upon one's definition of pregnancy; perhaps 30 to 50 percent of all fertilized ova do not survive long enough to become recognized pregnancies.

[2] The accounting should also show that approximately 10 to 15 percent of couples are involuntarily sterile.

[3] In the United States, 15 percent of all marital births between 1965 and 1970 were classified as unwanted and an additional 29 percent were classified as timing failures (U.S. Commission on Population Growth and the American Future, 1972). In England and Wales, 16 percent of women who gave birth during 1969 to 1970 reported that they "were sorry that [the pregnancy] had happened at all" (Bone 1973). Knowledge, attitude, and practice surveys in thirty-five developing countries show that about 70 percent of women say they want no more children after the fourth (Nortman 1974a). These fragmentary data support the inference that the magnitude of unwanted births and timing failures throughout the world ranges from about 25 to 45 percent of all births—the proportions utilized here.

Table 1.1  Table of Accounts of Human Reproduction, ca. 1975

| | |
|---|---|
| Number of conceptions | 335 million[a] |
| Spontaneous abortions and stillbirths | 160 million[a] |
| Recognized | 50 million[a] |
| Unrecognized | 110 million[a] |
| Induced abortions | 50 million[a] |
| Live births | 125 million |
| Premature | 20 million |
| Normal delivery | 105 million |
| Infant deaths | 14 million |
| Deaths between age 1 and age 15 (at present mortality rates) | 18 million |
| Chromosomal abnormalities, hereditary biochemical disorders, and major congenital defects at birth | 6 million[b] |
| Maternal deaths (death attributable to pregnancy and delivery including induced abortion) | 0.2 million[a] |
| Maternal morbidity and relatively serious complications of pregnancy and delivery | 25 million[a] |

[a]Data insufficiency means that these are very approximate estimates.

[b]Based on rough estimates of 700,000 chromosomal anomalies, 1,400,000 hereditary biochemical disorders, and 4,000,000 major congenital defects, as derived from the data reported by Milunsky 1973.

Source: These estimates, which are presented only as suggestive orders of magnitude, appear to be consistent with the available fragmentary data. We are indebted to Karin Edstrom, M.D., of the Maternal and Child Health Unit of the World Health Organization, Geneva, and to Dorothy Nortman and Christopher Tietze, M.D., of the Population Council, New York, for generous assistance in developing these estimates.

caps that are either untreatable or require intensive sophisticated health resources.[4]

• Perhaps 3 to 4 million were seriously retarded and most will remain so as long as they live.

• An unknown but large number were malnourished, pre- or postnatal-

[4]This may well be an underestimate; some studies have shown congenital abnormalities as high as 7.1 percent of pregnancies (detected by age two in a sample of births to women enrolled in the Health Insurance Plan of New York) (Shapiro and Abramowicz 1969). For a comprehensive review of available studies, see Milunsky 1973.

ly, with the resulting risk of mental retardation and chronic debilitating illness.

Of women who carried these pregnancies, at least two hundred thousand died before the gestational period was completed or during parturition; an unknown but large number suffered from maternal malnutrition, anemia, infection, or chronic illness, often with lifelong effects; and about 25 million experienced serious illness or complications of pregnancy and delivery.

There is, of course, some overlap between these adverse pregnancy outcomes. But the malformed, retarded, and sick babies who survive face limited opportunities to participate meaningfully in the lives of their families and societies: Their ability to obtain an education will be more restricted than that of other children in their societies, as will their ability to learn a trade and engage in productive employment.

These crude estimates are only rough orders of magnitude, but they illustrate the importance of the human reproductive process for the health, well-being, and future functioning of individual women and children. If more adequate epidemiological data were collected, it is highly likely that the accounts would show even greater wastage of human reproductive effort. Biological imperatives make some reproductive wastage unavoidable, but additional knowledge and its application would reduce these risks to levels closer to the irreducible biological minimum and thereby enable more of the world's children to be born healthy and to lead productive lives.

These health and social consequences are not confined to the individual women and their babies; they also affect the functioning of families, communities, and nations. This is perhaps best illustrated in a developed country such as the United States. Consider, for example, the impact of pregnancy on the life chances of an unmarried American adolescent aged sixteen: She is immediately faced with the difficult choice of terminating the pregnancy or carrying it to term. If she chooses the latter course, she faces, in addition to the health risks already noted, other consequences: In all probability she will drop out of school and not complete her education. She will bear an out-of-wedlock child or rush into a precipitous marriage. Both courses will severely limit her chances to obtain employment and establish a career, and the marriage itself carries a 50 percent probability of ending in divorce. Youngsters in these circumstances pay dearly, both immediately and in terms of their long-term earning capacity. In the United States in 1973, nearly one in ten adolescent women found themselves in such circumstances. (In contrast, about one out of a hundred U.S. youngsters aged fifteen to nineteen had gonorrhea in 1973 (1155/100,000) and two out of every ten thousand had primary and secondary syphilis (19.2/100,000)—rates that were increasingly characterized as of "epidemic" proportions [U.S. Department of Health, Education and Welfare 1975].)

These problems facing individuals thus aggregate to a series of intractable social problems in industrialized nations: illegitimacy, early marriage, educational disadvantage, the perpetuation of poverty, family disruption, the need to divert considerable resources from the maintenance of health to the intensive care of the seriously ill and disabled. Each of these problems necessarily commands societal attention and requires societal resources that, in a country like the United States, account for a significant share of governmental expenditures. In developing nations that are not yet able to sustain extensive health and welfare programs, the consequences are expressed differently and are less visible at the national level, but they are nonetheless similar in principle.

Thus, human reproduction, which is essential to the survival of the species, can also exact high costs from both individuals and societies. These adverse consequences have multiple causes, ranging from our ignorance of many aspects of the basic biological processes of reproduction to the grinding poverty and malnutrition that afflicts much of the world and the inadequacies of our educational and health systems. In the complex web of interacting factors adversely affecting reproductive outcome, the inability of millions of individuals adequately to regulate pregnancy—to determine whether and when they will bear children—is critically important:

• Bearing children at too young or too old an age or too close together and having too many children are major factors contributing to one or more of the following adverse health outcomes of pregnancy:

infant and maternal mortality and morbidity, malformations, and congenital defects.
- Unplanned and unwanted pregnancies—estimated to comprise between 25 and 45 percent of pregnancies worldwide (see footnote 3)—are prevalent particularly among younger and older women, respectively. The younger and older ages of childbearing are precisely those at which the risks of adverse health outcomes are greatest and the socioeconomic consequences most acute.

These interrelations have been documented in detail in numerous studies[5], and improvements in the health of mothers and children resulting from more effective regulation of pregnancy can be demonstrated in many ways. To take but one example, the risk of mortality for the woman and of mortality or malformation for the fetus is generally related to the age of the mother. Figures 1.1a and 1.1b express the risks for various maternal ages compared to the level of risk for mothers aged twenty to twenty-four.

Because unwanted and mistimed births are concentrated at young and old maternal ages, the reduction of unintended fertility would tend to reduce the incidence of such age-related problems. For example, one estimate is that if all unwanted births[6] in the United States in the period 1965 to 1970 had been avoided, the incidence of serious sex chromosome defects such as Down's Syndrome would have been about 12 percent lower. If all births to women over age thirty-five had been avoided, a reduction in incidence of about 25 percent would have been expected (Teitelbaum 1972).

The effort to solve these varied problems proceeds on several fronts: Some lines of biomedical research attempt to find direct solutions for adverse health outcomes (such as the important work in perinatology aimed at developing better means of managing threat-

---

[5] See, for example, the excellent reviews of the literature in Wray 1971, Omran 1971, World Health Organization 1970, Nortman 1974b, and Siegel and Morris 1974. Sixteen studies of the effects of birth interval on mortality, morbidity, prematurity, growth, and intelligence of children and on maternal health are reviewed in Wray; the studies were done in the United States, England and Wales, India, Colombia, and Thailand.

[6] Defined as unwanted according to the definition employed in the 1965 National Fertility Study in the United States. See Ryder and Westoff 1971.

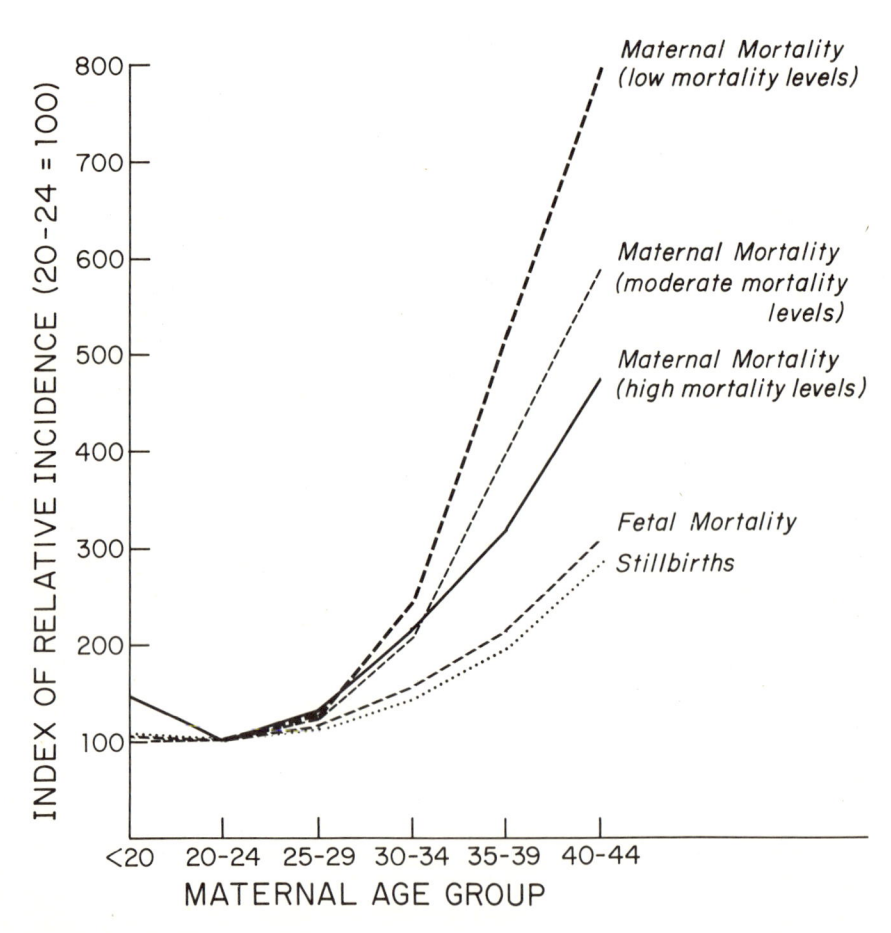

**Figure 1.1a** Index of relative incidence of selected adverse outcomes of pregnancy, by maternal age (incidence of each outcome for age group 20-24 = 100). Source: Calculated on basis of median index numbers for age group from Nortman 1974b, tables 2, 3, 4, 5, 7, 8, 9, and 11a.

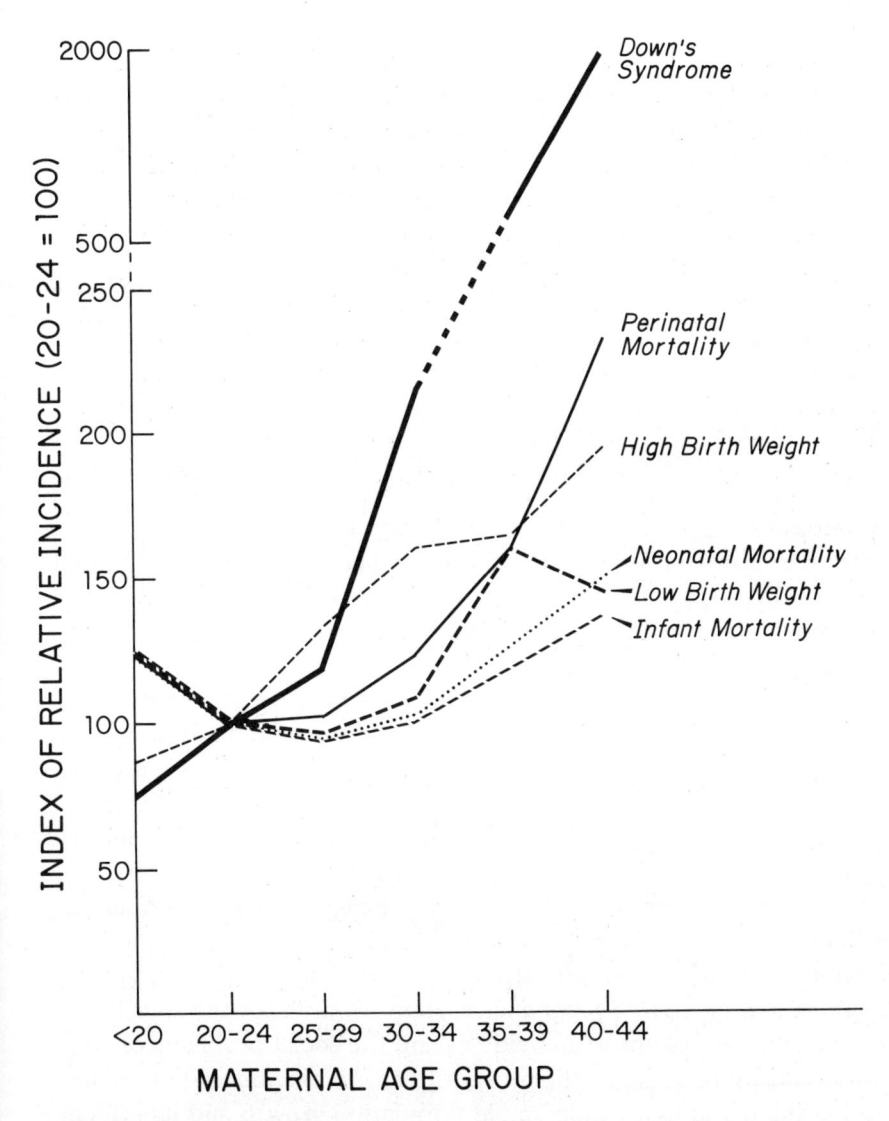

Figure 1.1b Index of relative incidence of selected adverse outcomes of pregnancy, by maternal age (incidence of each outcome for age group 20-24 = 100). Source: Calculated on basis of median index numbers for age group from Nortman 1974b, tables 2, 3, 4, 5, 7, 8, 9, and 11a.

ened pregnancies). Other researchers seek to unravel the mysteries of genetic development and to determine the causes of and remedies for retardation. Meanwhile, as part of their overall efforts to improve the well-being of their citizens, nations attend to the not insignificant health, social, and economic consequences of children already born. And individual men and women, in both developing and industrial nations, cope with what life brings them.

In the last fifteen years, another branch of biomedical research, the reproductive sciences and contraceptive development, has emerged. This branch seeks indirectly to affect all of these health, social, and economic consequences of the reproductive process. This is the worldwide effort to develop more effective, safer, and simpler techniques for regulating fertility. The reproductive sciences branch complements (and to some extent overlaps) ongoing research in perinatal development, genetics, and endocrinology. Unlike these other branches, however, it seeks to add to fundamental biomedical knowledge findings that are particularly applicable to the task of discovering improved ways for human beings safely to control the reproductive process.

Unwanted and unplanned pregnancies, and the negative outcomes associated with them, are, or can be, largely preventable if more individuals are able to regulate their fertility. This branch of reproductive research, aimed at the discovery of improved means of fertility regulation, would thus contribute greatly to reducing the incidence of the adverse social, health, and economic consequences of pregnancy and to ameliorating the resulting aggregate societal problems.

*Such a development would also have another important effect: It would contribute to fertility decline by reducing or eliminating unplanned and unwanted pregnancies.* Here research for the solution of the personal problems of individuals and the social problems of societies links up to some of the most critical and overarching problems facing the world as a whole—rapid population growth and its effects on economic development in the Third World and poverty in the industrial nations; environmental degradation; and, indeed, the very quality of human life.

Volumes have been written about population problems and there are lively debates among scientists and policy makers about their determinants and consequences. For our purposes it will be sufficient merely to outline the main parameters of the world's current demographic situation.

World population is presently growing at an annual rate of 2 percent, a rate without precedent in history. From the appearance of humankind on earth up to early historical times, the rate of increase in the human population averaged almost zero—perhaps +0.002 percent per annum over the long term. By the mid-1700s the average annual rate of growth had increased to about 0.3 percent. From that period onward the rate has rapidly accelerated, reaching 1 percent in the 1950s and doubling again to 2 percent only twenty years later.[7]

Growth rates are far from uniform among and within nations. They are much higher in developing than in industrial countries; and within nations, they are typically higher among the poorest and least educated. In developing countries, which comprise more than two-thirds of the world's population, annual growth rates average 2.5 percent and range as high as 3.4 percent. These are extraordinarily high by all historical standards, including the rates of growth in Europe during the nineteenth and twentieth centuries. The sources of this unprecedented growth are the marked declines in mortality in the developing world over the past several decades, combined with birth rates that are very high compared to the rates in Europe before the onset of its mortality decline.

This unprecedented rapid population growth is intimately related to problems of development, food supply, and the environment. The processes summarized in the term "development" represent complex and multifaceted transformations of the entire social order from the traditional agrarian society to the modern industrialized society.

---

[7] A continuation of this 2 percent growth rate into the long term is a logical absurdity. Such a continuation would imply that:
in just over three hundred years the entire land area of the earth would have a density of New York City today;
in six hundred years there would be one person per square meter of land area; and
in twelve hundred years the weight of world population would be greater than the mass of the earth.

Although there is disagreement concerning which components of change are most important for the development process, most scholars agree that traditional societies are characterized by predominant patterns of rural settlement; heavy concentration of the labor force in agriculture, fishing, or forestry; low per capita income; and restricted educational opportunities, especially for women.

Rapid population growth can have major negative impacts upon the rate of change in each of these correlates of underdevelopment. In theory, the rate of population growth could be slowed either by increasing levels of mortality or declining levels of fertility. Because every nation and social system is committed to policies and programs to reduce mortality levels, the prospects for reducing growth rates depend, in practice, on the reduction of fertility. Fertility decline would speed the development process by reducing the burden of child dependency, which in the short run would permit a higher level of productive investment and make less difficult the provision of employment opportunities. In the long run, reduction of fertility would restrain the overall level of growth in population size and density and reduce the increase of social demands that these imply.

These relationships are reflected in many ways. For example, although the past decades have seen substantial improvements in nutritional levels around the world, severe shortages of food in the last two years have led to widespread famine and malnutrition in some parts of the world. These shortages are due only in part to population increase, with climatological, economic, and political factors playing more important roles. But rapid population growth contributes to the increasing difficulty of feeding the human population. Increased cultivation of less fertile land and more intensive cultivation of present farm land involves major inputs of capital, fertilizer, and technology. The balance of food supply and demand has become more delicate, so that apparently minor fluctuations in climate and productivity can set off complex reactions of price changes and supply deficiencies. Substantial advances can still be made, and indeed must be made, if the world is to feed a doubled population by 2010. But beyond a doubling of agricultural productivity, the prognosis for further increases in food production become less optimistic.

In many developing nations these effects are dramatic and urgent. Comparable effects also can be traced on efforts to improve living standards of high growth subgroups within industrial nations. The difficulties that rapid growth imposes on development in developing nations are paralleled in industrial nations by the difficulties that rapid growth imposes on the poorest and least educated. In the urbanized individual circumstances prevalent in industrial countries, the direct and indirect costs of many children are higher than in the rural agricultural context of many developing countries. The earning power of both father and mother can be negatively affected by many children, and the direct investment required per child is much higher than in the developing countries (Kuznets 1974).

In industrial nations the impact of rapid growth is often discussed in terms of the deterioration of the environment, and there is considerable dispute as to the relative importance for environmental degradation of population growth, energy consumption practices, waste disposal, or industrial production. In developing countries with low levels of industrialization and high rates of population growth, there is less dispute that population growth is more important than industrial processes in leading to environmental degradation. This is particularly true in such areas of the developing world as parts of the Indian Subcontinent where rapid defoliation of forest land is occurring to provide fuel for the burgeoning population. Yet industrial development is a declared goal of most leaders of developing countries, so successful efforts toward this goal will likely increase the environmental problems posed by industrial practices.

In many nations, therefore, there are cogent reasons to believe that fertility decline would have beneficial effects. Ansley Coale (1973) has succinctly summarized three broad preconditions for such declines:

Fertility must be within the calculus of conscious choice; potential parents must consider it an acceptable mode of behavior to balance advantages and disadvantages before deciding to have another child. Reduced fertility must be seen as advantageous to individual couples. Effective techniques of fertility reduction must be available and known by the couples.

A complex web of social, psychological, economic, medical, and environmental factors are thus intertwined in what we have come to call "population problems" and the problems of adverse reproductive outcome. The solutions will necessarily also be complex, involving basic changes in individual behavior, health, and educational systems and social and economic structures. The development of improved fertility control methods does not envision a technological panacea; neither the problems of rapid population growth nor those of adverse reproductive outcome will be solved simply by a technological discovery. It does envision that by facilitating the realization of Coale's third condition, technological change can have an important impact on the interrelated processes that contribute to the severity of these problems. Improved control of fertility in both developing and industrial nations, leading to more rapid fertility decline than would otherwise be the case, can intensify the effects of other social programs and accelerate the pace of change. Moreover, the costs of even an optimum research effort in this field are modest compared to the costs of other development or social programs, a significant consideration in a world where resources are always limited.

Reproductive research related to fertility regulation has thus emerged as a critically important area of scientific inquiry. The ultimate objective is to contribute to improvement of human welfare through an improved fertility control technology. But progress toward that goal is dependent upon knowledge of the human reproductive process; we can learn to regulate safely and efficiently only what we understand. As a result, much of the work in the field necessarily consists of fundamental studies in the reproductive sciences and its findings are relevant to a wide range of health problems associated with childbearing.

This report seeks to assess the progress of this research effort thus far, to illuminate what has been learned in the last fifteen years, to identify persisting gaps in the knowledge base as well as promising lines of further investigation, and to outline the scope of a worldwide program commensurate with the size of the task. Because reproduction is a social and cultural as well as biological process, the task is formidable, as is evident from consideration of the limitations of the current contraceptive array and the requirements imposed on the technology by the diversity of the human reproductive process itself.

## References

Berelson, B. 1971. Report of the president: 18-35 in place of 15-45? In *Population Council, Annual Report*, pp. 19-27. New York: Population Council.

Bone, M. 1973. *Family planning services in England and Wales.* London: Her Majesty's Stationery Office.

Coale, A. 1973. The demographic transition reconsidered. In *Proceedings of the IUSSP International Population Conference*, pp. 53-72. Liege: International Union for the Scientific Study of Population.

Kuznets, S. 1974. Fertility differentials between less developed and developed regions: Components and implications. Discussion Paper No. 217. Yale University, Economic Growth Center.

Milunsky, A. 1973. *The prenatal diagnosis of hereditary disorders.* Springfield, Illinois: Charles C Thomas.

Nortman, D. 1974a. Population and family planning programs: A factbook. *Reports on Population/Family Planning*, no. 2, 6th ed.

Nortman, D. 1974b. Parental age as a factor in pregnancy outcome and child development. *Reports on Population/Family Planning*, no. 2, 6th ed.

Omran, A. R. 1971. *The health theme in family planning.* Chapel Hill: Carolina Population Center, monograph no. 16.

Ryder, N. B., and Westoff, C. F. 1971. *Reproduction in the United States: 1965.* Princeton: Princeton University Press.

Shapiro, S., and Abramowicz, M. 1969. Pregnancy outcome correlates identified through medical record-based information. *American Journal of Public Health* 59(9): 1629-1650.

Siegel, E., and Morris, N. M. 1974. Family planning: Its health rationale. *Am. J. Obst. Gynecol.* 118:995-1004.

Teitelbaum, M. S. 1972. Some genetic implications of population policies. In U.S. Commission on population growth and the American future. *Research reports, demographic and social aspects of population growth*, vol. I, ed. C. F. Westoff and W. Parke, Jr., pp. 493-503. Washington, D.C.: Government Printing Office.

United Nations. 1974. ECONF 60/CBA/14.

U.S. Commission on Population Growth and the American Future. 1972. *Population and the American future.* Washington, D.C.: Government Printing Office.

U.S. Department of Health, Education and Welfare. Center for Disease Control. 1975. *VD fact sheet 1974.* DHEW publication no. (CD) 75,8195. Washington, D.C.: Government Printing Office.

World Health Organization. 1970. *Health aspects of family planning.* Technical report series no. 442. Geneva: World Health Organization.

Wray, J. D. 1971. Population pressure on families: Family size and child spacing. In National Academy of Sciences, *Rapid population growth: Consequences and policy implications*, pp. 434-45. Baltimore: Johns Hopkins Press.

# 2 The Uses and Limits of Current Contraceptive Technology

Whether one's perspective is limited to the personal effects of high fertility on individuals or its aggregate effects on nations and the world community, it seems evident that reduced fertility levels are a desirable, even urgent, objective. Some degree of fertility decline would be expected as a consequence of social transformation. But in many nations and subgroups within nations, high fertility itself retards the process of social transformation. If fertility could be reduced concomitantly with other efforts at development, it would have a multiplier effect on the overall process. To attempt directly to reduce fertility levels, many nations have instituted national programs to educate individuals about the importance of fertility regulation for their own well-being and to give them greater access to existing contraceptive technology. Because the existing technology has many limitations, however, research is also underway in many nations to discover new and improved techniques for fertility regulation.

It is important to recognize that the two approaches are entirely complementary and indeed, necessary. A considerable degree of fertility decline would be expected to result from more vigorous family planning programs that distribute the current technology more efficiently throughout the world. At the same time, new methods, particularly those better adapted to mass application in nations lacking sufficient health resources, would make a large difference in the worldwide effort. At any level of motivation to regulate fertility, more effective and acceptable methods that are suitable for less costly and sophisticated delivery systems would increase the effectiveness of fertility regulation and the magnitude of the reduction in fertility.

Current fertility control methods can conveniently be classified as conventional contraceptives (diaphragms, condoms, foams, creams, and so forth), hormonal contraceptives, intrauterine devices (IUDs), voluntary sterilization, and induced abortion.[1] Oral contraceptives and current IUDs are entirely the result of reproductive research, which has also yielded important improvements in procedures for

---

[1] The classification omits rhythm, withdrawal, and other folk methods used in some parts of the world that are rarely provided through organized family planning programs, physicians, or commercial distributors.

abortion and female sterilization. These technological changes in the last fifteen years have brought about a "contraceptive revolution" (Ryder and Westoff 1971) in industrial nations with long histories of contraceptive practice and were major stimulants to the initiation of family planning programs in developing nations in which fertility regulation was either minimal or dependent on folk methods.

**Impact of the New Technology**

The pill was introduced in the United States in 1960 and IUDs were introduced two years later. Within ten years nearly three in five married couples were using pills, IUDs, or voluntary sterilization, the three most effective methods (Westoff 1972). These changes were among the principal factors responsible for a 36 percent reduction in unwanted births among U.S. couples during the last half of the 1960s, and even greater reductions among subgroups that had previously experienced the highest rates of unwanted fertility (Ryder and Westoff 1972). Not only was the overall efficacy of contraceptive practice substantially improved, but an increasing proportion of U.S. couples initiated contraceptive practice earlier in their family-building years.

Similar changes, though differing in some details, occurred in England and Wales (Bone 1973), Australia (Borrie 1974), and other industrial nations. The pill had become the second most popular method in Great Britain by 1970, although it was used by a somewhat smaller proportion of couples than in the United States (as was contraceptive sterilization). The pill was also used by a greater proportion of younger English and Welsh women, presaging future changes in contraceptive patterns.

The effect of the new technology in developing nations is illustrated in other ways. Of sixty-six such countries that either have adopted official policies to reduce population growth or provide support for family planning programs, all but one adopted the policy or initiated the program *after* the new technology began to become available in 1960 (Nortman 1974). The limitations of conventional contraceptive methods had previously been so extensive as to discourage many health and governmental leaders from undertaking programs, regardless of their interest in curbing population growth. The new technology encouraged, for the first time, serious consideration of the initiation of large-scale programs. This was particularly true of the IUD; the pill was at first regarded as too expensive and unsuitable for mass application, and some governments were concerned over its safety.[2] As one observer put it in 1967, the IUD (and inferentially, the new technology as a whole) "by giving national programs some hope of success . . . stimulated a wholly new level of effort, improved the morale of family planning workers from the top down, and most importantly, brought about the development of family planning organizations in a form and magnitude not previously known" (Berelson 1969, p. 365).

By the early 1970s, the IUD and the pill had become the predominant methods used by new acceptors in all but seven of forty-four national programs in Asia, Africa, and Latin America for which data were available. Only the programs in India, Bangladesh, and Singapore relied predominantly on conventional methods and sterilization; and only the programs in Pakistan, Nepal, Jamaica, and Trinidad depended primarily on conventional methods alone (Nortman 1974). Similar patterns are reported in the International Postpartum Program, a demonstration project initiated in 1966 by the Population Council to test the feasibility of maternity-centered delivery of family planning services. This project, which links programs in twenty-one nations, had provided contraception to more than a million women by the beginning of 1973. In this program, 68 percent of acceptors used the IUD and 29 percent used the pill (Sivin 1974). There were wide variations in these percentages from country to

---

[2] Referring to the initiation of large-scale family planning program activities in India and Pakistan around 1965, Finkle states, "No development was more significant in making 1965 the landmark year for family planning than the introduction of a new contraceptive technology known as the intrauterine contraceptive device (IUD). . . . It was later realized that the IUD was not the ideal contraceptive that would enable the two countries to sharply curtail fertility rates in a matter of years. . . . Ironically, if it had not been for the unreal expectations attributed to the IUD, neither country probably would have invested the resources and manpower in family planning that it did in 1965" (Finkle 1971, pp. 263-264).

country, implying that the choice of method may have been significantly influenced by the preferences of program staffs; but the results nevertheless suggest the importance of the new methods in developing nations. In this program, as in other reports from both developing and industrial nations, pill acceptors tend to be younger and to have fewer children than IUD acceptors (Sivin 1974; Nortman 1974).

The new technology has thus spurred the growth of conscious efforts to introduce modern contraception in the poorer nations. Despite the limited funds made available to support these efforts, the programs have contributed to a widespread increase in public knowledge and awareness of family planning methods and to important increases in contraceptive use (Mauldin 1975).

The impact of the programs appears to depend on their scope. All five classes of fertility control methods (conventional, orals, IUDs, sterilization, and abortion) are generally available in the People's Republic of China and South Korea, and four are generally available in Puerto Rico, Singapore, Taiwan, and Thailand. Most of these countries experienced very rapid fertility declines, ranging from 26 to 43 percent, during the 1960s; the exception is Thailand, which adopted its national policy in 1970 and had a more modest fertility decline of 14 percent since 1960[3] (Mauldin 1975). In countries where programs offered fewer methods, the decline in fertility has been smaller.

## Limitations

The experience in both industrial and developing nations thus suggests that millions of men and women have substantially altered their fertility regulation practices as a result of the improved contraceptive technology that emerged in the last fifteen years. In certain industrialized nations, the use of modern contraceptives backed up by substantial numbers of abortions is capable of lowering crude birth rates to a level of 15 per 1,000. Such results have not been possible in

[3] For data demonstrating greater fertility declines in countries with more energetic family planning programs, also see Ravenholt and Chao 1974.

developing nations, where the social, economic, and cultural circumstances are markedly different from industrial nations and where the health services infrastructure is particularly inadequate for the wide distribution of medically supervised methods of fertility regulation. Based on these considerations, the necessity for new techniques adapted to the conditions of developing countries is fairly widely understood.[4]

What is not so widely understood are the limits of current contraceptive technology even in industrial nations such as the United States and the extent to which the regulation of fertility in developed nations is dependent on the widespread practice of abortion because of those limits. The pill and the IUD represent major improvements over conventional contraceptive methods. But they have limitations with respect to efficacy, safety, acceptability, and continuity of use that at the least make more difficult and expensive the delivery of family planning services and probably also limit the extent of effective regulation of fertility that can be expected with current technology.

Data from the 1970 National Fertility Study in the United States provide an overview of these difficulties even in a highly contracepting, affluent nation with high levels of literacy and relatively well developed health service, transportation, and communication networks. The study found that one-third of married contraceptors who do not want another child fail within five years to prevent an unwanted conception (Ryder 1973). The extent of contraceptive failure was markedly lower among younger women who married during the 1960s, and more than half of this decrease was attributable to more frequent adoption of the pill by these women (Ryder 1973), foreshadowing improved contraceptive efficacy in the future as the older marriage cohorts leave the childbearing ages. Despite these indicators of progress toward effective regulation of fertility, the study found the overall use-effectiveness of current contraceptive methods

[4] A review of the current status of family planning programs concludes, "An improvement in the technology of fertility control, in the form of an attractive new substance, device, or mode of administration (e.g., vaccination) would make a non-trivial difference, particularly if it had high continuity and did not require medical intervention" (Freedman and Berelson 1976, p. 13).

to be considerably lower than would theoretically be expected. The standardized proportions of married women who fail to delay a wanted pregnancy or to prevent an unwanted pregnancy within twelve months of initiation of use of various contraceptive methods were as follows (Ryder 1973):

| Method | Percent Failing |
| --- | --- |
| Pill | 6 |
| IUD | 12 |
| Condom | 18 |
| Diaphragm | 23 |
| Foam | 31 |
| Rhythm | 33 |
| Douche | 39 |

These high failure rates in actual use explain, at least in part, the study's findings that 15 percent of all births to U.S. married couples between 1965 and 1970—a minimum of 2,650,000 births—were unwanted at time of conception and an additional 29 percent were unplanned (U.S. Commission on Population Growth and the American Future 1972). (These data, based on a sample of married women only, omit unintended births to unmarried women, which are not inconsequential.) They also help explain the high incidence of induced abortion in the United States (without which the number of unintended *births* would be even greater). In 1973, the year in which the U.S. Supreme Court invalidated state laws barring or restricting abortion in the first and second trimesters, more than 745,000 legal abortions were reported in the United States. Based on trends in the last part of 1973 and the first quarter of 1974, the number was projected to rise to 892,000 in 1974 (Weinstock et al. 1975). Even though there remain wide disparities in the geographic availability of legal abortion and the reported statistics are regarded as underestimates, the ratio of reported legal abortions in the United States reached 238 abortions per 1,000 live births in 1973 and was expected to climb still higher in 1974.

Comparable detailed information on contraceptive use and failure rates is not available for other industrial nations, but abortion ratios are. They provide some indication that unintended pregnancy remains a problem of epidemic proportion even in industrial nations with highly contracepting populations, although the highest ratios are reported in countries in which contraceptive use is limited (Tietze and Murstein 1975, tables 2A-D[5]).

| Country | Year | Legal Abortions Per 1,000 Live Births |
| --- | --- | --- |
| Bulgaria | 1974 | 828 |
| Canada | 1973 | 145 |
| Czechoslovakia | 1973 | 287 |
| Denmark | 1974 | 344 |
| England and Wales (residents) | 1974 | 176 |
| Finland | 1973 | 392 |
| German Democratic Republic | 1973 | 626 |
| Hungary | 1974 | 518 |
| Japan | 1974 | 308 |
| Norway | 1972 | 192 |
| Poland | 1971 | 236 |
| Sweden | 1974 | 285 |
| U.S.A. | 1973 | 238 |
| Yugoslavia | 1968 | 642 |

Because both contraception and abortion are means to prevent unintended births, improved contraceptive efficacy will, in the long run, tend to reduce the incidence of abortion. The data from the table imply that nations such as Bulgaria, Yugoslavia, and several other Eastern European countries, which have high abortion ratios, could reduce the incidence of abortion by more systematic distribution of existing contraceptive technology. *But they also suggest that given the limits of current contraceptives, there will be a considerable need for abortion even in nations using the technology maximally.*[6]

[5] All ratios are computed from officially reported abortion data, except for the United States, which is derived from a nationwide survey of provider agencies.

[6] Findings of a computer simulation imply that even with highly effective use of current contraceptive methods, the abortion ratio is likely to be at least 100 per 1,000 births. See Tietze and Bongaarts 1975.

The search for an improved technology is thus in part a search for a means to avoid abortion as a means of regulating fertility.

**Safety and Acceptability**

The high failure rates observed with such methods as the pill and the IUD, which have high theoretical effectiveness, present a paradox explained in large part by their limitations with respect to safety, acceptability, and continuity of use. The relationships are complex in general but the safety factors (coupled with irritating side effects, which pose less risk) are sufficiently problematic to make these methods unacceptable to some women and to induce levels of anxiety that militate against regular use by others. Taken together, these factors also make more costly and complicated the task of building an effective system for the distribution of modern contraceptive methods.

Since oral contraceptives were introduced in 1960, they have been subject to perhaps closer medical evaluation than any other drug. Based on these studies, scientists with the information necessary for the sophisticated weighing of benefits against risk are generally agreed that they "are highly effective and generally safe for most women; that the risk of developing serious illness as a consequence of taking the pill is small; that mortality associated with the orals is of a very low order of magnitude, much lower than that associated with pregnancy;... [and] that there are significant health benefits associated with [their] use" (Connell 1975, p. 62).[7] But the orals are also associated with troublesome minor side effects and rare but serious medical risks, both of which contribute to persistent anxieties among physicians and users, particularly because the orals are used primarily by normal, healthy women over long periods of time. The use of some (but not all) oral contraceptives during the period immediately following birth has negative effects on lactation, which is an important source of nutrition in many countries. As a result, physicians in some nations regard the orals as unacceptable—they were only approved for use in the Indian family planning program in 1975, for

[7] See also Royal College of General Practitioners 1974.

example—and in others, the medical consensus is that the orals' "long-term effects must continue to be monitored closely to safeguard women who use this method of contraception" (Connell 1975, p. 62).

There are very real difficulties in attempting definitively to assess the safety of any medication. When side effects are low in incidence, an extraordinarily large number of cases need to be monitored to pinpoint rare events. Many of the minor side effects associated with use of the orals, such as nausea, menstrual irregularities, melasma, or chloasma, are not life threatening, but they are sufficiently distressing to affect the acceptability and/or continuity of use of these medications for many women. The complications of greatest concern to women using the orals and to physicians are their possible carcinogenicity, their association with thromboembolic disease, and their potential relationship to liver, gallbladder, renal, and heart disease; hypertension; and possible fetal damage. Studies have shown that the risk of mortality from thromboembolism, while still small, is greater among users of orals than among other women. The orals have also been shown to produce an elevation of blood pressure in some women and the incidence of hypertension increases with duration of use. Recent studies have shown that older users of oral contraceptives, particularly women aged forty and older who have predisposing factors, are at greater risk of both fatal and nonfatal myocardial infarction than nonusers. There is no evidence of a greater risk of cancer among users, but investigations of this issue are continuing because of the concern stemming from the association between steroids and carcinogenesis in animal studies and because of the known length of induction time of malignant changes in the human (fifteen to twenty-five years). The restoration of fertility after discontinuation of use, possible teratogenesis in cases of contraceptive failure, and metabolic disorders also continue to be investigated.

As a result of these uncertainties, the U.S. Food and Drug Administration has contraindicated use of the orals by women with thromboembolic disorders, impaired liver function, carcinoma of the breast, estrogen-dependent neoplasia, or genital bleeding, or who are pregnant. Regulatory agencies in some other nations have promulgated similar restrictions.

The limitations associated with intrauterine devices are somewhat different than those associated with the orals. The devices are frequently not tolerated by women and in rare cases, carry the risk of serious consequences. As with the pill, there are non-life-threatening side effects, such as bleeding and pain, that are sufficiently distressing to make the devices unacceptable to many women. The relationship between IUDs, or particular devices, and infection has recently become an issue that is not yet fully resolved. A 1974 study by the Center for Disease Control of the U.S. Public Health Service suggested an increased incidence of septic midtrimester abortions among women who became pregnant while using a shield-shaped device. Although not conclusive, these epidemiological data led to a decision to discontinue distribution of this specific device and to a Food and Drug Administration examination of infection rates with all IUDs (1974). The Central Medical Committee of the International Planned Parenthood Federation (1975) similarly decided to urge its member organizations to discontinue use of the device and the manufacturer has withdrawn it from the market. Subsequent laboratory data have established that a feature of the specific device implicated in the CDC study, which is not present in other devices, could account for the passage of bacteria into the uterus (Tatum et al. 1975). However, observations of clinical investigators suggest that Fallopian tube pathology secondary to mild infection may be more frequent among users of an IUD than among nonusers (Wright 1974). The data are preliminary and inconclusive and the subject is receiving continuing attention.

Factors such as these have had two principal direct consequences: First, they sustain the prevailing belief that distribution of the orals and the IUDs must be dependent on relatively expensive physician-based delivery systems so that side effects can be monitored and evaluated. Second, they contribute to a pervasive anxiety over safety among many physicians and women that has reduced the acceptability of these methods and increased discontinuation rates.

Many physicians believe that the uncertainties associated with oral contraceptives require not only that the drugs be prescribed by physicians, but that users return every six months or year for check-up examinations. A 1973 study showed, however, that oral contraceptives may be legally sold without prescription in nineteen of forty-five industrial nations and are readily available over the counter in twelve of the remaining twenty-six (Black 1974). The necessity for a medical delivery system for the orals has been questioned in recent years, particularly in developing nations where rates of maternal mortality and morbidity are high and physicians and health personnel are in short supply (Atkinson et al. 1974). In such countries, the balance of benefits against risks supports nonclinical distribution because it will clearly be a long time before sufficient health resources are available to provide services to the whole population. In response to these initiatives, community-based distribution plans are being tested in several countries. The same question has recently been raised in the United Kingdom (*The Lancet* 1974; Smith 1974) and, by inference, in other industrial nations where the regulation of fertility in some subgroups (such as adolescents) seems to be intractable as long as oral contraceptives are dependent on physician prescription. The discussion thus far has been useful in compelling a more accurate calculation of the actual benefits and risks associated with the alternative modes of distribution, but continuing doubts about the pill's safety have undoubtedly influenced the course of the debate. It seems probable that official sanction of nonclinical delivery systems will be limited until safer methods are found, despite the fact that such modes of distribution are clearly necessary in many countries.

There is no simple way to assess the influence of these concerns on acceptability. As noted above, both the pill and the IUD are important methods in industrial and developing nations. But there is no way to determine from the program statistics available how many women did *not* come to the program because of concern over safety or how many women could *not* be reached because of the requirements of a medically based delivery system. The view that concern over safety and discomfort plays an important role in limiting the acceptability of the modern methods is supported by data on the high discontinuation rates associated with both the pill and the IUD. Nor is concern over safety limited to the pill and the IUD. In recent years unresolved health-related issues have been raised about other current fertility regulation methods—male and female sterilization, abortion, and even spermicidal agents. These concerns vary in severity, but they contribute to a high level of anxiety in many nations concerning the regulation of fertility.

Since 1960, it is estimated that 150 million women throughout the world have at one time or another used the pill (Connell 1975), but only about 50 million are currently using it (Piotrow and Lee 1974).[8] Both figures can only be regarded as approximations, and some women who began to use the pill during the last fifteen years undoubtedly discontinued as a result of planned pregnancy, menopause, sterilization, or other reasons indicating that they were no longer at risk of unwanted pregnancy. But the magnitude of the difference between ever-users and current users suggests that discontinuation rates are high even among women who continue to be at risk. In the United States, analysis of data from the 1970 National Fertility Study indicated that 16 percent of new pill users in 1970 discontinued during the first three months and 41 percent discontinued in the first year (Westoff and Jones 1975). In three organized U.S. programs for low income persons, between 16 and 26 percent of patients discontinued use of the IUD by the end of twelve months, compared to 25 to 45 percent of all pill users (Tietze and Lewit 1971). The patterns are similar in developing nations. Data from programs in twelve developing nations and the International Postpartum Program show standardized twelve-month discontinuation rates for IUD users ranging from 23 to 42 percent and for pill users ranging from 31 to 59 percent (Sivin 1974). Because the pill and the IUD (like other contraceptives) are only effective when used regularly, these findings go a long way toward explaining why their high theoretical effectiveness is not translated directly into an even more rapid reduction of fertility.

Thus, the current technology cannot be regarded as adequate to meet individual *or* societal needs in either industrial *or* developing

---

[8] There are, of course, great difficulties in preparing such estimates, which necessarily are only approximations. For example, 26 to 40 percent of the estimated current users are in the People's Republic of China, where there are no official published statistics.

nations. What is needed is an array of contraceptive methods that require less complex distribution systems; that are perceived by users as safer, less discomforting, and more convenient than current methods; and that are suited to the diverse requirements imposed by the diversity of conditions throughout the world.

The magnitude of the task remaining can be gauged by the results of the survey of world needs in family planning conducted by the International Planned Parenthood Federation in 1971 (IPPF 1974). Of approximately 500 million women throughout the world at risk of unwanted pregnancy, an estimated 70 percent were using no contraceptive method at all. Although expanded and more vigorous program efforts could bring the current technology to many of these women, "almost certainly more effective contraception will result over a 5-10 year period from a major advance in technology than from information and persuasive efforts" (Berelson 1969, p. 367).

## References

Atkinson, L.; Castadot, R.; Cundros, A.; and Rosenfield, A. G. 1974. Oral contraceptives: Consideration of safety in nonclinical distribution. *Studies in Family Planning* 5:242-249.

Berelson, B. 1969. National family planning programs: Where we stand. In *Fertility and family planning: A world view*, ed. S. J. Behrman, L. Corsa, and R. Freedman. Ann Arbor: University of Michigan Press.

Black, T. R. L. 1974. Oral contraceptive prescription requirements and commercial availability in 45 developing countries. *Studies in Family Planning* 5:250-254.

Bone, M. 1973. *Family planning services in England and Wales.* London: Her Majesty's Stationery Office.

Borrie, W. D. 1974. Australia. In *Population policy in developed countries*, ed. B. Berelson, pp. 270-293. New York: McGraw-Hill.

Connell, E. 1975. The pill revisited. *Family Planning Perspectives* 7:62-71.

Finkle, J. L. 1971. Politics, development strategy and family planning programs in India and Pakistan. *Journal of Comparative Administration* 3:259-295.

Food and Drug Administration. Ad Hoc Ob-Gyn Advisory Committee. 1974. Report on safety and efficacy of the Dalkon shield and other IUDs. Mimeographed. Washington, D.C.

Freedman, R., and Berelson, B. 1976. The record of family planning programs. *Studies in Family Planning*, no. 7:1-40.

International Planned Parenthood Federation. 1974. *Survey of world needs in family planning.* London: I.P.P.F.

———. Central Medical Committee. 1975. Central Medical Committee decisions—April 1975.

Mauldin, W. P. 1975. Family planning programs and fertility decline in developing countries. *Family Planning Perspectives* 7:32-38.

Nortman, D. 1974. Population and family planning programs: A factbook. *Reports on Population/Family Planning,* no. 2.

The pill off prescription? 1974. Editorial. *The Lancet* 2:933.

Piotrow, P. T., and Lee, C. M. 1974. Oral contraceptives—50 million users. *Population Report,* series A (1).

Ravenholt, R. T., and Chao, J. 1974. Availability of family planning services: The key to rapid fertility reduction. *Family Planning Perspectives* 6:217-223.

Royal College of General Practitioners. 1974. *Oral contraceptives and health—an interim report from the oral contraception study of the Royal College of General Practitioners.* New York: Pitman Medical.

Ryder, N. B. 1973. Contraceptive failure in the United States. *Family Planning Perspectives* 5(3):133-142.

Ryder, N. B., and Westoff, C. F. 1971. *Reproduction in the United States: 1965.* Princeton: Princeton University Press.

———. 1972. Wanted and unwanted fertility in the United States: 1965 and 1970. In *Demographic and social aspects of population growth,* ed. C. F. Westoff and R. Parke, Jr. Washington, D.C.: Government Printing Office.

Sivin, I. 1974. *Contraception and fertility change in the International Postpartum Program.* New York: The Population Council.

Smith, M. 1974. Distribution and supervision of oral contraceptives. *Br. Med. J.* 4:161.

Tatum, H. J.; Schmidt, F. H., Phillips, D.; McCarthy, M.; and O'Leary, W. M. 1975. The Dalkon shield controversy. *J.A.M.A.* 231(7):716-717.

Tietze, C., and Bongaarts, J. 1975. Fertility rates and abortion rates: Simulations of family limitation. *Studies in Family Planning* 6:114-120.

Tietze, C., and Lewit, S. 1971. The IUD and the pill: Extended use effectiveness. *Family Planning Perspectives* 3(2):53-55.

Tietze, C., and Murstein, M. J. 1975. Induced abortion: A factbook. *Reports on Population/Family Planning,* no. 14, 2nd edition.

U.S. Commission on Population Growth and the American Future. 1972. *Population and the American future.* Washington, D.C.: Government Printing Office.

U.S. Public Health Service, Center for Disease Control. 1974. Current trends—IUD safety: Report of a nationwide physician survey. *Morbidity and Mortality Weekly Report* 23:225.

Weinstock, E.; Tietze, C.; Jaffe, F. S.; and Dryfoos, J. G. 1975. Legal abortions in the United States since the 1973 Supreme Court decisions. *Family Planning Perspectives* 7:23-31.

Westoff, C. F. 1972. The modernization of U.S. contraceptive practice. *Family Planning Perspectives* 4(3): 9-12.

Westoff, C. F., and Jones, E. F. 1975. Discontinuation rates of the pill and the IUD in the United States, 1960-1970. *Mt. Sinai J. Med.* 42:384-390.

Wright, N. H. 1974. Unsuspected pelvic infection discovered at tubal ligation: Relationship to use of intrauterine contraception. *International Conference on Intrauterine Contraception.* Cairo: Population Council.

# 3 The Human Requirements of an Adequate Technology for Fertility Regulation

The control of human fertility is by no means a modern invention. Some individuals in nearly all societies have practiced fertility control throughout known history (Himes 1963), and anthropological evidence suggests that early human hunting and gathering societies maintained societal control upon aggregate fertility levels via regulation of marriage and by traditional means of contraception, abortion, and infanticide (Dumond 1975). What *are* new phenomena are the near-universal use of contraception in some societies and the systematic application of scientific knowledge and research toward the development of methods suitable for the rational control of fertility.

Because the process of reproduction is preeminently sociocultural as well as biological in nature, the task of developing new contraceptive techniques is a formidable one. Certain criteria, especially those of *medical safety* and *contraceptive effectiveness*, must be met by all contraceptive methods. In addition, intrinsic properties favorable to *continuity of use* are highly desirable, but such continuity should also be balanced by ready *reversibility*.

In addition to these universally important criteria, contraceptive methods must meet requirements for fertility control that vary widely with the stage of the person's life cycle, the culture in which he or she lives, and the availability of social and medical services. Indeed it is likely that there will never be an "ideal" contraceptive, for the requirements on contraceptive technology are as diverse as the human situation itself.

Table 3.1 presents one listing of acceptability criteria that vary widely among individuals and within and among societies.

With a list of criteria as diverse as this one, it is easy to see how a method that is heavily preferred in one setting can be unpopular (or even illegal) in another. Such divergent demands upon contraceptive methods exist not only among societies but also within them. Indeed, during each individual's lifetime, different types of contraceptives may be needed, depending upon life circumstances and stage of the reproductive life cycle. As an illustration of the diversity of demand experienced by each individual, four[1] stages of the reproduc-

---

[1] The age of menarche (the onset of menstruation and reproductive potential) is also a factor affecting the life of every woman. In Western Europe, the age of menarche has declined about four months per decade between 1830 and 1960 (Tanner 1962). The sources of this decline are not yet clear, though it does ap-

Table 3.1 Criteria of Personal and Social Acceptability of Contraceptive Methods

| Criterion | Attributes Required |
| --- | --- |
| Personal (motivational) | Simplicity and ease of use, independence from sex act, freedom from "nuisance" side effects, one-time administration for long-lasting effect, protection of privacy, male and/or female application, appropriateness to stage of life cycle. |
| Cultural | Suitability of method to local customs, modesty, resistance to superstition, concern with menstruation. |
| Religious | Catholic preference favoring rhythm method and opposing mechanical and chemical methods, Islamic attitudes toward menstruation as they relate to methods that change menstrual patterns, opposition of several religions to abortion. |
| Sexual | Real or perceived method effects upon libido and pleasure. |
| Medical | Preference for mode of administration (pills vs. injections), non- or positive effect on normal physiological functions. |
| Organizational and logistic | Method's reliance or nonreliance on supply lines, need or lack thereof for doctors (usually male), positive diffusibility of method through word of mouth, appropriateness of method for self-administration, ease of access to method. |
| Economic | Costs of method to individuals and/or society. |
| Political | Degree of welcome given to products from outside, symbolisms attached by society to different methods and hence their vulnerability to attack. |
| Philosophical | Degree to which method fosters freedom of choice of reproductive experience (including coerciveness or noncoerciveness of administration, effectiveness in preventing involuntary pregnancy, ready reversibility). |

Source: Adapted from Freedman and Berelson 1976, p. 14.

tive life cycle may be described:
1. Premarital[2]
2. Delay (postmarital, pre-first birth)
3. Spacing (post-first birth, before completion of fertility)
4. Completion of fertility

This categorization is, of course, an arbitrary one, abstracted from the continuum of the life cycle. The particular forms in which the stages are expressed will vary depending on the social customs and family formation patterns of different nations, and indeed some stages may be inapplicable in some circumstances. The differing characteristics of contraceptives most suitable for each of these stages are presented in table 3.2. Premarital exposure to risk of pregnancy, for example, varies widely from culture to culture. In some (such as China, by all reports) premarital sexual activity is rare. In others (such as Western Europe and North America) premarital intercourse is more common and frequent, though still far less so than within marriage. Hence where contraceptive application is related to intercourse the method may be somewhat more acceptable in the premarital stage than in the delay and spacing stages (with their greater frequency and regularity of intercourse). In many societies premarital pregnancy is subject to severe negative sanctions, hence contraception at this stage of life must be highly effective when used on an occasional basis. Indeed, in such circumstances, postcoital forms of fertility control are particularly attractive, given the irregular and infrequent exposure to risk. Premarital methods must be highly acceptable and convenient to use because persons in this stage (as well as in the later delay stage) are being introduced to contraception and will develop patterns of attitudes and practice that may last throughout their reproductive lives. Access to medical contraceptive services is often restricted for unmarried persons in societies with severe nega-

---

pear that improved nutrition is an important factor (Frisch and McArthur 1974). In most developed countries the age of marriage is relatively high, and these declines in the age of menarche have therefore had little impact upon fertility after marriage. But in developing countries such as India, where the age at marriage is often less than fifteen, similar declines in age at menarche would imply increased years to exposure to pregnancy.

[2] "Marital" as employed here includes stable consensual union.

Table 3.2 Characteristics of Contraceptives Related to Life-Cycle Stage

| Stage | Characteristics[a] |
|---|---|
| 1. Premarital | Relatively irregular and infrequent exposure. Intercourse-related methods (particularly postcoital) somewhat more acceptable than in delay and spacing stages. Serious consequences for contraceptive failure. Limited knowledge of and access to fertility control. Limited independent access to medical system, hence nonmedical delivery is preferable. Reversibility highly important. |
| 2. Delay: Postmarital, Pre-first birth | Frequent exposure. Relatively moderate consequences for contraceptive failure. Relatively short period of protection required. Methods where application is independent of intercourse are highly desirable. High acceptability and convenience important. Method delivery via medical system is less undesirable than in premarital stage due to readier access to medical system. Reversibility highly important. |
| 3. Spacing: Post-first birth, Precompletion | Frequent exposure. Moderate consequences for contraceptive failure. Long time span (as sum of separate birth intervals) of protection required. Reversibility somewhat less important than in delay stage. |
| 4. Completion of wanted fertility | Long time span of protection required. Less frequent exposure than in delay and spacing stages. Serious consequences for contraceptive failure. Intercourse-related methods somewhat more acceptable than in delay and spacing stages. Acceptability and convenience less important than in earlier three stages. Reversibility less important than in three previous stages. |

[a] All are *average* characteristics and need not apply to any particular individual in any stage.

tive sanctions, so methods suitable for distribution outside the formal medical system are preferred. Finally, reversibility of effect is a highly important attribute in the premarital stage.

In the delay stage,[3] following marriage but before the first birth, there is frequent exposure to risk of pregnancy. However, the level of effectiveness required in this stage may be somewhat less than in the premarital stage because the social and other costs of contraceptive failure are considerably lower following marriage. The length of protection required is relatively short because most couples do not delay the first pregnancy in marriage for more than a few years. The higher frequency and regularity of sexual activity implies that a method independent of intercourse is preferable, but it must have a high degree of acceptability and convenience. Delivery other than through the medical system is preferable in many nations, though the importance of this factor may be less than in the premarital stage given the lack of negative sanction on intercourse within marriage. Finally, most couples intend to have children at a later date, so reversibility is very important.

Similar criteria apply in the spacing stage as in the delay stage. The effectiveness of the method need not be as high as in the delay stage, however, because a contraceptive failure during the spacing stage has the relatively less significant impact of compressing the desired birth interval between wanted children. The length of protection required, however, is longer in the spacing stage (as the sum of separate birth intervals) than in the delay stage. Reversibility is somewhat less important because at least one child has already been born.

Following the birth of the last wanted child, births must be prevented effectively over a long time span. Exposure to the risk of pregnancy may be somewhat less than in the delay and spacing stages, but the cost of contraceptive failure increases sharply upon the arrival of the last wanted child. Hence, methods used in the completion stage must be highly effective for a long period of time. Because fecundity is likely to decline along with exposure to risk of

---

[3] In many countries, especially in the developing world, there may in fact be no delay stage, but rather a desire to achieve fertility as soon after marriage as possible.

pregnancy, methods related to the act of intercourse (such as condom and diaphragm) become slightly more attractive during this stage, though techniques separate from sexual activity are still much to be preferred. The acceptability and convenience of the method are less important than in the earlier stages of the reproductive cycle. Finally, reversibility becomes less important upon the birth of the last wanted child.

Hence at each of these four stages of reproductive life, the needs of couples place differing demands upon contraceptive technology. Another dimension of human diversity of demand is presented by the heterogeneity of personal, social, religious, cultural, and economic circumstances, as discussed previously. Individuals in all social settings experience changed requirements as they proceed through the life cycle; so these two dimensions of diversity act as multipliers of variability, presenting the reproductive scientist with a very large range of demands. It is for this reason that an "ideal contraceptive" is an unlikely development, but the realistic researcher can nonetheless aspire to the development of an array of methods with differing and diverse attributes, which collectively provide an adequate scientific response to the needs of human diversity.

There are uniformities as well. *All societies interested in reduction in aggregate rates of population increase must be actively concerned with the acceptability[4] and continuity[5] of contraceptive use. At any given level of motivation the intrinsic properties of the method will affect the degree of acceptance achieved initially and the continuity of use thereafter.* In order to have an important effect upon aggregate fertility, the overall levels of acceptability and continuity must be as large as (or even larger than) has recently been experienced in those developing countries with relatively advantageous socioeco-

[4] Empirical measurement of the comparative acceptability levels of alternative forms of contraception presents a formidable methodological challenge. There are few settings in which a full range of contraceptive methods are freely and equally available, and the socioeconomic and cultural factors thought to affect acceptability mean that data from one setting cannot be readily generalized.

[5] That is, the degree to which contraception is in continuous use, either through regular administration of methods with short-term effects (condom or diaphragm) or through continuous activity of methods needing only occasional administration (IUD, injectables, sterilization).

nomic settings and strong family planning programs. It follows that in those countries with less favorable socioeconomic conditions, even higher levels of acceptability and continuity are likely to be required of contraceptive methods if they are to make notable impacts upon aggregate fertility rates.[6] Yet none of the available methods combines high levels of both acceptability and continuity. As an illustration of this point table 3.3 categorizes on a two-dimensional grid the available modern methods of fertility control (oral contraceptives, IUDs, sterilization) and more traditional methods (diaphragm, rhythm, condom). The vertical dimension is acceptability, the horizontal, continuity.

As indicated in this overly simplified table, oral contraceptives (currently used by about 50 million women) have proven to be highly acceptable in a variety of socioeconomic and cultural settings, but also have low levels of continuity. The IUD (currently used by about

Table 3.3 Acceptability and Continuity of Use of Current Contraceptive Methods

|  | Continuity | | |
|---|---|---|---|
| Acceptability | High | Medium | Low |
| High | – | – | Orals |
| Medium | Sterilization | IUD | – |
| Low | – | – | Condom<br>Diaphragm<br>Rhythm |

Source: Based on Ross 1975. Although this table is a generalization based on worldwide experience, there are exceptions in some countries.

---

[6] If there are trade-offs between initial acceptability and eventual continuity, it is evident that the most dramatic short-term effects can be achieved via high acceptability; but in the long term the level of continuity of use is equally important.

10 to 15 million women) has somewhat better continuity experience, but also tends to be somewhat less acceptable. Sterilization has very high continuity given its nonreversibility, but in part because of this property it has only moderate acceptability. Traditional contraception tends to suffer both from low acceptability and low continuity. Abortion, perhaps the most widely practiced form of fertility control, does not appear on this table because it is difficult to characterize its level of acceptability (often official opposition in a context of widespread private use) and continuity (data on frequency of abortion are notoriously deficient). It is worth noting, however, that efficient control of fertility today does require the use of abortion; all available methods experience failure rates that would otherwise result in substantial unwanted fertility (Tietze and Bongaarts 1975).

The gaps in the upper left-hand portion of table 3.3—that denoting a method of fertility control characterized by high acceptability *and* high (or even medium) continuity—is a poignant reminder of the lacunae in our knowledge and technology. Until these gaps are filled by a variety of methods, the prospects for substantial impact upon aggregate fertility levels in the less favorable socioeconomic settings of the developing world will remain daunting.

Thus what is needed is a broader range of effective methods that are appropriate for the great variety of personal, social, cultural, religious, and economic circumstances of a diverse and pluralistic world and that are acceptable to a larger number of persons than current methods and will be used more regularly by them. This is the challenge of reproductive research aimed at improving contraceptive technology to meet human needs.

## References

Dumond, D. E. 1975. The limitation of human population: A natural history. *Science* 187:713-721. Also reprinted in *Population: Dynamics, ethics, and policy*, ed. P. Reining, and I. Tinker, pp. 83-90. Washington, D.C.: American Association for the Advancement of Science, 1975.

Freedman, R., and Berelson, B. 1976. The record of family planning programs. *Studies in Family Planning* 7:1-40.

Frisch, R. E., and McArthur, J. W.

1974. Menstrual cycles: Fatness as a determinant of minimum weight for height necessary for their maintenance or onset. *Science* 185:949-951.

Himes, N. E. 1963. *Medical history of contraception.* New York: Gamet Press, Inc.

Ross, J. (Population Council). 1975. Personal communication.

Tanner, J. M. 1962. *Growth and adolescence.* Oxford: Blackwell Scientific Publications.

Tietze, C., and Bongaarts, J. 1975. Fertility rates and abortion rates: Simulations of family limitation. *Studies in Family Planning* 6:114-120.

# II Major Advances and New Opportunities

# Introduction to Part II

The operation of the scientific process in the area of fertility control is neither the acme of efficiency nor the "Final Solver of All Problems." As is evident to all who read, it works by fits and starts, it leads to errors that must be painfully reviewed and critically rejected, it often dwells lengthily on minutiae, overstresses conditional findings, and for long periods fails to illumine factual obscurities. Nonetheless, it gets ahead, haltingly perhaps, but inevitably. The mystery and wonder of conception becomes describable in terms of gametes and their movements, in terms of fertilization reactions and the operation of replication mechanisms, in terms of oviduct chemistry and hormonal regulation. In each of these is also mystery and wonder, for there is still more to discover than we now know. But in the blazing or flickering light of what we do know a priori judgments and willful prejudices fade. And our considered and tested knowledge offers a firm basis for what we can and should do.

Gregory Pincus
*The Control of Fertility*

In the following pages account is taken of many of the major advances in knowledge of reproductive phenomena both in the female and the male achieved during the periods before and after 1960. Although it is the advances made since 1960 that are of particular pertinence to this review, these are presented against a background of earlier development to show how progress in this field has been intensified in recent years. The division of scientific research into pre- and post-1960 periods is made for expository purposes only; clearly it must be recognized that scientific knowledge is gained by accretion. Any scientific discovery has its roots in earlier work.

In order to discuss the prospects of developing new methods of fertility control, some of the major discoveries of the last fifteen years, especially in the human reproductive process, will be considered with special emphasis on those steps that appear to be particularly susceptible to hormonal, pharmacological, and immunological interference. This information was gathered from forty-one essays commissioned specifically for this review. The primary aim of this survey was to obtain from active investigators their personal assessments of where the field of investigation stands in respect to major accomplishments, knowledge gained, and concepts formulated. The essayists were also asked to point out the possibilities for interfering with the reproductive process and to identify problems in urgent need of intensive study and early solution. It was felt to be of equal importance to specify (to the extent possible) what is *not* known as well as what is known.

The essays are available in a companion volume. The biological and methodological advances in reproductive research, the present gaps of knowledge, and contraceptive possibilities are summarized in tabular form in volume II. In order to make this review accessible to a wider audience, an introduction to the field written with a minimal reliance on the specialized terminology of the reproductive sciences is provided in appendix B of this volume.

# 4 The Female Reproductive System

## Section 1. The Era Prior to 1960

### Ovary

**The Ovarian Hormones**
The ovary had been studied extensively in slaughter animals and some wild animals by the turn of the century and was well recognized to exhibit a cyclic pattern of anatomical changes (figure 4.1). Two stages were recognized: The first, made especially conspicuous by the presence of large blistery follicles of whitish color and filled with translucent fluid, was succeeded by a remarkable shift to the presence of large fleshy bodies (corpora lutea) protruding from the surface of the ovary. These cyclic changes in the ovary were also known to be correlated with changes seen in the reproductive tract. The mechanism linking the two was unknown but was thought to be due to some kind of internal secretion or "generative ferment" produced by the ovary and carried to the female reproductive tract by the bloodstream (Knauer 1900; Halban 1900; Heape 1905; Marshall and Jolly 1908; Hammond and Marshall 1914). The concept was prophetic. In 1923, immature rats and mice were treated with aspirated porcine follicular fluid and promptly exhibited precocious vaginal estrus and at autopsy the uterus was found to be greatly enlarged (Allen and Doisy 1923). From that time forward, the ovary was known not to be just a source of eggs, but an endocrine gland as well.

Succeeding chemical work showed this activity to be an oil soluble substance and the race was on to isolate the hormone and identify its structure. Estrone was isolated in crystalline form by Doisy, Veler, and Thayer (1929 and 1930), and estriol by Marrian (1930). The structure of the hormone secreted by the ovary, 17-$\beta$ estradiol, was not obtained until 1936 and was based on the extraction of more than four tons of porcine ovaries by MacCorquodale, Thayer, and Doisy (1936a, 1936b).

The isolation of progesterone was achieved in 1934 (Butenandt; Slotta, Ruschig, and Fels; Allen and Wintersteiner) and its structure

---

This chapter by Dr. Roy O. Greep summarizes the essays commissioned for the review pertaining to female reproductive research.

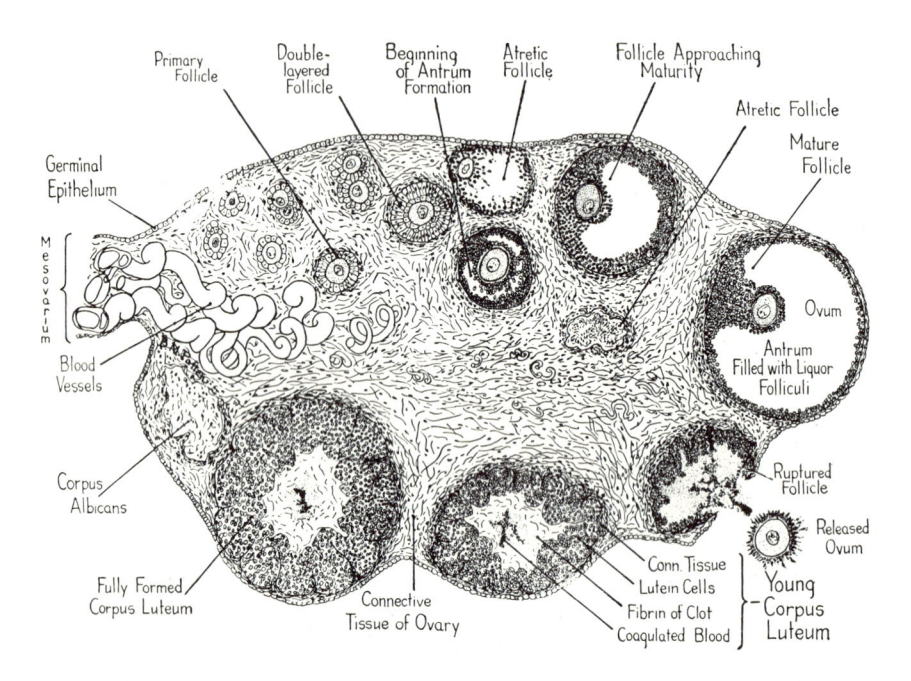

**Figure 4.1** A schematic representation of follicular growth, maturation and ovulation, and corpus luteum formation. Follicular atresia and luteolysis are also illustrated.
Source: Slightly modified from Patten's *Human Embryology*, 2nd Edition, 1956. Used with permission of the McGraw-Hill Book Company.

determined in the same year (Butenandt and Schmidt). The next thirty years witnessed an explosion of interest in steroid biochemistry by both industrial and academic investigators. The biosynthetic pathways and the metabolism of all the steroid hormones including those from the adrenal and testes dominated research not only in reproduction but in the entire field of endocrinology. It was an extremely productive era for steroid hormone biochemistry and a period of spectacular advances in the treatment of endocrine dyscrasias resulting from deficiencies or excesses of the steroid hormones or their altered metabolism. Early in the period, investigators stopped referring to estrogens and androgens as "female" and "male" sex hormones respectively because it was found that they were not sex specific nor even organ specific. Androgens were found in females and estrogens in males. Although the ovarian follicles are the primary source of estrogen and the corpora lutea are the primary source of progesterone, the ovarian follicles produce some progesterone and the corpora lutea some estrogen.

## The Biosynthesis of Ovarian Steroid Hormones

The details of the metabolic pathways for the biosynthesis of the ovarian steroid hormones were a mystery until the biosynthetic scheme was elaborated for adrenal steroids in the late 1940s and early in the 1950s. The techniques so developed could then be applied to ovarian tissue.

The first insights into the metabolic steps were accomplished by perfusions of isolated ovaries (Werthessen, Schwenk, and Baker 1953) and possible intermediates were identified by isolation from ovarian follicular fluid (Short 1961). The pathways demonstrated in pregnant animals (Heard et al. 1956) and women (Werkin et al. 1957) involved both acetate and cholesterol in the biosynthesis of estrogen and progesterone and provided the clues for later work.

Tracer techniques using radioactive steroid precursors: acetate, cholesterol, pregnenolone, progesterone, androstenedione, and testosterone were next employed with isolated ovarian follicles, corpora lutea, and stroma as well as whole ovaries (see Ryan 1963a; 1963b for overview).

The unraveling of five major steps cracked the secret.

1. Acetate conversion to cholesterol demonstrated in the liver and adrenal provided the basis for formation of the steroid nucleus that is common to all steroid hormones. This was also demonstrated in the ovary (Ryan and Smith 1961).

2. Cholesterol side chain cleavage to pregnenolone and progesterone (Staple, Lynn, and Gurin 1956) provided the basis for the subsequent formation of all other steroid hormones. Side chain cleavage was first elaborated with the adrenal and placenta and subsequently the corpus luteum (Marsh, Mason, and Savard 1961).

3. The conversion of pregnenolone to progesterone, a crucial intermediary step first described in 1951 (Pearlman, Cerceo, and Thomas) provided the basis for the conversion of $\Delta 5$ to $\Delta 4$ steroids.

4. Conversion of the $C_{21}$ steroids, pregnenolone and progesterone, to androgens by 17-hydroxylation and side chain cleavage was demonstrated in adrenal (Solomon, Vande Wiele and Lieberman 1956) and testes (Slaunwhite and Samuels 1956).

5. Finally, aromatization of androgens to estrogens by tracer techniques in the whole ovary (Baggett et al. 1956; Wotiz et al. 1956) and then at substrate levels in the placenta provided the last link to total biosynthesis (Ryan 1959; Smith and Ryan 1961; Ryan and Smith 1961).

When the total biosynthetic scheme was first put together, it was demonstrated that the rate and amount of conversion increased as biosynthesis progressed toward the end product estrogen(s). The pathway was essentially in one direction and the intermediates could be identified by isolation at each step along the way. The quantitation came later by measuring ovarian vein secretion and production rates.

The formation of specific steroids could then be linked to ovarian subcompartments: follicle, stroma, and corpus luteum, correlated with gonadotropin stimulation of each of these ovarian units and ultimately related to the age (childhood, adult reproductive period, and menopause) and endocrine state of the ovary.

**Maturation of Oocytes**

All the oocytes that the ovary will ever have, approximately a million, are present at birth. Vast numbers of these undergo atresia during infancy and childhood so that only about three hundred thousand are left by the age of seven. Throughout normal reproductive life only approximately three hundred fifty to four hundred reach maturity and are ovulated. The remainder dwindle in number with each new menstrual cycle as a result of further atresia. Oocyte maturation, marked by two divisions that reduce the number of chromosomes by half, starts even before birth but becomes arrested early in the first division. Maturation is not resumed until the ovum is ready to be ovulated. The stimulus to ovum maturation is generally thought to be due to the ovulatory surge of luteinizing hormone (LH), but follicle stimulating hormone (FSH) may also be involved.

The interesting question is what holds oocyte maturation in check. It appears that the granulosal cells may be responsible because maturation proceeds promptly when the oocytes are removed and placed in ordinary culture media but not when they are cultured in follicular fluid (Nekola and Smith 1974). Thus it appears that the granulosa cells produce an oocyte inhibitory factor; but the nature of this substance, if it exists at all, is unknown.

It is of immediate contraceptive importance to determine what chemical signals are responsible for the initiation, inhibition, and completion of the maturational preparation of the ovum for fertilization.

## Discovery of the Pituitary and Placental Gonad Stimulating Hormones

Probably the most far reaching discovery ever made in reproductive biology was the revelation in the mid-twenties that the male and female reproductive systems are under the functional control of the anterior hypophysis. The first firm evidence was obtained by demonstrating in weanling rats and mice that implants of fresh anterior hypophysial tissue induced precocious sexual maturation and marked enlargement of the ovaries (Smith 1926; Zondek and Aschheim 1926). This electrifying breakthrough was quickly bulwarked by the reverse demonstration that removal of the pituitary gland without injury to the brain resulted in profound atrophy and total loss of function by the reproductive system in both sexes (Smith 1927; 1930). A few earlier discoveries and many later ones of great significance have been made, but none had quite the impact on the field nor led to such explosive exploitation as this long-delayed exposure of the mechanism controlling the histophysiology of the testis and ovary. Although the pituitary body is located at some distance from the gonads, the evidence proved it to be the source of blood-borne gonad stimulating hormone(s).

Following hard on the heels of this great discovery, gonad stimulating hormones were found in the blood and urine of postmenopausal women (Zondek 1930) and pregnant women (Aschheim and Zondek 1927) and in the serum of pregnant mares (Cole and Hart 1930). The first of these proved to be of pituitary origin, but the latter two were shown to be placental secretions.

Strong physiological evidence was marshalled in the early thirties in support of the postulation that the pituitary secretes not one single gonad stimulating hormone governing all aspects of gonadal function, but two hormones, FSH and LH or testis interstitial cell stimulating hormone (ICSH) (Fevold, Hisaw, and Leonard 1931). The basis of this concept rested on the chemical fractionation of these two activities from crude extracts of acetone dried pituitary powder obtained from sheep. Although definitive chemical confirmation was not to be forthcoming until nearly forty years later, the two-hormone concept dominated all research thinking and application during that interval.

The primate reproductive cycle is comprised of three phases: a follicular phase, a midcycle ovulatory phase, and a luteal phase. The follicular phase is characterized by the growth and maturation of one or more follicles (depending on the species) accompanied by the secretion of estrogen and generalized proliferative activity in the reproductive tract. The change-over from the follicular to the luteal phase is entirely dependent upon the ovulation of one or more follicles and their conversion into luteal bodies or corpora lutea. The luteal phase terminates in menstruation and the initiation of growth of a new set of follicles to start the next cycle.

FSH was established as the sole agent required for the growth of ovarian follicles to the stage of maturity (of large fluid-filled structures bulging from the surface of the ovary and ready to undergo rupture, see figure 4.1). An increased secretion of LH participates in the final preovulatory enlargement of the follicle and appears to be primarily responsible for the rupture of the follicular wall and release of the free-floating ovum. Although it is likely that under normal circumstances both FSH and LH participate in the ovulating process and experimental ovulation can be induced by either hormone acting alone, LH has far greater ovulating potency than FSH and is generally and justifiably regarded as the ovulating hormone. The secretion of estrogen by the maturing follicles is mainly under the control of LH and is carried out by the encapsulating thecal layer (Greep, van Dyke, and Chow 1942). Under in vitro conditions the granulosa has also been shown to be capable of steroid synthesis (Ryan 1962; 1963; Ryan and Short 1965; Channing 1969; Erickson and Ryan 1975).

The hormonal control of the formation, life span, and function of the corpus luteum has posed some difficulties. There is no doubt that LH is responsible in all species for the luteinization of the follicle and formation of a luteal body. What causes the corpus luteum to secrete progesterone has been less clear cut. The luteotropic stimulus varies somewhat between species. The rat and mouse are notable examples of animals that require prolactin to induce luteal function (Astwood 1941; Evans et al. 1941). The hamster appears to require a "luteotrophic complex" of FSH, prolactin, and a trace of LH (Greenwald 1967). The rabbit requires only estrogen (Hammond and Robson 1951). During pregnancy in some species, including the rat and mouse, luteal support is provided by placental luteotropins with the result that the pituitary gland can be removed without interrupting the pregnancy. In other animals, such as the guinea pig, sheep, and monkey, the placenta also takes over the progesterone secreting function of the corpus luteum so that the pituitary and/or the ovary can be removed without aborting the conceptus.

On the matter of bringing about the cessation of function and demise of the corpus luteum, firm evidence of general significance has proved hard to come by. As far back as 1923 (Loeb) it was conclusively shown that removal of the uterus in guinea pigs greatly prolonged the life of the corpora lutea. This has since been shown to hold true for several other species, but not for monkeys or humans. Subsequent work has shown that a uterine implant will hasten luteolysis in those species in which hysterectomy has the reverse effect. The uterus clearly produces a positive luteolytic factor in some species, and this will be referred to later.

Prolactin is, in many ways, a maverick among the several tropic hormones secreted by the anterior lobe of the hypophysis. It has no common role in the animal kingdom. It is the only hormone secreted by the pituitary in the absence of vascular contact with the hypothalamus. It is also the only pituitary hormone to have been crystallized and the only gonadotropin that is not a glycoprotein.

The hormonal activity that bears the name prolactin was discovered in 1928 as the hormone inducing milk secretion in rabbits (Stricker and Grueter). In 1933 the hormone stimulating the formation of crop milk in pigeons and named prolactin (Riddle, Bates, and Dykshorn 1933) turned out to be the same hormone that stimulates milk secretion in animals. In 1937 it was further purified and obtained in crystalline form (White, Catchpole, and Long).

The functional virtuosity of prolactin throughout the vertebrate kingdom and evolutionary history is astounding. More than eighty effects among the vertebrates have been identified by Nicoll (1974). It appeared among the fishes as a regulator of osmolality that enables some species to migrate between fresh and salt water habitats. In certain of the amphibia it became the water drive hormone that brought land-dwelling forms back to water to breed. Nothing is known of its role, if any, in the reptiles; but in birds prolactin turned up as the brooding or nesting factor with a marked influence on behavior. In birds and mammals, it acts also to suppress the secretion of other pituitary gonadotropins. When it turned out that LH could induce the formation of corpora lutea in rats and mice but not support their secretory function, prolactin was found to be the needed factor during pseudopregnancy and early pregnancy. In mammals, prolactin appears to have no critical, but perhaps a permissive, function in the male; in females it is required for lactation and has a naturetic function during pregnancy. Under special circumstances prolactin has also been shown to have a luteolytic action (Malven 1965).

## Factors Influencing the Secretion of Pituitary Gonadotropins

**Sex Steroids**
That the pituitary does not function autonomously but is linked interdependently to the hormones produced by the gonads was clearly demonstrated in 1930 (Meyer et al.) by showing that the estrogenic hormones of ovarian origin suppress the gonad stimulating functions of the pituitary. Pituitary gonadal interrelationships were further clarified in 1932 (Moore and Price) by showing that the action of exogenous pituitary gonad stimulating hormones was not suppressed by estrogen treatment. This proved for the first time that the gonadal hormones suppress reproductive functions not by inactivating the pituitary hormones or blocking their action on the gonads, but by reducing their output.

This was the beginning of what came to be known as the negative feedback concept of regulating the functions of the pituitary gland. The evidence was incontestably fortified by the earlier finding that loss of the secretory functions of the gonads, as by gonadectomy or the menopause, was followed by marked hypersecretion of the pituitary gonad stimulating hormones (Engle 1929; Evans and Simpson 1929).

A clear-cut negative feedback role for progesterone trailed that of estrogen. It came with the demonstration in 1937 (Makepeace, Weinstein, and Friedman) that the administration of progesterone blocked postcoital ovulation in rabbits. The logical extension of this observation to the possibility of controlling fertility was to lie dormant for a period of sixteen years. By hindsight this may seem an oversight, but it should be noted that the progestins then available were active only by subcutaneous injection and that contraception had not yet become a pressing societal problem.

As early as 1932, it was observed that, contrary to the earlier demonstration of the suppressive action of estrogen on the pituitary, there were circumstances in which estrogens appeared to enhance the pituitary output of gonadotropins. This was especially true in rats approaching puberty. In these animals estrogen was claimed to induce precocious ovulation and luteinization of the ovaries. With what can only be termed a stroke of prescience, it was postulated (Hohlweg and Junkman 1932) that the actions of the gonadal hormones on the pituitary were mediated through what was termed a "sex center" in the brain. This concept was unfortunately taken lightly, disregarded, or ridiculed; but history was to make amends some thirty-five years later.

The female pituitary functions in a cyclic manner; the male pituitary is notably acyclic. In 1936, evidence was produced by Pfeiffer suggesting that the pituitary might be sexually bipotential. This belief was based on the fact that the male pituitary could be made to function in a cyclic manner and the female pituitary could similarly be made to function acyclically. Thus, male rats castrated at birth and later bearing an ovarian graft in the anterior eye chamber showed cyclic ovarian function, whereas intact females bearing a testis graft from birth failed to display cyclic vaginal smears. This concept of the bisexuality of the pituitary was soon challenged by experiments involving transplantation of the pituitary gland. Although pituitaries transplanted to ectopic sites failed to function, those autografted into the empty sella (Greep 1936) or underneath the median eminence (Harris and Jacobsohn 1952) of the brain functioned normally. Greep also found that pituitaries exchanged between sexes functioned not in accordance with the sex of their origin, but in accordance with the sex of the host. Such evidence made it clear that sexuality was being imposed on the pituitary. The site of sexual bipotential has been shown to reside not in the pituitary but in the brain, namely, in the hypothalamic area.

### Neural Influences on Reproductive Functions

That the regulation of the gonadotropic functions of the anterior pituitary might not be controlled entirely by the feedback actions of the gonadal secretions (for review see Donovan 1966) was surmised from experiments in the early thirties showing that by artificially lengthening the daily photoperiod (length of daylight) the recrudescence of gonadal functions could be hastened in seasonal breeders.

Birds and most mammals studied responded to extensions of the photoperiod, but some mammals that breed during the fall responded to reductions in the photoperiod. This clearly suggested that the gonadal functions of the pituitary are also subject to some form of neural control (figure 4.2). Buttressing this view were the observations that in the rabbit, ovulation was definitely linked to the physical stimulus of mating, and in the rat mechanical stimulation of the uterine cervix induced prolongation of the functional life of the corpora lutea characterized as pseudopregnancy.

The observations led not only to a futile search for nervous pathways to the anterior pituitary but also to attempts to excite pituitary secretion by electrical stimulation of these presumptive pathways. The early findings were paradoxical. The anterior pituitary appeared to have no neural connections to the brain, yet electrical stimulation evoked a secretory response when a current was passed through the head (Marshall and Verney 1936) or when the electrodes were placed in the central nervous system just above, but not in, the pituitary (Markee, Sawyer, and Hollinshead 1946). Many of these studies were carried out on rabbits, which ovulate only after mating. Here the effectiveness of electrical stimulation of pituitary function could be determined with certainty by the induction of ovulation and corpora lutea formation in nonmated rabbits. In unanesthetized rabbits bearing indwelling electrodes in the hypothalamus, it was shown that ovulation could be induced at will by electrical stimuli delivered via remote control (Harris 1948a).

Evidence favoring the neural control of reproductive functions accumulated in great abundance and variety in the period roughly from 1935 to 1945. Electrolytic lesions located in many different regions of the hypothalamus produced variable but repeatable alterations in ovarian functions. The effects of electrical stimuli were similarly area specific and depended upon the stereotaxic placement of the electrodes.

At the halfway mark of the present century, the presence of an adrenergic and a cholinergic component of the ovulatory process in rabbits was revealed by showing that drugs blocking brain function also block ovulation following mating (Markee, Everett, and Sawyer

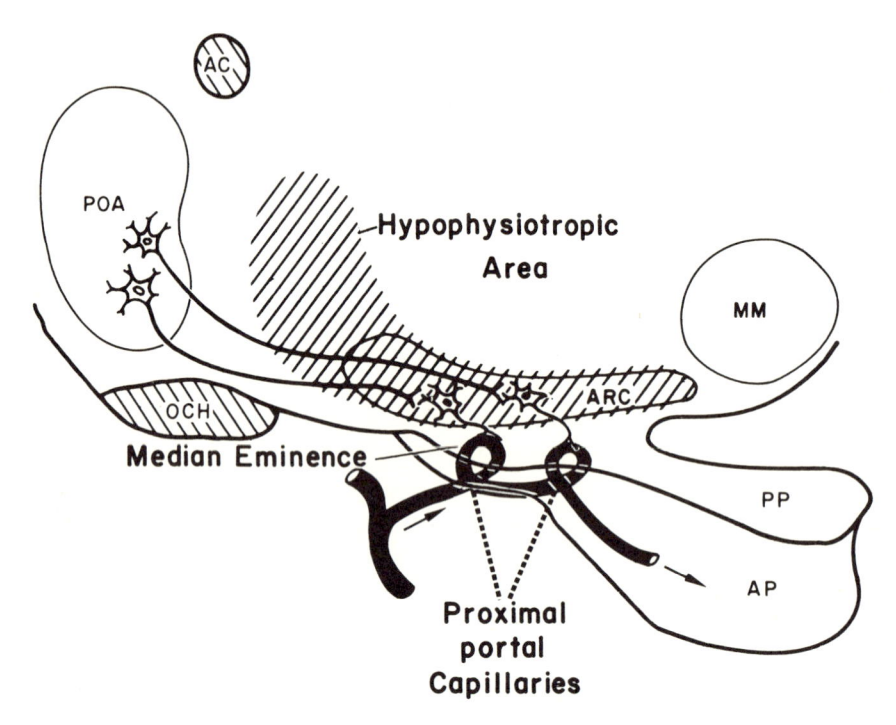

Figure 4.2 A schematic representation of the neuroendocrine mechanism controlling the gonad stimulating functions of the anterior pituitary. Endocrine neurons in the basal hypothalamus and median eminence area produce luteinizing hormone releasing factor, which is transported to the pituitary via the hypophysial portal vascular system.
Source: C. H. Sawyer 1975. Courtesy S. Karger AG Basel.

1952). Dibenamine, an adrenergic blocking agent, inhibited ovulation when given within the first hour after mating; and atropine, a cholinergic blocker, had a similar action when given within the first few minutes after coitus.

## The Hypophyseal Portal System and the Neurovascular Mechanism Regulating Pituitary Functions

In 1930 Popa and Fielding made a chance observation based on the dissection of human cadavers whose blood vessels had been filled with dye. They saw a portal system that connects the hypophysis with the base of the brain. Although it was first suggested that this system drained from the pituitary to the brain, the direction of flow was soon established to be from brain to pituitary (Wislocki and King 1936). This new portal of entry to the pituitary soon led to speculation that neural stimuli may be conveyed to the pituitary by means of neurohumoral signals reaching the pituitary via this portal system.

The evidence favoring neural influence on the pituitary function was put together by Green and Harris (1947) and Harris (1948b) as the neurovascular concept (figure 4.2). Although it was to take another quarter-century to identify the elusive neurohumoral signals that effectuate the neurovascular mechanisms, the impact of the enunciation of this concept on research in reproductive biology was indeed revolutionizing. A decade later, these hypothalamic neurohumors were shown to bring about the almost immediate release of stored pituitary hormones and were termed "releasing factors."

## Inhibition of Ovulation by Orally Active Progestational Steroids

The year 1953 is notable for a very important event. The era of *oral* contraception was ushered in by the demonstration that a number of synthetic progestational compounds including norethynodrel (Searle) synthesized by Frank Colton (1951) and norethisterone (Syntex) (synthesized by Carl Djerassi 1951) were active inhibitors of ovulation in mated rabbits when given by mouth (Pincus and Chang 1953). The testing of norethynodrel in sexually active women was initiated in 1956 and its contraceptive efficacy was established beyond question (Pincus et al. 1958). Because the norethynodrel was found to contain an estrogen that was contributing to the contraceptive action, the "pill" was formulated to contain an added amount of an orally active estrogen, mestranol. That the combination of orally active progestin and estrogen controlled fertility by inhibiting ovulation was inferred from an abundance of clinical evidence (Rock, Garcia, and Pincus 1957) including no midcycle shift in basal body temperature and no excretion of pregnanediol, a breakdown product of progesterone, during the second half of the menstrual cycle.

### Heightened LH Output at Midcycle

Follicular rupture and release of an ovum occurs spontaneously in primates and most other species, so the important question to be answered was what triggers off this important event. Because an injection of LH induced ovulation in all species studied, it was assumed that some mechanism must exist for the acute discharge of LH prior to ovulation. However, no assay system available was sufficiently sensitive to measure the pituitary gonadotropins in the bloodstream during reproductive life in any species. Nevertheless, it did prove possible to measure LH in concentrates of twenty-four-hour urine samples collected daily during the human menstrual cycle. A 1958 study by McArthur, Ingersoll, and Worcester revealed for the first time an elevated presence of LH at or near the time of ovulation and presaged the spectacular midcycle events later to be revealed by the more sensitive radioimmunoassay.

### Introduction of an Improved Bioassay for FSH

For twenty-three years following the first publication on FSH, progress in purification of the hormone and in critically assessing its concentration in the pituitary gland and body fluids under varying experimental and clinical conditions was severely hampered by the lack of a suitably sensitive and specific bioassay system. This log jam was largely broken in 1953 by the introduction of the new Steelman-Pohley bioassay based on augmentation of ovarian weight in immature rats given an overdose of human chorionic gonadotropin (hCG). Although this valuable tool has now been replaced to a large extent by the more sensitive radioimmunoassay, it remains the bioassay of choice.

## Progesterone Feedback

Progesterone had long been accepted as an effective inhibitor of ovulation in rabbits and during lactation in women, so it came as somewhat of a surprise to learn during the forties that progesterone could also facilitate the induction of ovulation in rats (Everett 1940). That progesterone, like estrogen, has the capability of manifesting both a stimulatory and inhibitory role in reproductive physiology was an important conceptual advance (reviewed by Everett 1961b). The major gap left in our knowledge of these processes was whether these feedback actions of the gonadal steroids were being effected at the level of the pituitary or the hypothalamus or both. For further evidence see studies after 1960.

## Discovery of the "Critical Period" in Rats

An offshoot of the discovery of the profound influence of the photoperiod on sexual functions was the discovery that the estrous cycle in the laboratory rat is linked to the daily light-dark cycle. Thus it was found that ovulation could be delayed by the injection of an anesthetizing agent during the early afternoon on the day of proestrus but not at any other time of the cycle. Further study revealed that a preovulatory discharge of LH occurred between the hours of 2 and 4 P.M. on the day of proestrus. When this mechanism was blocked by anesthesia, proving that a neural component was involved, the process was delayed until the next day at the same two-hour period (Everett and Sawyer 1950 and 1953). This came to be known as the "critical period." With this vital piece of information, it became possible to study the neurohormonal control of ovulation far more effectively. The LH discharged during the critical period was shown to bring about the induction of ovulation some ten to twelve hours later.

## Uterus

The uterus provides a site for the attachment of the fetus to the mother, an environment for the development of the fetus during gestation, and a mechanism for its expulsion at term. To serve these purposes, the uterus, a functional unit, is composed anatomically of two unrelated components: a heavy muscular coat, or myometrium, and a thick glandular lining, the endometrium. Despite their extreme structural dissimilarities, both the myometrium and the endometrium respond to and are controlled by the same ovarian hormones, estrogen and progesterone. The uterus also responds in one way or another to other agents, such as oxytocin, prostaglandins, relaxin, catecholamines, drugs, and trauma.

Research on the uterus has been sporadic. Following the isolation of the ovarian hormones, estrogen and progesterone, in the late twenties and early thirties, there was much interest in learning how these affect the endometrium with respect to its structure, its receptivity to nidation, its loss during menses, and its ability to regenerate with each new menstrual cycle throughout reproductive life (for a review see Hisaw and Hisaw, Jr. 1961).

During the forties and early fifties, work on the uterus occupied the attention of but few investigators. Over the past twenty years interest shifted increasingly to the myometrium, currently a subject of considerable controversy and confusion, especially with respect to the mechanisms that prevent abortion and initiate parturition.

Growth of the uterus during puberty and as witnessed in the laboratory or clinic is primarily dependent upon the action of estrogen, the only hormonal stimulus known to induce cell proliferation in any and all parts of the uterus. Progesterone can only induce cytomorphosis as in the decidualization of the stroma or the conversion of the epithelium from proliferative to secretory activity. The effects of estrogen and progesterone on the contractility of the myometrium are much less firm. In general, contractility tends to be enhanced by estrogen and depressed by progesterone, but comparative studies on different species have shown that there are exceptions. Progesterone, regardless of whatever other roles it may play, is first and foremost the hormone of pregnancy. Its essentiality for the establishment and maintenance of pregnancy is absolute. It is produced in great quantities during at least the early stages of pregnancy and remains at a high level in the bloodstream during the whole of gestation in man but recedes to a very low level after the first trimester in the rhesus monkey. In both instances, it is believed to fulfill an obligatory role, but the mechanism of its action is most uncertain. In those species such as man, rhesus monkey, and guinea pig in which an established pregnancy is not interrupted by removal of the ovaries, the alternate source of progesterone has been shown to be the placenta. Removal or death of the fetus (or fetuses) does not disturb the course of pregnancy in these species so long as the placenta remains intact.

As indicated above, estrogen stimulates the growth of the mucosal lining of the uterus, the endometrium. It also primes the endometrium for the action of progesterone. Nowhere is the difference in the actions of estrogen and progesterone on the endometrium shown in more spectacular fashion than in the rabbit and the primate. In estrogenized rabbits, the luminal surface of the uterus exhibits smooth, longitudinal ridges and these are converted by progesterone into hundreds of thin plicae so that cross sections of the mucosa reveal a lacelike pattern (Corner and Allen 1929). This remarkable transformation of the appearance of the uterus in cross section became the basis of a widely used semiquantitative bioassay for progesterone or progesteronelike compounds known as progestins.

Similarly, in primates, the actions of estrogen and progesterone on the endometrium are unmistakably distinguishable. The long straight glands built by estrogen during the follicular phase become conspicuously tortuous during the luteal phase as a result of the action of progesterone (figure 4.3). Furthermore, the changes induced by estrogen and progesterone on the epithelium lining the lumen and all the glands are similarly hormone specific. With estrogen the epithelial cells are columnar with a subnuclear accumulation of large vacuoles containing glycogen. Progesterone moves these vacuoles to the supranuclear pole of the cell where they are secreted, giving the luminal border of the cells a frayed appearance. These changes in the histology of the endometrium provide a basis for diagnosing with certainty

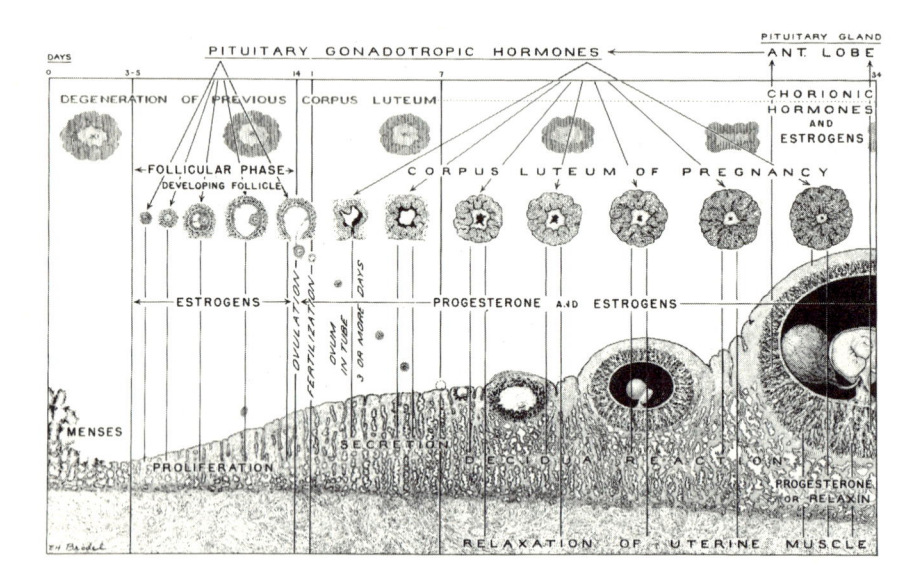

**Figure 4.3** Illustrates the sequence of ovarian events and the pituitary and ovarian hormonal actions involved in the control of the uterine endometrium during a fertile cycle.
Source: Hamilton, Boyd, and Mossman, *Human Embryology*. With permission.

whether the uterus is under the influence of estrogen or progesterone and also whether ovulation and corpus luteum formation has occurred. By the same means, it is possible to estimate the age of the corpus luteum in primates (Corner, Jr. 1956). These observations, now routine in fertility clinics, provided for the first time an indirect means of gaining insight relative to the state of ovarian function in women (for a review see Allen, Hisaw, and Gardner 1939).

## Hysterectomy

The first observed influence of the uterus on the life and function of the corpora lutea was made in the guinea pig by Loeb in 1923. Following hysterectomy, the corpora lutea were seen to persist for three to four months, during which time estrous cycles did not recur. A similar phenomenon was not found in hysterectomized women; the ovarian cycles continued in normal fashion (Whitelaw 1958). It is now well established that hysterectomized women have normal cyclic ovarian function and do not experience the climacteric prematurely as was earlier thought.

## Cervix

Much of what was known about the cervix in this early period stemmed from study of its microscopic anatomy and clinical experience rather than experimental observations. Such information was helpful in understanding the nature of the cervix; its role during insemination, pregnancy, and delivery; and the cyclic changes in its structure and function. Although the anatomy of the cervix varies enormously between species, it has some common characteristics, the most important of which is a mucinogenic epithelium. The cervical fluid serves as a barrier to the passage of sperm except at the time of estrus and mating. It also prevents at all times the passage of infectious bacteria into the uterine lumen. In humans, the cervical mucus varies in physical properties during the menstrual cycle and permits the passage of sperm through the cervical canal only at midcycle. This is the only function of the cervix that is of pertinence to this review. The fact that the cervical mucus can serve as a barrier to the penetration of sperm at other times during the cycle presents the possibility of rendering the mucus hostile to sperm movement during the fertile period. Tests developed to judge the physical and chemical properties, such as ferning, spinnbarkeit, and tackiness, of the mucus proved useful in determining the normalcy of the reproductive cycle.

## Section 2. The Modern Era (after 1960)

### Advances in Hormone Research

Hormones known to influence directly the gametogenic or secretory function of the gonads include, in addition to FSH, LH, hCG, and pregnant mare serum gonadotropin (PMSG), the "lactogenic" hormone prolactin (PRL). PRL must also be included because it has acquired a definite role in the regulation of luteal function in the rat and mouse and less certainly in a few other species including ferrets, sheep, monkeys, and humans.

### FSH, LH, and HCG

The nature and activities of the proteinaceous gonad stimulating hormones have been widely studied over the past forty years, but only in the past five years has substantial progress been made in understanding their chemical structure. The amino acid composition and primary structure (amino acid sequence) has been determined for FSH (Papkoff 1973; Shome and Parlow 1974a and 1974b), LH (Liu, Sweeney, and Ward 1970; Pierce et al. 1971; Sairam, Papkoff, and Li 1972; Papkoff et al. 1973; Shome and Parlow 1973), and hCG (Canfield et al. 1971; Bahl et al. 1972; Bahl, Carlsen, and Bellisario 1973). Examination of the tertiary (three-dimensional or conformational) structure of the gonadotropins is being undertaken using nuclear magnetic resonance, optical rotatory dispersion, and circular dichroism; but application of X-ray diffraction awaits their crystallization. Only some speculative information is available concerning the conformation of the gonadotropins (Bewley, Sairam, and Li 1972; Ward et al. 1973); therein lies a new frontier in this area. Such information may be of great value in determining the structure-function relationships of the gonadotropins and in developing antagonists.

The extraordinary feature of the primary structure of FSH, LH, and hCG is that each hormone is comprised of two dissociable polypeptide chains or subunits of approximately equal molecular weight. The subunits designated as $\alpha$ and $\beta$ are nonidentical and lightly bonded together. They can be dissociated by exposure to low pH, urea, or guanidine and isolated by countercurrent distribution. Most unexpectedly, the $\alpha$ subunits of FSH, LH, and hCG have proven to be interchangeably identical. The $\beta$ subunits differ from their natural $\alpha$ counterparts between hormones and somewhat between species for any given hormone. They are, therefore, both hormone specific and species specific and are the determinants of both biological and immunological activity. Even so, large segments of the amino acid sequence of the $\beta$ subunits from different hormones are homologous. These advances have added much to our knowledge of the chemical structure of the gonadotropins but have not altered significantly the previously established information concerning their physiological activity.

The individual subunits have very little, if any, biological activity by themselves; but when the subunits of either FSH, LH, or hCG are recombined, the activity of the native hormone is largely or fully restored. When the $\beta$ subunit of one hormone is combined with the $\alpha$ subunit of another hormone to form a hybrid molecule, activity is also restored. In these instances the activity expressed is that of the hormone from which the $\beta$ subunit was derived.

The variation in primary structure between the $\beta$ subunits of different hormones and of a given hormone from different species confers the immunogenecity that has led to the development of radioimmunoassays for the gonadotropins. The extreme sensitivity of these assay systems has been of inestimable value in studying the rate of secretion and quantitative presence of the gonadotropins in body tissue and fluids.

Carbohydrate moieties are present in different quantities in FSH, LH, and hCG and also in their $\alpha$ and $\beta$ subunits. The exact location of these moieties in the amino acid sequence has been determined for LH and hCG and their subunits. These glycoprotein hormones all contain the same monosaccharides, fucose, manose, galactose, glucosamine, and sialic acid. hCG has the highest content of carbohydrate complexes and also of sialic acid, the bulk of which is in the $\beta$ subunit. Reduction of the sialic acid content by treatment with neuraminidase lowers or destroys the biological activity of hCG and FSH but not LH as tested in vivo. Other evidence indicates that the loss of sialic acid does not preclude binding nor does it interfere with the activity of the gonadotropins in in vitro systems, so it appears that the sialic acid serves to slow the metabolism of the hormones in vivo and extend their half-life (Van Hall et al. 1971).

Several recent studies have brought to light the fact that FSH can occur in different forms in a single species. For instance, it was observed that the FSH obtained from the pituitaries of ovariectomized rhesus monkeys does not behave chemically the same as FSH found in the bloodstream (Peckham et al. 1973). A similar phenomenon was observed in castrated male rats treated with testosterone (Bogdanove, Nolin, and Campbell 1975). Under both of these circumstances the pituitary synthesizes an FSH molecule of larger size, increased biological activity, and longer half-life. A similar phenomenon has not been observed with respect to any of the other gonadotropins; but heterogeneity of hormones (prohormones and so-called big hormones, little hormones, and so forth), such as is found in connection with insulin and parathyroid hormone, is well known and it will be surprising if similar findings are not made with respect to other gonadotropins. The structure of hormones can only be determined after a long purification procedure and may not reflect the form in which they are synthesized or secreted or exist in the bloodstream. The eyes of investigators are only now being opened to the complexities with which they are being confronted in this area of research.

Prolactin

Up until about the past five years, prolactin had received much less attention than the other pituitary hormones. Research was considerably handicapped by lack of a satisfactory quantitative assay system. As mentioned before, there was even doubt until recently that the human pituitary secreted prolactin. The isolation of ovine prolactin was accomplished in 1957 (Li) and the sequence of its 198 amino acids determined in 1969 (Li et al.). It is a straight chain molecule, does not possess subunits, and is not a glycoprotein. Attention then turned to human pituitaries as source material from which prolactin was not only identified as a chemically distinct entity but also from which most of the amino acid sequence has been worked out (Niall et al. 1973).

## The Isolation and Structure of the Hypothalamic Luteinizing Hormone Releasing Factor

Following the unequivocal demonstration of the presence of a luteinizing hormone releasing factor (LRF) in extracts of hypothalamic tissue that could stimulate the pituitary to release LH (McCann, Taleisnik, and Friedman 1960; Harris 1961) hypothalamic fragments were collected from 5 million sheep and 2.5 million pigs in a herculean effort to isolate this elusive agent. In 1971 the LRF was characterized as a decapeptide (figure 4.4), first from pig (Matsuo et al.) and a few months later from sheep (Burgus et al.). The amino acid sequence of LRF from these two species proved to be identical. LRF was immediately reproduced by total synthesis and is now available in essentially unlimited quantity. This decapeptide has been shown to stimulate the secretion of both LH and FSH (figure 4.5) (Schally et al. 1971; Kastin et al. 1972) and has been termed LH/FSH-RH for luteinizing hormone and follicle stimulating hormone releasing hormone or more generally GnRH for gonadotropin releasing hormone. For simplicity it will be referred to here as LRF. A separate FSH-RF has not been identified, but an abundance of physiological evidence suggests that such a substance may yet be brought to light. The synthetic LRF is already in use throughout the world as a diagnostic tool to ascertain the functional capacity of the pituitary to secrete gonadotropins. Its ability to induce ovulation in several species including man has been demonstrated and its clinical use in the correction of gonadotropin deficiencies of hypothalamic origin is now in the developmental stage (see Schally et al. 1973 for a review).

Synthetic analogs of the LRF decapeptide are being prepared on a wide scale. A few (super LRFs) have proven to be more potent than the native peptide, some by as much as fiftyfold (Fugino et al.

## STRUCTURAL FORMULA OF PORCINE AND OVINE LRF

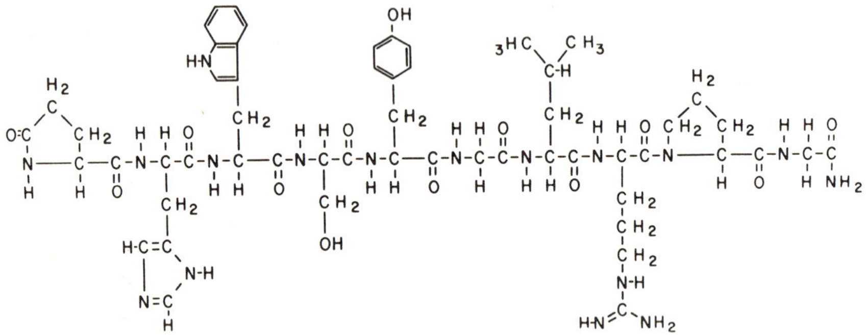

pGLU - HIS - TRP - SER - TYR - GLY - LEU - ARG - PRO - GLY - NH$_2$

**Figure 4.4** The amino acid sequence of luteinizing hormone releasing hormone (LRF).
Source: Guillemin 1972, with permission.

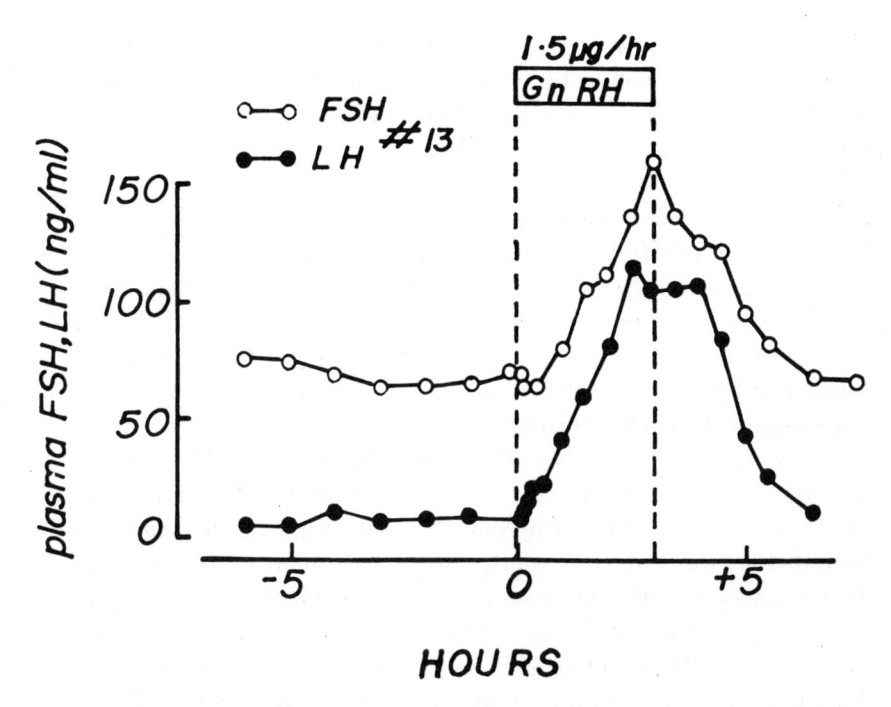

Figure 4.5 Showing elevation of plasma FSH and LH induced by a single injection of LRF (also termed GnRH).
Source: Jonas et al. 1973, with permission.

1974); others antagonize the action of LRF (Vale et al. 1972; Fugino et al. 1973); presumably by competitive inhibition. High hopes are held that this line of investigation will lead to a new means of contraceptive medication.

**Neural Mechanism Regulating the Release of Prolactin**
There is a total difference between the hypothalamic mechanisms controlling the release of gonadotropins on the one hand and prolactin on the other. The secretion of gonadotropins is effected by a factor that stimulates release, whereas the secretion of prolactin is controlled by a hypothalamic factor that inhibits release and is known as prolactin inhibiting factor (PIF).

Suckling is the most effective stimulus to prolactin secretion (Hwang, Guyda, and Friesen 1971) and is mediated by suppression of PIF through some unknown mechanism. Removal or neutralization of this factor by stalk sectioning or treatment with tranquilizers such as chlorpromazine or reserpine allows prolactin secretion to proceed uninhibited in laboratory animals and women (see Meites and Clemens 1972 and Clemens and Meites 1974 for a review). When the pituitary gland is removed and transplanted to the kidney capsules and thereby freed from hypothalamic influence, it proceeds to produce only prolactin. Similarly, pituitaries cultured in vitro synthesize (de novo) and release large quantities of prolactin and this can be quickly repressed by the addition of PIF or the catecholamine, dopamine (MacLeod and Lehmeyer 1972).

The drug perphenazine is an effective stimulus to prolactin secretion; and, obversely, the ergot alkaloids ergocornine and ergocryptine are effective blockers or prolactin secretion (Clemens and Meites 1974). PIF has not been isolated, but there is evidence to suggest that it is a very small molecule and it has been shown that the brain catecholamine, dopamine, fulfills all the requirements of a PIF (Schaar and Clemens 1974) but it is claimed to be ineffective when injected into the portal veins leading to the pituitary (Kamberi 1972). No prolactin releasing factor (PRF) has thus far been identified. Strangely, the secretion of prolactin has been shown to accompany the injection of thyroid stimulating hormone releasing factor (TRF) (Jacobs et al. 1971), but there is no firm basis for believing that TRF is the postulated PRF. These two responses to TRF can be elicited separately under appropriate circumstances. The physiological significance of this bizarre and intriguing partnership remains to be elucidated.

Studies on prolactin binding, prolactin receptors, and the mechanism of prolactin action are under intense study; but it is too early to draw conclusions. It is of primary importance to identify and chemically characterize PIF and settle the question as to the existence of a PRF. It has been observed that the response to LRF in women who are breast feeding is much reduced in comparison to those who are not. The implication is that high prolactin levels may be able to inhibit the preovulatory LH surge and prevent ovulation, but this has not been demonstrated. The daily administration of TRF does abolish the normal midcycle LH peak. Is this due to prolactin? Such questions cry out for answers and these can only come from tests on women.

Radioimmunoassay

No other single methodology has done so much to advance the study of reproductive endocrinology as has the development of the highly sensitive radioimmunoassay (RIA). RIA was born belatedly in the late fifties and within a decade became a routine procedure for nearly every hormone as well as a host of other antigenic substances. Its development depended on the availability of important immunochemical concepts developed much earlier. Indeed the first RIA, that for insulin reported by Yallow and Berson in 1960, arose as a consequence of their preceding studies designed to learn the biological fate of radioiodinated insulin in human diabetics treated with nonhomologous insulin. By 1963 the general concept that the competition between radiolabeled and nonlabeled chemicals for a binding protein (such as antibody) could be used to quantify the relative amounts of the two chemicals became an established method.

As the term RIA implies, one must have antigens, radiolabeled antigens, and binding proteins (antisera). Although the antigenic nature of the gonadotropins was recognized by Rowlands in 1938, the highly purified LH, FSH, prolactin, and hCG required for RIA were not available until many years later. The sex steroids, on the other hand, were available in pure form prior to World War II; but it was not until the late fifties when androgens (Erlanger et al. 1957) and estrogens (Sehon, Goodfriend, and Rose 1958) were coupled to proteins that the immunological properties of these compounds could be well defined (Lieberman et al. 1959). RIA also requires high affinity antisera that have been well characterized and radiolabeled antigens of high specific activity. Methods for protein iodination were available long before the first RIA; but it was not until 1963 when Greenwood, Hunter, and Glover introduced the chloramine-T procedure for iodination that RIA became routine.

At some point in the RIA analysis procedures radiolabel bound to the antiserum must be separated from that remaining free. The initial procedure of electrophoresis employed by Yallow and Berson (1959) was replaced: in 1962 by immunoprecipitation with gamma globulin (Utiger, Parker, and Daughaday); in 1965 by salt precipitation (Imura et al.) or absorption to charcoal (Herbert et al.); in 1967 by solid phase techniques (Catt and Tregear).

RIA procedures for the gonadotropins became available in the midsixties. Methods were developed for hCG by Paul and Odell (1964) and Wilde, Orr, and Bagshaw (1965); LH by Midgley (1966) and Odell, Ross and Rayford (1966); FSH by Faiman and Ryan (1967) and Midgley (1967); prolactin by Hwang, Guyda, and Friesen (1971) and Raud and Odell (1973).

In 1963 Murphy, Engelberg, and Pattee introduced a method of measuring sex steroids in plasma based on *competitive protein binding*, but the method lacked the desired specificity. This limitation was circumvented by Murphy in 1970 by first partially purifying the steroids in a chromatographic system. In the same year Niswender and Midgley eliminated a similar problem of nonspecificity of the steroid antisera by attaching the steroid to the protein antigen on a nonfunctional side group. They also increased the sensitivity of the assay by an improved procedure for radioiodination of the steroid hormones. Thus many steroids can now be measured directly in plasma with great accuracy.

## Changes in Levels of Reproductive Hormones during the Cycle

Profiles of the serum levels of FSH and LH measured daily through the sexual cycle in laboratory animals, domestic animals, and primates, including women, have revealed cyclic patterns that are characteristic for each hormone (figures 4.6, 4.7, and 4.14). One thing these profiles have in common between hormones and among all species studied is a sharp preovulatory spike for LH and a lesser one for FSH. The FSH peak may occur at the same time or slightly later than the LH peak. The blood levels of both hormones remain low and relatively constant at all other times during the cycle. In primates the FSH profile depicts a steady, though mild, rise during the early and midfollicular phase followed by a slow decline to a nadir coincident with the beginning of the midcycle LH peak (Knobil 1974; Yen et al. 1974).

Because the radioimmunoassay requires but minute amounts of blood, serial determinations have been done as often as every fifteen minutes over a period of several hours with surprising results. Instead of the blood levels remaining at a fairly constant level, a pulsatile pattern was observed (figures 4.8 and 4.9). The pattern for LH consisted of a sharp rise at approximately hourly intervals followed by a gradual fall that corresponds to the half-life decay pattern following an injection of LH. Thus, it was established that the secretion of LH is not steady but episodic (Dierschke et al. 1970; Yen et al. 1972). The pulsatile pattern appears to be imposed by episodic release of hypothalamic LRF (Blake and Sawyer 1974). This, however, has not yet been demonstrated and remains an important problem for future research.

The levels of FSH also show a mildly pulsatile pattern (figure 4.9). The FSH pulses may, but do not always, occur coincident with those of LH and are always of much less amplitude (Yen et al. 1973). The rise in FSH occurs more slowly than that for LH (figure 4.10). These differing patterns raise questions as to whether both hormones are released by a single releasing hormone and also whether they are secreted by a single cell type or by two different types. Present evidence suggests that both hormones are released by a single releasing hormone LH/FSH-RH (Schally et al. 1971) and derived from a common pituitary cell type.

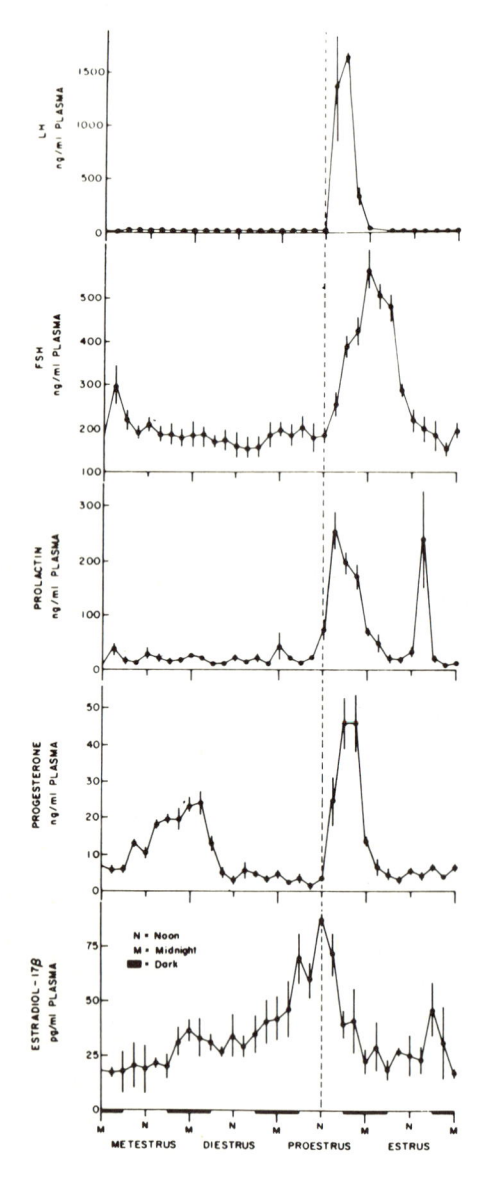

**Figure 4.6** Profiles of the plasma concentration of LH, FSH prolactin, progesterone, and estradiol-17β during the estrous cycle of the rat.
Source: Butcher, Collins, and Fugo 1974, with permission.

PLASMA LH-FSH IN NORMAL MENSTRUAL CYCLE

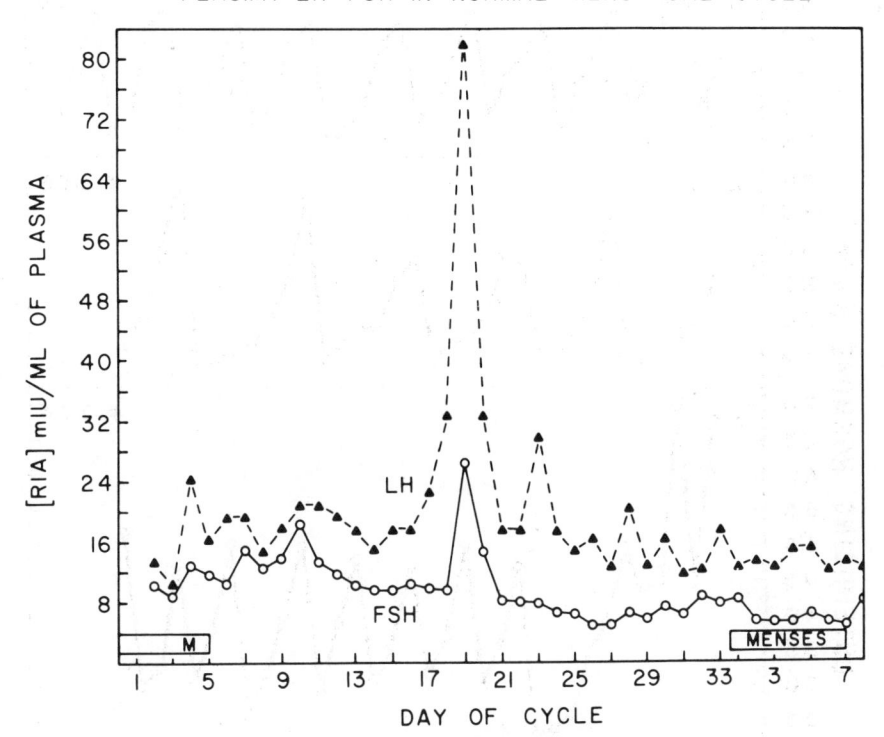

**Figure 4.7** Profiles of the plasma concentration of FSH and LH during the normal menstrual cycle of the human female.
Source: Guillemin 1972, with permission.

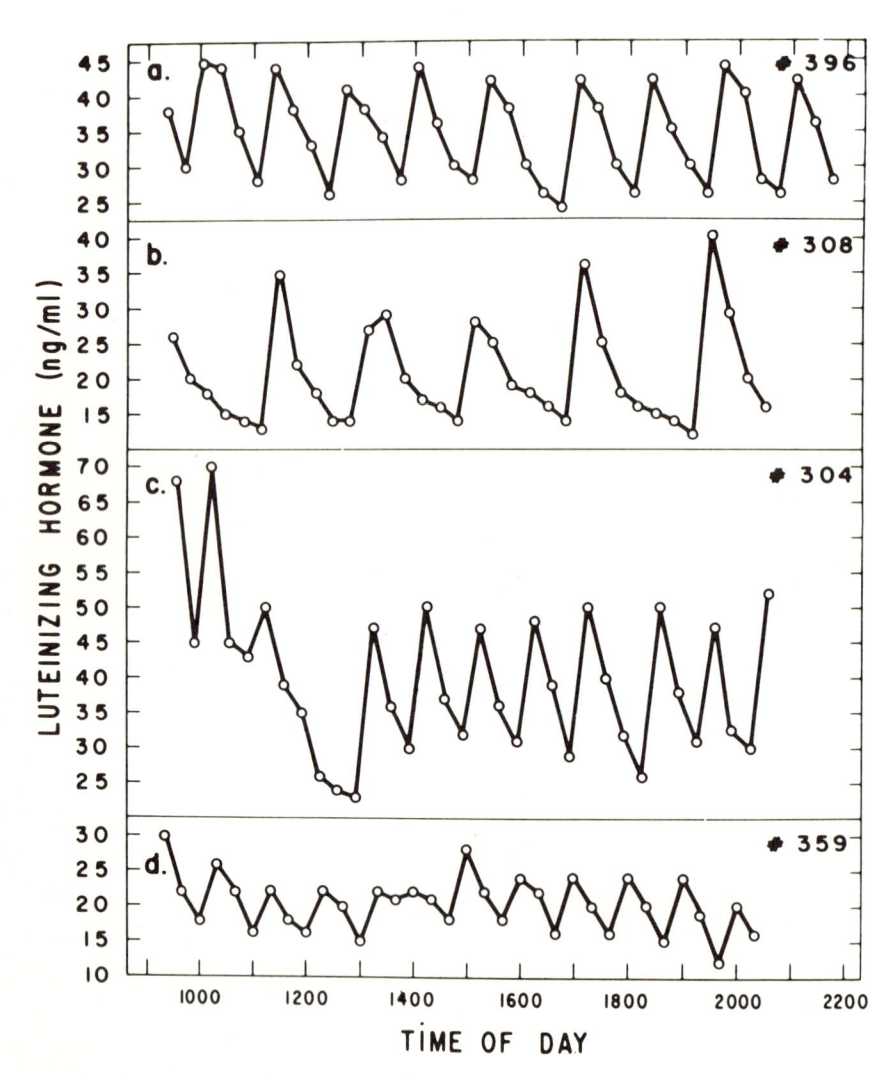

**Figure 4.8** Showing the pulsatile secretion of LH by four ovariectomized rhesus monkeys.
Source: Dierschke et al. 1970, with permission.

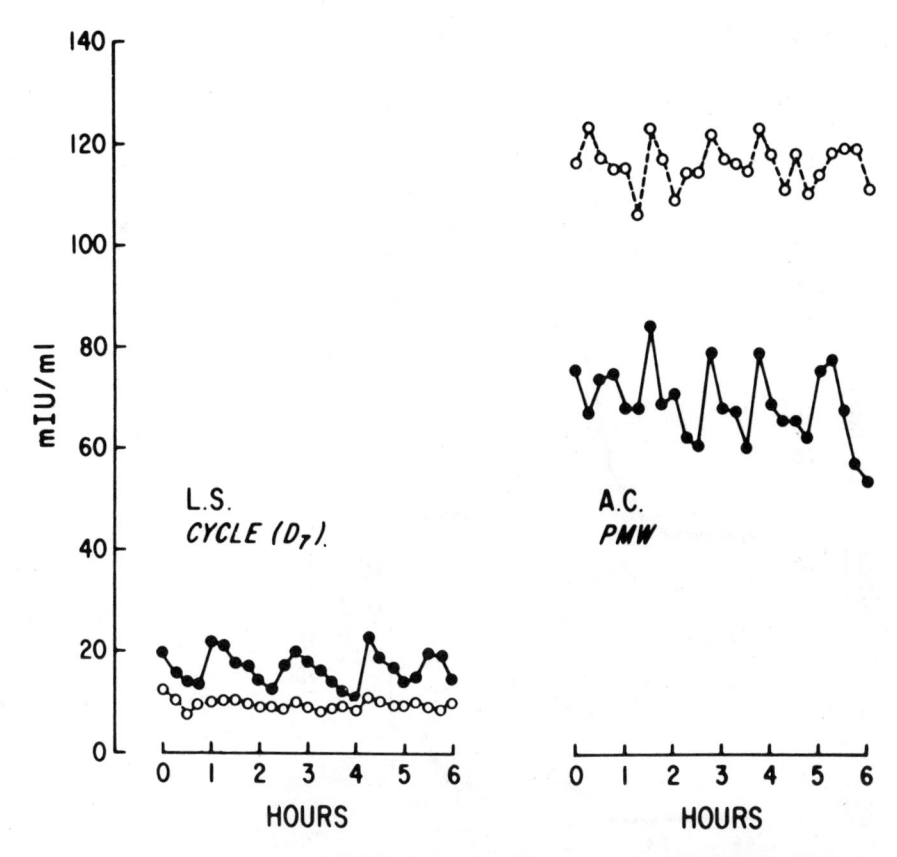

Figure 4.9 Showing the pulsatile secretion of FSH and LH in the human female on day 7 during a normal menstrual cycle and after menopause when the plasma level of gonadotropin is elevated.
Source: Yen et al. 1972, with permission.

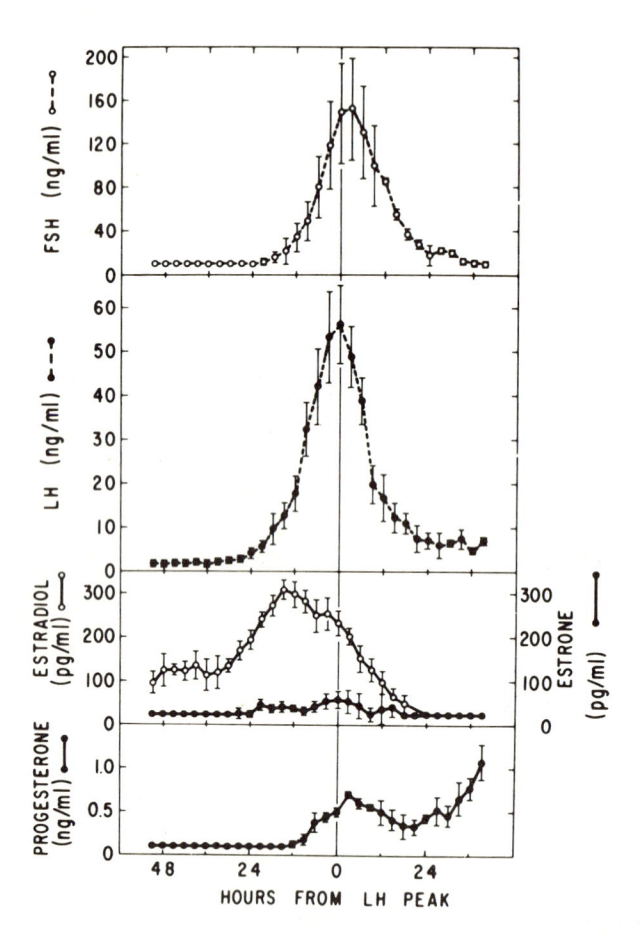

Figure 4.10 Profiles of the mean plasma concentration of FSH, LH, estradiol, estrone, and progesterone during the menstrual cycle in the rhesus monkey. Source: Weick et al. 1973, with permission.

Now that a sensitive radioimmunoassay for prolactin has been developed (Hwang, Guyda, and Friesen 1971), the blood levels can be followed on a closely timed basis. Prolactin levels in peripheral serum fluctuate in an irregular pulsatile manner before and after the ovulatory period (figure 4.6). A modest peak also occurs in unison with the preovulatory surge of LH and FSH in rats (Neill, Freeman, and Tillson 1971; Neill and Smith 1974; Butcher, Collins, and Fugo 1974) and primates (Friesen, forthcoming). Marked diurnal cycles in the blood levels of prolactin have been observed in pseudopregnant rats (Freeman et al. 1974) and man (Jacobs and Daughaday 1974) with elevated levels being reached during sleep.

### Estrogen and Progesterone

The ovarian secretion of estrogen and progesterone during the rodent estrous cycle and the primate menstrual cycle as reflected by their concentration in the peripheral bloodstream is shown in figures 4.10 and 4.14. The degree to which investigators in this field have been hampered by lack of assay procedures sensitive enough to measure quantitatively the minute concentration of reproductive hormones in the bloodstream will be evident from the fact that the measurement of the changes in blood levels of the hormones of the pituitary-gonad axis during an estrous or menstrual cycle has only become possible within the past few years (Ross et al. 1970; Knobil 1974; Schwartz 1974; Butcher, Collins, and Fugo 1974). The critical need of such information had been evident for about forty years and is an outcome of the perfection of assay techniques using hormone-specific antibodies. The data provide a few surprises, but in the main they are in general agreement with prior concepts based on other physiological studies and the measurement of some of the hormones or their metabolic breakdown products in concentrates of urine.

Plasma estrogen concentration in women climbs fairly steadily through the follicular phase to a peak at midcycle followed by a sharp decline during the stages of ovulation and early luteinization. A second mild rise during the midluteal phase eventuates in a final dropping off to base line values at the end of the cycle (figure 4.14). Blood progesterone values remain barely detectable through the follicular phase, start rising as ovulation becomes imminent, slacken off with rupture of the follicle, resume their climb to a temporary plateau (or double peak), and end with a final plunge to base line values a day or two before menstruation begins. There can be little doubt that the circulating levels of estrogen and progesterone reflect the degree of their stimulatory activity. In rats (figure 4.6) all five hormones exhibited a peak in plasma concentration on the day of proestrus. In addition, progesterone showed a slight rise during diestrus and prolactin a brief second peak on the day of estrus. It will be noted that the estrogen peak preceded the preovulatory LH surge and that the FSH peak rose more slowly and was more prolonged than the LH peak. The FSH peak is believed responsible for initiating the growth of a new set of follicles for the next cycle (Schwartz 1973; Neill and Smith 1974).

### Feedback Action of Ovarian Hormones

Whether the gonadal hormones exert their feedback action at the level of the pituitary or the brain or both remains a matter of long-standing controversy. Nonetheless, some important observations bearing on this matter have been made. Autoradiographs of animals injected with isotope-labeled estrogen show the anterior pituitary, median eminence, and hypothalamus (Pfaff 1968; Stumpf 1970; Kato 1970) to be heavily labeled, suggesting multiple sites of major feedback action. Many other extrahypothalamic areas possibly concerned with behavior or modulation of reproductive functions also show some labeling, the significance of which needs much additional study.

An abundance of recent evidence leaves no doubt that the gonadal steroids markedly influence the responsiveness of the pituitary to the action of LRF. The response to a given dose of LRF is known to be far greater in the preovulatory period (figure 4.11) than at any other time of the cycle in rodents, monkeys, and the human female (Yen et al. 1973 and 1975). This is believed to be due to an increase in the serum estrogen content. Although such evidence suggests that the preovulatory LH surge may be due to increased pituitary sensitivity to LRF and not to an increased output of LRF, there is substantial

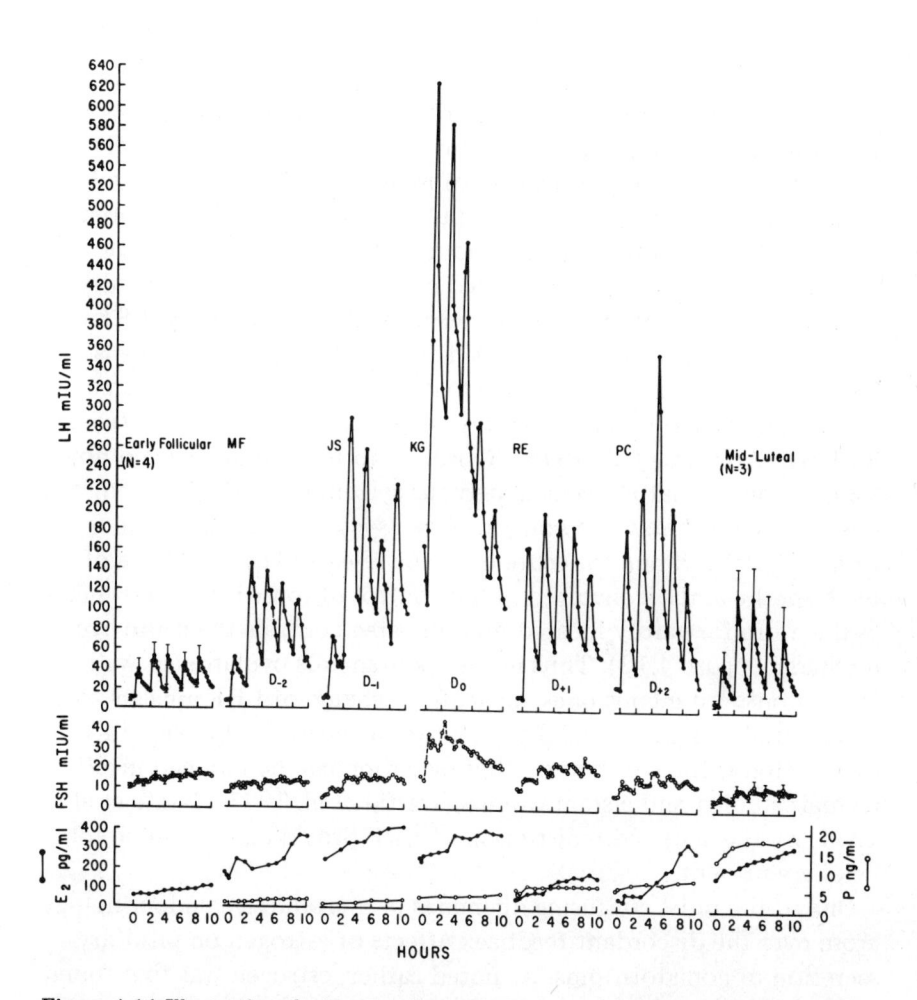

Figure 4.11 Illustrating the increased responsiveness to LRF at midcycle in the human female. Each LH spike represents the response to a single injection of a constant dose of LRF. Simultaneous responses in plasma FSH, estrogen, and progesterone are also shown.
Source: Yen et al. 1975. Courtesy of Academic Press.

evidence that the hypothalamus plays a more positive role in the induction of ovulation. Three lines of evidence support this view. First, in those animals like the rabbit and cat, which ovulate only after mating, it is the physical act of mating that triggers the ovulatory mechanism. Here, nervous impulses conveyed to the hypothalamus could only act by bringing about an acute discharge of LRF. This has not yet been demonstrated due to the technical difficulty of collecting blood from the portal vessels of these animals for LRF analysis. Second, with the help of radioimmunoassay, evidence is being accumulated that shows that LRF can be detected in women in the peripheral serum only at midcycle (Malacara, Seyler, and Reichlin 1972; Arimura et al. 1974). Third, in both spontaneous and nonspontaneous ovulators, central nervous system control of reproductive function is known to consist of two systems—one for tonic release of LRF and one for acute or cyclic release. The basal area of the hypothalamus is involved in the control of tonic secretion whereas the more forwardly located preoptic area controls the acute cyclic discharge (figure 4.12). Thus in the spontaneous ovulators as well as in the nonspontaneous ones, an acute discharge of LRF appears to be an essential component of the ovulatory mechanism. Backing this assumption is the observation that ovulation can be induced in all animals studied and also in women (Kastin et al. 1971; Zarate et al. 1974) by the acute administration of LRF providing ovulatable follicles are present.

One of the most confusing dilemmas in reproductive endocrinology arose over the discordant feedback effects of estrogen on pituitary secretion of gonadotropins. As noted earlier, estrogen was first found to have a powerful depressant action on the pituitary-gonadal axis in both males and females. However, when it became possible within the past few years to measure the blood levels of the gonadotropins and the gonadal steroids on a daily basis throughout the estrous (Schwartz 1969; Butcher, Collins, and Fugo 1974) or menstrual cycle (Midgley and Jaffe 1966; Ross et al. 1970; Hotchkiss, Atkinson, and Knobil 1971; Mishell et al. 1971; Abraham et al. 1972; Monroe, Jaffe, and Midgley 1972; Knobil 1974; Yen et al. 1975), it was discovered that the preovulatory LH peak was always preceded by a

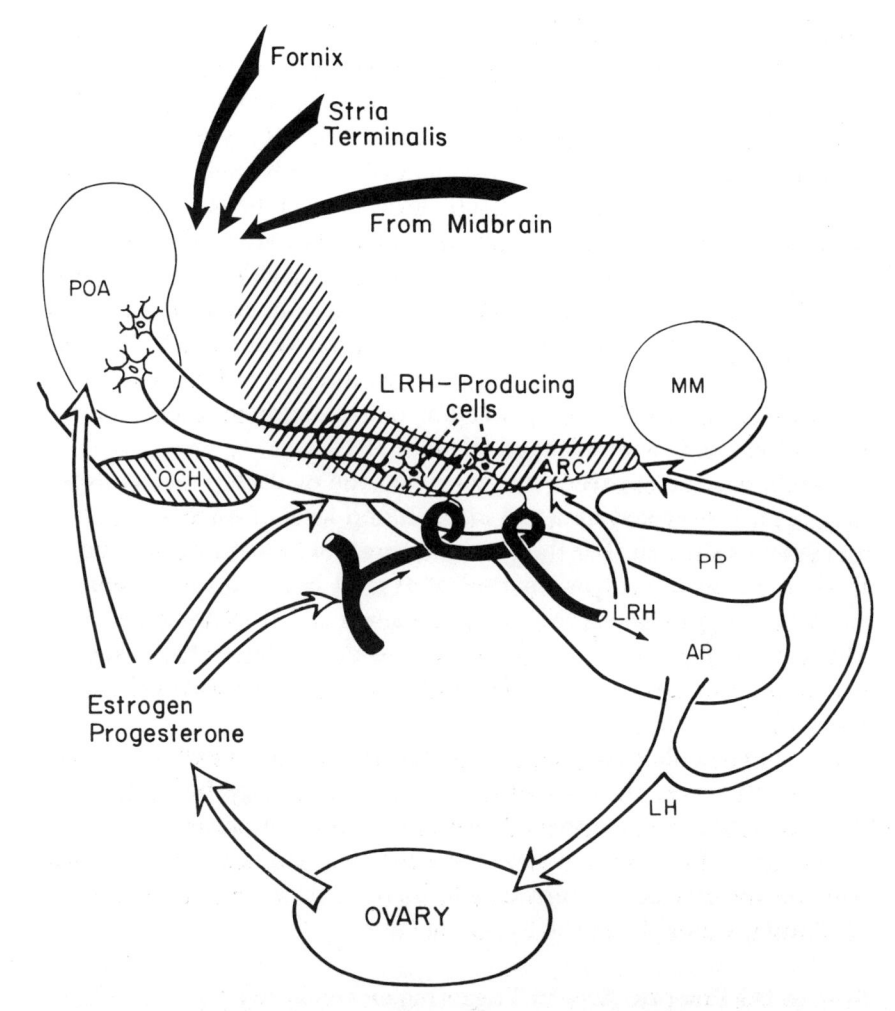

Figure 4.12 A schematic representation of the interrelationships between the brain, pituitary, and ovary. LRF produced by endocrine neurones and transported to the pituitary via the portal vessels stimulates the release of LH. This in turn stimulates the ovary to secrete estrogen and progesterone, both of which exert feedback reactions on the neural centers. LRF is known to influence sex behavior. The solid arrows represent the influence of other brain centers on mechanisms controlling reproductive functions. POA, preoptic area; ARC, arcuate nucleus; OCH, optic chiasma; MM, mammillary body; PP, posterior pituitary; AP, anterior pituitary.
Source: C. H. Sawyer, 1975. Courtesy S. Karger AG, Basel.

sharp rise in the level of estrogen (figure 4.10). It was subsequently demonstrated that acute estrogen treatment would lead to a surge in blood LH in sheep (Goding et al. 1969), monkeys (Yamaji et al. 1971), (figure 4.13) and women (Yen and Tsai 1972) identical to that seen at estrus or at midmenstrual cycle (see Knobil 1974 and Yen et al. 1975 for a review). Such elevations can be elicited in females at any time during the follicular phase but are followed by ovulation only when the estrogen treatment coincides with the presence of mature follicles. This action of estrogen in eliciting an outpouring of pituitary LH is referred to as "positive feedback." It does not occur in the presence of progesterone nor in intact males. An estrogen-induced LH surge can be elicited in castrated males following pretreatment with estrogen (Knobil 1974).

Thus it became apparent that whatever the ovulatory mechanism may be, it is triggered by a rise in the blood level of estrogen. Follow-up evidence showed that the estrogen threshold must not only be reached but sustained for a period of at least twelve hours (Yamaji et al. 1971) to bring about the ovulatory surge of LH (Knobil 1974). This wholly new concept and unexpected development has again added enormously to our understanding of the hormonal control of the ovulatory process.

The well-founded presumption is that the positive feedback action of estrogen acts at the level of the preoptic area or cyclic center of the neuroendocrine control of pituitary gonad stimulating function. As a matter of doubly assuring that ovulation will occur, the estrogen also acts directly on the pituitary to increase its sensitivity to the LRF forthcoming from the hypothalamus.

### Role of the Preoptic Area in Triggering an Ovulatory Discharge of LH

Although it was known that the blockage of spontaneous ovulation in nembutalized rats could not be overcome by electrical stimuli applied to the basal arcuate area of the hypothalamus, a chance observation based on slight forward misplacement of the electrodes revealed that ovulation could be induced in such rats when the electrodes were placed in the preoptic area (Everett 1961a).

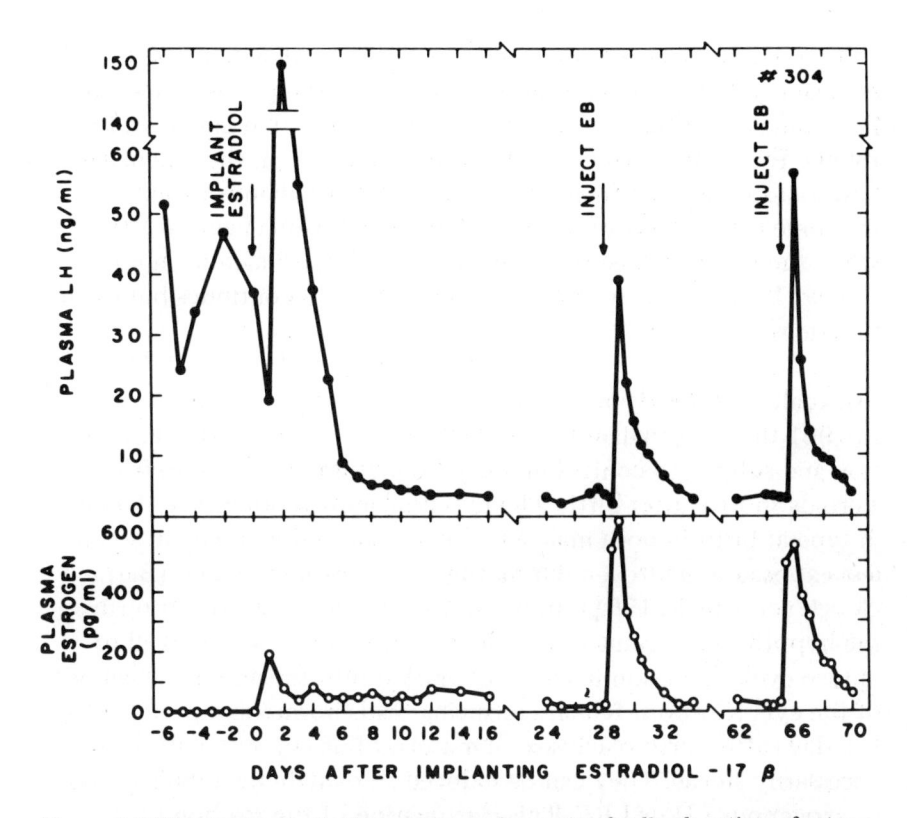

Figure 4.13 Illustrating both the positive and negative feedback action of estrogen on LH secretion in the female rhesus monkey. A silastic capsule containing estradiol-17$\beta$ (which can slowly leach out) was implanted into an ovariectomized monkey. The first effect of the implant was to trigger an LH surge (positive response). This was followed by a chronic negative feedback reaction that lowered the pituitary secretion of LH. The subsequent single injections of estradiol benzoate invariably elicited within 12 to 24 hours an LH surge that is similar in every respect to a normal midcycle preovulatory LH surge.
Source: Karsch et al. 1973, with permission.

Other studies have shown that the positive feedback action of estrogen and progesterone serves to stimulate the preoptic area (cyclic center) and effects the acute discharge of LH that is known to follow. By contrast, the negative feedback action of these gonadal hormones is believed to act on the basal hypothalamus to suppress the tonic center regulating gonadotropin secretion (figure 4.12). When the preoptic area is severed from the hypothalamus in rats (Halasz 1969), tonic secretion of gonadotropins continues but ovulation does not occur.

**Bisexuality of the Hypothalamus**
In 1961 the hypothalamus was identified as being bisexual in function and subject to control in rats by unnatural exposure to sex steroids shortly after birth. The hypothalamus is potentially female in type at birth in both male and female rats and will remain so unless exposed to androgen during the early postnatal period (Barraclough and Gorski 1961). In newborn male rats castrated at birth, the hypothalamus remains female in type and can be observed to induce cyclic function in an ovary grafted into the anterior chamber of the eye. Newborn females receiving testis hormone during the first few days after birth reach sexual maturity but never ovulate and are necessarily sterile. They can be induced to ovulate by administration of exogenous LH or LRF. This clarification of the relation of the sexual differentiation of the brain to the control of pituitary gonadotropic functions in the two sexes marked a major advance in our understanding of reproductive biology in mammals and opened a new approach to the study of the control of fertility (for a review see Barraclough 1967).

It has recently been demonstrated that the hypothalamus is capable of metabolizing androgens to estrogens via ring A reduction, aromatization, or hydroxylation (reviewed by Naftolin et al. 1975). This coupled with the fact that estrogens proved to be as effective as androgens in masculinizing the hypothalamus points to the possibility that even in males estrogen may participate in the sexual differentiation of the brain (Flores et al. 1973). Most of the work on brain differentiation has been done in rodents and uncertainty exists as to whether similar mechanisms are operative in primates. The difficulty lies in the fact that although differentiation occurs after birth in rodents, it most likely occurs during early pregnancy in animals with longer gestation periods. Exposure of the monkey fetus to androgen after forty days has resulted in some masculinization of behavior patterns in the female offspring, but they have normal menstrual cycles and are fertile. Given at earlier stages in pregnancy, androgens cause abortion. Much work needs to be done to determine whether the sexual differentiation of the brain of subhuman primate fetuses can be artificially altered as in rodents.

**Refining the Limits of the Critical Period**
As recently demonstrated (Takahashi et al. 1974), the precise limits of the two-hour critical period on the afternoon (2 to 4 P.M.) of proestrus are remarkable. Removal of the pituitary gland at any time up to 3:30 P.M. blocked ovulation completely; with removal at 3:45 P.M. the blockage was partial; but when the operation was performed at or after 4 P.M. ovulation occurred normally. The explanation is that by 3:30 P.M. the amount of LH discharged was not sufficient to bring about ovulation. A half hour later an ovulatory quantum was in the bloodstream and ovulation could no longer be blocked by removal of the pituitary. Other recent work has demonstrated that the signal for the LH discharge during the critical period begins at about 10 A.M. (Aiyer and Fink 1974). If the ovaries are removed at this time, the LH surge during the critical period does not occur. If an injection of estrogen is given immediately after ovariectomy, the surge occurs normally. Other evidence has suggested that the hypothalamus is attuned to respond daily but is effective only when the estrogen has reached a critical level, as on the day of proestrous (Fink and Aiyer 1974). Whether these events in the rat are applicable to other mammals is unknown, but it has been shown that sheep do not have a fixed critical period. It is likely, however, that something similar to the chain of events found in the rat will prove to be a common feature of the neuroendocrine ovulatory mechanism in spontaneous ovulators.

## Ovarian Binding of Gonadotropins, Receptor Proteins, and Mechanism of Hormone Action

Hormones as blood-borne chemical messengers are distributed ubiquitously throughout the body but evoke a response only by those cells capable of decoding the message. This is the mechanism whereby hormones achieve their specificity of action. The cells of organs that respond to a particular hormone are referred to as the target cells or target organs of that hormone. In order to be effective, the hormone must be taken up and bound to a tissue-specific receptor. The binding of hormone to receptor leads to an interaction and is but the first in a train of events that produces an overt response such as growth or secretion. Hormones are bound by target tissues in such infinitesimally small quantities that their presence cannot be detected by bioassay. It is necessary, therefore, to tag the hormone with a radioactive isotope and track the radioactivity. This can be measured in whole organs or tissue fractions and visualized in microscopic sections by autoradiography (Midgley 1973).

Labeled LH, FSH, or hCG bind to ovarian tissue with high affinity and specificity (LH, hCG: Lee and Ryan 1971, 1972, 1973; Ashitaka, Tsong, and Koide 1973; Rajaniemi and Vanha-Pertulla 1972; FSH: Presl et al. 1971; Rajaniemi and Vanha-Pertulla 1973). The receptor for these hormones is thought to be localized in the cellular membrane. The use of labeled hormone and autoradiography shows very clearly that LH binds to the luteal tissue in vivo (Rajaniemi and Vanha-Pertulla 1973) and in vitro (Lee and Ryan 1971; Channing and Kammerman 1974). It does not bind to granulosa cells until they are on the verge of undergoing luteinization (Channing 1970). The number of receptors for LH increases as the follicle nears maturity; this has been shown to be a function of the action of FSH (Zeleznik, Midgley, and Reichert 1974; Channing 1974; 1975). FSH also serves to increase the number of LH (or hCG) binding sites in the theca interna. FSH, on the contrary, binds only to granulosa cells in vivo. This is in agreement with the concept that FSH itself does not stimulate the theca interna and ipso facto is not a stimulus to estrogen secretion under in vivo conditions. Under in vitro conditions FSH does stimulate granulosa cells to secrete estrogen (Moon, Dorrington, and Armstrong 1975). Prolactin binds irregularly to corpora lutea of all ages and to some extent to the follicular granulosa of the rat ovary (Midgley 1973).

The cells of the corpus luteum have *specific receptors* not only for prostaglandins but also for LH and hCG, especially for their $\beta$ subunits. Hence, it is conceivable that chemically modified fragments could be obtained that would exhibit little, if any, biological activity, but that would still bind firmly to the luteal receptors and compete with the biologically active hormone for binding sites. What degree of saturation by these inert analogs would be needed to achieve luteolysis and how this could be brought about remains to be answered by future research.

**Ovulation in the Human Female**

Figure 4.14 shows the cyclic patterns of the hormones in women as measured by radioimmunoassay. The biologically active LH in plasma is believed to be reliably gauged by radioimmunoassays. The LH surge precedes ovulation by some twenty to thirty hours, and coincides with a significant rise in the plasma levels of $17\alpha$-hydroxyprogesterone, progesterone, and $20\alpha$-dihydroprogesterone. It is followed by a significant increase in basal body temperature. The so-called preovulatory surge in plasma estradiol and estrone is more variable and may occur a day before or after the LH surge or simultaneously with it. Knowledge of the reliability of these and other parameters in the *prediction*, or at least *detection*, of ovulation is of considerable importance because of the short fertilizable life of the freshly ovulated human egg, estimated to be not more than twenty hours. Because ovulation is soon followed by a period of "physiological infertility," a better knowledge of the exact time of ovulation might not only improve the efficiency of the so-called rhythm method of fertility regulation but could also be of use in the development of new methods of contraception. That is why it is believed that the development of a simple and reliable "kitchen" method for ovulation prediction or detection would represent a significant advance in fertility control.

A major midcycle surge in plasma LH and FSH is considered to

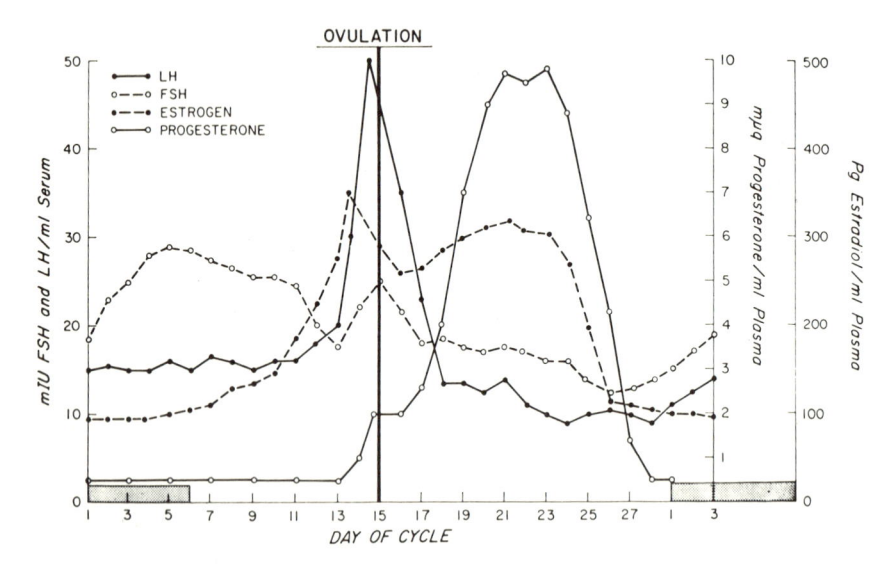

Figure 4.14 A model of the profiles of FSH, LH, estradiol and progesterone during the human menstrual cycle.
Source: Taymor et al. 1972, with permission.

have a high predictive value for an impending ovulation in untreated subjects; however, its diagnostic value is much less certain when it comes to the inhibition of ovulation, especially in women taking steroidal contraceptives. For the assessment of ovulation inhibition the assay of plasma progesterone (or urinary pregnanediol) is much more reliable.

Although a large number of compounds (such as certain depressants of the central nervous system, cholinergic and adrenergic blocking agents, inhibitors of catecholamine synthesis, monoamineoxidase inhibitors, antigonadotropic substances, and perhaps also releasing hormone analogs) are known to inhibit ovulation in experimental animals, thus far only steroidal preparations have been found suitable for large-scale use as contraceptives. Their mechanism of action may involve a spectrum of endpoints, but the currently used estrogen-progestogen contraceptive formulations do consistently inhibit ovulation. It has been reported several times that the various types of progestogen-only "minipills" do not invariably inhibit ovulation; the proportion of normal ovulatory cycles in such subjects remains to be ascertained in carefully designed large-scale studies.

## The Formation and Demise of the Corpus Luteum

In recent years much has been learned concerning the mechanism controlling the transformation of follicular granulosa cells into lutein cells from studies conducted in vitro. Cells harvested from preovulatory follicles in pigs, mares, monkeys, and women all undergo spontaneous luteinization in culture without the addition of LH or FSH to the medium; they have already been sufficiently exposed to these hormones in vivo. Granulosa cells from medium-sized follicles do not luteinize unless FSH and/or LH are added to the medium. Cells from small follicles do not luteinize unless they are first primed with both FSH and LH. Under normal in vivo conditions the follicles are exposed to both FSH and LH at all times during the cycle. Although as previously indicated the primary action of FSH on ovarian morphology is to stimulate the growth and maturation of follicles and the primary function of LH is to induce ovulation and luteinization, these functions are always carried out in the presence of both hormones. Their actions can and have been studied independently but not with assurance that such observations reflect normal physiology. Although the follicular granulosa cells accumulate LH receptor sites and begin to bind LH as they near maturity, they do not luteinize until the ovum has been expelled during the ovulatory process. According to a thesis advanced by Nalbandov's group, the ovum or its enveloping cluster of cells produces a factor that inhibits luteinization (El-Fouly et al. 1970).

Luteinization, whether it occurs in vivo or in vitro is in all instances accompanied by the secretion of progesterone. This can be increased in vitro by the addition of either LH or adenosine 3',5'-cyclic monophosphate (cyclic AMP) (Marsh and Savard 1966; Channing 1973). The latter is believed to result from the interaction of LH and its receptor protein.

Although LH was well known to be capable of inducing ovulation and luteinization in women, doubt existed as to what pituitary hormonal stimulus, if any, was necessary for the secretion of progesterone. In 1971, the issue was settled by the demonstration that human LH alone initiated and maintained luteal function in a woman in the complete absence of the pituitary (Vande Wiele et al. 1970).

The major unsolved problem with respect to the human corpus luteum is what causes it to regress and cease functioning at the end of the cycle. The answer to this long-standing problem could have great contraceptive significance. About the only firm knowledge we have concerning the luteal demise is that it is not due, as was widely anticipated, to withdrawal of LH (see figure 4.10).

Diethylstilbestrol and ethinyl estradiol in high doses are used in several countries experimentally for so-called *postcoital contraception*. The treatment appears to be effective when given within seventy-two hours following exposure but fails to interfere with gestation once implantation has taken place. For the time being, this treatment is associated with a variety of unpleasant side effects, cannot be repeated often, and should absolutely be regarded as an emergency measure rather than an established routine method. However, the approach as such is promising and it is at least possible that the same effect can be achieved by the use of other types of steroids exhibiting slight or no side effects. Although the postcoital administration of estrogens in high doses depresses the secretion of progesterone, the mechanisms of their action is not thought to be luteolysis but rather interference with tubal transport. Studies on the mechanism of action of this type of fertility regulation might be very rewarding.

Plasma progesterone levels are also greatly suppressed by the daily oral administration of high doses of contraceptive progestins such as norgestrel, norethynodrel, norethindrone, chlormadinone acetate, and medroxyprogesterone; but they have not proved effective as postcoital agents in preventing pregnancy.

In those species (guinea pig, sheep, pig) in which hysterectomy prolongs life of the corpus luteum, there is clear-cut evidence that the uterus produces a *luteolytic factor*. In sheep, this factor has been identified as prostaglandin $F_{2\alpha}$ produced by the uterus, passed by countercurrent transmission from the uterine vein to the closely applied ovarian artery, and carried directly to the corpus luteum (McCracken et al. 1972). Exogenously administered prostaglandins are also luteolytic in rats (Phariss and Wyngarden 1969; Behrman, Macdonald, and Greep 1971). How the prostaglandins effect luteal involution and loss of function has not been determined. Prostaglandins have, otherwise, been shown to stimulate cyclic AMP formation and progesterone synthesis when added to incubating slices of rat or bovine lutein tissue (see Behrman and Caldwell 1974). Although the prostaglandins cause abortion in monkeys (Kirton, Phariss, and Forbes 1970) and have been widely used as abortifacients during the midtrimester in women (Karim and Filshie 1970), there is no evidence that disruption of luteal function is involved in the termination of pregnancy by prostaglandins in women.

**The Ovarian Stroma**
An intriguing question is the function of the ovarian stroma in the reproductive cycle. It is capable of secreting steroids, mainly androgens; but there is evidence that it may be a major source of progestins in the rabbit (Hilliard, Archibald, and Sawyer 1963; Hilliard, Penardi, and Sawyer 1967). However, studies of binding affinity for

gonadotropins have thus far failed to show significant binding to stromal or interstitial tissue.

Restoration of Fertility
The normal sequence of events leading to ovulation had been so successfully replicated by 1958 that it had become possible to induce ovulation in many fertile women by treatment with human menopausal gonadotropin (a strongly follicle stimulating extract of the urine of postmenopausal women) for seven to ten days followed by acute injection of a high dose of LH, usually in the form of a potent luteinizing hormone from the placenta known as human chorionic gonadotropin of hCG. Failure to understand the proper dosages of FSH and hCG has occasionally resulted in superovulation and multiple births in women treated with "fertility drugs." Although the treatment is imperfect, it must be recognized that this has, for the first time in history, made it possible for many infertile women to become pregnant and give birth to normal offspring. The number of instances of excessive ovarian enlargement and multiple ovulations, although not yet eliminated, has been greatly reduced by adjustment of dosage and duration of treatment (see chapter by Gemzell in volume II of this series).

Reproductive Immunoendocrinology

Antisera to Gonadotropins
As an experimental tool, the neutralization of a specific hormone by immunologic means has many advantages for the study of reproductive processes. Antibodies produced in rabbits against hCG have been shown to be capable of blocking ovulation and lowering progesterone secretion in monkeys (Moudgal, MacDonald, and Greep 1971). They also inhibit implantation in rats and produce abortion when administered during the first half of pregnancy (Loewit, Badawy, and Laurence 1969; Loewit and Laurence 1969; Madhwa Raj and Moudgal 1970). This demonstration that LH is essential for the maintenance of luteal function in the pregnant rat was unexpected because previous evidence had shown that luteal function, at least in the pseudo-

pregnant rat, is dependent upon another hormone, prolactin. Active immunization of monkeys against conjugates of estrogen has also been shown to result in cessation of menstrual cycles and long-lasting sterility (Ferin et al. 1974).

### Antisera to Releasing Factors

Recent work has demonstrated that both natural and experimentally induced ovulation can be inhibited by passive immunization with antisera to LRF, the luteinizing hormone releasing factor of hypothalamic origin. Similarly, animals actively immunized against an LRF conjugate show the same severe impairment of structure and loss of secretory function of the gonads as is seen after removal of the pituitary (Schally et al. 1973; Takahashi and Yoshinaga 1974). This confirms, in agreement with earlier work on transplantation of the pituitary to sites away from the base of the hypothalamus, that the pituitary acting in vivo cannot secrete FSH or LH in the absence of LRF.

The most promising avenue for the employment of immunizing procedures for the control of fertility would appear to be in the development of antibodies to some antigen of the conceptus. The antigen of choice at this writing is the $\beta$ subunit of hCG or some fragment of same (Talwar 1974 and Talwar and associates 1976). Such a procedure would presumably have no effect on the normal reproductive cycle and no side effects. It should only result in inhibition of the establishment of pregnancy and this has already been demonstrated in baboons (Stevens 1973 and 1974). Small-scale testing in women is already underway (Talwar 1974).

### Oviduct

### Transport of Eggs and Sperm

Although the oviduct has been studied rather intensively over the past fifteen years (for a review see Hafez and Blandau 1969 and chapters 34 through 38, *Handbook of Physiology*, section 7, volume 2, part 2, 1973), it is doubtful that there is any other component of the female reproductive system about which so many questions remain. The oviduct is now regarded as a dynamic and complex structure, but for many decades it was looked upon as an inert conduit where sperm meets egg and the resulting zygote moves haltingly downstream to the uterus. Most of the early observations centered on the microscopic anatomy of the different segments of the oviduct, such as the fimbriated infundibulum at the distal end, the ampulla, the ampullary-isthmic junction, the isthmus, and the intramural segment or uterotubal junction (figure 4.15). Many years ago it was noted by visual inspection that at or near the time of ovulation, the funnel-shaped opening of the oviduct moved closer to the ovary and was described as "massaging" its surface. Such movement is thought to be due to contraction of smooth muscle fibers in the oviducal ligaments. In any event, the pick-up of released ova is very efficient and occurs within minutes following follicular rupture. The ova are fertilized in the oviduct and pass quickly to the ampullary-isthmic junction, where they are held in check for a matter of three days while the uterus is being prepared for their reception. During their stay in the oviduct the fertilized ova continue to undergo cleavage and blastocyst formation (figure 4.15).

The mechanism of transport of sperm and ova through the oviduct has been the subject of many studies in a wide variety of animals, but there is still little definitive information. Most of the cilia beat toward the uterus, but in the rabbit a longitudinal band of cilia beat in the opposite direction and may aid distal migration of sperm. It is, however, becoming increasingly evident that the primary factor in the transport of both sperm and ova is the segmental and peristaltic contractions of the investing layers of smooth muscle fibers. In summary, the functions of the oviduct may be broadly described as transport of gametes and provision of a suitable fluid environment for the sperm, the ovum, the fertilization process, and the newly created zygote.

Ciliogenesis
Without question, the most striking observation made on the oviduct in recent years is the cyclic regeneration and loss of cilia with each recurrent menstrual cycle in the rhesus monkey (Brenner 1969).

Figure 4.15 Showing schematically ovarian follicular development, ovulation, corpus luteum formation, and processes involved as the ovum passes through the oviduct to become embedded in the wall of the uterus. 1) Freed ovum (in second meiotic metaphase) with one polar body and cumulus cells; 2) Fertilization, dispersal of cumulus cells. Formation of second polar body; 3) Showing pronuclei of ovum and sperm; 4) First cleavage in process; 5,6,7) Additional cleavage stages; 8) Morula; 9 & 10) Free blastocyst; 11) Beginning of implantation. Source: R. L. Dickinson, *Human Sex Anatomy*. William and Wilkins Co., Baltimore, 1933.

During the luteal phase, the cilia on the fimbria and to some extent the ampulla are completely shed except for a basal body lying deep beneath the luminal surface of the cells. The same loss occurs after ovariectomy. With the initiation of a new cycle, and after treating ovariectomized animals with estrogen, the cilia reappear (the same phenomenon has recently been found to occur in cats). Ciliogenesis is, therefore, estrogen dependent. Loss of cilia during the luteal phase is not due to the action of progesterone per se but to its suppression of estrogen receptors.

Regeneration of the extremely complex microanatomy of each cilium from a tiny basal body has been elucidated in minute detail by means of the electron microscope. The mechanism directing this fantastic reconstruction of hundreds of motile cilia on single cells is a total mystery. A similar phenomenon has not been observed in the oviduct of women, but it has long been known that the epithelium does undergo an estrogen-dependent change in height and appearance during each cycle and after gonadectomy or the menopause. The epithelial cells become maximally columnar at the time of ovulation and recede to a low cuboidal state before the cycle ends. With total loss of gonadal function, the cells undergo extreme involution.

**Luminal Fluid**

Since 1960 much attention has focused on the oviducal fluid as a medium for the biological processes that occur in the oviducts. These processes include capacitation of sperm, fertilization, and cleavage. Oviducal fluid has been shown to be formed through the processes of selective transudation from the blood plasma and exocrine secretion by the oviducal epithelium. Under normal circumstances such fluid also receives contributions from the peritoneal fluid, follicular fluid, and uterine fluid. Various devices have been developed for the collection of oviducal fluid from monkeys on a daily basis throughout the cycle. The volume is dependent upon estrogen stimulation and is maximal at the time of ovulation (Mastroianni, Shah, Abdul-Karim 1961). Progesterone has a suppressive action on volume but, again, this may be due to blocking the action of estrogen. Numerous enzymes, proteins, carbohydrates, minerals, electrolytes, and other constituents of oviduct fluid have been identified. This fluid is a perfect medium for fertilization and cleavage but only at the time of ovulation. Newly fertilized mouse ova have special requirements for their energy source (Brinster 1965; Biggers, Whittingham, and Donahue 1967). They cannot utilize glucose but must rely on its breakdown products such as pyruvate or lactate. Such substrates are provided by the oviduct. Although little is known about the difference in fluid composition in different segments of the oviducts, it is a requirement that they meet the changing needs of the fertilized ovum as it transcends the tube (figure 4.15). The study of the composition of oviduct fluid is of fundamental importance in developing suitable media for in vitro fertilization and culture of early embryos. With full knowledge of these requirements, it may be a short step to arresting any one of the critical biological processes that occur in this organ. Moreover, it is firmly established that the timing of the passage of the cleaving ovum through the oviduct is absolutely critical. If the passage could be hastened or delayed, even by as much as one day, a newly formed blastocyst would not be likely to find a favorable uterine environment for its continued development.

Uterotubal Junction

Most of the progress made in the study of the uterotubal junction has concerned its morphology and its role as a gateway to the passage of sperm to the oviduct. The morphological complexity of this portion of the female genital tract varies enormously between species (Hafez 1973). In most litter-bearing animals, it serves also as a reservoir for sperm. Although it has been known for many years that the uterotubal junction limited the number of sperm entering the oviduct, only in recent years has this been quantitatively documented. The ability of the junction to exercise selective control over the passage of motile and dead or nonmotile sperm or inert particles also varies between species. In rats inseminated with a mixture of sperm from several species, the junction will admit the passage of only rat sperm. The junction in man does not select between human sperm and inert particles. The junction does not appear to delay the passage of ova into the uterus. It does, nevertheless, have that potential by virtue of its investing musculature.

The uterotubal junction has been studied extensively as a possible site for occluding the lumen of the female genital tract. Various means, such as the infusion of sclerosing agents, insertion of plugs or adhesives, and cauterization, have been tried with qualified success. Occlusion of the junction can occur in normal cycling women and is a common cause of infertility. Additional research on the uterotubal junction is urgently needed. Some of the avenues for important study include (1) use of drugs that may close the lumen through their action on the musculature, (2) determination of the mechanism whereby the junction exercises control over the passage of sperm, (3) clarification of the effects of ovarian hormones on the structure and function of the junction, and (4) perfection of a simple and reliable means of occluding the lumen of the junction.

## Uterus

Evidence that the uterus may exert a regressive (luteolytic) influence on the corpus luteum has stimulated much interest in uterine physiology during the past fifteen years. Some excitement was fired by the findings that, as in the guinea pig, removal of the uterus in pigs, sheep, and cattle led to persistence of the life and function of the corpora lutea for at least the normal length of gestation (see Anderson 1973 for a review). Moreover, it was observed that the duration of pseudopregnancy in hysterectomized rats and hamsters was extended by at least a week. Although in each instance the blood level of progesterone failed to rise as in pregnancy, it did remain at about the maximal level seen in the nonpregnant cycling animal of the same species. Moreover it was observed that in unilaterally hysterectomized pigs and sheep, the corpora lutea underwent regression only on the side with the intact uterine horn. Similarly, unilateral uterine transplants proved to have a deleterious influence on the corpora lutea in the adjacent ovary. These effects were observed only with transplants of the endometrium.

With such solid evidence for the uterine origin of a luteolytic factor, extracts of uterine tissues were prepared and uterine flushings were collected. These proved not to be luteolytic by in vivo tests with one exception; extracts of hamster uteri collected on days 6 and 7 of pseudopregnancy shortened the duration of pseudopregnancy in hamsters significantly (Mazur and Wright 1968). Also, in an in vitro system, porcine uterine flushings collected during the late luteal phase induced luteolysis and depressed progesterone synthesis by luteal tissue in vitro but not in vivo (Schomberg 1969; Schomberg, Campbell, and Wiltbank 1969). All the evidence agrees that only the endometrium is essential for expression of the uterine luteolytic activity, whatever its nature may be. That the uterus has no adverse influence on luteal function in monkeys was shown by the observation that hysterectomy does not alter the normal pattern of cyclic progesterone secretion (Neill, Johansson, and Knobil 1969).

As mentioned elsewhere in this report, it is a virtual certainty that in sheep the uterine luteolytic factor is prostaglandin $F_{2\alpha}$ (McCracken et al. 1972). There is, in addition, suggestive evidence that the prostaglandins may also be luteolytic in a few other species (Phariss, Tillson, and Erickson 1972) including the subhuman primates (Auletta, Speroff, and Caldwell 1973), and women (Lehman et al. 1972). What has been firmly established is that the prostaglandin content of the blood and amniotic fluid increases considerably prior to and during labor, that exogenous prostaglandins stimulate myometrial contractions directly, and that prostaglandins can be used to induce premature labor or to facilitate delivery at term.

## The Mechanism Initiating Parturition

The most interesting and unexpected recent development with respect to parturition was unquestionably the finding that in sheep initiation of the delivery mechanism rests with the fetus. That the fetus in any species can determine when it will be born was, to state the obvious, a departure from conventional thinking. The breakthrough came with the observation that surgical removal of the fetal hypophysis led to delayed delivery (Liggins et al. 1973). Administration of adrenocorticotropin hormone (ACTH) or adrenocortical glucocorticoids also brought on premature delivery, an event that could be inhibited by progesterone. What this implies is that the fetal pituitary stimulates the fetal adrenals to secrete hormones that reduce the placental synthesis of progesterone. Whether parturition follows as a direct result of "progesterone withdrawal" is doubtful. An accompanying increase in estrogen and prostaglandins most likely participate in the initiation of labor in the ewe. The mechanism that initiates labor in women remains controversial and obscure (Porter and Finn 1976).

Cervix
In recent years interest has centered on the chemical composition of the cervical mucus and the peculiar linear alignments of its macromolecules at midcycle. This arrangement facilitates and may even direct the migration of sperm through the cervical canal. It also appears that the cervical crypts may serve as reservoirs for the storage and protracted release of sperm (see Moghissi 1973).

There is much current interest in devising a simple chemical test of cervical mucus that would produce a positive reaction only at or near the time of ovulation. Tests have been developed for this purpose, but none has proven to have the required degree of reliability for wide acceptance.

**Some Problems Remaining to Be Solved, with Particular Reference to Conception Control**

The Gonadotropins: Structure, Binding, Receptors, and Mechanism of Action
Because the ovaries are totally nonfunctional in the absence of the pituitary gonad stimulating hormones, it is of the utmost importance in terms of fertility control to understand how these hormones (FSH, LH, and prolactin) carry out their respective functions and how their own production is regulated.

Although the primary amino acid sequence of the LH molecules from several species is known, little is known about the three-dimensional structure of the molecule. The disulphide bonds must be established and the tertiary, quaternary, and carbohydrate structures identified in order to determine how the molecule interacts with its target organ receptors. Evidence suggests that the carbohydrate residues of hCG are not important for biological function, but they are important for stabilization of the hormone in circulation. Similar studies must be carried out with LH.

Because LH and FSH are heterogeneous glycoproteins and because the structure of the molecules in circulation have not been elucidated, there is a great need to confirm the plasma levels of FSH and LH as measured by radioimmunoassay and by bioassay using newly devised highly sensitive assays. There is also a need to examine critically the most sensitive bioassays, including the "redox" assay. Further along this line there is a requirement for standardized, universally accepted radioimmunoassay systems for FSH and LH. This would necessitate distribution of common, well-characterized antisera along with highly purified species-specific LH and FSH. Inasmuch as our understanding of reproductive function is necessarily dependent on reliable quantitative estimates of circulating hormone levels, this need to improve the assay systems is urgent.

Although the kinetics of LH and FSH clearance rates have been "studied," knowledge of how the hormones are metabolized is lacking. Inasmuch as circulating levels of LH and FSH are dependent on the addition and removal of these hormones from the blood, knowledge of hormone metabolism is as important as determining secretion rates.

It is well documented that FSH and LH are stored in granules in the anterior pituitary, but the mechanisms of synthesis, storage, and release are not understood. What are the roles of cyclic AMP, calcium, prostaglandins, protein kinases, and LRF in this process? Is there more than one gonadotropin releasing factor? How do the gonadal steroid hormones modulate LH and FSH secretion in response to LRF? It is not clear how FSH and LH are secreted at different rates throughout the estrous or menstrual cycles. It was postulated that identification of the cell types responsible for LH and FSH storage would clarify this problem. As it appears that this question of differential release cannot be answered by further work in this area, new approaches must be taken to solve the perplexing dilemma of differential release. Better procedures are needed for studying LH and FSH secretion from pituitary cells in vitro.

Recent studies suggest that the nocturnal release of LH during sleep may play a key role in the development of puberty. This very ill defined area should be studied further.

The mechanism of action of LH in stimulating steroidogenesis and causing ovulation is still far from understood. Quantitative measurements of cellular processes must be made under physiological conditions. For example, cyclic AMP is postulated to mediate the LH stimulation of steroidogenesis; nonetheless, LH can stimulate steroidogenesis without stimulating cyclic AMP accumulation if physiological levels of LH are added to Leydig cells. Much higher levels of LH must be used to stimulate cyclic AMP synthesis. The relationship between gonadotropin stimulation and prostaglandin synthesis is also unclear. Although a large number of studies have described binding of LH to its receptor and the receptor has even been "isolated," it is not known how the hormone-receptor complex mediates hormone action. Why are only 1 percent of the receptors filled for complete stimulation of steroidogenesis? Even less is known about the manner in which LH stimulates ovulation.

LH is present in the plasma at all phases of the cycle and remains at low "tonic" levels except at midcycle, when it increases tenfold or more. This surge induces ovulation, luteinization, and corpus luteum formation. Further research is needed to determine whether the binding sites for LH are different in quantity or character at this time.

Although the preliminary structure of human FSH has been determined, it needs to be confirmed. The structure of FSH molecules from other species needs to be determined. The considerations of the three-dimensional structure need to be established.

Differences between pituitary FSH and circulating FSH have recently been detected. Are these differences related to the metabolism of the hormone or are they related to the functional form of the molecule? This question must be answered.

The relatively low number of binding sites in ovarian tissue for

FSH may be accounted for by endogenous FSH competing with the labeled FSH used to determine binding characteristics. The fact that FSH has not been found to bind to thecal cells may or may not be a valid observation. Also, FSH has been implicated in early follicular growth, yet present studies have indicated binding to only the larger antral follicles. Further studies on the binding of FSH to granulosa and thecal cells as a function of the state of maturation of the follicle will be required before a clear picture of correlation between FSH binding and biologic action is obtained.

The amino acid sequence of human prolactin needs to be completed and fragments of the hormone tested for biologic activity. The identification of pleomorphic forms of prolactin such as are produced by tumor cells needs clarification.

Although prolactin binding sites have been identified in mammary tissue, they have also been found to be more widespread and are especially abundant in the uterus. Why? Prolactin can be both luteotropic and luteolytic in the rat; but its effect, if any, on luteal function in other mammals is unclear.

Prolactin tends to suppress the secretion of FSH and LH and may be a factor in the suppression of ovulation during lactation. This has contraceptive significance and is in urgent need of clarification.

## Neuroendocrinology of Reproduction

The great progress that has been made in clarifying the role of the central nervous system in regulating the gonad stimulating functions of the anterior pituitary have raised perhaps undue expectation that new means of fertility control might soon be forthcoming. What these spectacular new findings have done is to open an immense new field to scientific exploitation. Nonetheless, serious attempts are being made to develop new means of controlling fertility at the neuroendocrine level. These attempts involve mainly efforts to synthesize analogs of LRF that might block the action of LRF and thus inhibit ovulation or be active orally so that ovulation could be therapeutically induced and the fertile period prescribed.

In this field there is need for such basic knowledge as the mapping of the nerve pathways and interconnections in the hypothalamic area. Also, where in the endocrine neurones is LRF synthesized and how does it enter the primary plexus of the portal system? It is known that the synaptic transmission of signals to these endocrine neurones is effected by catecholamines or similar substances collectively referred to as neurotransmitters. Therein lies a vast new area for the study of the action of pharmacological agents on the hypothalamic control of reproductive functions.

Immunology of Reproduction
The immunological approach to fertility control is currently receiving a great deal of attention. The possibility of developing an immunity to pregnancy is very real. An essential condition for such a "vaccination" is that this should be done with a completely specific "foreign" alloantigen that should be rendered immunogenic without resorting to noxious adjuvants. This immunization should result in the formation of specific antibodies directed exclusively to the biologically active principle in question. There is great need for a suitable animal model in which the same effect could be elicited and in which the necessary developmental work could be carried out. To develop such a vaccine, furthermore, would necessitate providing satisfactory answers to questions such as toxic reactions, if any, the length of time the antibodies produced would be effective, the need for periodic reimmunization, and the reversibility and possible teratogenic complications during the period of declining antibody titers. Although this is a formidable list of criteria to be met, preliminary evidence suggests that at least part of them can be fulfilled, thanks to recent studies with fragments of the $\beta$ subunit of hCG. Such substances occur normally in the female organism only in pregnancy. Immunologically characteristic portions of them can be produced synthetically, and they can be rendered immunogenic by coupling to tetanus toxoid or by diazotization. There is an animal model, the baboon, in which the above approach has already produced evidence of efficacy. Although it is recognized that this is a singularly complicated field of research that will require considerable time and effort before it will result in a vaccine for mass use, it should be borne in mind that the potential payoff of this approach is considerable.

## Ovary

*Oocyte Maturation:* Oocyte maturation factors, known to be normal constituents of the ovaries of marine organisms, appear to play a role in the mammalian ovary as well. Studies in the rat show that the normal arrest of oocyte development during the first meiotic division may be the result of a specific maturation inhibitor produced by the primary ovarian follicle. The reactivation of oocyte maturation as the follicle nears the time of ovulation appears to be due to the midcycle surge of LH, which carries the maturation process through to the metaphase of the second meiotic division. Completion of meiosis depends upon fertilization of the ovum. These processes can be studied in follicles that are maintained in organ culture and it would obviously be of importance to understand how in nature the arrest of oocyte maturation can be maintained for from twelve to forty years.

*Atresia:* The factors responsible for follicle atresia during embryonic life, childhood, and even within each cycle continue to elude researchers. Speculation formerly centered around those factors responsible for stimulating ovulation (FSH and LH). Insufficient estrogen necessary for follicle maturation was also implicated. More recently it has been shown that the mature follicles are relatively more resistant to atresia, perhaps due to the degree of luteinization and slight steroid synthesizing enzymatic activities. Increased FSH stimulation is involved in ovarian hypertrophy and super ovulation—phenomena that might lend themselves to answer questions governing follicular atresia and rescue of follicles from atresia.

The mass induction of atresia as an irreversible means of fertility regulation has few serious advocates. Although some have predicted that prolonged use of contraceptive steroids may induce either premature depletion or abnormal maintenance of the egg reservoir of ovaries, the accumulating evidence based on long-term usage does not support either of these assumed outcomes.

## Luteolysis and Antiprogestins

A very promising approach to the control of conception rests with the induction of premature luteolysis. Certain *prostaglandin analogs* with a relatively long half-life appear to induce at least a temporary reduction in circulating progesterone levels and bind very firmly to specific receptors in the human corpus luteum. More and more potent analogs are being synthesized and screened for a luteolytic effect in laboratory animals, and it is at least possible that orally active prostaglandin analogs can be developed that will be effective in the human female. Preliminary studies suggest, furthermore, that certain orally active *plant extracts* (for instance the Mexican plant *Montanoa tuberosa*, called Zoapatle) are capable of suppressing circulating progesterone levels. Infusions prepared from this plant are said to induce abortion also in the second trimester of gestation because of a sustained *oxytocic* effect (oxytocin is a posterior pituitary hormone that causes the uterine muscles to contract). It is not known at this time whether the luteolytic and oxytocic effects are due to the same or different principles. A synthetic drug (Centchroman) developed in India is also said to interfere with luteal function. Because the corpus luteum is a temporary and renewable bodily organ, it can be adversely affected by immunologic, cytotoxic, or other pharmacological means with impunity; a new structure will develop in the next cycle.

In terms of sheer effectiveness and probable innocuousness, there is no other systemically active contraceptive that has quite the appeal and promise of an antiprogestin. Effective antiestrogens and antiandrogens have been synthesized and are available in quantity, but no really promising breakthrough has been obtained with respect to an antiprogesterone. Such a substance would not disturb the secretion of progesterone but would block its action and could be the basis of an ideal end-of-the-month menses regulator and contraceptive.

### Reproductive Tract

*Oviduct:* It is clear that the oviduct produces specific secretory materials, but our knowledge of their chemical nature and function is shamefully limited. Only the rabbit has been more than superficially studied, although similar secretory material has been noticed in many mammalian species including monkey and man. Are the secretory products necessary and if so, why? What is their chemical makeup? Can they be altered or eliminated specifically? The physiologic and biochemical roles of these substances need to be defined because their regulation might lead to a fertility control method specific to the oviduct level.

There are cleavage inhibiting agents in preovulatory oviduct fluid of the rabbit. Are such agents responsible for the specific timing of embryo cleavage? Are they found in other species, especially humans? Are they specific to each species or are they universal inhibitors? What is the mechanism of their inhibition? What is their chemical nature? Can they be induced to remain at high levels in the oviduct for contraceptive activity?

Delivery of the embryo to the uterus at specifically the right time is required. Are the fluid constituents so programmed that small or great alterations are required to impede normal development? What are the most crucial constituents? How are they produced or regulated and can they be altered?

The biochemical relationship between the zygote and oviduct function in the proximity of the zygote is just beginning to be studied. However, very preliminary work, anatomical observations, and thymidine incorporation suggest that important communication processes are at work. We should learn what mediates messages between embryo and oviduct, what is produced or released by the embryo, what is produced (and how) by the oviduct and how it is released, and what possible constraints could be directed at either the embryo or oviduct.

Several pharmacologic agents have been identified as being capable of entering the oviducal lumen, but many others do not. If agents are to be used to control oviduct secretions or embryonal development, we have to know about the mechanism of transport of agents from the blood to the fluid. What local delivery systems could be developed? What agents are most likely to be effective?

Little information is available on the oviducal environment of Homo sapiens. Some information is available in the subhuman primate but many of the essential processes are not known.

We need to emphasize the relationship between fluid flow and transport. Would alteration of fluid flow affect embryo transport? What role does fluid dynamics play in gamete transport? Would alteration of fluid flow cause ectopic pregnancy? What pharmacological agents would alter fluid flow?

## Uterotubal Junction

There is urgent need for greater understanding of the changes occurring in the anatomy of this structure during the menstrual cycle and of its role in controlling the passage of sperm and the entry of the ovum into the uterus. Studies are also needed to clarify the effect of the oral contraceptives and pharmacological agents on the uterotubal junction. Numerous attempts are being made to develop a satisfactory method of occluding this passageway by mechanical means and by the application of sclerosing agents or cauterization.

## Uterus

It is well known that the uterus remains in a receptive state for the attachment of a developing fertilized ovum for only a very brief period and that this coincides with the presence and readiness of the ovum for implantation. This is termed the period of *sensitivity* and at all other times the uterus is in a state that is *refractory* to nidation by the ovum. The control of uterine sensitivity is well known and depends upon a well-timed sequence and shifting balance in the relationship of estrogen and progesterone. Any disturbance of these conditions will prevent the uterus from becoming receptive even though ovulation and fertilization may have occurred. The minipill and the progestin-medicated IUDs and vaginal rings all subject the uterine endometrium to the influence of progestins on an abnormally sustained basis and induce sterility.

Because quiescence of the uterine myometrium is also essential to the establishment and maintenance of gestation, the availability of any myometrial stimulant might also serve to forestall or disrupt a pregnancy. In this connection the search for substances such as Zoapatle and Centchroman mentioned earlier should be intensified as possible postcoital or once-a-month contraceptives.

*Cervix:* The role of the cervix in the control of conception centers to a large extent on its capacity to secrete a mucin hostile to penetration by spermatozoa. The composition and character of the cervical secretion undergo marked changes during the menstrual cycle in the human female and in most primates studied. During all but the midcycle ovulatory phase the mucin is a thick glairy substance that acts as a barrier to the passage of sperm. At midcycle the mucin changes to a thin watery fluid that seems actually to facilitate penetration by sperm migrating under their own flagellary motility.

## Vagina

It has been demonstrated that contraceptive steroids are readily absorbed through the vaginal wall into the bloodstream. Progestins in medicated vaginal rings are believed to exert their contraceptive action by reaching the uterine endometrium (and oviduct as well) through the peripheral circulation. The vagina as a route for the administration of contraceptive agents has distinct advantages in terms of accessibility, convenience, and periodicity of treatment. Additional studies are needed to ascertain the tolerance of the vaginal mucosa to the presence of medicated vehicles.

## Concluding Comments

Our understanding of the reproductive process has made great strides over the past fifty years. That much remains to be learned is simply a reflection of the fact that the development, structure, regulation, and metabolism of the reproductive system is enormously complex. The attainment of a truly advanced understanding of this system is a long-term proposition, but it is to be noted that the rate of acquisition of new knowledge has been accelerating rapidly over the past dozen years. There can be no doubt that the study of reproductive phenomena has now entered a phase of exponential growth and productivity, largely as a result of recognized need and farsighted support by both public and private donor agencies.

Concern over the capacity of the human species to increase its numbers was first expressed by Malthus at the beginning of the past century. His well-known prognostications were not taken seriously and did not materialize for reasons that he could not anticipate. Natural forces and crude though effective fertility controlling practices served to keep the growth of the population within reasonable bounds well into the present century. With the lowering of the death rate due to the introduction of life-saving drugs and numerous other improvements in health care during the second quarter of this century, the procreative capacity of mankind again became a matter of concern. Hunger and malnutrition became widespread. Problems relating to spoilage of the environment and exhaustion of many natural resources were causing concern. Reduction of the birth rate became a matter of increasing urgency. Later, the introduction of two new and effective means of inhibiting conception (the pill and the IUD) along with the demonstrated realities as well as the limitations of the rhythm method, plus the successful use of medical agents to overcome infertility served to heighten public awareness and focus attention on the need for expanded research in this area of biomedical science.

Pressure for more financial support of reproductive research was obviously mounting during the late fifties and early sixties; but this pressure continued to be offset, although to a lessening extent, by the traditional constraints such as taboos, religious views, and prudery. Be that as it may, the late sixties were to witness a notable expansion of support and research effort. The resulting chronicle of achievements as related in this chapter has been nothing short of phenomenal. Problems that a few years ago seemed unlikely to be solved within this century have succumbed to scientific assault in domino fashion. The determination of the chemical structure of one pituitary hormone was followed in rapid succession by the elucidation of the structure of every known gonadotropin. Similarly, the isolation and chemical characterization of one hypothalamic "releasing" factor was followed by a succession of like accomplishments in respect to several other neuroendocrine factors. This listing of emerging denouements could be extended to nearly every aspect of research on both the male and female reproductive systems.

It has not been possible, nor is it the intent, to cover all the significant accomplishments in this brief account. Only the surface has been skimmed. Volumes have been written on some of the topics merely touched upon here. Moreover, any degree of familiarity with the literature will reveal that there exists a vast number of reported observations that cannot yet be fitted into the functional scheme contrived by nature over millions of years of evolution.

What is important is to recognize that great progress has been made in understanding the basic features of the reproductive process in a relatively short period of time and that the potential for further gains has never been stronger than it is today. Trained manpower and the facilities are in place as a basis on which to build. There remains only the provision of adequate financial support in order to develop a research effort commensurate in scope with the magnitude of the problem being addressed. To this end, it should be noted that modern gains in the control of human procreation have freed millions of women from the burden of unwanted or unplanned pregnancies and raised man's hopes for the welfare of future generations. The choice is between the development of acceptable and effective means of controlling conception on a massive and enduring scale or the endangerment of the future of humankind. The time to choose is now.

## References

Abraham, G. E.; Odell, W. I.; Swerdloff, R. S.; and Hopper, K. 1972. Simultaneous radioimmunoassay of plasma progesterone, 17-hydroxyprogesterone, and estradiol-17$\beta$ during the cycle. *J. Clin. Endocrinol.* 34:312-318.

Aiyer, M. A., and Fink, G. 1974. The role of sex steroid hormones in modulating the responsiveness of the anterior pituitary gland to luteinizing hormone releasing factor in the female rat. *J. Endocrinol.* 62:553-572.

Allen, E., and Doisy, E. A. 1923. An ovarian hormone: a preliminary report on its localization, extraction and partial purification and action in test animals. *J.A.M.A.* 81:819-821.

Allen, E.; Hisaw, F. L.; and Gardner, W. U. 1939. The endocrine functions of the ovaries. In *Sex and internal secretions*, 2nd ed., ed. E. Allen, C. Danforth, and E. A. Doisy, pp. 452-629. Baltimore: Williams and Wilkins.

Allen, W. M., and Wintersteiner, O. 1934. Crystalline progestin. *Science* 80:155.

Anderson, L. L. 1973. Effects of hysterectomy and other factors on luteal function. In *Handbook of Physiology*, sect. 7, vol. II, part 2, ed. R. O. Greep, pp. 69-86. Bethesda: American Physiological Society.

Arimura, A.; Kastin, A. J.; Schally, A. V.; Saito, M.; Kumasaka, T.; Taoi, Y.; Nishi, N.; and Ohkura, K. 1974. Immunoreactive LH-releasing hormone in plasma: midcycle elevation in women. *J. Clin. Endocrinol. Metab.* 38:510-513.

Aschheim, S., and Zondek, B. 1927. Hypophysenvorderlappenhormone und ovarialhormone im Harn von

Schwangeren. *Klin Wochschr.* 6:1322.

Ashitaka, Y.; Tsong, Y. Y.; and Koide, S. S. 1973. Distribution of tritiated human chorionic gonadotropin in super-ovulated rat ovary. *Proc. Soc. Exp. Biol. Med.* 142:395-397.

Astwood, E. B. 1941. Regulation of corpus luteum function by hypophysial luteotrophin. *Endocrinology* 28:309-319.

Auletta, F. J.; Speroff, L.; and Caldwell, B. V. 1973. Prostaglandin $F_{2\alpha}$ induced steroidogenesis and luteolysis in the primate corpus luteum. *J. Clin. Endocrinol. Metab.*

Baggett, B.; Engel, L. O., Savard, K.; and Dorfman, R. I. 1956. The conversion of testosterone-3-$C^{14}$ to $C^{14}$-estradiol-17β by human ovarian tissue. *J. Biol. Chem.* 221:931-941.

Bahl, O. P.; Carlsen, R. B.; and Bellisario, R. 1973. Human chorionic gonadotropin amino acid sequence of the α and β subunits and the nature and location of the carbohydrate units. In *Endocrinology*, p. 667-672. Amsterdam: Scow. Excerpta Medica. International Congress Series 273.

Bahl, O. P.; Carlsen, R. B.; Bellisario, R.; and Swaminathan, N. 1972. Human chorionic gonadotropin: amino acid sequence of the α and β subunits. *Biochem. Biophys. Res. Commun.* 48:416-422.

Barraclough, C. A. 1967. Modifications in reproductive function after exposure to hormones during the prenatal and early postnatal period. In *Neuroendocrinology*, vol. 2. ed. W. F. Ganong and L. Martini, pp. 61-99. New York: Academic Press.

Barraclough, C. A., and Gorski, R. A. 1961. Evidence that the hypothalamus is responsible for androgen-induced sterility in the female rat. *Endocrinology* 68:68-79.

Behrman, H. R., and Caldwell, B. V. 1974. Role of prostaglandins in reproduction. In *Reproductive physiology*, ed. R. O. Greep, pp. 63-94. London: Butterworth.

Behrman, H. R.; Macdonald, G. J.; and Greep, R. O. 1971. Regulation of ovarian esters: Evidence for the enzymatic sites of prostaglandin-induced loss of corpus luteum function. *Lipids* 6:791-796.

Bewley, T. A.; Sairam, M. R.; and Li, C. H. 1972. Circular dichroism of ovine interstitial cell stimulating hormone and its subunits. *Biochem.* 11:932-936.

Biggers, J. D.; Whittingham, D. G.; and Donahue, R. P. 1967. The pattern of energy metabolism in the mouse oocyte and zygote. *Proc. Natl. Acad. Sci. (U.S.)* 58:560-567.

Blake, C. A., and Sawyer, C. H. 1974. Effect of hypothalamic deafferentation on the pulsatile rhythm in the plasma concentrations of luteinizing hormone in ovariectomized rats. *Endocrinology* 94:730-736.

Bogdanove, E. M.; Nolin, J. M.; and Campbell, G. T. 1975. Qualitative and quantitative gonad-pituitary feedback. *Recent Prog. Horm. Res.* 31:567-619.

Brenner, R. M. 1969. Renewal of oviduct cilia during the menstrual cycle of the rhesus monkey. *Fertil. Steril.* 20:599-611.

Brinster, R. L. 1965. Studies on the

development of mouse embryos *in vitro*. IV. Interaction of energy sources. *J. Reprod. Fertil.* 10:227-240.

Burgus, R.; Butcher, M.; Amoss, M.; Ling, N.; Monahan, M.; Rivier, J.; Fellows, R.; Blackwell, R.; Vale, W.; and Guillemin, R. 1971. Structure moleculaire di facteur hypothalamique (LRF) d'origine ovine controlant la secretion de l'hormone gonadtrope hypophysaire de luteinization. *C.R. Acad. Sci.* (Paris) 273:1611-1613.

Butcher, R. L.; Collins, W. E.; and Fugo, N. W. 1974. Plasma concentration of LH, FSH, prolactin, progesterone and estradiol 17β throughout the 4-day estrous cycle of the rat. *Endocrinology* 94:1704-1708.

Butenandt, A. 1934. Neure Ergebnisse auf dem Gebiet der Sexualhormone. *Wein. Klin. Wchenschr.* 47:934-936.

Butenandt, A., and Schmidt, J. 1934. Uberfuhrung des pregnandiols im corpus-luteum-hormons. *Bei. Chem. Ges.* 67:1901-1903.

Canfield, R. E.; Morgan, F. J.; Kammerman, S.; Bell, J. J.; and Agosto, G. M. 1971. Studies on human chorionic gonadotropin. *Recent Prog. Horm. Res.* 27:121-156.

Catt, K. J., and Tregear, G. W. 1967. A solid phase radioimmunoassay in antibody coated tubes. *Science* 158:1570-1572.

Channing, C. P. 1969. Studies on tissue culture of equine ovarian cell types: Pathways of steroidogenesis. *J. Endocrinol.* 43:403-414.

———. 1970. Influences of the *in vivo* and *in vitro* hormonal environment upon luteinization of granulosa cells in tissue culture. *Recent Prog. Horm. Res.* 26:589-622.

———. 1973. Regulation of luteinization in granulosa cell cultures. In *The regulation of mammalian reproduction*, ed. S. Segal, R. Crozier, P. A. Corfman, and P. G. Condliffe, pp. 509-518. Springfield: Charles C Thomas.

———. 1974. Temporal effects of LH, hCG, FSH and dibutyryl 3'5' cyclic AMP upon luteinization of rhesus monkey granulosa cells in culture. *Endocrinology* 94:1215-1223.

———. 1975. Follicle stimulating hormone stimulation of $^{125}$I-human chorionic gonadotropin binding in procine granulosa cell cultures. *Proc. Soc. Exp. Biol. Med.* 149:238-241.

Channing, C. P., and Kammerman, C. 1974. Binding of gonadotropins to ovarian cells. *Biol. Reprod.* 10:179-198.

Clemens, J. A., and Meites, J. 1974. Comparative mammalian studies of the control of prolactin secretion. In *Lactogenic hormones, fetal nutrition and lactation*, ed. J. B. Josemovich, M. Reynolds, and E. Coco, pp. 111-142. New York: John Wiley & Sons.

Cole, H. H., and Hart, G. H. 1930. Potency of blood serum of mares in progressive stages of pregnancy in effecting sexual maturity of immature rats. *Am. J. Physiol.* 93:57-68.

Corner, G. W., Jr. 1956. The histological dating of the human corpus luteum. *Am. J. Anat.* 98:377-392.

Corner, G. W., and Allen, W. M. 1929.

Physiology of the corpus luteum. II. Production of a special uterine reaction (progestational proliferation) by extracts of the corpus luteum. *Am. J. Physiol.* 88:326-339.

Dierschke, D. J.; Bhattacharya, A. N.; Atkinson, L. E.; and Knobil, E. 1970. Circhoral oscillations of plasma LH levels in the ovariectomized rhesus monkey. *Endocrinology* 87:850-853.

Djerassi, C.; Rosenkranz, G.; Iriarte, J.; Berlin, J.; and Romo, J. 1951. Steroids. XII. Aromatization experiments in the progesterone series. *J. Am. Chem. Soc.* 73:1523-1527.

Doisy, E. A.; Veler, C. D.; and Thayer, S. A. 1929. Folliculin from the urine of pregnant women. *Am. J. Physiol.* (abstr.) 90:329-330.

————. 1930. The preparation of the crystalline ovarian hormone from the urine of pregnant women. *J. Biol. Chem.* 86:499-509.

Donovan, B. T. 1966. The regulation of the secretion of follicle-stimulating hormone. In *The pituitary gland*, vol. II, ed. G. W. Harris and B. T. Donovan, pp. 49-98. Berkeley: University of California Press.

El-Fouly, M.; Cook, B.; Nekola, M.; and Nalbandov, A. V. 1970. Role of the ovum in follicular luteinization. *Endocrinology* 87:288-293.

Engle, E. T. 1929. Effect of daily transplants of anterior lobe from gonadectomized rats on immature test animals. *Am. J. Physiol.* 88:101-106.

Erickson, G. F., and Ryan, K. J. 1975. The effect of LH/FSH, dibutyryl cyclic AMP and prostaglandins in the production of estrogens by rabbit granulosa cells *in vitro*. *Endocrinology* 97:108-113.

Erlanger, B. F.; Borek, F.; Beiser, S. M.; and Lieberman, S. 1957. Steroid-protein conjugates: 1. Preparation and characterization of conjugates of bovine serum albumin with testerone and with cortisone. *J. Biol. Chem.* 228:713-727.

Evans, H. M., and Simpson, M. E. 1929. Comparison of anterior hypophyseal implants from normal and gonadectomized animals with reference of their capacity to stimulate immature ovary. *Am. J. Physiol.* 89:371-380.

Evans, H. M.; Simpson, M. E.; Lyons, W. R.; and Turpeinen, K. 1941. Anterior pituitary hormones which favor the production of traumatic uterine placentoma. *Endocrinology* 28:933-945.

Everett, J. W. 1940. The restoration of ovulatory cycles and corpus luteum formation in persistent-estrous rats by progesterone. *Endocrinology* 27:681-686.

————. 1961a. The preoptic region of the brain and its relation to ovulation. In *Control of ovulation*, ed. C. A. Villee, pp. 101-112. New York: Pergamon Press.

————. 1961b. The mammalian female reproductive cycle and its controlling mechanisms. In *Sex and internal secretions*, vol. I, 3rd ed. Ed. W. C. Young, pp. 497-555. Baltimore: Williams and Wilkins.

Everett, J. W., and Sawyer, C. H.

1950. A 24-hour periodicity in the LH-release apparatus of female rats disclosed by barbiturate sedation. *Endocrinology* 47:198-218.

———. 1953. Estimated duration of the spontaneous activation which causes release of the ovulating hormone from the rat hypophysis. *Endocrinology* 52:83-92.

Faiman, C., and Ryan, R. J. 1967. Radioimmunoassays for human follicle stimulating hormone. *J. Clin. Endocrinol. Metab.* 27:444-447.

Ferin, M.; Dryenfurth, I.; Cowchock, S.; Warren, M.; and Vande Wiele, R. L. 1974. Active immunization to 17β-estradiol and its effects upon the reproductive cycle of the rhesus monkey. *Endocrinology* 94:765-776.

Fevold, H. L.; Hisaw, F. L.; and Leonard, S. L. 1931. The gonad-stimulating and the luteinizing hormones of the anterior lobe of the hypophysis. *Am. J. Physiol.* 97:291-301.

Fink, G., and Aiyer, M. S. 1974. Gonadotropin secretion after electrical stimulation of the preoptic area during the oestrous cycle of the rat. *J. Endocrinol.* 62:589-604.

Flores, F.; Naftolin, F.; Ryan, K. J.; and White, R. J. 1973. Estrogen formation by isolated perfused rhesus monkey brain. *Science* 180:1074-1075.

Freeman, M. E.; Smith, M. S.; Nazian, S. J.; and Neill, J. D. 1974. Ovarian and hypothalamic control of the daily surges of prolactin secretion during pseudopregnancy in the rat. *Endocrinology* 94:875-882.

Friesen, H. G. Forthcoming. Prolactin. In *Frontiers in reproduction and fertility control*, ed. R. O. Greep. Cambridge: MIT Press.

Fugino, M.; Shinogawa, S.; Obayashi, M.; Kobayashi, S.; Fukuda, T.; Yamazaki, F.; and Nakayama, R. 1973. Further studies on the structure-activity relationships in the C-terminal part of the luteinizing hormone-releasing hormones. *J. Med. Chem.* 16:1144-1146.

Fugino, M.; Yamazaki, F.; Kobayashi, S.; Fukuda, T.; Shinogawa, S.; Nakayama, R.; White, W. F.; and Rippel, R. H. 1974. Some analogs of luteinizing hormone releasing hormone (LH-RH) having intense ovulations inducing activity. *Biochem. Biophys. Res. Commun.* 57:1248-1256.

Goding, J. R.; Catt, K. J.; Brown, J. M.; Kaltenback, C. C.; Cummings, I. A.; and Mole, B. J. 1969. Radioimmunoassay for ovine luteinizing hormone. Secretion of luteinizing hormone during estrus and following estrogen administration in the sheep. *Endocrinology* 85:133-142.

Green, J. D., and Harris, G. W. 1947. The neurovascular link between the neurohypophysis and adenohypophysis. *J. Endocrinol.* 5:136-146.

Greenwald, G. S. 1967. Luteotropic complex of the hamster. *Endocrinology* 80:118-137.

Greenwood, F. C.; Hunter, W. L.; and Glover, J. J. 1963. The preparation of [131]I-labeled growth hormone of high specific activity. *Biochem. J.* 89:114-123.

Greep, R. O. 1936. Functional pituitary grafts in rats. *Proc. Soc. Exp. Biol. Med.* 34:754-755.

Greep, R. O.; van Dyke, H. B.; and Chow, B. F. 1942. Gonadotropins of the swine pituitary. *Endocrinology* 39:635-649.

Guillemin, R. 1972. Physiology and chemistry of the hypothalamic releasing factors for gonadotropins: A new approach to fertility control. *Contraception* 5:1-20.

Hafez, E. S. E. 1973. Anatomy and physiology of the mammalian uterotubal junction. In *Handbook of physiology*, ed. R. O. Greep, sect. 7, vol. II, part 2, pp. 87-96. Washington, D.C.: Physiological Society.

Hafez, E. S. E., and Blandau, R. J. 1969. *The Mammalian Oviduct*. Chicago: University of Chicago Press.

Halasz, B. 1969. The endocrine effects of isolation of the hypothalamus from the rest of the brain. In *Frontiers on neuroendocrinology*, vol. I, ed. W. F. Ganong and L. Martini, pp. 307-342. London: Oxford University Press.

Halban, J. 1900. Uber den Einfluss der Ovarien auf die Entwicklung der Genitales. *Monatschr. F. Geburtsh u. Gynak.* 12:496-506.

Hammond, J., and Marshall, F. H. A. 1914. The functional correlation between the ovaries, uterus and mammary glands in the rabbit, with observations on the oestrous cycle. *Proc. R. Soc. (London)*, ser. #B. 87:422-440.

Hammond, J. Jr., and Robson, J. M. 1951. Local maintenance of the rabbit corpus luteum with oestrogen. *Endocrinology* 49:384-389.

Harris, G. W. 1948a. Electrical stimulation of the hypothalamus and the mechanism of neural control of the adenohypophysis. *J. Physiol.* 107: 418-429.

————. 1948b. Neural control of the pituitary gland. *Physiol. Rev.* 28: 139-179.

————. 1961. The pituitary stalk and ovulation. In *Control of ovulation*, ed. C. A. Villee, pp. 56-74. New York: Pergamon Press.

Harris, G. W., and Jacobsohn, D. 1952. Functional grafts of the anterior pituitary gland. *Proc. R. Soc. (London)*, ser. B. 139:263-276.

Heape, W. 1905. Ovulation and degeneration of ova in the rabbit. *Proc. R. Soc. (London)*, ser. B. 76:260-268.

Heard, R. D. H.; Bligh, E. G.; Cann, M. C.; Jellinck, P. H.; O'Donnell, V. J.; Rao, B. G.; and Webb, J. L. 1956. Biogenesis of the sterols and steroid hormones. *Recent Prog. Horm. Res.* 12:45-70.

Herbert, V.; Lau, K.; Gottlieb, C. W.; and Bleicher, S. J. 1965. Coated charcoal immunoassay of insulin. *J. Clin. Endocrinol. Metab.* 25:1375-1384.

Hilliard, J.; Archibald, D.; and Sawyer, C. H. 1963. Gonadotropic activation of preovulatory synthesis and release of progestin in the rabbit. *Endocrinology* 72:59-66.

Hilliard, J.; Penardi, R.; and Sawyer, C. H. 1967. A functional role for 20α hydroxypregn-4-en-3-one in the rabbit. *Endocrinology* 80:901-909.

Hisaw, F. L., and Hisaw, F. L., Jr. 1961. Action of estrogen and progesterone on the reproductive tract of lower primates. In *Sex and internal secretions*, vol. I, 3rd ed., pp. 556-590. Baltimore: Williams and Wilkins.

Hohlweg, W., and Junkman, K. 1932. Die hormonal-nervöse Regulierung der Funktion des Hypophysenvorderlappens. *Klin. Wchnschr.* 11:321-323.

Hotchkiss, J.; Atkinson, L. E.; and Knobil, E. 1971. Time course of serum estrogen and luteinizing hormone (LH) concentration during the menstrual cycle of the rhesus monkey. *Endocrinology* 89:177-183.

Hwang, P.; Guyda, H.; and Friesen, H. 1971. A radioimmunoassay for human prolactin. *Proc. Natl. Acad. Sci.* (U.S.) 68:1902-1906.

Imura, H.; Sparks, L. L.; Grodsky, G. M.; and Forsham, P. H. 1965. Immunological studies of adrenocorticotrophic hormone dissociation of biologic and immunologic activities. *J. Clin. Endocrinol. Metab.* 25:1361-1369.

Jacobs, L. A., and Daughaday, W. H. 1974. Physiologic regulation of prolactin secretion in man. In *Lactogenic hormones, fetal nutrition and lactation*, ed. J. B. Josimovich, M. Reynolds, and E. Cobe, pp. 351-378. New York: John Wiley and Sons.

Jacobs, L. S.; Snyder, P. J.; Wilber, J. F.; Utiger, R. D.; and Daughaday, W. H. 1971. Increased serum prolactin after administration of synthetic thyrotropin releasing hormone (TRH) in man. *J. Clin. Endocrinol.* 33:996-998.

Jonas, H. A.; Salamonsen, L. A.; Burger, H. G.; Chamley, W. A.; Cummings, I. A.; Findlay, J. K.; and Goding, J. R. 1973. Release of FSH after administration of gonadotropin-releasing hormone or estradiol to the anestrous ewe. *Endocrinology* 92: 862-865.

Kamberi, I. A. 1972. *Recent Prog. Horm. Res.* 28:281.

Karim, S. M., and Filshic, G. M. 1970. Therapeutic abortions using prostaglandin $F_{2\alpha}$ *Lancet* 1:157-158.

Karsch, A. J.; Dierschke, D. J.; Weick, R. F.; Yamaji, T.; Hotchkiss, J.; and Knobil, E. 1973. Positive and negative feedback control by estrogen of luteinizing hormone secretion in the rhesus monkey. *Endocrinology* 92:799-804.

Kastin, A. J.; Schally, A. V.; Gual, C.; and Arimura, A. 1972. Release of LH and FSH after administration of synthetic LH-releasing hormone. *J. Clin. Endocrinol.* 34:753-776.

Kastin, A. J.; Zarate, A.; Midgley, A. R.; Canales, E. S.; and Schally, A. V. 1971. Ovulation confirmed by pregnancy after infusion of porcine LH-RH. *J. Clin. Endocrinol. Metab.* 33: 980-982.

Kato, J. 1970. Estrogen receptors in the hypothalamus and hypophysis in relation to reproduction. In *Hormonal steroids*, pp. 764-773. Amsterdam: Excerpta Medica Congress Series No. 219.

Kirton, K. T.; Phariss, B. B.; and Forbes, A. D. 1970. Luteolytic effects of $PGF_2$-alpha in primates. *Proc. Soc. Exp. Biol. Med.* 133:314-316.

Knauer, E. 1900. Die ovarian transplantatin. *Arch. F. Gynak.* 60:322-376.

Knobil, E. 1974. On the control of gonadotropin secretion in the rhesus monkey. *Recent Prog. Horm. Res.* 30:1-36.

Lee, C. T., and Ryan, R. J. 1971. The uptake of human luteinizing hormone (hLH) by slices of luteinized rat ovaries. *Endocrinology* 89:1515-1523.

Lee, C. Y., and Ryan, R. J. 1972. Luteinizing hormone receptors: Specific binding of human luteinizing hormone to homogenates of luteinized rat ovaries. *Proc. Natl. Acad. Sci.* (U.S.) 69:3520-3523.

_____. 1973. Luteinizing hormone receptors in luteinized rat ovaries. In *Receptors for reproductive hormones*, ed. B. W. O'Malley and A. R. Means, pp. 419-430. New York: Plenum Press.

Lehman, F.; Peters, F.; Breckwoldt, M.; and Bettendorf, G. 1972. Plasma progesterone levels during infusion of prostaglandin $F_{2\alpha}$ in the human. *Prostaglandins* 1:269-277.

Li, C. H. 1957. Studies on pituitary lactogenic hormone. XVII. Oxidation of the ovine hormone with performic acid. *J. Biol. Chem.* 229:157-163.

Li, C. H.; Dixon, J. S.; Lo, T-B; Pankov, Y. A.; and Schmidt, K. D. 1969. Amino acid sequence of ovine lactogenic hormone. *Nature* 224:695-696.

Lieberman, S.; Erlanger, B. F.; Beiser, S. M.; and Agate, F. J. 1959. Steroid-protein conjugates: Their chemical, immunochemical and endocrinological properties. *Recent Prog. Horm. Res.* 15:165-196.

Liggins, G. C.; Fairclough, R. J.; Grieves, S. A.; Kendall, J. Z.; and Knox, B. S. 1973. The mechanism of initiation of parturition in the ewe. *Recent Prog. Horm. Res.* 29:111-150.

Liu, W. K.; Sweeney, C. M.; and Ward, D. N. 1970. The -COOH and $-NH_2$ terminal peptides of ovine luteinizing hormone α-subunit. *Res. Commun. Chem. Pathol. Pharmacol.* 1:214-222.

Loeb, L. 1923. The effect of extirpation of the uterus on the life and function of the corpus luteum in the guinea pig. *Proc. Soc. Exp. Biol. Med.* 20:441-443.

Loewit, K. K., and Laurence, K. A. 1969. Termination of pregnancy in the rat with rabbit antibovine LH serum. *Fertil. Steril.* 20:679-688.

Loewit, K.; Badawy, S.; and Laurence, K. 1969. Alteration of corpus luteum function in the pregnant rat by antiluteinizing serum. *Endocrinology* 84:244-251.

McArthur, J. W.; Ingersoll, F. W.; and Worcester, J. 1958. Urinary excretion of interstitial cell stimulating hormone by normal males and females of various ages. *J. Clin. Endocrinol.* 18:460-469.

McCann, S. M.; Taleisnik, S.; and Friedman, H. M. 1960. LH-releasing activity in hypothalamic extracts. *Proc. Soc. Exp. Biol. Med.* 104:432-434.

MacCorquodale, D. W.; Thayer, S. A.; and Doisy, E. A. 1936a. The crystalline ovarian hormone. *Proc. Soc. Exp. Biol. Med.* 32:1182.

_____. 1936b. The isolation of the principal estrogenic substance of liquor folliculi. *J. Biol. Chem.* 115: 436-448.

McCracken, J. A.; Carlson, J. C.; Glew, M. E.; Goding, J. R.; Baird, D. T.; Green, K.; and Samuelsson, B. 1972. Prostaglandin $F_2$ alpha identified as a luteolytic hormone in sheep. *Nature (New Biol.)* 238:129-134.

MacLeod, R. M., and Lehmeyer, J. E. 1972. Regulation of synthesis and release of prolactin. In *Lactogenic hormones*, ed. G. E. W. Wolstenholme and J. Knight, pp. 53-76. Edinburgh and London: Churchill Livingston.

Madhwa Raj, H. G., and Moudgal, N. R. 1970. Hormonal control of gestation in the intact rat. *Endocrinology* 86:874-889.

Makepeace, A. W.; Weinstein, G. L.; and Friedman, M. H. 1937. The effect of progestin and progesterone on ovulation in the rabbit. *Am. J. Physiol.* 119:512-516.

Malacara, J. M.; Seyler, L. E., Jr.; and Reichlin, S. 1972. Luteinizing hormone releasing factor activity in peripheral blood from women during the mid-cycle luteinizing hormone ovulatory surge. *J. Clin. Endocrinol.* 34: 271-278.

Malven, P. V. 1965. Morphological regression of the copora lutea in prolactin-treated hypophysectomized rats. *Anat. Rec.* 151:381 (abstr).

Markee, J. E.; Everett, J. W.; and Sawyer, C. H. 1952. The relationship of the nervous system to the release of gonadotropin and the regulation of the sex cycle. *Recent Prog. Horm. Res.* 7:139-157.

Markee, J. E.; Sawyer, C. H.; and Hollinshead, W. H. 1946. Activation of the anterior hypophysis by electrical stimulation in the rabbit. *Endocrinology* 38:345-357.

Marrian, G. F. 1930. The chemical nature of crystalline preparations. *Biochem. J.* 24:1021-1030.

Marsh, J. M.; Mason, N. R.; and Savard, K. 1961. An *in vitro* action of gonadotropin. *Fed. Proc.* 20:187 (abstr).

Marsh, J. M., and Savard, K. 1966. The stimulation of progesterone synthesis in bovine corpora lutea by adenosine 3'5'-monophosphate. *Steroids* 8:133-148.

Marshall, F. H. A., and Jolly, W. A. 1908. On the results of heteroplastic ovarian transplantation as compared with those produced by transplantation in the same individual. *Q. J. Exp. Physiol.* 1:115-120.

Marshall, F. H. A., and Verney, E. B. 1936. The occurrence of ovulation and pseudopregnancy in the rabbit as a result of stimulation of the central nervous system. *J. Physiol.* 86: 327-336.

Mastroianni, L. Jr.; Shah, U.; and Abdul-Karim, R. 1961. Prolonged volumetric collection of oviduct fluid in the rhesus monkey. *Fertil. Steril.* 12:417-424.

Matsuo, H.; Nair, R. M. G.; Arimura, A.; and Schally, A. V. 1971. Structure of the procine LH- and FSH-releasing hormone. I. The proposed amino acid sequence. *Biochem. Biophys. Res. Commun.* 43:1334-1339.

Mazer, R. S., and Wright, P. A. 1968.

A hamster uterine luteolytic extract. *Endocrinology* 83:1065-1070.

Meites, J., and Clemens, J. A. 1972. Hypothanlamic control of prolactin secretion. *Vitam. Horm.* 30:165-221.

Meyer, R. K.; Leonard, S. L.; Hisaw, F. L.; and Martin, S. J. 1930. Effect of oestrin on gonad stimulating power of the hypophysis. *Proc. Soc. Exp. Biol. Med.* 27:702-704.

Midgley, A. R. Jr. 1966. Radioimmunoassay: A method for human chorionic gonadotropin and human luteinizing hormone. *Endocrinology* 79:10-18.

─────── . 1967. Radioimmunoassay of human follicle stimulating hormone. *J. Clin. Endocrinol. Metab.* 27:295-299.

─────── . 1973. Autoradiographic analysis of gonadotropin binding to rat ovarian tissue sections. In *Receptors for reproductive hormones*, ed. B. O'Malley and A. Means, pp. 365-378. New York: Plenum Press.

Midgley, A. R., Jr., and Jaffe, R. B. 1966. Human luteinizing hormone in serum during the menstrual cycle: Determination by radioimmunoassay. *J. Clin. Endocrinol. Metab.* 26:1375-1381.

Mishell, D. R.; Nakamura, R. M.; Crosignani, P. G.; Stone, S.; Kharma, K.; Nagata, Y.; and Thorney-Croft, I. H. 1971. Serum gonadotropin and steroid patterns during the normal menstrual cycle. *Am. J. Obstet. Gynecol.* 111:60-65.

Moghissi, K. S. 1973. Composition and function of cervical secretion. In *Handbook of physiology*, ed. R. O. Greep, sect. 7, vol. II, part 2, pp. 25-48. Washington, D.C.: Physiological Society.

Monroe, S. E.; Jaffe, R. B.; Midgley, A. R., Jr. 1972. Regulation of human gonadotropins XII. Increase in serum gonadotropins in response to estradiol. *J. Clin. Endocrinol. Metab.* 34:342-347.

Moon, Y. S.; Dorrington, J. H.; and Armstrong, D. T. 1975. Stimulatory action of follicle-stimulating hormone on estradiol-17$\beta$ secretion by hypophysectomized rat ovaries in organ culture. *Endocrinology* 97:244-247.

Moore, C. R., and Price, D. 1932. Gonad hormone functions and reciprocal influence between gonads and hypophysis with its bearing on problem of sex hormone antagonism. *Am. J. Anat.* 50:13-71.

Moudgal, N. R.; MacDonald, G. J.; and Greep, R. O. 1971. Effect of hCG anti-serum on ovulation and corpus luteum formation in the monkey. *J. Clin. Endocrinol. Metab.* 32:579-581.

Murphy, B. E. P. 1964. The application of the property of protein-binding to the assay of minute quantities of hormones and other substances. *Nature* 201:679-682.

─────── . 1970. Methodological problems in competitive protein-binding techniques. *Acta Endocrinol.* 64 (Suppl) 147:37-56.

Murphy, B. E. P.; Engelberg, W.; and Pattee, C. J. 1963. A simple method for the determination of plasma corticoids. *J. Clin. Endocrinol. Metab.* 23:293-300.

Naftolin, F.; Ryan, K. J.; Davies, I. J.; Reddy, V. V.; Flores, F.; Petro, Z.; Kuhn, M.; White, R. J.; Takooka, Y.; and Wolin, L. 1975. The formation of estrogens by central neuroendocrine tissues. *Recent Prog. Horm. Res.* 31: 295-315.

Neill, J. D., and Smith, M. S. 1974. Pituitary-ovarian interrelationships in the rat. In *Current topics in experimental endocrinology*, vol. 2, ed. V. H. T. James and L. Martini, pp. 73-106. New York: Academic Press.

Neill, J. D.; Freeman, M. E.; and Tillson, S. H. 1971. Control of the proestrus surge of prolactin and luteinizing hormone secretion by estrogens in the rat. *Endocrinology* 89:1448-1453.

Neill, J. D.; Johansson, E. D. B.; and Knobil, E. 1969. Failure of hysterectomy to influence the normal pattern of cyclic progesterone secretion in the rhesus monkey. *Endocrinology* 84: 464-465.

Nekola, M. V., and Smith, D. M. 1974. Steroid inhibition of mouse oocyte maturation *in vitro*. Proceeding 55th Annual Meeting, The Endocrine Society. Abst. No. 220, p. A-165.

Niall, H. D.; Hogan, M. L.; Tregear, G. W.; Segre, G. V.; Hwang, P.; and Friesen, H. 1973. The chemistry of growth hormone and the lactogenic hormones. *Recent Prog. Horm. Res.* 29:387-404.

Nicoll, C. 1974. Prolactin. In *Lactogenic hormones, fetal nutrition and lactation*, ed. J. B. Josimovich, M. Reynolds, and E. Cobo, pp. 69-84. New York: John Wiley and Sons.

Niswender, G. D. and Midgley, A. R., Jr. 1970. Haptene-radioimmunoassay for steroid hormones. In *Immunologic methods in steroid determinations*, ed. F. G. Peron and B. V. Caldwell, pp. 149-173. New York: Appleton-Century.

Odell, W. D.; Ross, G. T.; and Rayford, P. L. 1966. Radioimmunoassay of human luteinizing hormone. *Metabolism* 15:287-289.

Papkoff, H. 1973. Studies on the structure of ovine follicle stimulating hormone. In *Endocrinology*, pp. 596-600. Amsterdam: Excerpta Medica. International Congress Series 273.

Papkoff, H.; Sairam, M. R.; Farmer, S. W.; and Li, C. H. 1973. Studies on the structure and function of interstitial cell-stimulating hormone. *Recent Prog. Horm. Res.* 29:563-588.

Paul, W. E. and Odell, W. D. 1964. Radiation inactivation of immunological and biological activities of hCG. *Nature* 203:979-980.

Pearlman, W. H.; Cerceo, E.; and Thomas, M. 1951. The conversion of $\Delta^5$-pregnen-3$\beta$-Ol-20-one to progesterone by homogenates of human placental tissue. *J. Biol. Chem.* 208:231-239.

Peckham, W. P.; Yamaji, T.; Dierschke, D. J.; and Knobil, E. 1973. Gonadal function and the biological and physicochemical properties of follicle-stimulating hormone. *Endocrinology* 92: 1660-1666.

Pfaff, D. W. 1968. Autoradiographic localization of radioactivity in the rat brain after injection of tritiated sex hormones. *Science* 161:1355-1356.

Pfeiffer, C. A. 1936. Sexual differences of the hypophyses and their determination by the gonads. *Am. J. Anat.* 58:195-225.

Pharris, B. B., and Wyngarden, L. J. 1969. The effect of prostaglandin $F_{2\alpha}$ on the progestogen content of ovaries from rats. *Proc. Soc. Exp. Biol. Med.* 130:92-94.

Pharris, B. B.; Tillson, S. A.; and Erickson, R. R. 1972. Prostaglandins in luteal function. *Recent Prog. Horm. Res.* 28:51-73.

Pierce, J. G.; Liao, T. H.; Howard, S. M.; Shome, B.; and Cornell, J. S. 1971. Studies on the structure of thyrotropin: Its relationship to luteinizing hormone. *Recent Prog. Horm. Res.* 27:165-206.

Pincus, G., and Chang, M. C. 1953. The effects of progesterone and related compounds on ovulation and early development in the rabbit. *Acta Physiol. Latinoam.* 3:177-183.

Pincus, G.; Rock, J.; Garcia, C-R.; Rise-Wray, E.; Paniaga, M.; and Rodriguez, I. 1958. Fertility control with oral medication. *Am. J. Obstet. Gynecol.* 75:1333-1346.

Popa, G. T., and Fielding, U. 1930. A portal circulation from the pituitary to the hypothalamic region. *J. Anat.* 65: 88-91.

Porter, D. G., and Finn, C. A. Forthcoming. Present status and future prospects of research in the biology of the uterus. In *Frontiers in reproduction and fertility control*, ed. R. O. Greep. Cambridge: MIT Press.

Presl, J.; Figarova, V.; Pospisil, J.; Wagner, V.; and Horsky, J. 1971. Evidence for human chorionic gonadotropin binding sites in some organs of the immature female rats. Effects of non-labelled chorionic gonadotropin and follicle stimulating hormone on the distribution of $^{125}I$ labelled human chorionic gonadotropin. *Folia Morphol.* 19:171-176.

Rajaniemi, H., and Vanha-Pertulla, T. 1972. Specific receptor for LH in the ovary: Evidence by autoradiography and tissue fractionation. *Endocrinology* 90:1-9.

———. 1973. Attachment of the luteal plasma membranes: An early event in the action of luteinizing hormone. *J. Endocrinol.* 57:199-206.

Raud, H. R., and Odell, W. D. 1973. Studies of the measurement of bovine and porcine prolactin by radioimmunoassay and by systemic pigeon crop-sac bioassay. *Endocrinology* 88:991-1002.

Riddle, O.; Bates, R. W.; and Dyskhorn, S. W. 1933. The preparation, identification and assay of prolactin—a hormone of the anterior pituitary. *Am. J. Physiol.* 105:191-216.

Rock, J.; Garcia, C-R.; and Pincus, G. 1957. Synthetic progestins in the normal human menstrual cycle. *Recent Prog. Horm. Res.* 13:323-339.

Ross, G. T.; Cargille, C. M.; Lipsett, M. B.; Rayford, P. L.; Marshall, J. R.; Strott, C. A.; and Rodbard, D. 1970. Pituitary hormones in women during spontaneous and induced cycles. *Recent Prog. Horm. Res.* 26:1-48.

Rowlands, I. W. 1938. Selective neutralization of the luteinizing activity of gonadotropic extracts of pituitary by anti-sera. *Proc. R. Soc. (London)* ser. B. 126:76-86.

Ryan, K. J. 1959. Biological aromatization of steroids. *J. Biol. Chem.* 234:268-272.

―――――. 1963a. Synthesis of hormones in the ovary. In *The ovary.* International Acad. of Path. Monograph no. 3. ed. H. Grady, pp. 69-83. Baltimore: Williams and Williams.

―――――. 1963b. Biogenesis of estrogens. In *Proc. Fifth International Congress of Biochemistry*, vol. 7, pp. 381-394. Oxford: Pergamon Press.

Ryan, K. J., and Short, R. V. 1965. Formation of estradiol by granulosa and theca cells of the equine ovarian follicle. *Endocrinology* 76:108-114.

Ryan, K. Y., and Smith, O. W. 1961. Biogenesis of estrogens by the human ovary. I. Conversion of acetate-1-$C^{14}$ to estrone and estradiol. *J. Biol. Chem.* 236:705-709.

Sairam, M. R.; Papkoff, H.; and Li, C. H. 1972. The primary structure of ovine interstitial cell-stimulating hormone. I. The α-subunit. *Arch. Biochem. Biophys.* 153:554-571.

Sawyer, C. H. 1975. Some recent development in brain-pituitary-ovarian physiology. *Neuroendocrinology* 17:97-124.

Schaar, C., and Clemens, J. A. 1974. The role of catecholamines in the release of anterior pituitary prolactin *in vitro. Endocrinology* 95:1202-1212.

Schally, A. V.; Arimura, A.; Kastin, A. K.; Matsuo, H.; Beba, Y.; Redding, T. W.; Nair, R. M. G.; Debeljuk, L.; and White, W. F. 1971. Gonadotropin releasing hormone: one polypeptide regulates secretion of luteinizing and follicle-stimulating hormones. *Science* 173:1036-1308.

Schally, A. V.; Kastin, A. J.; Arimura, A.; Coy, D.; Coy, E.; Debeljuk, L.; and Redding, T. W. 1973. Basic and clinical studies with luteinizing hormone-releasing hormone (LH-RH) and its analogues. *J. Reprod. Fertil.* (suppl) 20:119-136.

Schomberg, D. W. 1969. Effects of pig uterine flushings on ovarian tissues *in vitro* and *in vivo*. In *Proceedings of the third international congress of endocrinology*, ed. C. Gual and F. J. Gebling, pp. 916-920. Amsterdam: Excerpta Medica Foundation.

Schomberg, D. W.; Campbell, J. A.; and Wiltbank, G. D. 1969. Effects of uterine flushings from pigs on porcine granulosa cells growing in tissue culture. *Advan. Biosci.* 4:429-440.

Schwartz, N. B. 1969. A model for the regulation of ovulation in the rat. *Recent Prog. Horm. Res.* 25:1-43.

―――――. 1973. Mechanism controlling ovulation in small mammals. In *Handbook of physiology*, sect. 7, *Endocrinology*, vol. II, ed. R. O. Greep, pp. 125-142. Washington, D.C.: American Physiological Society.

―――――. 1974. The role of FSH and LH and of their antibodies in follicle growth and in ovulation. *Biol. Reprod.* 10:236-272.

Sehon, A. H.; Goodfriend, L.; and Rose, B. 1958. Preparation of an estrone-protein conjugate. *133rd Am. Chem. Soc. Meeting.* San Francisco, 1958, p. 19 (abstr).

Shome, B., and Parlow, A. F. 1973. The primary structure of the hormone-specific, beta subunit of human pituitary luteinizing hormone (hLH). *J. Clin. Endocrinol. Metab.* 36:618-621.

―――. 1974a. Human follicle stimulating hormone (hFSH): First proposal for the amino acid sequence of the α-subunit of human luteinizing hormone (hLHα). *J. Clin. Endocrinol. Metab.* 39:199-202.

―――. 1974b. Human follicle stimulating hormone: First proposal for the amino acid sequence of the hormone-specific, β-subunit (hFSH). *J. Clin. Endocrinol. Metab.* 39:203-205.

Short, R. V. 1961. Steroids present in the follicular fluid of the mare. *J. Endocrinol.* 22:153-163.

Slaunwhite, W. R., and Samuels, L. T. 1956. Progesterone as a precursor of testicular androgen. *J. Biol. Chem.* 220:341-352.

Slotta, K. H.; Ruschig, H.; and Fels, E. 1934. Reindarstellung der hormone aus dem corpus luteum. *Bio. Chem. Ges.* 67:1270-1273.

Smith, O. W., and Ryan, K. J. 1961. Biogenesis of estrogens by the human ovary: The conversion of androstene-dione-4-C-14 to estrone and estradiol in high yield. *Endocrinology* 69:869-872.

Smith, P. E. 1926. Hastening development of the female genital system by daily homoplastic pituitary transplants. *Proc. Soc. Exp. Biol. Med.* 24:131-132.

―――. 1927. The disabilities caused by hypophysectomy and their repair. *J. A. M. A.* 88:158-161.

―――. 1930. Hypophysectomy and replacement therapy in the rat. *Am. J. Anat.* 45:205-274.

Solomon, S.; Vande Wiele, R.; and Lieberman, S. 1956. The *in vitro* synthesis of 17-hydroxyprogesterone and $\Delta^4$-androstene-3, 17-dione from progesterone by bovine ovarian tissue. *J. Am. Chem. Soc.* 78:5453-5454.

Staple, E.; Lynn, W. S.; and Gurin, S. 1956. An enzymatic cleavage of the cholesterol side chain. *J. Biol. Chem.* 219:845-851.

Steelman, S. L., and Pohley, F. M. 1953. Assay of the follicle-stimulating hormone based on the augmentation with human chorionic gonadotropin. *Endocrinology* 53:604-616.

Stevens, V. C. 1973. Immunization of female baboons with hapten-coupled gonadotropins. *Obstet. Gynecol.* 42:496-506.

―――. 1974. Fertility control through active immunization using placenta proteins. In *Immunological approaches to fertility control*, 7th Karolinska Symposium, pp. 357-369. Stockholm: Karolinska Institute.

Stricker, P., and Grueter, F. 1928. Action du lobe anterieur de l'hypophyse sur la montee laiteuse. *Compt. Rend. Soc. Biol.* 99:1978-1980.

Stumpf, W. E. 1970. Estrogen-neurons and estrogen neuron systems in the periventricular brain. *Am. J. Anat.* 129:207-217.

Takahashi, M., and Yoshinaga, K. 1974. Active immunization of rats with luteinizing hormone-releasing hormone. In *Proc. The Endocrine Society Meeting*, p. 80, abstr. 49.

Takahashi, M.; Ford, J. J.; Yoshinaga, K.; and Greep, R. O. 1974. Induction of ovulation in hypophysectomized rats by progesterone. *Endocrinology* 95:1322-1326.

Talwar, G. P. 1974. In *Immunological approaches to fertility control*, 7th Karolinska symposium, pp. 370-371. Stockholm: Karolinska Institute.

Talwar, G. P. and Associates. 1976 (Sixteen related papers) Contraception 13:129-268.

Taymor, M. L.; Berger, M. J.; Thompson, I. S.; and Karam, K. 1972. Hormonal factors in human ovulation. *Am. J. Obstet. Gynecol.* 114:445-453.

Utiger, R.; Parker, M. L.; and Daughaday, W. H. 1962. Studies on human growth hormone: A radioimmunoassay for human growth hormone. *J. Clin. Invest.* 41:254-261.

Vale, W.; Grant, G.; Rivier, J.; Monahan, M.; Amoss, M.; and Blackwell, R. 1972. Synthetic polypeptide antagonists of the hypothalamic luteinizing hormone releasing factor. *Science* 176:933-934.

Vande Wiele, R. L.; Bogumil, J.; Dryenfurth, I.; Ferin, M.; Jewelewicz, R.; Warren, M.; Rizkallah, T.; and Mikhail, G. 1970. Mechanisms regulating the menstrual cycle in women. *Recent Prog. Horm. Res.* 26:65-104.

Van Hall, E. V.; Vaitukaitis, J. L.; Ross, G. T.; Hickman, J. W.; and Ashwell, G. 1971. Immunological and biological activity of hCG following progessive desialylation. *Endocrinology* 88:456-464.

Ward, D. N.; Reichert, L. W., Jr.; Liu, W-K.; Nahm, H. S.; Hsia, J.; Lamkin, W. M.; and Jones, N. S. 1973. Chemical studies of luteinizing hormone from human and ovine pituitaries. *Recent Prog. Horm. Res.* 29:533-556.

Weick, R. F.; Dierschke, D. J.; Karsch, F. J.; Butler, W. R.; Hotchkiss, J.; and Knobil, E. 1973. Periovulatory time courses of circulating gonadotropic and ovarian hormones in the rhesus monkey. *Endocrinology* 93:1140-1147.

Werkin, H.; Plotz, J.; LeRoy, G. V.; and Davis, E. M. 1957. Cholesterol—a precursor of estrone *in vivo*. *J. Am. Chem. Soc.* 79:1012-1013.

Werthessen, N. T.; Schwenk, E.; and Baker, C. 1953. Biosynthesis of estrone and β estradiol in the perfused ovary. *Science* 117:380-381.

White, A.; Catchpole, H. R.; and Long, C. N. H. 1937. A crystalline protein with high lactogenic activity. *Science* 86:82-83.

Whitelaw, R. G. 1958. Ovarian activity following hysterectomy. *Br. J. Obstet. Gynaecol.* 65:917-932.

Wilde, C. E.; Orr, A. H.; and Bagshaw, K. D. 1965. A radioimmunoassay for human chorionic gonadotropins. *Nature* 205:191-192.

Wislocki, G. B., and King, L. S. 1936. The permeability of the hypophysis and the hypothalamus to vital dyes, with a study of the hypophysial vascular supply. *Am. J. Anat.* 58:421-472.

Wotiz, H. H.; Davis, J. W.; Lemon, H. H.; and Gut, M. 1956. Studies in steroid metabolism V: The conversion of testosterone-4-C$^{14}$ to estrogens by human ovarian tissue. *J. Biol. Chem.* 222:487-495.

Yallow, R. S., and Berson, S. A. 1959. Assay of plasma insulin in human subjects by immunological methods. *Nature* 184:1648-1649.

―――. 1960. Immunoassay of endogenous plasma insulin in man. *J. Clin. Invest.* 39:1157-1175.

Yamaji, T.; Dierschke, D. J.; Hotchkiss, J.; Bhattacharya, A. N.; Surve, A. H.; and Knobil, E. 1971. Estrogen induction of LH release in the rhesus monkey. *Endocrinology* 89:1034-1041.

Yen, S. S. C.; Lasley, B. L.; Wang, C. F.; LeBlanc, H.; and Silber, T. M. 1975. The operating characteristics of the hypothalamic-pituitary system during the menstrual cycle and observations of biological action of somatostatin. *Recent Prog. Horm. Res.* 31:321-357.

Yen, S. S. C.; Rebar, R.; Vaneneberg, G.; Ehara, Y.; and Siler, T. 1973. Pituitary gonadotropin responsiveness to synthetic LRF in subjects with normal and abnormal hypothalamic-pituitary-gonadal axis. *J. Reprod. Fertil.* (suppl) 20:137-160.

Yen, S. S. C., and Tsai, C. C. 1972. Acute gonadotropin release induced by exogenous estradiol during the mid-follicular phase of the menstrual cycle. *J. Clin. Endocrinol. Metab.* 34: 298-305.

Yen, S. S. C.; Tsai, C. C.; Naftolin, F.; Vandenberg, G.; and Ajabor, L. 1972. Pulsatile patterns of gonadotropin release in subjects with and without ovarian function. *J. Clin. Endocrinol. Metab.* 34:671-675.

Zarate, A.; Canales, E. S.; Soria, J.; Kastin, A. J.; Schally, A. V.; and Gonzales, A. 1974. Induction of ovulation with synthetic luteinizing hormone-releasing hormone. In *Recent Progress in Reproductive Biology*, ed. Crosignani and James, pp. 811-826. New York: Academic Press.

Zeleznik, A. J.; Midgley, A. R.; and Reichert, L. E. 1974. Granulosa cell maturation in the rat: Increased binding of human chorionic gonadotropin following treatment with follicle-stimulating hormone *in vivo*. *Endocrinology* 95:818-825.

Zondek, B. 1930. Über die Hormone des Hypophysenvorderlappens I. Wachstumshormon Follikelreifungshormone (Prolan A, Luteinisierungshormon) Prolan B, Stoffwechselhormone. *Klin. Wochschr.* 9:245-248.

Zondek, B., and Aschheim, S. 1926. Über die Funktion des Ovariums. *Z. Geburtshiffe Gynaekol.* 90:372-376.

# 5 The Male Reproductive System

## Section 1. Studies of the Male Reproductive Tract before 1960

In 1677 a student named Hamen observed with Leeuwenhoek's simple microscope motile "animalcules" in human semen. Although Leeuwenhoek considered these to be involved in reproduction, a majority of his contemporaries dismissed them as parasites or symbiotic protozoa. Their significance continued to be a subject of vigorous debate for nearly two centuries. By ingenious filtration experiments Spallanzani in 1780 demonstrated that their removal from semen prevented fertilization but neither he nor others correctly interpreted this observation and the belief that the *spermatozoa* were parasites, unrelated to generation, persisted until Kolliker (1841) traced their origin to the testis and LaValette (1865) clearly showed that they were highly specialized animal cells possessing a nucleus and cytoplasm. All doubt concerning their significance was finally dispelled by Barry's observation (1843) that the spermatozoon enters the egg and by Oscar Hertwig's demonstration (1875) that fertilization involves the union of the sperm nucleus with that of the ovum, which had been discovered by von Baer in 1827. Thus two centuries elapsed between the first description of the male gamete and the final acceptance of its essential role in reproduction.

Without the benefit of a microscope, Regnier de Graaf in 1668 wrote a classical account of the structure of the *testicle* in which he described it as consisting largely of minute convoluted tubules (figure 5.1). Some two centuries later these came to be called *seminiferous tubules* when they were found to be the site of spermatozoa formation. Tracing the sequence of changes in the germ cells during development of the spermatozoa in the lining of the seminiferous tubules was the work of a number of discerning microscopists in the late 1880s (La Valette 1865; Benda 1898; von Ebner 1888; Meves 1899). In addition to the germ cell line, highly branched supporting cells were identified by Sertoli (1865), after whom they were later named *Sertoli cells* (von Ebner 1888). These were originally described as individual cells but subsequent investigators insisted upon their syn-

This chapter by Dr. Don W. Fawcett summarizes the essays commissioned for the review pertaining to male reproductive research.

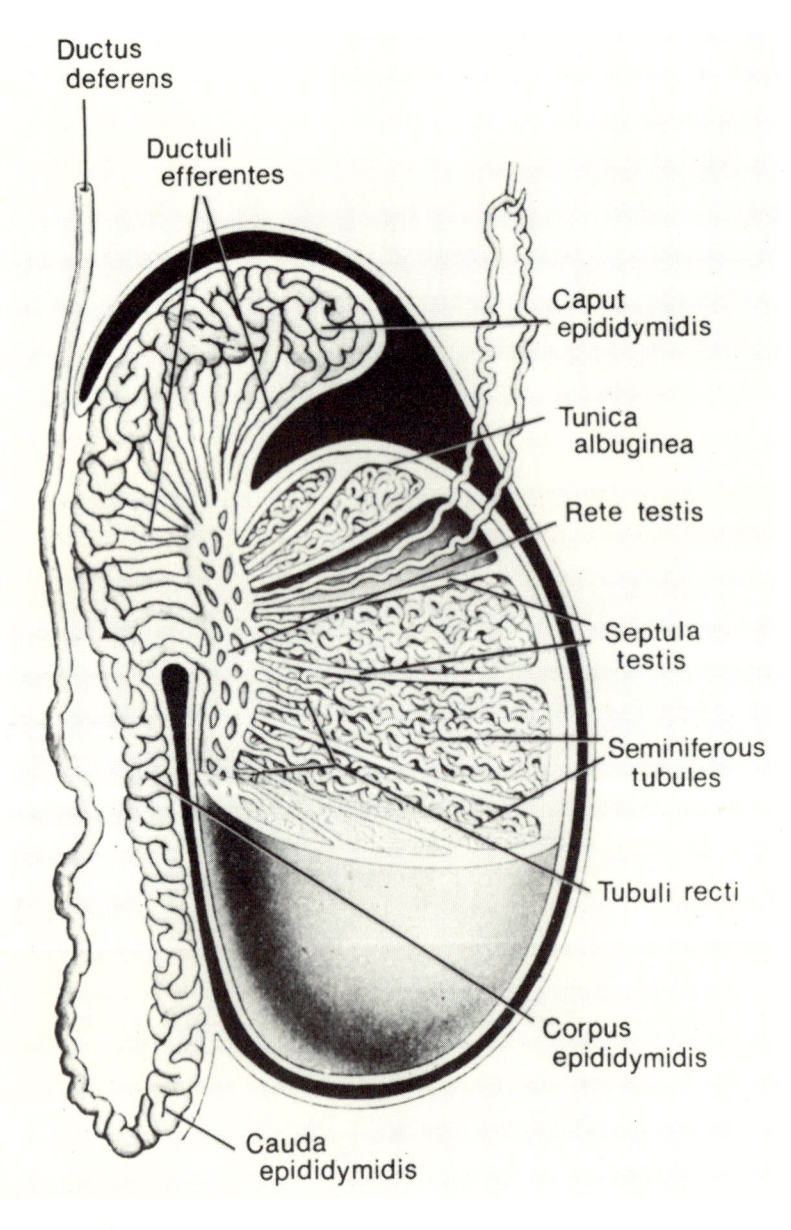

Figure 5.1 Cutaway diagram of the architecture of the human testis and its excurrent duct system.
Source: Drawing modified from W. J. Hamilton, *Textbook of Human Anatomy*. London, Macmillan and Co., 1957.

cytial nature (von Ebner 1871; Rolshoven 1940, 1945). All agreed that they subserved a mechanical supportive role and possibly a nutritive function for the developing germ cells that are arrayed along their sides and occupy deep recesses in their apex.

The seminiferous epithelium presented a confusing picture to early cytologists who were unable to resolve clearly its cell boundaries. Relying heavily on size and nuclear differences, they recognized several categories of cells representing successive steps in the evolution of spermatozoa (figure 5.2). Small rounded cells situated at the periphery of the tubules between the bases of the Sertoli cells were identified as the stem cells from which all subsequent stages of spermatogenesis were derived. These were called *spermatogonia* (La Valette 1865) and two kinds were later distinguished on the basis of nuclear staining characteristics. One having sparse, highly dispersed chromatin was designated *Type A* and the other, possessing a coarser chromatin pattern, was called *Type B* (Regaud 1901; Allen 1918).

Van Benedin (1875) discovered that the male and female nuclei that unite in the fertilized ovum each contains half as many chromosomes as the nongerminal body cells. It was soon found that in the course of development of both ova and spermatozoa the number of chromosomes characteristic of the species is reduced by half in two successive atypical cell divisions—a process called *meiosis*. In the male, clusters of large cells in the seminiferous epithelium immediately above the spermatogonia have nuclei with a conspicuous chromosomal pattern. These *primary spermatocytes* are in one phase or another of a protracted preparation for the first meiotic division. Smaller *secondary spermatocytes* resulting from their division are present for only a short time before they divide to give rise to *spermatids* containing half the original chromosomal complement. No further divisions of the germ cells take place, but the spermatids undergo a complex process of differentiation by which they are transformed into the highly specialized spermatozoa. The entire developmental process from spermatogonia to release of spermatozoa is called *spermatogenesis* and that part comprising the metamorphosis of the spermatids is called *spermiogenesis*.

The principal morphogenetic events of spermiogenesis were well

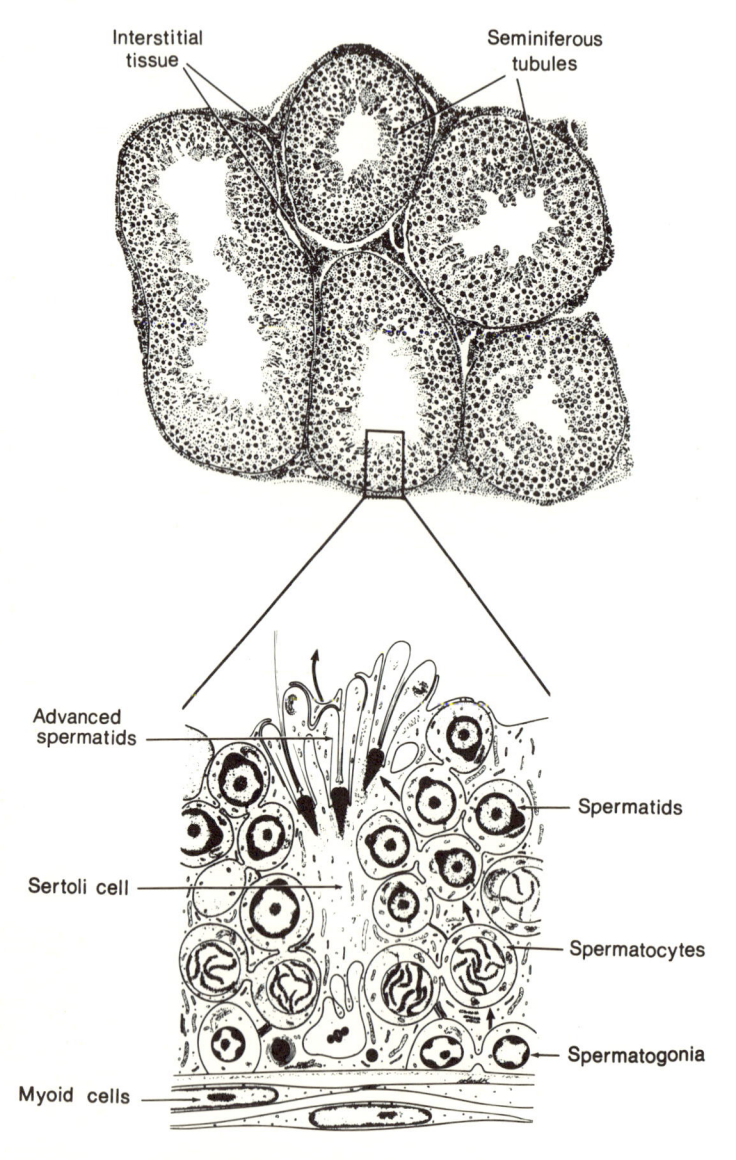

Figure 5.2 Seminiferous tubules and interstitial tissue as seen in histological sections. Below is a simplified drawing of the arrangement of the cellular components of the seminiferous epithelium as seen at higher magnification.
Source: Above, from Moore and Quick, *Am. J. Anat.* 34:317, 1924. Lower figure modified after D. W. Fawcett in *Male Fertility and Sterility*, eds. R. E. Mancini and L. Martini, Academic Press Inc., London, 1974.

documented by microscopists around the turn of the century. In brief they consist of outgrowth of a flagellum from one of a pair of centrioles situated near the plasmalemma of the spermatid (Meves 1899); the development, in the Golgi apparatus, of an acrosomal cap over one pole of the nucleus (Meves 1899; Gatenby 1918; Bowen 1920); the condensation of the chromatin and shaping of the nucleus; the investment of the base of the flagellum by a sheath of mitochondria (Benda 1898; Meves 1899); the casting off of the excess spermatid cytoplasm; and finally the release of the spermatozoon into the lumen of the tubule.

In the 1950s the cyclic nature of spermatogenesis and the mechanism of stem cell renewal received renewed investigative attention. Two morphologically distinct types of spermatogonia had been recognized by earlier investigators, but the use of radiolabeled thymidine and radioautography made it possible to identify in rodents at least six successive cell generations produced by a series of mitotic divisions of the Type A spermatogonium (Leblond and Clermont 1952a; Roosen-Runge and Giesel 1950; Ortovant 1958). These were designated Types $A_1$, $A_2$, $A_3$, $A_4$, Intermediate, and Type B. The division of Type B spermatogonia yields daughter cells that are preleptotene spermatocytes. These move off the basal lamina of the epithelium and enter meiosis.

Application of the then newly introduced periodic-acid Schiff staining reaction for carbohydrates made it possible for Leblond and Clermont (1952b) to stain selectively the acrosome and its precursors in the Golgi apparatus of the spermatid and to follow the evolution of this sperm organelle with much greater precision than had been possible with classical histological methods. Fourteen distinct steps in spermatid development were defined for the rat on the basis of the configuration of the developing acrosome. Then by carefully identifying the state of differentiation of the other germ cell types associated with each step in spermatid development, Leblond and Clermont (1952) described fourteen different associations of cells that could be defined as distinct *stages of spermatogenesis* (figure 5.3). The *cycle of the seminiferous epithelium* could then be defined as the series of changes taking place in any given area of the epithelium

170                                    Major Advances and New
                                       Opportunities

| | CELL TYPES | | | | | |
|---|---|---|---|---|---|---|
| I | A | In | P • | 1 • | 15 | |
| II | A + | In + | P | 2 | 16 | |
| III | A + | In + | P | 3 | 16 | |
| IV | A | In | P | 4 | 17 | |
| V | A + | B + | P | 5 | 17 | |
| VI | A | B | P | 6 | 18 | |
| VII | A + | R + | P + • | 7 • | 19 | |
| VIII | A + | R | P | 8 | 19 | |
| IX | A | L | P | 9 | | |
| X | A + | L | P | 10 | | |
| XI | A +▲ | L • | P | 11 | | |
| XII | A ▲ | L-Z | P • | 12 | | |
| XIII | A + | Z | Di | 13 | | |
| XIV | A | Z • | II D •  II P | 14 | | |

(STAGES OF THE CYCLE)

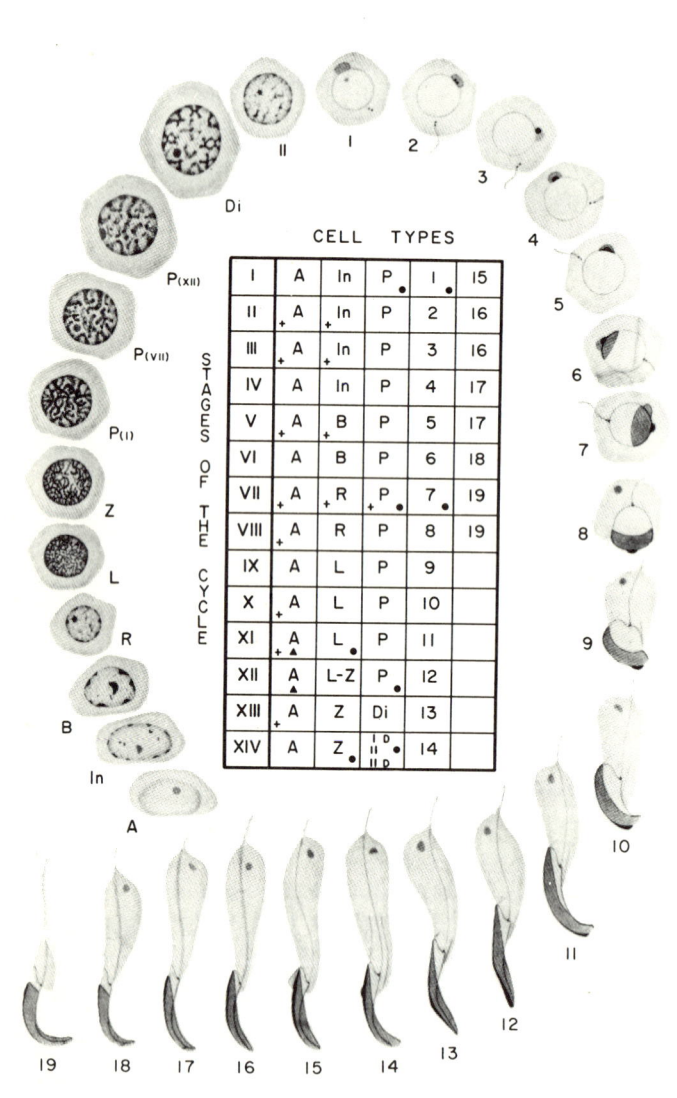

**Figure 5.3** Drawing of the steps of spermatogenesis in the rat. Lettering: A, In, B, successive types of spermatogonia; R, resting spermatocyte; L, leptotene spermatocyte; Z, zygotene spermatocyte; $P_{(I)}$, $P_{(VII)}$, $P_{(XII)}$, early, mid-, and late pachytene spermatocytes. The Roman numerals in brackets indicate the stages of the cycle in which they are found; Di, diplotene; II secondary spermatocyte; 1-19 steps of spermiogenesis.

The table in the center gives the cellular composition of the stages of the cycle of the seminiferous epithelium.

Source: Y. Clermont, *Am. J. Anat.* 111:111, 1962.

between two successive appearances of the same association of cell types. By means of radioautography after administering a pulse of tritiated thymidine, the duration of one cycle was established as twenty-three days in the rat and the entire course of spermatogenesis was found to occupy four cycles or forty-eight days (Clermont and Leblond 1953; Clermont, Leblond, and Messier 1959).

The progress in our understanding of the spermatogenic function of the seminiferous tubules was paralleled by a growing interest in the endocrine functions of the testis. In 1850 Franz Leydig observed clusters of epithelioid cells in the interstices among the seminiferous tubules (figure 5.2); but these attracted little attention until Bouin and Ancel (1903), on largely circumstantial and morphological evidence, interpreted them as an interstitial gland of the testis and prophetically attributed to these cells the function of producing secretions important for the development of sexual instincts and the maintenance of the secondary sexual characteristics of the male. The experimental studies that finally established the endocrine function of the testis extended over the first two decades of this century and involved many investigators. Extracts of testis were shown to promote growth of the comb in capons (McGee, Jukn, and Domm 1928) and this observation later became the basis for an important bioassay method for male sex hormone. Castration was found to result in atrophy and cytological regression of the prostate and seminal vesicles, and these effects could be reversed by injection of extracts of testis (Moore 1928; Moore, Hughes, and Gallagher 1930). Major advances in endocrinology were the first crystallization of androgen from extracts of urine and testis by Butenandt (1931) and the synthesis of androsterone (Ruzicka et al. 1934) and of testosterone (Butenandt and Hanisch 1935), which soon followed.

The essential role of the pituitary in the maintenance of the spermatogenic function of the testis was clearly demonstrated by the regression of both the Leydig cells and the seminiferous epithelium after hypophysectomy (Smith and Engle 1927; Zondek and Aschheim 1927) and by their cytological and functional restoration by pituitary implantation (Smith 1930) or administration of pituitary extracts (Greep, Fevold, and Hisaw 1936; Greep 1937). Fractionation of pituitary extracts into *follicle stimulating hormone* (FSH) and *luteinizing hormone* (LH) was accomplished by Fevold, Hisaw, and Leonard (1931). In addition to its effect on the ovary, FSH was found to stimulate development of the seminiferous tubules of immature or hypophysectomized male rats; whereas, LH acted upon the interstitial tissue and resulted secondarily in enlargement of the tubules and the accessory glands of male reproduction (Greep, Fevold, and Hisaw 1936). Qualitatively normal spermatogenesis could be maintained in hypophysectomized adult rats by testosterone (Nelson 1936), thus confirming the view that the effects of LH on gametogenesis in the male were a consequence of its stimulation of the Leydig cells to secrete testosterone.

At about the same time the effects of the gonadal hormones on the pituitary and the principle of feedback control of the release of gonadotropins in the male were established by studies of Moore and Price (1932) in which injection of androgens into intact male rats proved to be harmful rather than beneficial to the testis. It was correctly perceived that the excess male hormone had suppressed pituitary gonadotropin secretion below the level necessary to maintain normal testicular activity. This inference was later substantiated by observation of an elevated LH content of pituitaries of castrated rats, which could be lowered by administration of testosterone (Greep and Jones 1950).

A feedback control of FSH was postulated but proved difficult to establish without more sensitive bioassay systems. The question as to what testicular hormone might be responsible for feedback control remained unanswered. A hypothetical substance, *inhibin*, was postulated to originate in the seminiferous tubules (McCullagh 1932; McCullagh and Walsh 1935), but efforts to isolate and characterize such a substance were unsuccessful and existing bioassay systems failed to yield consistent evidence of elevation of FSH in the blood or the pituitary following castration.

Influences on the functions of the testis other than those mediated by hormones received some attention. It had long been known that in those species that normally have scrotal testes, failure of the testes to descend during development resulted in incomplete differentiation of the seminiferous tubules and consequent sterility (Felizet and Branca 1898). In an admirable series of experiments, Moore and his coworkers clearly established the temperature sensitivity of spermatogenesis and the thermoregulatory function of the scrotum. Surgical replacement of the testis in the abdominal cavity, insulation of the scrotum to reduce evaporative heat loss, or application of heat to the scrotum all resulted in degenerative changes in the seminiferous epithelium and ultimate sterility (Moore 1924a, 1924b, 1926a, 1926b). The endocrine function of the testes under all of these conditions was retained but significantly reduced. A vascular adaptation that facilitates maintenance of lower testicular temperature was recognized (Harrison 1949). En route to the testis the spermatic artery is in extensive contact with the surrounding pampiniform plexus of veins. This topographical relationship constitutes a countercurrent system providing for loss of heat from the arterial blood to the cooler venous return, thus precooling the blood entering the testis. The temperature sensitive metabolic or biosynthetic pathway in spermatogenesis has yet to be identified. Furthermore, in the evolution of mammals it is by no means clear what advantage, if any, was gained by having spermatogenesis take place in a scrotum 4 to 5° C below core body temperature. Why a few homeothermic mammals and all birds are exempted from this limitation and carry out spermatogenesis at 37 to 40° C remains a mystery.

The excretory ducts of the male gonads were traced in dissections by gross anatomists of the eighteenth century, but their microscopic structure appears to have been neglected until the middle of the nineteenth century. The spermatozoa formed in the seminiferous tubules are carried in dilute suspension through short straight tubules (*tubuli recti*), which converge upon the *rete testis*, a network of thin-walled channels in the axis of the testis, or near its posterior surface, depending upon the species (figure 5.4). From the rete, a dozen or more small efferent ducts (*ductuli efferentes*) conduct the sperm and testicular fluid to the *epididymis*, an organ consisting of a single highly convoluted duct (*ductus epididymidis*) that forms a compact organ closely applied to the posterior surface of the testis. The duct, some seven meters long in man, straightens out at the lower pole of

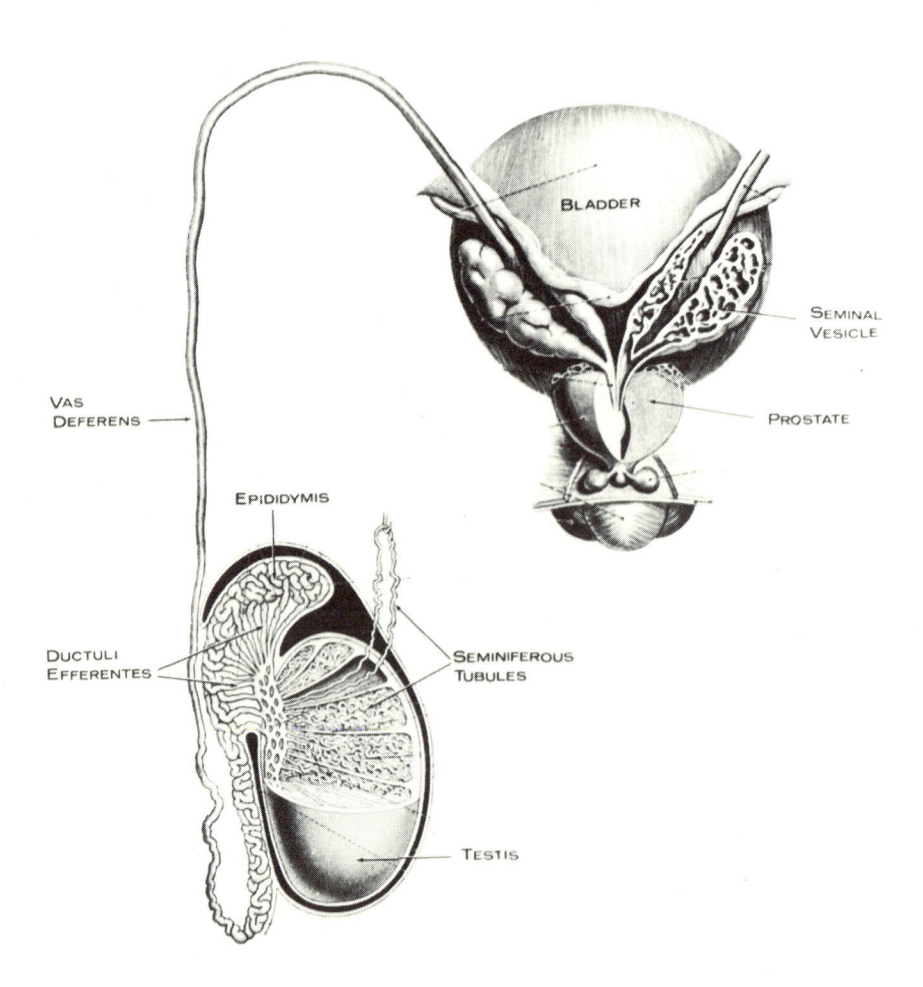

Figure 5.4 Composite drawing of the excurrent ducts and accessory glands of the male reproductive tract.
Source: After W. J. Hamilton, *Textbook of Human Anatomy*, London, Macmillan and Co., 1957.

the testis and continues as the *vas deferens (ductus deferens)*. This slender duct with a thick muscular wall passes upward out of the scrotum entering the pelvis in the groin. There it courses retroperitoneally toward the midline and terminates in a slender *ejaculatory duct* that opens into the urethra as it passes through the prostate gland just below the urinary bladder (figure 5.4). Becker (1857) first described motile cilia on the epithelium lining the ductuli efferentes and stereocilia in the epididymal duct and took note of regional differences in height of the lining epithelium. In addition to the columnar *principal cells*, small pyramidal *basal cells* were identified. The numerous vacuoles and granules in the principal cells were generally interpreted as manifestations of secretory activity (Regaud 1901; Benoit 1926). Others, finding that the epithelial cells took up injected dyes and carbon particles from the duct lumen, placed more emphasis upon a possible absorptive function of the epithelium. The relative importance of absorption and secretion in the epididymis remains unresolved.

The spermatozoa were found to undergo further maturation and gradually to acquire fertilizing capacity as they passed through the epididymis. However, these changes were believed to be inherent in the spermatozoa and simply dependent upon the passage of time rather than upon any specific secretory function of the epididymal duct (Young 1929, 1931). From the histological observation of increased numbers of spermatozoa in successive segments of the duct, it was concluded that this organ simply concentrated the relatively dilute sperm suspension reaching it from the testis and served as a site of accumulation and storage of spermatozoa (Nicander 1957; Reid and Cleland 1957). Physiological and biochemical studies of the epididymis were very few, but the glycerophosphorylcholine of semen was shown to be a product of its epithelium (Dawson and Rowlands 1959). The prevailing view of the storage function of the epididymis was to change dramatically after 1960.

Opening into the excurrent ducts are a number of accessory glands of the male reproductive tract (figure 5.4). These include the *seminal vesicles*, a pair of saccular glands on the posterior aspect of the bladder that open into the ejaculatory ducts, and the *prostate*, a firm compact gland surrounding the first portion of the urethra immediately below the neck of the urinary bladder and discharging its secretion through multiple small ducts that open into the prostatic urethra. The *bulbourethral glands* secrete via short ducts into the membranous urethra just below the prostate. Early studies of these principal accessory glands of male reproduction were, for the most part, comparative morphological descriptions of their remarkable variations in different mammalian species. They were known to be involved to varying degree in the formation of the seminal plasma, but their respective contributions to this fluid were poorly understood until ingenious surgical procedures developed for the dog (Huggins et al. 1939) permitted collection of prostatic fluid and made it possible to relate the secretory activity of the gland to the endocrine status of the animal. These studies led to extensive analyses of the chemical constituents of the secretions stored in the accessory glands and present in the seminal plasma (Mann 1945, 1954; Mann, Davies, and Humphrey 1949; Williams-Ashman 1954; Williams-Ashman and Banks 1954). A number of substances were demonstrated that are either absent or present only in very low concentrations elsewhere in the body. Acid phosphatase, produced in the prostate and present in human semen (Krutscher and Wolbergs 1935), became a useful index of circulating androgen levels. Abnormally high values in the blood proved to be a reliable diagnostic sign of osteoblastic metastases of carcinoma of the prostate (Gutman, Sproul, and Gutman 1936). The demonstrated effectiveness of the female hormone *estrogen*, acting via suppression of gonadotropin release and diminished androgen production, proved to be an important advance in the hormonal control of prostatic cancer (Huggins 1956). Zinc was found to occur in the prostate and in its secretions in up to twenty times its concentration in other organs (Mawson and Fischer 1957; Gunn and Gould 1956). What enzymes or other molecules it is associated with and what their function may be remains unclear. Among other unusual compounds in the secretions of the accessory glands were fructose, sorbitol, inositol, amino sugars, citric acid, polyamines, ergothionine, and prostaglandins. Fructose, produced in the seminal vesicles of man and many other mammals, was found to be the principal natural energy source of ejaculated spermatozoa (Mann 1954). The amount present in semen proved to be strictly controlled by testicular hormones (Mann and Parsons 1950) and this became the basis of a sensitive method for assessing the endogenous production of androgens in man as well as in domestic animals (Landace and Laughead 1951). Prostaglandins were first identified and purified from the secretions of the male accessory glands (von Euler 1936; Bergstrom 1949). Although the widespread occurrence of prostaglandins and their multiple biological activities have since been thoroughly documented, their function in the seminal plasma remains obscure.

To summarize the state of our knowledge prior to 1960, it can be said that the general architecture of the testis had been described at the light microscope level and the gonadal and hypophyseal hormones had been identified and purified. The cyclic nature of spermatogenesis was recognized and its stages had been defined for the common laboratory rodents. The principal structural components of the spermatozoon and the sequence of cytological changes involved in its development had been described at the light and electron microscope levels. The sensitivity of spermatogenesis to heat and the thermoregulatory functions of the scrotum were appreciated. The dependence of spermatogenesis upon pituitary gonadotropins and the feedback modulation of gonadotropin release by gonadal hormones had been established in broad outline. The biochemical analysis of semen had identified a number of biologically active compounds and had traced their origin to specific accessory glands, but the functional significance of most of these substances remained unexplained. The epididymis was regarded as a relatively inactive organ for storage of spermatozoa, a view that was to change dramatically in the ensuing decade.

## Section 2. Advances in Reproductive Biology of the Male since 1960 and Unresolved Problems

### Organization of the Seminiferous Epithelium and the Blood-Testis Barrier

In the classical period of cytology that followed the introduction of well-corrected lenses for the compound microscope and the introduction of aniline dyes as biological stains, a multiplicity of complex methods of tissue preservation and coloration was developed. Several were capable of revealing mitochondria. A different set of methods was required to impregnate and visualize the Golgi apparatus, and still others were needed for optimal observation of nuclear structure. Thus the prevailing concept of the cytology of the seminiferous epithelium was a composite pieced together from the application of multiple empirical techniques, each selective for particular cell organelles or inclusions. No single method permitted simultaneous observation of all the cell components or the study of their relations to one another. The thickness of the sections that could be cut from tissue embedded in paraffin or celloidin resulted in a confusing superimposition of component parts of adjacent cells, and the optics of the light microscope did not permit a clear resolution of cell boundaries. Therefore, the images of the complex seminiferous epithelium were, at best, confused; and the information derived from its study was imprecise and subject to varying interpretations.

After 1950 the availability of the electron microscope, the introduction of improved methods of chemical fixation and embedding in plastics, and the design of ultramicrotomes capable of cutting slices of tissue as thin as a tenth of a micrometer revolutionized the study of tissues and organs. For the first time reproductive biologists were able to observe in a single preparation all the membranous organelles of cells as well as their protein, polysaccharide, and lipid inclusions and to record exquisitely sharp images with two hundred times the resolution of the compound microscope and at magnifications up to four hundred thousand times. The rapid exploitation of this powerful new instrument in the early 1960s yielded a rich harvest of new structural information and brought fundamental changes in our concept of the organization of the seminiferous epithelium.

The Sertoli cells, which were previously believed by many to form a syncytium, were shown instead to be individual cells uniformly spaced on the basal lamina with the germ cells occupying expanded intercellular spaces between them (Burgos and Fawcett 1955; Brokelmann 1963). These supporting cells cease to proliferate at about the time of puberty but persist throughout the lifetime of the individual. The germ cells, on the other hand, are a constantly renewing population with their stem cells at the base of the epithelium and more advanced stages at successively higher levels (figure 5.2). As they develop, the germ cells are slowly displaced upward along the sides of the supporting cells. When they approach the tubule lumen as spermatids, their nucleus condenses, the cell body elongates, and they establish a new relationship to the Sertoli cells in which they occupy deep recesses in the apical cytoplasm.

The occurrence of two cell populations, one fixed and the other constantly moving upward, is a feature peculiar to the seminiferous epithelium and the continually changing topographical relationships between the germ cells and their supporting cells create some unique problems of cell coherence and cell communication that may have important implications for the control of spermatogenesis. The cells of other columnar epithelia in the body have, on their lateral surfaces, local specializations of the cell membrane that maintain cohesion and provide for communication between cells. The function of maintaining cell attachment is subserved by a beltlike zone of membrane fusion (the *zonula occludens*) that encircles the apical ends of the cells and also by attachment plaques (*desmosomes*) scattered over their lateral surfaces. The communication between cells that is essential for physiological integration of the activities of the entire epithelium is maintained through specialized sites of close membrane apposition called *communicating junctions* (*gap junctions*), which permit passage of ions and small molecules from cell to cell. In addition to providing low resistance electrical coupling of cells, these gap junctions are also sites of very firm attachment of the opposing membranes and thus contribute to the mechanical stability of the epithelium as well as to its functional integration. In the seminiferous epithelium the necessity for upward mobility of the developing germ cells with respect to the stationary population of supporting cells does not permit the establishment of enduring surface specializations for cell attachment such as are found in other epithelia. Thus, in the seminiferous tubules, neither typical desmosomes nor gap junctions are found on the interfaces between Sertoli cells and germ cells in the upper two-thirds of the epithelium. It appears therefore that in this epithelium freedom of the germ cells to move upward has been maintained at the sacrifice of communicating junctions that would normally be expected to provide for electrotonic and metabolic coupling of the germ cells to their supporting cells (Fawcett 1974, 1975b). However, a unique type of occluding junction (figure 5.5) has been described between adjacent Sertoli cells near the base of the seminiferous epithelium (Flickinger and Fawcett 1967; Nicander 1967). The finding of these junctions accounted for contemporary observations on the exclusion of blood-borne dyes from the seminiferous epithelium (Kormano 1967, 1968) and the results of physiological experiments which revealed that a wide range of substances injected into the bloodstream rapidly appeared in the lymph but was not detectable in the fluid collected from the rete testis (Setchell 1967; Setchell, Voglmayr, and Waites 1969). Several morphological studies using electron opaque probes of the extracellular space showed that substances do penetrate the base of the seminiferous epithelium and may fill the intercellular cleft surrounding the spermatogonia, but deeper penetration is prevented by occluding junctions between Sertoli cells (Fawcett, Leak, and Heidger 1970; Dym and Fawcett 1970; Neaves 1973a). It was concluded therefore that these junctions constitute the morphological basis of the blood-testis permeability barrier (figure 5.6). Situated between Sertoli cell processes that overarch the spermatogonia, the junctions also divide the epithelium into a *basal compartment* containing the stem cells of spermatogenesis and an *adluminal compartment* containing the more advanced stages of germ cell development (Dym and Fawcett 1970). These junctions first appear at puberty concurrently with the initial establishment of a barrier, the onset of fluid secretion, and the development of a lumen in the seminiferous tubules. These temporally integrated developmental events are evidently dependent upon the formation of occluding junctions near the base of the epithelium (Vitale-Calpe, Fawcett, and Dym 1973).

The development of the method of freeze-fracturing (Moor 1964; Branton 1965) made it possible for cell biologists to split membranes in half and to examine their internal differentiations in high-resolution replicas. Application of this technique to studies of the Sertoli

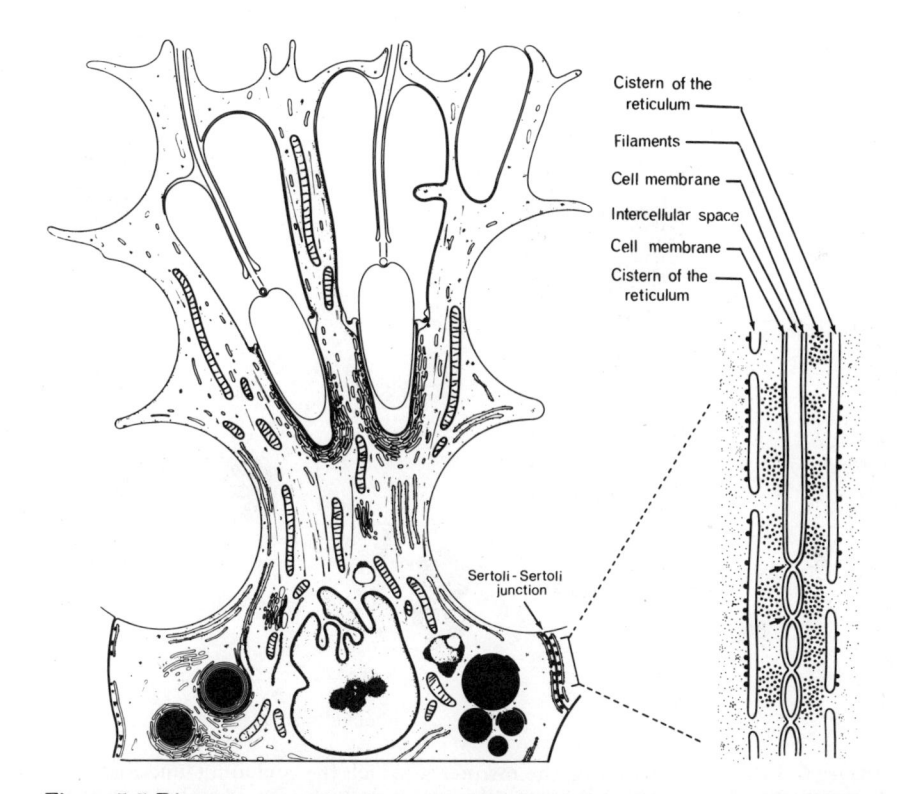

Figure 5.5 Diagram presenting the typical location and the components of the junctional complex between Sertoli cells. The opposing membranes are fused at multiple sites indicated by arrows. The subjacent cytoplasm contains bundles of filaments coursing parallel to the cell surface. Deep to these in each cell are cisternae of the endoplasmic reticulum bearing ribosomes on the side toward the cell body.
Source: D. W. Fawcett in *Handbook of Physiology*, Endocrinology V, Chapter 2. American Physiological Society, Washington, D.C., 1975.

Occluding Sertoli junction

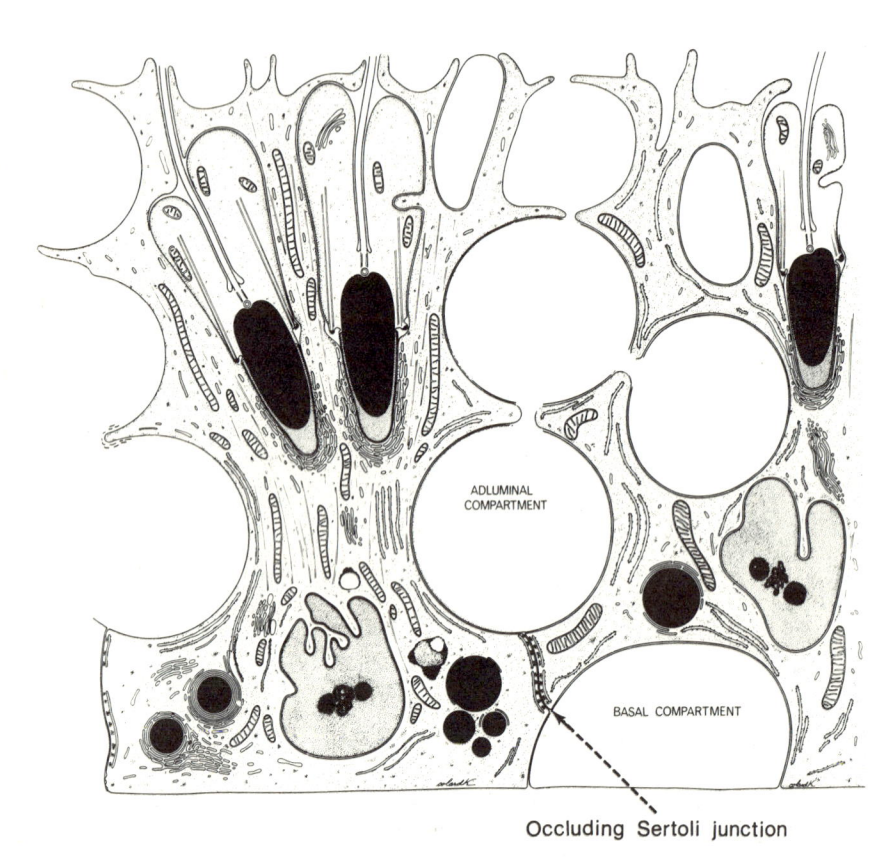

**Figure 5.6** Drawing illustrating the manner in which the occluding junctions between Sertoli cells divide the seminiferous epithelium into a basal compartment occupied by the spermatogonia and preleptotene spermatocytes and an adluminal compartment containing more advanced stages of the germ cell population. The occluding Sertoli-Sertoli junctions are the principal component of the blood-testis barrier.
Source: D. W. Fawcett in *Handbook of Physiology*, Endocrinology V, Chapter 2. American Physiological Society, Washington, D.C., 1975.

cell junctions revealed twenty to fifty parallel rows of intramembrane particles coursing circumferentially around the base of the Sertoli cells. Each of these particle rows is associated with a line of obliteration of the intercellular space and fusion of the opposing cell membranes (Gilula, Fawcett, and Aoki 1976). Exposure of the seminiferous tubules to hypertonic solutions, which rapidly open the blood-brain barrier and dissociate the junctions of other epithelia, does not open the blood-testis barrier. Thus the Sertoli junctions seem to constitute one of the tightest and most resistant epithelial permeability barriers in the body.

Spermatogenesis is divisible into clearly defined stages or cell associations occupying successive segments of up to a millimeter in length along the seminiferous tubules (Perey, Clermont, and Leblond 1961). Thus an explanation is needed for the precise coordination of developmental events over considerable distances. The apparent absence of communicating junctions between Sertoli cells and germ cells, or between neighboring syncytial clusters of germ cells, would seem to eliminate such membrane specializations as the basis for the integration of the cellular activities within a given stage. However, in freeze-fractured preparations, atypical gap junctions have been observed intercalated between the particle rows of the extensive occluding junctions between Sertoli cells (Gilula, Fawcett, and Aoki 1976; Nagano and Suzuki 1976a,b). These provide for ionic and metabolic coupling of the supporting cells. The coordination of germ cell development may therefore be attributable to communication among the Sertoli cells that create the milieu for germ cell development in the adluminal compartment. The compartmentation of the seminiferous epithelium that is achieved by the occluding Sertoli junctions serves to isolate the germ cells from the general extracellular space of the testis, permitting the supporting cells to maintain in the adluminal compartment a microenvironment favorable for the continuing differentiation of the germ cells (Dym and Fawcett 1970; Fawcett 1974). Analysis of testicular fluid obtained by cannulation of the rete testis (Voglmayr, Waites, and Setchell 1966; Waites and Einer-Jensen 1975) or by micropuncture of the tubules (Tuck et al. 1970) shows it to be rich in potassium, poor in protein, and essentially devoid of glucose. But it contains relatively high concentrations of androgens and estrogens, inositol, and certain amino acids, notably glutamic and aspartic (Setchell 1974). It is evident that the unusual composition of the fluid bathing the germ cells constitutes a very special milieu, but it is not yet clear which components are essential for germ cell differentiation, which are waste products, and which are destined to exert their effects farther along the male reproductive tract.

It is speculated that maintenance of a permeability barrier at the base of the epithelium may be essential for its secretory function—making it possible to create a standing osmotic gradient in the adluminal compartment that would tend to move fluid across the epithelium into the lumen of the tubules (Setchell 1970; Fawcett 1975b). The intraepithelial barrier is also important for protection of the germ cell line from blood-borne noxious agents and for impounding within the tubules antigenic products of postmeiotic cells that might otherwise reach the bloodstream and induce an autoimmune response (Johnson 1970, 1973).

Despite the evidence for unusual stability of the Sertoli cell junctions, it is obvious that the occluding Sertoli junctions cannot endure unchanged for longer than one cycle of the seminiferous epithelium, for then it becomes necessary for the next generation of spermatocytes to move from the basal to the adluminal compartment. A major gap in our understanding of the dynamics of the seminiferous epithelium is how this translocation takes place. It could involve transient dissolution of the Sertoli junctions, upward movement of the next generation of spermatocytes followed by reformation of the occluding junctions below them. Alternatively, there might be a progressive dissociation of the lines of membrane fusion above the germ cells and a concurrent formation of new rows of attachment between Sertoli cell processes interposed between the ascending germ cells and the basal lamina. The latter mechanism would provide for maintenance of the barrier at all times during the transition. There is, as yet, no adequate basis for choice between these alternatives, but experiments involving use of electron opaque probes of the extracellular space favor a gradual upward movement of the germ cells without interruption of the permeability barrier (Aoki and Fawcett 1975; Gilula, Fawcett, and Aoki 1976). The local control of this orderly process of modification of the Sertoli junctions presents a challenging unsolved problem, but it seems inescapable that the dissolution and formation, or opening and closing, of the junctions at exactly the appropriate stage of the cycle must somehow depend upon signals emanating from the germ cells when they reach a certain stage of their differentiation.

The significance of the blood-testis barrier for development of new methods of population control resides in the fact that potential antispermatogenic compounds must be able either to dissociate the Sertoli junctions or to achieve their effects by acting upon those cells that are outside of the barrier (Fawcett 1975a). It has been shown that a few antifertility agents given systemically do appear in the rete testis fluid (Waites and Setchell 1969), but it could not be ascertained from these studies whether the compounds traversed the epithelium by permeating the extracellular clefts of the epithelium or whether their path was transcellular. Morphological studies suggest that at least one compound, urea, which is not excluded by the barrier, enters the Sertoli cells and probably does not penetrate or disrupt the Sertoli junctions (Gilula, Fawcett, and Aoki 1976). The observation that a potent nonsteroidal antiandrogen causes regression of accessory glands without impairment of spermatogenesis (Neri et al. 1972) suggests that this compound is excluded by the barrier.

The observations of the past decade on the structural organization of the seminiferous epithelium have profoundly changed our thinking about the milieu in which spermatogenesis takes place and about the possible mechanisms involved in its regulation. Before 1960, explanations were sought almost solely in terms of the endocrine environment of the germ cells. Now one must also consider the topographical relations within the seminiferous epithelium, the short-range interactions between its cell types, and the membrane specializations for cell-to-cell communication and for isolation of developmental events by occlusion of extracellular pathways. Whereas formerly the Sertoli cells were assigned little more than a mechanical supportive role, the realization that they maintain the blood-testis barrier, form the walls of the adluminal compartment, and create a special fluid environment in which the germ cells differentiate has elevated these cells to a position of primary importance in spermatogenesis and has made them the subject of intensive biochemical investigation.

## Spermatogenesis

### Syncytial Nature of the Germ Cells and the Mechanism of Sperm Release

Among the unexpected early results of electron microscopy on the seminiferous epithelium was the observation that groups of developing germ cells are joined together by intercellular bridges that result from incomplete cytokinesis in the mitotic and meiotic divisions of spermatogensis. It was concluded that the synchrony of differentiation of the male germ cells is attributable, at least in part, to their syncytial nature (Fawcett, Ito, and Slautterback 1959; Fawcett 1961). Bridges were first observed connecting groups of spermatocytes and of spermatids, but subsequently they were also found joining chains of spermatogonia (Dym and Fawcett 1971; Huckins and Oakberg 1971; Gondos and Zemjanis 1970). It was difficult in these early studies to ascertain the total number of cell bodies that were joined in any given group because only a few bridges were included in a single thin section of tissue. But as methods improved, it became apparent that the size of the syncytia had been greatly underestimated (Dym and Fawcett 1971). It is now widely accepted that mitotic divisions of some of the primitive spermatogonia result in separate daughter cells that contribute to maintenance of the stem cell pool. In other stem cell divisions the daughter cells remain joined by a bridge; and in the period of continuing spermatogonial proliferation that follows, cytokinesis is always incomplete resulting in a lengthening chain of cells that remain interconnected by bridges (figure 5.7). The same applies to the subsequent meiotic divisions of the spermatocytes. From the number of divisions known to occur in the course of spermatogenesis, the estimated number of conjoined spermatids would be at least 512. For technical reasons, it has not been

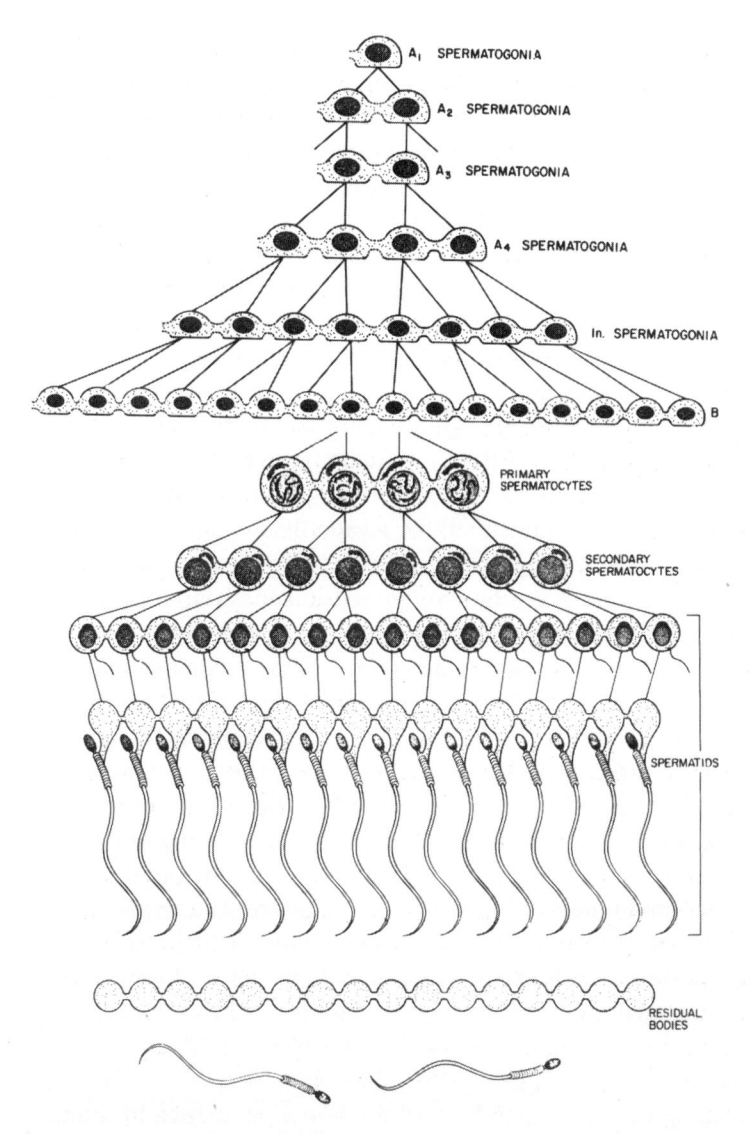

Figure 5.7 Schematic representation of the syncytial nature of the mammalian germ cells. Cytokinesis is incomplete in all but the earliest spermatogonial divisions, resulting in expanding clones of germ cells that remain joined by intercellular bridges.
Source: D. W. Fawcett in *The Developmental Biology of Reproduction*, eds. C. L. Marhert and J. Papanconstantinou. Academic Press, Inc., New York, 1975.

possible to ascertain whether this theoretical number is ever achieved. There is evidence from radiation experiments that if one member of a chain of spermatogonia is sublethally damaged, the entire group of connected cells may degenerate (Huckins and Oakberg 1971). This does not rule out the possibility that in the unirradiated testis there might be random degeneration of individual cellular units that would permit survival of the reminder of the chain. It is known that up to 25 percent of germ cells normally degenerate in the course of spermatogenesis. If degeneration in chains of cells is not necessarily an all-or-none phenomenon, random cell death would create gaps in the chains and reduce the upper limit of size of the syncytia. Recent studies on serial sections parallel to the base of the epithelium have shown at least 80 conjoined spermatocytes where 128 might have been predicted on theoretical grounds (Moens and Go 1971; Moens and Hugenholz 1975). Available evidence thus indicates that despite substantial losses from spontaneous degeneration, very large numbers of cells remain connected in syncytial clusters. The phenomenon of incomplete cytokinesis in a normal developmental process has few parallels in nature. In spermatogenesis its original interpretation as a device for maintaining synchrony of differentiation still seems valid, although there may be other consequences.

The syncytial nature of the germ cells also has important implications for the dynamics of cell movement within the epithelium and for the mechanism of sperm release. The ascent of preleptotene spermatocytes from the basal to the adluminal compartment obviously does not involve independent movement of single cells, but of long chains linked by intercellular bridges. Because there is little to suggest that the spermatocytes are actively motile, it seems likely that their upward displacement requires synchronous modification of the occluding junctions between a number of Sertoli cells, interposition of processes of the supporting cells between the preleptotene spermatocytes and the basal lamina, and coordinated active change in shape of the base of the Sertoli cells. This complex maneuver could hardly be accomplished without cell-to-cell communication among Sertoli cells to integrate their activities.

No less remarkable is the mechanism by which individual spermatozoa are finally separated from large syncytial groupings of spermatids at the time of sperm release (figure 5.8). The intercellular bridges that join the cell bodies together after the second meiotic division persist throughout differentiation of the spermatids. At the beginning of spermiation, a lobule of excess cytoplasm remains attached to the neck region of each of the hundreds of nascent spermatozoa. These lobules, representing the bulk of the cytoplasm of the spermatid cell bodies, are still joined together by the original intercellular bridges. As movement of the apex of each Sertoli cell extrudes each future spermatozoon into the lumen, the slender strand connecting the neck region to its lobule of residual cytoplasm becomes increasingly attenuated and ultimately breaks, releasing the spermatozoon and freeing it from the large group of conjoined residual bodies that are retained within the epithelium (Fawcett and Phillips 1969a; Burgos, Sacerdote, and Russo 1973). The germ cells seem to be relatively passive in a process of release that must involve complex motor activity on the part of the Sertoli cells. The very large numbers in any one cohort of synchronously differentiating spermatids makes it inevitable that the group extends far beyond the limits of any one supporting cell and probably spans the apices of many. Therefore, the synchronous release of the entire cohort almost certainly requires the coordinated activity of many Sertoli cells.

Electron microscopic studies in the past two decades have thus established that the cohorts of developing male germ cells are syncytial throughout spermatogenesis. These studies have also radically changed interpretations of the mechanism of sperm release. The complex process by which individual spermatozoa are separated from the germ cell syncytia now clearly emerges as a vulnerable step in spermiogenesis that might lend itself to interference. The accumulated morphological evidence that motor activity of the Sertoli cell is involved in spermiation as well as in the upward movement of the next generation of spermatocytes into the adluminal compartment focuses new attention upon the supporting cells as an appropriate target of efforts at fertility control. A pharmacological agent that would selectively impair Sertoli cell motility or interrupt communication and cooperative interaction among the supporting cells might very well result in infertility.

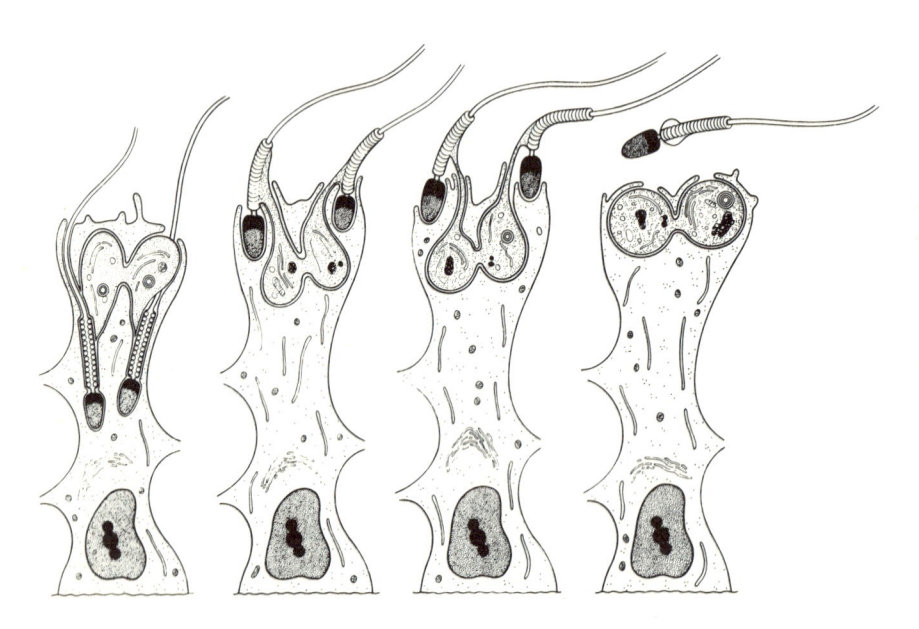

**Figure 5.8** Diagram of the stages of sperm release. The conjoined cell bodies of the advanced spermatids are retained in the epithelium while the nucleus, neck region, and tail are gradually extruded into the lumen. The narrow stalk connecting the neck region with the cell body becomes increasingly attenuated and finally gives way. Individual spermatozoa are thus separated from the syncytial cell bodies.
Source: D. W. Fawcett in *Regulation of Reproduction*, S. J. Segal et al., eds. Charles C. Thomas, Springfield, 1973.

## Stem Cell Renewal and the Spermatogenic Cycle

The pioneering studies of Clermont and Leblond in the 1950s described a set of histologically recognizable "cell associations" in the seminiferous epithelium that correspond to fourteen "stages" of spermatogenesis in the rat. Similar histological and autoradiographic analyses of the spermatogenic cycle were soon carried out on several laboratory and domestic animal species: ram (Ortovant 1956); guinea pig (Clermont 1960); rabbit (Swierstra and Foote 1963); bull (Hochereau 1963). The human testis proved to be more difficult to analyze for, unlike the other species, no single cell association extends around the entire circumference of the seminiferous tubule. Instead, a given cell association may occupy only a small sector of the epithelium and three or more different cell associations (stages of spermatogenesis) may be found in the same cross section of a tubule. Intermingling of the different cell associations at their boundaries further complicates the task of defining their limits. Nevertheless, it was possible to establish that the human seminiferous epithelium is organized on the same principle as in rodent species but with only six cell associations, which form a mosaic of irregular patches in the lining of the tubules (Clermont 1963). The length of the cycle was found to be sixteen days and the total duration of spermatogenesis was estimated to be sixty-four days. This important contribution dispelled the earlier erroneous impression that there was in man no cycle of the seminiferous epithelium comparable to that of rodents. The way was thus opened for more precise quantitative investigations of human spermatogenesis.

Spermatogenesis continues throughout the adult life of the male, so it is obvious that there must be continual renewal of the stem cells. The mechanism of this renewal has been a subject of controversy. It was originally proposed for the rat that stem cell divisions at the beginning of each cycle produced two kinds of daughter cells: *Type A* spermatogonia that were committed to enter spermatogenesis and develop into spermatozoa and spermatogonia that remained dormant until the subsequent cycle, when they, in turn, divided producing a new generation of committed Type A spermatogonia and uncommitted stem cells (Clermont and Leblond 1953). The subsequent recognition of several distinct categories of A spermatogonia ($A_1$, $A_2$, $A_3$, $A_4$) led to the proposal of alternative schemes for stem cell renewal. Introduction of a technique for studying whole mounts of intact seminiferous tubules permitted observations that led to a new hypothesis (Clermont and Bustos-Obregon 1968). A new category of *"reserve stem cells"* was postulated (spermatogonia $A_0$). These were believed to remain dormant for long periods unless called upon to enter the spermatogenic process as a result of destruction of the *renewing stem cells* (spermatogonia $A_1$ through $A_4$). The proliferation of these renewing spermatogonia was thought to result in large groups of some $A_4$ spermatogonia. The division of one of these latter was believed to produce two $A_1$ daughter cells, thus renewing the stem cell population, while the other seven $A_4$ cells divided to form *intermediate* spermatogonia, which in turn gave rise to *Type B* spermatogonia and subsequent phases of germ cell differentiation. This scheme did not gain wide acceptance because the observation that all but the most primitive spermatogonia are joined to others by intercellular bridges (Dym and Fawcett 1971) made it seem unlikely that one member of a syncytium of eight could behave differently from the others and, by its division, contribute two individual daughter cells to the stem cell pool.

This interpretation was soon succeeded by a new hypothesis of stem cell renewal (Huckins 1971) that continues to gain favor among students of spermatogenesis (figure 5.9). According to this scheme all spermatogonia are assigned to one of three categories: (1) stem cells (spermatogonia $A_s$), (2) two or more generations of proliferating spermatogonia designated $A_{pr}$ (paired) or $A_{al}$ (aligned) depending upon the number of interconnected cells, or (3) differentiating spermatogonia (types $A_1$, $A_2$, $A_3$, $A_4$, $A_{In}$, and B). The stem cells (spermatogonia $A_s$) are believed to occur singly and are randomly distributed throughout the length of the tubules. Their division either results in two separate $A_s$ daughter cells that renew the stem cell compartment or their cytokinesis may be incomplete, resulting in paired daughter cells joined by an intercellular bridge (spermatogonia $A_{pr}$). These latter begin a series of synchronous divisions resulting in chains of interconnected cells of increasing length (spermatogonia $A_{al}$). The

# The Male Reproductive System

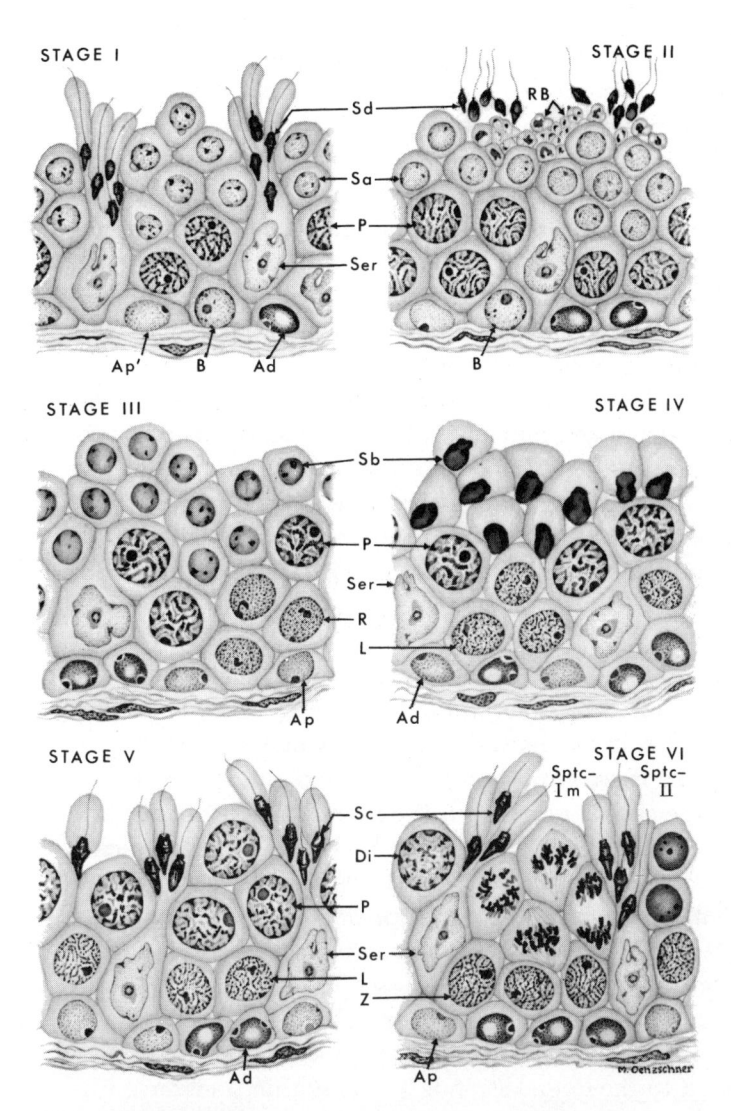

**Figure 5.9** Diagram of the six recognizable cell associations or stages of the cycle of the human seminiferous epithelium. Ser, Sertoli cell; $Ad$ and $Ap$, dark and pale type A spermatogonia; $B$, type B spermatogonia; $R$, resting primary spermatocyte; $L$, leptotene spermatocyte; $Z$, zygotene spermatocyte; $P$, pachytene spermatocyte; $Di$, diplotene spermatocyte; $Sptc$-$Im$, primary spermatocyte in division; $Sptc$-$II$, secondary spermatocyte in interphase; $Sa$, $Sb$, $Sc$, $Sd$, spermatids in various stages of differentiation; $RB$, residual bodies of Regaud.
Source: Y. Clermont, *Am. J. Anat.* 112:35, 1963.

proliferative activity of these cells results in a ninefold increase in the population of undifferentiated spermatogonia (Huckins 1971). The chains of aligned spermatogonia are then transformed into $A_1$ spermatogonia, which proceed to differentiate and undergo synchronous divisions at particular stages of the spermatogenic cycle, giving rise to $A_2$, $A_3$, $A_4$, Intermediate, and B spermatogonia. This scheme of stem cell renewal and spermatogonial proliferation in the rat appears to be in good agreement with ultrastructural studies on the clonal nature of the spermatogonia. Stem cell renewal in the human testis has yet to be worked out in comparable detail. It may be considerably simpler than in the rat, but it seems likely that the same principles apply. We remain ignorant of the biochemical events that take place in the successive generations of so-called "differentiating" spermatogonia. And the factors that determine when a stem cell will divide and whether the progeny will renew the stem cell pool or take the path of differentiation remain important gaps in our knowledge.

### Meiosis and Synthetic Activities during Spermatogenesis

Much remains to be learned about *meiosis*, the special type of cell division that occurs in the germ cells of all sexually reproducing organisms. Two of the most important events in this process are the linear pairing of the chromosomes and the interchange of segments between the homologous chromatids. A landmark morphological advance in our understanding of meiosis was the discovery of a unique ribbonlike structure, the *synaptonemal complex*, running longitudinally along the interface between paired homologous chromosomes in primary spermatocytes (Moses 1956, 1968; Fawcett 1956). It consists of a pair of dense lateral strands about 60 nm in thickness and approximately 100 nm apart and a slender central element connected to the lateral strands by many thin transverse filaments. This complex has been found in the meiotic chromosomes of all higher plants and animals in which chromosomal synapsis and crossing-over occur and is credited with providing point-for-point pairing between the homologous chromosomes (Stubblefield 1973). A recent technical advance permits display of the full length of the synaptonemal complexes of all the chromosomes of germ cells and promises to yield valuable new information on chromosomal behavior in meiosis (Moses, Counce, and Paulsen 1975).

A prominent DNA containing body at the periphery of the nucleus in pachytene spermatocytes, traditionally called the "sex vesicle," has been found to be a single heterochromatic mass containing the sex chromosomes. It is now called the *XY body*. RNA can be demonstrated in the nucleolus, which is often associated with its surface but not in the XY body itself (Solari and Tres 1967). Tritiated uridine is not incorporated into the nucleolus or the XY body, suggesting that synthesis of ribosomal RNA is suppressed during meiotic prophase (Monesi 1965). In electron micrographs two linear densities can be identified within the XY body, a shorter one corresponding to the axis of the Y chromosome and a longer one corresponding to that of the X chromosome. The two come together at one end to form a short synaptonemal complex that terminates on the nuclear envelope (Solari and Tres 1970).

Autoradiographic studies have yielded interesting and important information on RNA and protein synthesis by the germ cells (Monesi 1962, 1965; Kierszenbaum 1974, Kierszenbaum and Tres 1974). After a relatively low rate of RNA synthesis in early prophase of meiosis, it rises to a peak in midpachytene only to fall again to very low levels in late pachytene and diakinesis. The RNA molecules synthesized in meiosis remain associated with the chromosomes for an extraordinarily long time and are then released into the cytoplasm at diakinesis, contributing the bulk of the RNA present in the spermatid. Genetic inactivation occurs relatively early in spermatid differentiation and there is essentially no RNA synthesis thereafter. This means that the various protein structural constituents of the spermatozoon formed during the two-week period of spermiogenesis must either be synthesized from very long-lived messengers formed during meiosis or they are simply assembled from pools of macromolecular subunits synthesized days or weeks earlier than their appearance as formed elements in the spermatids. This appears to be a unique feature of spermiogenesis; in most differentiating somatic cell systems the nuclei do not become inactive and therefore transcription to produce new messenger, transfer, and ribosomal RNA goes on concurrently with cytoplasmic synthesis and assembly of the products.

A more detailed understanding of the molecular events and their genetic control during meiosis and spermiogenesis will require development of methods for the segregation of single cell types from the heterogeneous population that comprises the seminiferous epithelium. Only when this is accomplished will it be possible to study the biochemistry of spermatogenesis. Encouraging initial steps toward this objective have been reported in studies on the transformation of histones to basic proteins of higher arginine content, which takes place late in spermiogenesis concurrently with condensation of the spermatid nucleus. Basic proteins were isolated from highly enriched fractions of round spermatids and residual bodies obtained by velocity sedimentation of cell suspensions of mouse, guinea pig, and rabbit testes. Round spermatids and residual bodies contained proteins that comigrated on gels with calf thymus histone. On the other hand, sperm chromatin contained a single basic protein with a high arginine and cysteine content. The replacement of lysine-rich histones by arginine-rich histones previously inferred from autoradiographic studies was thus substantiated by biochemical analyses on the isolated cell types (Bellvé and Romrell 1974; Bhatnagar, Bellvé, and Romrell, 1976). Major interspecific differences in the amino acid composition of this protein are found (Bellvé, Anderson, and Hanley-Bowdoin 1976).

### Morphogenesis of Spermatozoa

Ultrastructural studies of spermiogenesis since 1960 have resulted in a voluminous literature describing in considerable detail the morphogenesis of spermatozoa in many mammalian species. To review here that rich harvest of new structural information would serve no useful purpose. A few of the salient features will suffice.

The earlier ambiguities concerning the origin of the *acrosomal cap* were resolved by the demonstration that it arises in the Golgi apparatus of the spermatid in a manner comparable, in nearly all respects, to the development of a secretory granule in a glandular cell (Burgos and Fawcett 1966; de Kretser 1969; Ploen 1973). These observations clarified the function of the Golgi apparatus and contributed importantly to the current concept of the acrosome as a secretory organelle that releases its enzyme-rich content in response to specific stimuli in the vicinity of tubal ova.

The *manchette*, described by classical cytologists as a cylindrical, sleevelike structure extending caudally from the nucleus of elongating spermatids, was found to be composed of large numbers of microtubules in parallel array. Indeed, the first description of cytoplasmic microtubules and the first hint that they play a role in the determination of cell shape came from early electromicroscopic studies of spermiogenesis (Burgos and Fawcett 1955). The microtubules of the manchette polymerize adjacent to the caudal pole of the spermatid nucleus and subsequently become aggregated to form a hollow cylinder attached to a perinuclear differentiation of the cell membrane. The elongation of the spermatid is attributed to flow of cytoplasm back along the manchette. This brings the cell membrane at the anterior part of the cell into close apposition with the acrosomal membrane and transports the bulk of the cytoplasm into lobules surrounding the proximal portion of the flagellum. The manchette is one of several transient organelles that develops in the course of spermiogenesis and then disappears without leaving any residue in the mature spermatozoon.

The process of condensation of chromatin and shaping of the nucleus in the course of spermiogenesis has been investigated by both morphological and biochemical methods. Early in the process, slender filaments of DNA and protein aggregate into coarser strands that become progressively thicker and more dense. The ultrastructure of the chromatin in later stages is difficult to study in thin sections because of the density and close packing of its subunits. But freeze-cleaving studies have revealed a lamellar structure of the condensed chromatin with the layers parallel to the broad surfaces of the flattened nucleus. Associated with these morphological changes is a remarkable transformation in the basic proteins of the nucleus. The lysine-rich histones typical of somatic cell nuclei disappear from the nucleus of late spermatids and accumulate in the residual cytoplasm. At the same time, arginine-rich proteins are synthesized and become associated with the condensing chromatin. Thus, late in spermiogenesis spermatid nucleohistones are transferred to the residual bodies and are replaced by protaminelike proteins of low molecular weight that exhibit species specificity (Bellvé and Romrell 1974). It has been suggested that the shaping of the sperm nucleus may result from a characteristic pattern of aggregation of the complexes of DNA and specific basic proteins (Fawcett, Anderson, and Phillips 1971).

Other morphological studies have defined the structure and origin of the *connecting piece* that joins the condensed nucleus to the outer dense fibers of the flagellum. Its cross-striated fibers appear to arise by self-assembly under the inductive influence of the centrioles at the base of the flagellum (Fawcett and Phillips 1970). The origin of the axoneme by assembly of nine doublet microtubules on the template provided by the nine triplet microtubules in the wall of the centriole has been described but is still poorly understood. The development of the *outer dense fibers* of the flagellum by accretion to a ridgelike radial appendage on the nine doublets of the axoneme has been followed in electron micrographs (Fawcett and Phillips 1970).

Developmental biologists are familiar with inductive phenomena in organogenesis wherein extracellular matrices or diffusable products of certain cells initiate or regulate the differentiation of other cells. At the present time we know almost nothing about inductive phenomena in organellogenesis at the subcellular level. How do centrioles induce assembly of the connecting piece? What initiates polymerization of tubulin to form the axoneme? What terminates the process and limits tail length? What triggers the aggregation of mitochondria around the base of the flagellum to form the mitochondrial sheath of the middle piece? What determines how many of the total complement of spermatid mitochondria will be used in construction of a midpiece that is remarkably constant in length for any given species? These are but a few of the challenging fundamental questions that should fire the imagination of cell biologists and reproductive biologists in the coming decade.

In the period covered by this report, the descriptive phase of the analysis of sperm development has been largely completed. We know the principal morphogenetic events in transformation of spermatids into spermatozoa and the temporal sequence in which they occur.

But our ignorance of the chemical processes and of the control mechanisms involved in initiation, regulation, and termination of each of the developmental events is profound. We envisage the formation of cellular structures from cytoplasmic pools of macromolecules by self-assembly, assembly by end-addition, assembly by accretion to a preformed template; but these are empty phrases coined to identify vague concepts that have yet to be given meaning by more precise definition and by searching experimental analysis. These unanswered questions are not "trivial" or "academic" in the context of contraception, for obviously interference with any one of the long train of interdependent sequential events of differentiation leading to the spermatozoon would result in infertility. But clearly we must understand the molecular mechanisms before we can hope to devise specific methods for their suppression.

## Contraception by Agents Acting upon the Multiplication and Differentiation of the Germ Cells

In seeking antifertility agents affecting spermatogenesis, certain principles must be borne in mind. The process of spermatogenesis depends upon continual renewal by mitotic division of a population of spermatogonia. Any agent that would destroy these stem cells would result in irreversible infertility, which is not the goal of contraceptive research. After the series of spermatogonial divisions leading to formation of primary spermatocytes, these enter meiosis. During the long prophase of the first meiotic division, lasting for several days, reduplication of the genetic material of the germ cells occurs and a complex pairing and interchange of genetic material takes place between chromosomes. This is a crucial period from a genetic point of view and one in which the chromosomes are especially vulnerable to damage. Therefore one would like to avoid drugs that might affect this phase of spermatogenesis because of the possibility of mutation or other alteration of the genetic material that might result in abnormal offspring. The exclusion of spermatogonia and spermatocytes as targets for drug action leaves as relatively safe candidates for pharmacological interruption only those later stages of spermatogenesis when the spermatids are being transformed into spermatozoa. An acceptable antispermatogenic drug should therefore have a high degree of selectivity for the later stages of spermatogenesis and should certainly not be mutagenic or destructive of the stem cells. No drug currently available adequately fulfills these criteria. However, several that have shown some initial promise have been extensively studied in animals and a few have had human trials. These drugs usually have not been developed primarily as contraceptives but were products of a quest for cancer chemotherapeutic agents, antibiotics, or amoebocides. In the course of their routine animal testing, scientists noted their antispermatogenic effects and their potential as antifertility agents was then pursued as a secondary objective.

A variety of compounds that act like ionizing radiation (radiomimetic drugs) are used in the treatment of cancer. These drugs include nitrogen mustards, ethyleneimine derivatives, and mono- and diesters of methanesulphonic acid. These compounds in low doses also interrupt spermatogenesis by destroying spermatogonia or arresting their division (Jackson 1959, 1970). At higher doses they appear also to affect spermatocytes and spermatids. They are cumulative in their effects, highly toxic to bone marrow, potentially mutagenic and therefore hold little promise as antifertility agents.

Somewhat more promising is a group of heterocyclic compounds—*nitrofurans*, *thiophenes*, *bis(dichloroacetyl) diamines*, and *dinitropyrroles*. These have been shown to inhibit spermatogenesis in animals with no evidence of mutagenesis and with complete restoration of fertility after withdrawal of the drug. The nitrofurans and thiophenes at doses that suppress spermatogenesis in man were found to have undesirable toxic side effects and have not been pursued further (Nelson and Bunge 1957). The dinitropyrroles (ORF 1616) are highly effective in rats, inhibiting spermatogenesis at the pachytene stage of meiotic prophase. Infertility can be maintained in rats by administering the compound once every four weeks (Patanelli and Nelson 1964). Toxic effects during testing in dogs prevented the compound from coming to clinical trials. Nitroimidazoles are reported to be highly effective in stopping spermatogenesis in experimental animals, acting at the primary spermatocyte stage. The tubules are virtually depopulated of germ cells with prolonged treatment. The mechanism of action of the compound is not known, but the evidence suggests an interference with the protein biosynthetic pathways of the testis (Patanelli 1975). This class of compounds may deserve further study.

An interesting group of compounds are the diamines, originally synthesized in a search for effective amoebocides. Oral antispermatogenic activity was demonstrated in rats, monkeys, and dogs with complete recovery after cessation of the drugs (Beyler et al. 1961; Drobeck and Coulston 1962; Nelson and Patanelli 1965). Administered to human volunteers, $N, N^1$-bis(dichloroacetyl)-1,8-octane diamine (WIN 18,446) was well tolerated for extended periods and resulted in complete suppression of sperm production. Fertility returned upon discontinuation of treatment (Heller, Flageolle, and Matron 1963). Histological examination of testicular biopsies suggested that the principal effects were on the spermatids, although there was some damage to spermatocytes as well. The high hopes raised by these initial favorable results suffered a discouraging setback when it was discovered that the drug is incompatible with alcohol, possessing a distressing Antibuse-like effect. This has prevented its further trial as an agent for fertility control in men.

Because the diamines fulfill some of the requirements for an acceptable antispermatogenic agent, it seemed worthwhile to carry out electron microscopic studies of their effects upon the seminiferous epithelium of animals in order to learn more about their site of action at the subcellular level and to assess the degree of their selectivity for the relatively safe later stages of spermiogenesis (Flores and Fawcett 1972; Fawcett 1975a). When administered to guinea pigs in relatively small doses over a two-week period, the drug permits quite normal development of spermatids up to the cap phase of acrosomal development. However, the subsequent complex morphogenetic forces within the acrosome and nucleus that normally result in acquisition of the definitive shape of the sperm head seem to be especially vulnerable to the effects of the drug. There is a remarkable and highly variable distortion of the acrosome and the nucleus tends to roll up around its long axis. Many of the bizarre, misshapen sperm that result are not released but are retained within the epithelium and undergo degeneration there. Those that are released are apparently incapable of fertilization. The Sertoli cells, although not seriously damaged, exhibit considerable vacuolization and it is not clear whether the sperm retention is simply due to the abnormal shape of the spermatozoa or whether it may also be a consequence of impaired Sertoli cell function.

The highly characteristic pathology of the late spermatids observed after treatment of guinea pigs with the diamines is not specific for this drug. Almost identical distortions occur after brief cryptorchidism (Ploen 1973) or after active immunization with testis antigens (Nagano and Okumura 1973). The deleterious effects of all three conditions are not confined to the shaping of the sperm head but seem also to involve the Sertoli cells, interfering with germ cell escalation and sperm release. It is evident from these and other studies that the Sertoli cells are not as resistant as they were formerly thought to be, but indeed they are sensitive to a variety of chemical and physical agents. The recent increase in our awareness of their functional importance and their responses to injury suggests that drugs may yet be found that will indirectly inhibit spermatogenesis or prevent sperm release by a selective action upon supporting cells. Such an action would allay anxiety over possible genetic effects of antispermatogenic drugs directly affecting the germ cells.

The rational development of antispermatogenic agents will require a prior knowledge of biochemical processes in germ cell differentiation that might be selectively inhibited. The acquisition of such knowledge has been largely prevented by our inability to attribute biochemical events to specific cell types in the highly heterogeneous cell population of the testis. Only recently has there been significant progress in separation of the components of the testis. It was shown some ten years ago that, with patience, the tubules can be separated manually from the interstitial tissue in certain rodents (Christensen and Mason 1965). This made it possible to differentiate the steroidogenic functions of these two major compartments of the organ. Within the past five years a method originally devised for separation of the cell types of the blood or bone marrow (Miller and Phillips 1969) has been modified and applied to the testis (Lam, Furrer, and Bruce 1970; Meistrich 1972). This procedure involves dissociating the seminiferous epithelium after enzymatic digestion of extracellular components and layering the resulting mixed suspension of cells onto a density gradient of bovine serum albumen. During sedimentation at unit gravity, cells distribute themselves in distinct layers according to differences in their size. Early efforts to segregate the cell types of the seminiferous epithelium by this method were frustrated by the syncytial nature of the germ cells. Syncytia were fragmented into uninucleate, binucleate, or multinucleate units whose varying size resulted in the same germ cell type distributing itself in different fractions. The preparation of the initial cell suspension proved to be a critical step in avoiding this. As improved dissociation procedures have been developed, this problem has been minimized and it is now possible to separate fractions of residual bodies, round spermatids, and pachytene spermatocytes in 85 to 90 percent purity (Romrell, Bellvé, and Fawcett 1976). Ingenious protocols have also been devised for obtaining cultures of myoid and Sertoli cells by taking advantage of their differing size and differing capacity to adhere to a plastic substrate in vitro (Steinberger et al. 1975). Such cultures are now being used to study the synthetic capacities of these cell types and their response to hormones in vitro. Biochemical studies of the testis have thus progressed in the past decade from analysis of the whole organ to study of its segregated endocrine and exocrine components to investigations of single cell types. These methodological advances promise at last to make the complex process of germ cell differentiation accessible to biochemical analysis at the cellular and subcellular levels. When such analyses have been achieved, the rational development of safe antispermatogenic agents may become an attainable goal.

## Structure and Function of the Peritubular and Interstitial Tissue of the Testis

### The Lamina Propria or Boundary Tissue of the Seminiferous Tubules

Interest in the peritubular tissue of the testis resides in the belief that it may influence access of metabolities to the seminiferous epithelium and that it generates the shallow peristaltic movements that help to transport spermatozoa toward the rete testis and epididymis. In the testis of oligospermic and azoospermic men, the peritubular tissue is also the site of the pathological changes resulting in "hyalinization" of the tubules.

Electron microscopic studies since 1960 have considerably modified interpretations of the organization of the boundary tissue and have documented significant differences among mammalian species. The peritubular tissue was historically considered comparable to the connective tissue lamina propria of mucous membranes, and the peristaltic movements of the tubules were attributed to contractility of the Sertoli cells. Electron microscopy has now revealed in the rodent testis a single continuous layer of peritubular epithelioid cells possessing abundant cytoplasmic filaments and other cytological characteristics of smooth muscle (Clermont 1958; Ross 1967; Ross and Long 1966; Fawcett, Leak, and Heidger 1970). The view that these cells, and not the Sertoli cells, are responsible for tubule contractions rapidly gained acceptance. Because the polygonal, squamous form of these cells and their epithelioid organization were incompatible with their designation as "smooth muscle," the term *myoid cell* was suggested. Definitive evidence for their contractility has recently come from identification of their cytoplasmic filaments as actin by their specific binding of heavy meromyosin (Toyama 1975). The myoid layer does not seem to be innervated but has an inherent rhymicity that is affected by oxytocin (Niemi and Kormano 1965).

The single layer of myoid cells around the seminiferous tubules of rodents is bounded on its outer surface by a sheet of extremely attenuated cells comprising the parietal endothelium of peritubular lymphatic sinusoids. In larger species, the lymphatics do not form peritubular sinusoids and the boundary tissue consists of multiple cell layers. The cells nearest the epithelium have all the characteristics of contractile cells, but in the more peripheral layers there is a progressive diminution of filaments as the outermost cells come to resemble fibroblasts. The significance of this gradient in the degree of differentiation for contractility deserves further study (Fawcett 1973). An intact pituitary-testicular endocrine system is required for normal differentiation of the myoid cells (Bressler and Ross 1972;

Hovatta 1972). In the testes of infertile men there is often a marked increase in the abundance of collagen fibers between the cellular layers of the boundary tissue and an accumulation of material resembling that of the basal lamina (Bustos-Obregon and Holstein 1973; de Kretser, Kerr, and Paulsen 1975). It would be of considerable clinical interest to determine whether the pathology of the seminiferous epithelium in these azoospermic men is primary or secondary to disturbed function of the cells in the thickened peritubular tissue.

**The Organization of the Interstitial Tissue**
In histological sections of rodent testis the tubules are surrounded by a clear zone. Before 1960 this was usually interpreted as a shrinkage artifact, but electron micrographs of testes fixed by vascular perfusion have now clearly shown that the seminiferous tubules are surrounded by a labyrinthine system of peritubular lymphatic sinusoids. These are lined by a thin layer of endothelium applied, on one side, to the myoid layer of the tubules and, on the other, to clusters of Leydig cells disposed along the walls of blood vessels that are centrally situated in the intertubular spaces (Fawcett, Leak, and Heidger 1970). The Leydig cells in these species are thus interposed between the blood vessels on one side and the peritubular lymphatic sinusoids on the other and doubtless release androgen into both. The lumen of the sinusoids provides a plane of cleavage that makes it possible to tease out the tubules of rodent testes for biochemical study (Christensen and Mason 1965). Such a separation is not possible in larger species, including man, where centrally placed lymphatic vessels (figure 5.10) replace the peritubular sinusoids (Fawcett, Leak, and Heidger 1970; Fawcett, Neaves, and Flores 1973). In the absence of this natural plane of cleavage, seminiferous tubules cannot be separated without extensive Leydig cell contamination.

The main thrust of ultrastructural studies on the organization of the interstitial tissue has been to direct more attention to the dependence of the spermatogenic function of the testis upon the close topographical relationship of the Leydig cells to the seminiferous tubules. The association of Leydig cells with blood vessels and the level of blood-borne androgen are probably of much less importance

206  Major Advances and New
Opportunities

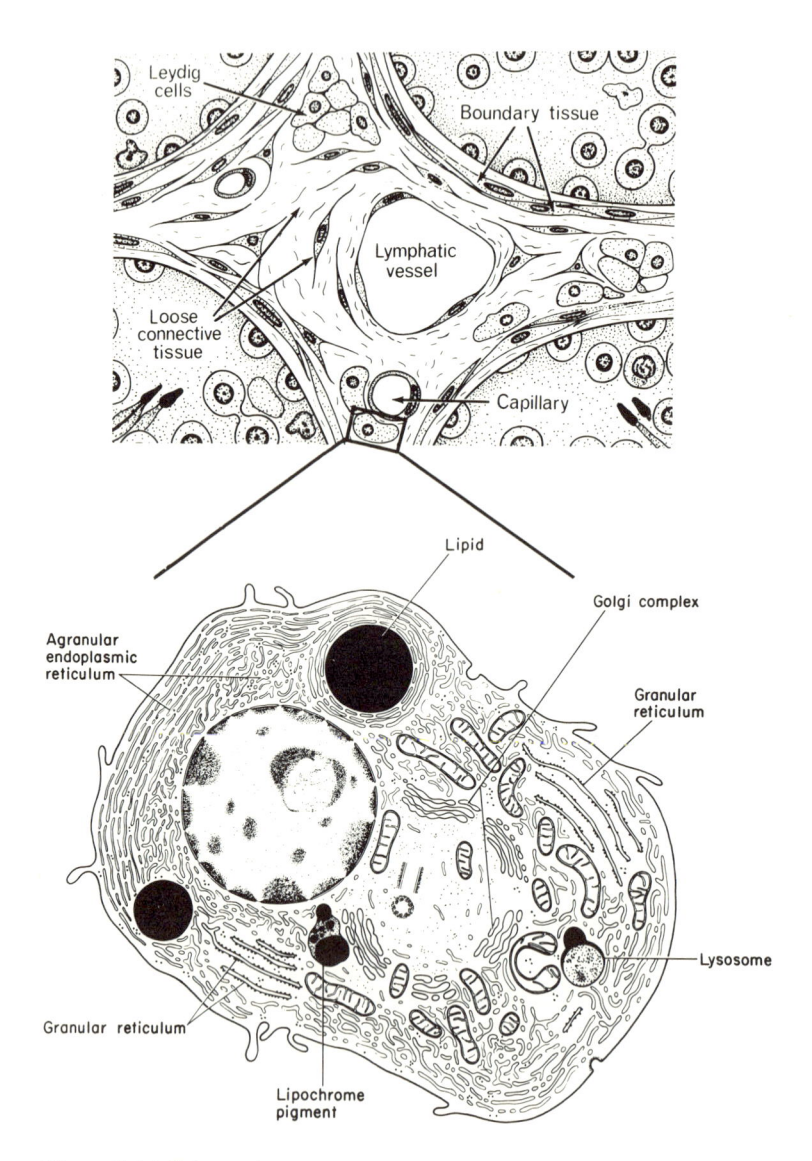

Figure 5.10 Schematic representation (above) of the arrangement of the blood and lymph vessels and cellular and extracellular components of an intertubular space in the human testis.
Source: Fawcett, *Schering Workshop on Contraception, the Masculine Gender*, Advances in Biosciences 10, Pergamon Press-Vïeweg, London, 1973.

Shown below is a diagram of the structure of a typical Leydig cell as seen with the electron microscope.
Source: Fawcett et al. *Advances in Hormone Research* 25, 1969. Academic Press, New York.

for spermatogenesis than is the accumulation of androgen in relatively high concentration in the peritubular lymphatics or in the fluid of the edematous loose connective tissue of the interstitial tissue.

The development of a technique for cannulating the rete testis of experimental animals (Voglmayr, Waites, and Setchell 1966) has given us new insight into the fluid environment of spermatogenesis and the vehicle of sperm transport. The concentration of androgen in the rete testis fluid in some species was found to be nearly as high as that in the spermatic vein (Cooper and Waites 1974) and as much as fifteen times the concentration in peripheral blood (Harris 1973). These findings have served to emphasize the intratesticular function of androgen. Investigative attention in the past was directed mainly to the role of androgen acting at a distance to maintain the accessory glands and other extratesticular manifestations of maleness, but now there is a greater awareness that the primary function of the Leydig cells is to maintain in and around the tubules the high local concentration of androgen that is required to sustain spermatogenesis. In this context, the close proximity of the Leydig cells to the germinal epithelium and their relationship to the blood and lymph vascular systems take on new interest and importance.

## The Ultrastructure of Leydig Cells and the Biosynthesis of Androgens

Although the Leydig cells have been considered the probable source of testicular androgens since the beginning of this century (Bouin and Ancel 1903), the seminiferous tubules could not be excluded as possible contributors. Biochemical investigation of the biosynthesis of testicular androgens was long hampered by the small partial volume of Leydig cells and the heterogeneity of the cell types contributing to centrifugal fractions obtained from homogenates of whole testes. Much of the recent progress in this area has resulted from the discovery that in the rat, interstitial tissue can be separated from tubules for in vitro experiments on steroid metabolism (Christensen and Mason 1965) and from the finding that transplantable Leydig cell tumors exhibit many of the physiological properties of normal Leydig cells, including testosterone secretion and responsiveness to luteinizing hormone (Moyle and Armstrong 1970).

In the past decade, the attribution of androgen synthesis to the Leydig cells has been greatly strengthened by several different lines of investigation. A histochemical method for detection of an essential enzyme in the biosynthesis of testosterone (3β-hydroxysteroid dehydrogenase) localized the enzyme in the Leydig cells (Baillie, Ferguson and Hart 1966). An interstitial tissue fraction of rat testis was shown to be far more efficient than the tubule fraction in the conversion of exogenous progesterone to testosterone (Christensen and Mason 1965). The interstitial tissue also synthesized testosterone from endogenous precursors in vitro whereas the tubules do not (van der Molen et al. 1973). Thus it has been concluded that if the tubules make any contribution to androgen synthesis, it is quantitatively insignificant compared to that of the Leydig cells.

Descriptive studies of the fine structure of the Leydig cells have defined a set of cytological features characteristic of steroid-secreting cells in general and have provided the morphological basis for relating specific cytoplasmic organelles to particular steps in the biosynthesis of steroids (Christensen and Fawcett 1961; Fawcett, Long, and Jones 1969). The most conspicuous cytoplasmic organelle of the Leydig cell is the smooth endoplasmic reticulum—an extensive network of membrane-limited tubules and fenestrated cisternae that extends throughout the cell body (figure 5.10). In some areas these tubular elements are continuous with parallel aggregations of cisternae bearing ribosomes on their surface. These limited areas of rough endoplasmic reticulum and their associated ribosomes are concerned with the protein synthesis required for the maintenance and renewal of cytoplasmic constituents. The much more extensive smooth reticulum, on the other hand, contains most of the enzymes necessary for the synthesis of cholesterol and its conversion to androgenic steroid hormones. When Leydig cells are homogenized for biochemical analysis, a very large proportion of the microsome fraction is derived from fragmentation of the smooth reticulum, whereas the contribution of the rough reticulum is relatively small (Christensen and Gillim 1969; Aoki and Massa 1976).

Cholesterol is believed to be the precursor of all testicular androgens (Hall 1970) and it is synthesized within the testis from acetate

(Morris and Chaikoff 1959; Gerson, Shortland, and Dunckley 1964). Most of the enzymatic steps in its synthesis are found in the microsome fraction. Cholesterol destined for conversion to androgenic steroids may be synthesized in advance and stored as cholesterol esters in the lipid droplets commonly found in Leydig cell cytoplasm. Synthesis of testosterone first requires retrieval of cholesterol by hydrolysis of the ester. This important step is stimulated by luteinizing hormone of the pituitary gland (Moyle 1970). Administration of LH results in a dramatic reduction of lipid droplets in electron micrographs of Leydig cells and a significant depletion of cholesterol ester as determined by biochemical analysis (Pokel, Moyle, and Greep 1972; Aoki 1970; Merkow et al. 1968).

The mechanism by which gonadotropin stimulates testosterone secretion has been clarified by autoradiographic and biochemical studies. It has been shown that radioiodine-labeled LH binds to the membrane of Leydig cells (Mougdal, MacDonald, and Greep 1971; Desjardin 1975) where it activates adenyl cyclase, which in turn causes intracellular accumulation of cyclic adenosine monophosphate (CAMP) (Murad, Strauch, and Vaughn 1969) and release of active kinase. These intracellular conditions trigger hydrolysis of cholesterol esters, increasing available precursor and stimulating steroidogenesis by accelerating the rate limiting first step in the biosynthetic processing of the cholesterol (Hall and Eik-Nes 1964).

The first step in the conversion of cholesterol to steroid hormones involves cleavage of its side chain in a series of enzymatic steps yielding pregnenolone (figure 5.11). Side chain cleavage takes place in the mitochondria and particularly in the inner mitochondrial membrane (Toren et al. 1964; Moyle 1970). The Leydig cell mitochondria are unusual in their internal structure in that the inner membrane forms tubular or vesicular invaginations into the matrix instead of the foliate cristae commonly found in other cell types. Whether this unusual inner membrane configuration is specifically related to the capacity of these mitochondria for cholesterol side chain cleavage is not clear; but LH, which influences this important rate limiting step in the biosynthesis of testosterone, also induces subtle changes in the internal structure of the mitochondria (de Kretser 1967; Aoki 1970; Neaves 1975).

210  Major Advances and New
Opportunities

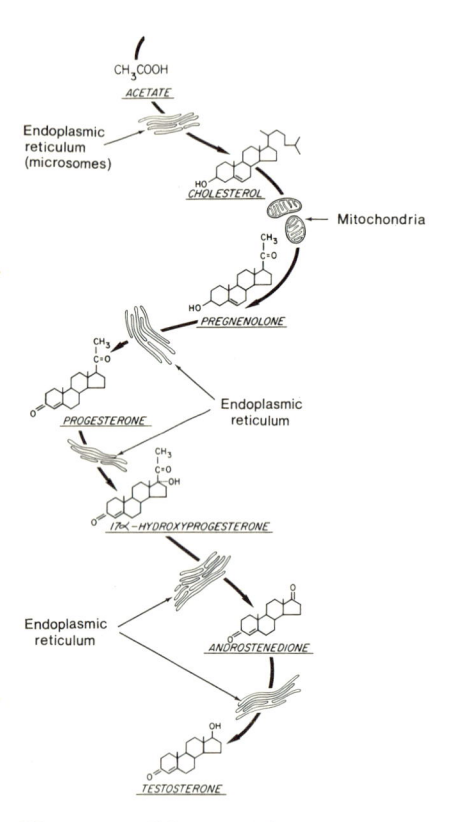

Figure 5.11 Scheme of the successive steps in the synthesis of testosterone from acetate, with indication of the cytoplasmic organelle of the Leydig cell in which each step is believed to take place.
Source: After M. Dym in *Histology*, ed. R. O. Greep and L. Weiss, 4th Ed. McGraw-Hill.

The pregnenolone resulting from cholesterol side chain cleavage is converted to progesterone by 3β-hydroxysteroid dehydrogenase, which is then reduced to 17α-hydroxyprogesterone (by 17α-hydroxylase) and this in turn is converted to androstenedione (by $C_{17}$-$C_{20}$ lyase). Reduction of androstenedione (by 17β-hydroxysteroid dehydrogenase) yields testosterone. All the enzymes involved in conversion of pregnenolone to testosterone are located in the endoplasmic reticulum of the Leydig cells and hence in the microsome fraction of homogenates (Murota et al. 1966; Tamaoki et al. 1969). The important role of the smooth reticulum in testosterone synthesis is supported by the observation that in patients whose testicular microsome fractions exhibit decreased conversion of progesterone to testosterone the Leydig cells exhibit poorly developed endoplasmic reticulum (Murota, Shikita, and Tamaoki 1966). Chronic stimulation of the Leydig cells of immature animals by chorionic gonadotropin increases the activities of enzymes in the pathway from pregnenolone to testosterone and also markedly increases the quantity of smooth endoplasmic reticulum seen in electron micrographs (Shikita and Hall 1967; Aoki 1970).

Correlation of structure and function for the other organelles and inclusions of steroid secreting cells is less satisfactory. They have a prominent, centrally situated Golgi complex that responds to tropic hormone stimulation by an increase in size; but at present the functional significance of this response is not clear (Long and Jones 1967; Fawcett, Long, and Jones 1969). In other glandular cells this organelle has an important function in segregating and concentrating the product in secretory droplets or granules. In steroid secreting cells, however, the product is not stored in visible granules or vesicles and the function of the Golgi apparatus of the Leydig cells and the mechanisms of release of their product represent significant gaps in our knowledge.

Thus, in summary, it can be said that the past decade and a half has seen the accumulation of incontestable evidence that the Leydig cells of the testis are the main source of androgen in the male and that luteinizing hormone is the principal regulator of their function. There have been rapid and important advances in the correlation of the structure of Leydig cells with their function. The biosynthetic pathway of androgenic steroids has been worked out in considerable detail. The effects of gonadotropin on the chemical reactions controlling availability of precursor and on rate limiting steps in androgen synthesis have been identified. We know, in general, where the principal enzymatic steps take place in the cytoplasmic organelles; and it has become increasingly evident that fertility requires a concentration of testosterone within the testis that is very much higher than the blood levels that are adequate for maintenance of the accessory glands, libido, and other manifestations of maleness. All of this information is clearly essential for the rational development of contraceptive strategies that depend upon reducing testosterone synthesis below the level required for maintenance of spermatogenesis.

## Control of Spermatogenesis and Biochemistry of the Seminiferous Tubules

It is generally accepted that spermatogenesis is dependent upon LH stimulation of the Leydig cells to maintain a high local concentration of testosterone around the tubules, but the mode of action of androgen on spermatogenesis remains obscure. It has been demonstrated that testicular tissue can form dihydrotestosterone (DHT) from testosterone (Folman, Haltmeyer, and Eik-Nes 1972; Rivarola, Podesta, and Chemes 1972) and it is possible that DHT is the androgen that is active in the tubules, as has been shown to be the case for other target tissues (Wilson 1972). Whether the active androgen acts upon the Sertoli cells or directly on the germ cells has yet to be determined.

### Effects of FSH on the Seminiferous Tubules
Nearly forty years have passed since it was first proposed by Greep and Fevold (1937) and Greep, Van Dyke, and Chow (1942) that in the male, luteinizing hormone acts upon the Leydig cells and that follicle stimulating hormone affects the seminiferous epithelium. This postulate has withstood the test of time, but the supporting evidence for its two parts has been unequal. The role of LH in the control of Leydig cell function has been firmly established, but the

role of FSH in the tubules has remained controversial. For many years this was attributable to the fact that preparations of FSH free of LH contamination were not available. As preparations of greater purity were developed, some of the previously reported effects of FSH could no longer be confirmed. The demonstration that spermatogenesis can be maintained in hypophysectomized adult rats by administration of exogenous testosterone alone (Walsh, Cuyler, and McCullagh 1934; Ahmad, Haltmeyer, and Eik-Nes 1973; Ewing, Stratton, and Desjardins 1973) also cast serious doubt upon the importance of FSH for fertility in the male. Many reproductive biologists concluded that FSH had no detectable effect upon the testis of hypophysectomized mature rats. There was, however, strong evidence that the initiation of spermatogenesis in the immature animal and its restitution in hypophysectomized animals after testicular regression both required the presence of FSH (von Berswordt-Wallrabe and Neumann 1968; Steinberger 1971). Recent experiments involving injection of antisera to highly purified FSH into immature rats have shown a highly significant effect on testicular weight and tubule diameter (Raj and Dym 1976). Thus there seems to be little doubt of the importance of FSH in testicular maturation, but whether it is essential in the adult rodent or in man remains unsettled.

Until recently, hypotheses as to the hormonal regulation of spermatogenesis were based almost exclusively on morphological observations after hormonal deprivation or administration. There has now been encouraging progress in defining the biochemical mechanisms of FSH action. Tritium-labeled FSH has been shown to bind specifically to the tubules of the testis and it was found that the immature testis binds more than the adult testis (Means and Vaitukaitis 1972). The binding is mainly to membranes, suggesting that FSH may exert its effect without entering cells, as has been shown to be true for several other peptide hormones (Means 1973). The hormone stimulates adenyl cyclase activity, resulting in intracellular accumulation of cyclic AMP and activation of protein kinase. This is followed by an increase in the rate of protein, RNA, and phospholipid synthesis (Means and Hall 1967, 1971).

These biochemical investigations on testicular homogenates could not identify the target cell of FSH action, but a number of morphological studies have implicated the Sertoli cell. There were early reports of hypertrophy of these cells after FSH administration to hypophysectomized animals (Murphy 1965). More recently, ferritin-labeled FSH has been localized on the Sertoli cells by electron microscopy and similar localization has been obtained by the fluorescent antibody technique (Castro, Seiguer, and Mancini 1970; Castro, Alonso, and Mancini 1972). Isolated Sertoli cells have been shown specifically to bind iodinated FSH but not LH, and the cells responded to FSH stimulation with a dramatic increase in cyclic AMP (Steinberger et al. 1975). Thus there is now compelling evidence that the Sertoli cell is one of the targets, and probably the principal target, of FSH activity in the testis.

**Androgen-Binding Protein and Cytoplasmic Androgen Receptor**
The biochemical functions of Sertoli cells that are activated by FSH are now beginning to come to light. A specific *androgen-binding protein (ABP)* with a molecular weight of about 86,000 has been found in the testis (Vernon, Kopec, and Fritz 1973; Ritzen et al. 1973; Hansson et al. 1973), in rete testis fluid (French and Ritzen 1973), and in the head of the epididymis (Ritzen et al. 1973). It is speculated that the accumulation of androgen in high concentrations in the seminiferous epithelium may depend upon the presence of androgen-binding protein in the tubular compartment. Androgen-binding protein disappears from the testis after hypophysectomy (Vernon, Kopec, and Fritz 1973; Steinberger, Steinberger, and Sanborn 1973; Hansson et al. 1973) and it is restored after administration of FSH. Because ABP was present in normal, or greater than normal, concentration after germ cell destruction by radiation, it was inferred that it is produced by the Sertoli cells (Hagenas, Ploen, and Ketzen 1975). This has now been verified by the demonstration that ABP is produced in vitro by cultures of isolated Sertoli cells and that such cultures are responsive to FSH stimulation (Fritz et al. 1974; Steinberger, Steinberger, and Sanborn 1975). The synthesis of ABP does not depend exclusively upon FSH, however, for it can also be in-

duced in hypophysectomized rats by LH or testosterone administration (Elkington, Sanborn, and Steinberger 1975).

Another significant advance in the biochemistry of the testis and epididymis has been the demonstration of a *cytoplasmic receptor protein* (CR) distinct from ABP. The receptor protein of epididymal and testicular epithelium has similar properties. As in other target organs that respond to steroid hormones, the receptor protein is believed to be responsible for translocation of androgen to the nucleus where it is bound to the chromatin and affects transcription (Blaquier 1971; Hansson et al. 1973; McLean et al. 1974).

Thus, after more than thirty years of controversy about the functions and site of action of FSH in the testis, research in several laboratories has now clearly established that FSH is bound specifically to Sertoli cells, where it stimulates cyclic AMP production, protein kinase activation, and protein synthesis. Among the proteins synthesized is ABP, which may play an essential role in maintaining in the seminiferous epithelium the high local concentration of androgen necessary for spermatogenesis. Whether this is an indispensable function of FSH remains an open question because of the finding that testosterone can also stimulate ABP production. The androgen-ABP complex is also carried downstream through the rete testis and efferent ducts to the epididymis, where it may be required for maintenance of the cytological differentiation of the epididymal epithelium (figure 5.12). Androgen receptor (CR) found in the cytosol of testis and epididymis appears to be an essential mediator of the androgenic stimulation of both organs.

Among the major gaps in our knowledge is the question of the respective roles and relative importance of FSH and testosterone in maintenance of normal levels of androgen-binding protein in the seminiferous tubules. Moreover, testosterone can freely diffuse into the tubules, so it remains to be demonstrated just what advantage is conferred by the formation of an androgen-ABP complex. Further studies are needed to localize androgen receptor (CR) to specific cell types within the seminiferous epithelium. Such information would help answer the question of whether testosterone stimulates spermatogenesis by a direct action on the germ cells or indirectly by affecting the Sertoli cells.

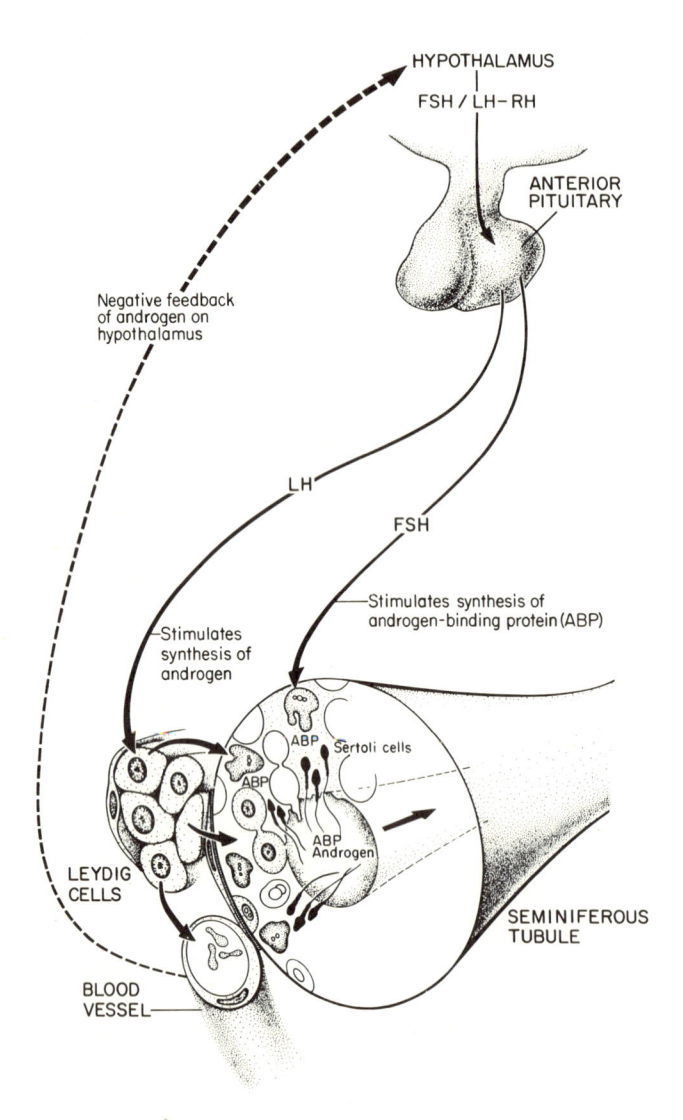

Figure 5.12 Schematic representation of the endocrine regulation of spermatogenesis. The high local concentration of androgen required is maintained by the Leydig cells in response to stimulation by luteinizing hormone (LH) from the pituitary. In addition to the well-established feedback of circulating androgen on LH levels, there is increasing evidence that FSH levels are similarly controlled by a hypothetical substance "inhibin" originating in the seminiferous tubules.
Source: D. W. Fawcett in *The Developmental Biology of Reproduction*, C. L. Markert and J. Papaconstantinou eds. Academic Press, New York, 1975.

## The Negative Feedback Control of FSH Release

The principle of negative feedback in the regulation of gonadotropin release has long been accepted, but the original proposal that gonadal steroids act directly upon the cells of the pituitary (Moore and Price 1932) has gradually been modified. The possibility of a direct effect upon the pituitary is not excluded, but in most instances the steroids seem to affect certain areas of the brain, which then act via the hypothalamus and hypothalamo-hypophysial portal vessels to accelerate or inhibit the release of gonadotropins. In the female, release of both LH and FSH is controlled by the blood levels of the steroid product of their respective target cells in the ovary. In the male, testosterone, the steroid product of the Leydig cells, controls release of LH; but a comparable feedback control of FSH by a product of the seminiferous epithelium has yet to be conclusively demonstrated.

A nonsteroidal, water-soluble hormone "inhibin" originating in the tubules was postulated over forty years ago (McCullagh 1932). Aqueous extracts of testis were reported to cause modification or cessation of the estrous cycle in normal female rats (McCullagh and Schneider 1940). The repeated demonstration that cryptorchidism, antispermatogenic compounds, vitamin A deficiency, and other factors that damage the seminiferous epithelium of experimental animals all result in an elevation in plasma levels of FSH was interpreted as strong indirect evidence for a negative feedback from the tubules. Men with germinal aplasia, or testicular damage from radiation or orchitis, also exhibit elevated FSH levels (Van Thiel et al. 1972; Saxena, Leyendecker, and Chin 1969; Franchimont et al. 1972). In the past, numerous efforts to isolate the hypothetical testicular factor inhibin responsible for FSH regulation have failed. Speculations as to the cell type producing it have varied. It was suggested by some that the late stages of germ cells were the source (Johnson 1970), but the results of more recent investigations indicate that it may be the absence of dividing spermatogonia that is associated with a rise in FSH (Lee et al. 1974a; Kruegar, Hodgen, and Scherins 1974).

Now after four decades of debate and inconclusive evidence, three laboratories have independently obtained "inhibin" preparations that selectively lower FSH without affecting LH. All three preparations were obtained from different sources—one by ultrafiltration from rete testis fluid (Setchell and Jacks 1975); one from semen (Franchimont, Chari, Hagelstein et al. 1975); and the third from extracts of bull testis (Lee et al. 1974b). There is general agreement that the substance is probably a polypeptide with a molecular weight less than 100,000. Efforts are now underway to purify, characterize, and sequence the active material. The work on inhibin is still hampered by lack of a dependable assay system, but this impediment to progress may soon be overcome. The recent recrudescence of interest in inhibin stems from the realization that if FSH is essential for spermatogenesis in man, administration of the substance responsible for its feedback control might produce infertility without impairment of libido.

**Suppression of Gonadotropin Release or Interference with Its Action as a Contraceptive Strategy**
There have been several attempts to induce infertility in men by taking advantage of the negative feedback regulation of gonadotropin release. The capacity of excess testosterone to lower plasma gonadotropin levels has been clearly established (Lee et al. 1972; Sherins and Loriaux 1973) and the suppression of LH is greater than that of FSH. Thus it was logical to expect that administration of excess testosterone might suppress release of LH, resulting in inhibition of spermatogenesis due to inadequate peritubular concentrations of testosterone. This expectation was borne out. Administration of testosterone proprionate daily or testosterone enanthate weekly resulted in infertility without suppression of libido or potentia (Heller et al. 1950; MacLeod 1965; Reddy and Rao 1972). Although the feasibility of this approach to fertility control has been demonstrated, it has not been actively pursued because the high dosage of testosterone required results in changes in lipoprotein metabolism and blood cell formation that might make long-term maintenance on this regime unsafe.

A synthetic derivative of ethinyl testosterone, Danazol, appears to act directly upon the Leydig cells to suppress their steroidogenesis and results in diminished sperm counts without marked decrease in libido (Sherins et al. 1971). Preliminary clinical trials have been carried out with Danazol combined with small doses of testosterone in the belief that they might act synergistically in the suppression of spermatogenesis. Sperm counts fell to infertile levels in two months with maintenance of normal libido (Shoglund and Paulsen 1973). The conclusion of these authors that Danazol plus some form of testosterone may prove to be a safe, effective contraceptive for men is not widely shared, but the principle underlying this approach is certainly deserving of further exploration.

Suppression of gonadotropins can also be achieved by administration of the female hormones, estrogens or progestogens. Administration of estrogens alone to men resulted in infertility but was accompanied by loss of both libido and potentia and was complicated by painful gynecomastia. Progestogens were also effective but had the same undesirable side effects (Heller et al. 1959).

More promising have been recent studies employing progestogens and testosterone in various combinations in the hope of achieving synergism in gonadotropin suppression while avoiding both loss of libido and gynecomastia. Implants of silastic capsules containing testosterone and others containing synthetic progestogens (norgestrienone or norethindrone) inhibited spermatogenesis for several months without depressing sexual drive or potency (Coutinho and Melo 1973; Frick and Bartsch 1973; Johansson and Nygren 1973). Fertility returned upon cessation of medication.

None of these combined methods has had approval by the regulatory agencies of the United States for general use and will not until extensive toxicological tests have been completed. However, a similar approach has been undertaken in England using existing oral hormone products that have been approved for sale in many countries. These products contain combinations of an orally effective androgen (methyltestosterone) and an estrogen (ethynyl-estradiol). In a small group of volunteers sperm numbers decreased to infertile levels in three months and returned to normal after cessation of treatment. No untoward side effects were noted (Briggs and Briggs 1974). More work is needed to assess possible long-term undesirable effects and to establish the most favorable estrogen or progestin-androgen ratios;

but the preliminary results justify some optimism that safe, reversible inhibition of spermatogenesis may be achieved by this approach.

The selective inhibition of FSH release would have an advantage over LH suppression in that it would not be complicated by loss of sexual desire due to impairment of testosterone production. If the current efforts to isolate and characterize inhibin are successful, this contraceptive strategy can be pursued. The rationale of this approach is that if androgen-binding protein is essential for maintenance of the high intratesticular and intraepididymal levels of testosterone, suppression of FSH release might remove the stimulus for ABP synthesis by Sertoli cells and result in levels of testosterone too low to support spermatogenesis or posttesticular sperm maturation. The prospect for success of this approach to contraception hinges upon the degree to which spermatogenesis in mature men is dependent upon FSH and this is not known. The recent finding that ABP production is not solely dependent upon FSH but can also be maintained by testosterone raises some doubt about the essentiality of FSH. It must be borne in mind, however, that synthesis of ABP is the only FSH-dependent Sertoli cell function discovered to date. There may be other FSH-dependent essential products or metabolic functions of the Sertoli cells of which we are still unaware.

The rapid advances in this area in the past five years have brought new concepts, challenging new problems, and new excitement to research on the reproductive biology of the male. If funding is maintained, we will certainly see continued rapid progress in our understanding of the control of spermatogenesis.

## The Structure and Function of the Excurrent Ducts of the Testis

The classical view of the excurrent ducts of the testis emphasized their role as conduits for transport and storage of spermatozoa. Relatively little credence was given to the possibility that their lining might have important physiological functions other than those related to transport. The ciliary beat of the lining of the ductuli efferentes was credited with helping propel the fluid contents toward the epididymis, and it was known that considerable resorption of water took place in this segment of the tract, thus concentrating the sperm suspension delivered to the epididymis (Benoit 1926; Mason and Shaver 1952). This absorptive function was shown to continue in the epididymis (Mason et al. 1952; Crabo 1965). The first clear-cut evidence that sperm survival in the epididymis depends upon androgen-stimulated functions of its lining epithelium came from experiments demonstrating that sperm remain viable on the operated side of unilaterally castrated animals several weeks longer than they do after bilateral castration (Benoit 1926) and that the viability of epididymal sperm in castrated animals can be maintained by administration of testis extracts (Moore and McGee 1928). Spermatozoa were found to be relatively infertile upon leaving the testis but gradually acquired fertilizing capacity as they passed slowly through the epididymal duct (Young 1931a,b; Nishikawa and Waida 1952). Whether these maturational changes were simply a time-dependent continuation of the development of spermatozoa or were dependent upon specific environmental conditions created by the lining of the epididymis was, for many years, a subject of debate. Research of the past decade has steadily accumulated new evidence for the thesis that androgens stimulate the epididymal epithelium to synthesize products required by the spermatozoa for their maturation and survival during storage (Orgebin-Crist 1968a,b; Orgebin-Crist, Danzo, and Davies 1975). This concept has stimulated the search for specific biochemical functions of the epididymis and has raised the hope that interference with its essential supporting functions might arrest the posttesticular maturation of spermatozoa and thus provide a promising approach to fertility control.

**Ultrastructural Studies of the Ductuli Efferentes and Ductus Epididymidis**
The presence of ciliated and nonciliated cells in the epithelium of the ductuli efferentes has been confirmed by electron microscopy and their ultrastructure has been described in detail (Monita 1966; Ladman 1967; Holstein 1969). These studies reveal very active endocytosis consistent with the absorptive function of the ductuli. The dense cytoplasmic granules interpreted by light microscopists as cytological

evidence of secretory activity contain acid phosphatase and are now believed to be lysosomes involved in intracellular digestive activity. After obstruction of the epididymal duct, the nonciliated cells of the ductuli efferentes may phagocytize and digest spermatozoa (Hoffer, Hamilton, and Fawcett 1973). There is at present no evidence of secretory activity in the ductuli efferentes, but their important role in fluid absorption and sperm transport are well documented.

Ultrastructural investigations of the epididymis have defined the cytological characteristics of the principal, basal, and apical cells of its epithelium (Nicander 1957a,b; Holstein 1969; Hamilton 1972) and have extended earlier histological studies showing that the epithelial lining of the duct exhibits several regions with different cytological characteristics. It is evident that biochemical studies based upon the crude subdivision of the organ into caput, corpus, and cauda do little to meet the need for precise correlation of function with microscopically identified segments of the duct (Glover and Nicander 1971; Hamilton 1972).

The principal cells of the epididymal epithelium have been most thoroughly studied by electron microscopy. Their exceptionally large Golgi complex and well-developed endoplasmic reticulum are characteristics usually found in very active secretory cells. On the other hand, the large surface area provided by their long stereocilia and the abundant evidence of uptake of fluid from the lumen by pinocytosis are features more typical of actively absorptive epithelia. These latter findings are consistent with physiological evidence that some 90 percent of the fluid leaving the testis is absorbed in the head of the epididymis (Crabo 1965; Waites and Setchell 1969). The remarkable development of the cell organelles usually involved in biosynthesis of proteins or glycoproteins makes it difficult to escape the conclusion that the principal cells are also secretory even though microscopic evidence of intracellular accumulation or release of a product is lacking (Hoffer, Hamilton, and Fawcett 1973). Histochemical studies of the excurrent duct system have demonstrated a variety of enzymatic activities in the epithelium, including glucose-6-phosphatase (Allen and Slater 1961), succinic dehydrogenases (Blackshaw and Samisoni 1967), $\beta$-hydroxybutyrate dehydrogenase (Niemi and Kormano

1965), and carbonic anhydrase (Cohen, Hoffer, and Rosen 1976). Thus it is evident from both ultrastructural and histochemical investigations that the epithelium of the excurrent ducts of the testis is not what would be expected of a simple transporting duct or passive storage organ. Instead it gives every indication of being metabolically active, biochemically versatile, and regionally specialized for specific functions. What those functions are remains largely unknown.

## Maturational Changes of Spermatozoa in the Epididymis

Significant changes take place in the spermatozoa of most mammals during epididymal transit. Early observations indicating that guinea pig sperm acquire the capacity for fertilization as they pass through the epididymis (Young 1931a,b) have now been extended to other species. In the rat, insemination with spermatozoa from the caput resulted in fertilization of only 8 percent of eggs, whereas similar numbers from the cauda fertilized over 90 percent (Blandau and Rumery 1964). In similar experiments on rabbits (Bedford 1966; Orgebin-Crist 1968a) and on hamsters (Horan and Bedford 1972) no eggs were fertilized by sperm taken from the caput. The acquisition of fertilizing capacity involves biochemical and physiological modifications, but in some species there are also morphological changes in the sperm head. These are especially prominent in guinea pig spermatozoa, where the acrosome becomes flexed at an angle of about 40° to the long axis of the nucleus while acquiring a crescentic thickening on one side and a deep concavity on the other. When these shape changes have occurred, rouleaux of a dozen or more spermatozoa are formed with the convexity of one acrosome fitting into the concavity of the next (Fawcett and Hollenberg 1963). Less conspicuous changes without rouleaux formation are reported for the chinchilla (Fawcett and Phillips 1969a) and the galago (Bedford 1973). In the relatively small acrosomes of the large domestic animals alterations of shape during epididymal transit are very slight or undetectable. In the human there appears to be no change of acrosomal contour during epididymal transit (Bedford 1973).

Changes in the character of the swimming movements of spermatozoa from different regions of the epididymis have been described by several investigators. Those removed from the head of the rat epididymis swim in circles with a relatively stiff midpiece; those taken from the tail exhibit normal progressive swimming (Blandau and Rumery 1964; Fray, Hoffer, and Fawcett 1972). A comparable maturation of flagellar function is reported for rabbit and guinea pig (Gaddum-Rosse 1968). A circling pattern of motility has not been described for human spermatozoa from the caput, but rapid forward progression does not occur until they have reached the middle of the body of the epididymis (Bedford, Calvin, and Cooper 1973). No structural changes have been described in the motor elements of the sperm tail that can be correlated with the acquisition of a mature pattern of motility. There is, however, biochemical evidence for formation of increased numbers of disulfide bonds in the tail fibers, and this might alter their flexibility (Calvin and Bedford 1971).

In spermiation a small residue of cytoplasm is left on the neck region of the spermatozoa when they break away from the spermatid syncytium in the seminiferous epithelium (Fawcett and Phillips 1969a). As the spermatozoa progress along the epididymal duct, this *cytoplasmic droplet* migrates from the neck back along the flagellum to the end of the midpiece where it gradually diminishes in size until it is no longer apparent. During its regression the membranous structures in its interior progressively diminish and are believed to provide phospholipid substrate for sperm metabolism. The capacity to hydrolyze phospholipids is acquired during posttesticular maturation (Scott and Dawson 1968) and the disappearance of the membranes of the droplet is correlated with a 25 to 50 percent loss of phospholipid content of the spermatozoa (Mills and Scott 1969; Paulos, Voglmayr, and White 1973; Voglmayr 1975). Thus the energy generating systems of spermatozoa seem to undergo both quantitative and qualitative changes during epididymal passage.

Changes in the charge on the cell membrane have also been reported. Spermatozoa from the caput epididymidis differ from those of the cauda in electrophoretic mobility (Bedford 1963) and in their binding of positively charged colloidal ferric oxide (Cooper and Bedford 1971).

The condensation of the chromatin in most mammalian species has progressed to a homogeneous, dense appearance before the spermatozoa leave the testis; but there is a significant increase in the number of disulfide bonds in the sperm head during passage through the epididymis. This is interpreted as an indication of continued stabilization of the chromatin (Calvin and Bedford 1971; Bedford, Calvin and Cooper 1973).

Thus, it has been established that mammalian spermatozoa are generally incapable of fertilization when they enter the epididymis, but they acquire this capacity as they slowly pass through the long epididymal duct. They concurrently undergo significant changes in metabolism and they may exhibit changes in pattern of motility, shape of the acrosome, position of the cytoplasmic droplet, degree of cross-linking of the nuclear chromatin, and surface charge. The nature of the epididymal contribution to these various aspects of sperm maturation is still not clear. The questions posed by Young some forty years ago remain unanswered. Do the sperm have an intrinsic capacity for maturation and merely find in the epididymis a favorable, but nonspecific, environment in which to complete the process? Or are the maturational changes dependent upon synthetic and secretory activities of specific regions of the duct? The answers may be different for the several components of the process and may vary from species to species. The influence of the epididymis appears to be essential in the rabbit (Orgebin-Crist 1967, 1968a). In the hamster, too, fertilizing ability does not develop when spermatozoa are retained in the proximal segments of the duct by ligation (Horan and Bedford 1972). On the other hand, the caudal portions of the duct may not be essential for sperm maturation in the human; men who have had epididymo-vasostomies are sometimes fertile even though the spermatozoa have bypassed the corpus and cauda of the epididymis. Much more work is needed to fill the numerous gaps in our understanding of the importance of epididymal function.

### Biochemical Functions of the Epididymis

The number of substances known to be synthesized by the epididymal epithelium is very small. *Glycerophosphorylcholine* was identified some twenty years ago, and the active incorporation of labeled phosphorous into this compound both in vivo and in vitro provided convincing evidence that it was synthesized by the epididymal epithelium (Dawson, Mann, and White 1957; Dawson and Rowlands 1959; Scott and Dawson 1968). Its biosynthetic pathway has not been completely worked out; its synthesis has not been localized to any particular cell type in the epithelium; and its physiological role remains unknown.

Also identified in the epididymis are *sialic acids*, a group of amino sugar containing carbohydrates associated with glycoproteins and glycolipids (Peyre and Laporte 1966; Rajalakshmi and Prasad 1968, 1969). Autoradiographic demonstration of rapid uptake of $^3$H-galactose or fucose into the Golgi complex of the epithelial cells and its later appearance in the lumen suggests that glycoproteins are synthesized in this organelle and subsequently secreted (Neutra and Leblond 1966; Bennett, Leblond, and Haddad 1974). A Golgi fraction isolated from bull epididymis was found to be rich in galactosyltransferase, an enzyme known to be involved in glycoprotein synthesis (Fleischer, Fleischer, and Ozawa 1969). There have been a number of speculations but no satisfactory explanation of the significance of sialoproteins in the epididymis. Their concentration falls following castration, and their production is stimulated by androgen administration (Prasad et al. 1973).

The steroids testosterone and dehydroepiandrosterone are present in the epididymis (White and Hudson 1968). For the most part these steroids are transported to the epididymis in the fluid from the rete testis (Waites and Setchell 1969a; Harris 1973), but there is suggestive evidence that the epididymis itself may have some capacity for synthesis of steroids. Hydroxysteroid dehydrogenases, enzymes that are essential for synthesis of androgenic steroids, have been demonstrated in the epididymis by both histochemical and biochemical analysis (McGadey, Baillie, and Ferguson 1966). It has been shown that the mouse epididymis in vitro can synthesize cholesterol from acetate (Hamilton, Jones, and Fawcett 1968; Hamilton 1971) and the in vitro synthesis of dehydroepiandrosterone from pregnenolone has been demonstrated for rabbit epididymis (Frankel and Eik-Nes 1968, 1970). Synthesis of small quantities of testosterone from various substrates by cell-free homogenates of rat epididymis has been reported (Inano, Machino, and Tamaoki 1969). To what extent epididymal steroid synthesis occurs in vivo and what its physiological significance might be are questions still unanswered.

*Carnitine*, a compound that, in other tissues is involved in the transport of certain fatty acids into mitochondria, is present in the epididymis in high concentrations (Marquis and Fritz 1965; Pearson and Tubbs 1967; Brooks et al. 1974). It was initially thought to be synthesized locally, but more recent work has found no evidence of local synthesis but has demonstrated its transport and intraluminal accumulation in the corpus and cauda in amounts up to seven times the concentration in plasma (Brooks, Hamilton, and Mallek 1973; Bohmer 1974). Its function in the seminal plasma is not known, but the observation that spermatozoa have exceptionally high activity of the enzyme carnitine acetyltransferase (Marquis and Fritz 1965) suggests that carnitine may influence the metabolism of spermatozoa during their storage in the epididymis. Various exogenous compounds are also known to be concentrated in seminal plasma (Mann 1968). The antifertility compound $\alpha$-chlorohydrin has been shown in autoradiographic studies to be concentrated in the corpus and cauda of the epididymis in much the same distribution as is carnitine (Crabo and Appelgren 1972).

In summary, the epididymal epithelium has been found to synthesize glycerophosphorylcholine and sialic acids and it may have a limited capacity for synthesis of steroids. It also is able to accumulate carnitine in high concentration in the duct lumen. How, or indeed whether, these substances contribute to maturation of spermatozoa during their transit, to their maintenance during storage, or to their activity after ejaculation are important gaps in our knowledge. There is no reason to assume that these compounds are the only products of epididymal synthesis and transport. Others of equal or even greater functional importance may be discovered in the future. The fact that certain exogenous compounds are also accumulated in the corpus and cauda of the epididymis suggests that it may be possible to take advantage of this property of the epithelium to achieve high intraluminal concentrations of antifertility agents directly deleterious to spermatozoa.

## Regulation of Epididymal Function

The dependence of the epididymis upon hormones of the testis has been recognized since the classical experiments of Benoit fifty years ago. The regression of the epididymis accompanying decline of androgen levels in seasonal breeding mammals (Wislocki 1949) and the reversal of postcastrational regression by administration of exogenous testosterone (Cavazos 1958; Maneely 1959) provided additional evidence for androgen regulation of epididymal function. Since 1960, studies of the morphological effects of castration or hypophysectomy on the epididymis have been extended to the electron microscope level. Either experimental procedure results in decreased duct diameter, diminution in epithelial height, reduction in size and number of stereocilia on the principal cells, reduction in size of their Golgi apparatus, and loss of endoplasmic reticulum from the apical cytoplasm (Hamilton, Jones, and Fawcett 1968; Orgebin-Crist and Davies 1974). Castration also results in a decrease in glycerophosphorylcholine, sialic acid, carnitine, and all other measurable biochemical functions of the epididymis. Relevant to the question as to the contribution of the epididymis to survival of the spermatozoa is the observation that epididymal spermatozoa lose fertilizing capacity and motility within three days after castration, but these functions are maintained for up to twelve days if the animal is injected with testosterone (Orgebin-Crist and Tichenor 1974; Lubicz-Nawrocki and Glover 1973; Lubicz-Nawrocki and Chang 1973).

Recent work has drawn attention to the special properties of the initial portion of the epididymal duct, which has a tall epithelial lining and an exceptionally rich blood supply (Hoffer, Hamilton, and Fawcett 1973; Brown and Waites 1972). In the rat it receives from the ductuli efferentes fluid rich in inositol and glutamic acid and containing testosterone in concentrations up to ten times that of the peripheral blood (Setchell, Dawson, and White 1968; White and Hudson, 1968; Cooper and Waites 1974; Harris 1973). Androgen-binding protein is present in highest concentration in the first part of the epididymis and falls off progressively in successive segments (Hansson, Ritzen, and French 1974). The epithelium lining this portion of the duct also possesses an androgen receptor protein in its cytoplasm

(Blaquier 1971; Hansson et al. 1973a). Testosterone taken up from the lumen is rapidly metabolized to dihydrotestosterone, which binds to the cytoplasmic receptor for transport to the nucleus. There it binds to the chromatin and by affecting transcription influences the rate of protein synthesis as measured by uptake of labeled amino acids (Blaquier and Callandra 1973).

Ligation of the ductuli efferentes to exclude testicular fluid results in a precipitous fall in androgen-binding protein in the caput epididymidis (Hansson, Ritzen, and French 1974) and in a marked decrease in diameter of the initial segments of the duct and in height of the epithelium accompanied by a striking cytological dedifferentiation of the principal cells (Fawcett and Hoffer 1976). This regression of the initial segments despite normal blood levels of androgen suggested that a high intraluminal concentration of testosterone might be required to maintain the cytological differentiation of the initial segment. However, administration of exogenous androgen in pharmacological doses that maintained blood testosterone levels at five to ten times normal failed to prevent regression of the initial segment after castration or ligation of the ductuli efferentes. It is concluded that although the cauda, corpus, and distal portions of the caput can be maintained by levels of androgen that normally circulate in the blood, the initial segments of the duct either have an obligatory requirement for high intraluminal androgen complexed to androgen-binding protein or they are in part dependent upon some other constituent of testicular fluid as yet unidentified (Fawcett and Hoffer 1976).

These and other recent findings represent significant advances in our understanding of the epididymis, but they leave many questions unanswered. It seems clear that transcription and protein synthesis in the epithelium of the first few segments of the epididymis are under control of androgen and possibly of other products of the testes, but we remain completely ignorant of the nature of the product synthesized by the duct or its role in the reproductive process.

**Control of Epididymal Function as an Approach to Contraception**
Spermatozoa first acquire the capacity to fertilize during their pas-

sage through the epididymis, so this organ would seem an ideal site of action for an antifertility agent designed to interrupt sperm maturation. Such an agent should also be relatively safe because the metabolically inert chromatin of the sperm nucleus is relatively resistant to chemical mutagenesis.

The discovery that the caput epididymidis is normally exposed to an unusually high concentration of intraluminal androgen suggested the possibility that an antiandrogen might interfere with the physiology of the epididymis at a dosage that would have little effect on spermatogenesis, libido, or accessory gland function. To pursue this strategy, *cyproterone acetate*, which acts by competitively inhibiting the action of testosterone on its target organs (Neumann, Potts, and Ryan 1970), was implanted in rats subcutaneously in silastic tubing. After several months of its slow continuous release, the testis weights and the accessory glands were reported to be normal but the epididymal duct showed definite regressive changes and the spermatozoa were not motile (Prasad, Singh, and Rajalakshmi 1970; Prasad et al. 1973). The changes were reversible on removal of the silastic capsules. These interesting results raised the hope of fertility control in the human male by selectively altering epididymal function by androgen deprivation. Clinical trials using oral administration are now in progress in countries where cyproterone acetate is approved for human use. Preliminary results point to a favorable result in inducing infertility, but the occurrence of a significant depression of sperm count as well as of sperm motility suggests that the drug is affecting spermatogenesis as well as epididymal maturation.

The compound $\alpha$-*chlorohydrin* has been found to produce reversible infertility in male rats (Ericsson and Youngdale 1970; Coppola 1969), guinea pigs (Ericsson and Baker 1970), hamsters, and monkeys (Kirton et al. 1970) without impairment of spermatogenesis. A posttesticular site of action was inferred from the rapid onset of sterility and from the fact that a much higher dose than that required to induce infertility in rats produces a highly localized lesion in the initial segment of the epididymis characterized by exfoliation of the epithelium and complete obstruction of the duct. It was postulated that this lesion resulted from an effect of the compound on the vasculature of the caput epididymidis (Ericsson and Baker 1970) but this has not been substantiated. More recent autoradiographic studies with labeled α-chlorohydrin have shown that the compound is concentrated in the lumen of the cauda epididymidis and appears to be associated with the spermatozoa (Crabo and Appelgren 1972). Current interpretations favor a direct action of a metabolite of α-chlorohydrin on epididymal spermatozoa possibly by blocking a specific step in the glycolytic pathway (Brown-Woodman, White, and Salamon 1975).

Alpha chlorohydrin has been considered too toxic for human trials. An analog, 1-amino, 3-chloro-2-propanol hydrochloride, has now been prepared (Coppola and Saldarini 1974). Resolution of its racemic mixture resulted in a remarkable separation of the toxicity and the antifertility effects when tested in small laboratory animals (Paul, Williams, and Cohen 1974). It was hoped that the l-isomer would prove sufficiently free of toxicity for human trials, but testing in monkeys disclosed unexpected and undesirable effects in the central nervous system. Other analogs may prove to be more promising.

Enthusiasm for the development of an antifertility agent with a posttesticular site of action continues to run high, but the rational development of new drugs will probably have to await the results of additional basic research on the physiology and biochemistry of the epididymis.

**The Spermatozoon and Fertilization**

We are approaching the three hundredth anniversary of the first description of the spermatozoon (1677) by Leeuwenhoek, the inventor of the simple microscope. Just as the discovery of the spermatozoon depended upon the development of the microscope, subsequent growth in our knowledge of its structure and behavior has depended in large measure upon advances in instrumentation. Progress in the design of light microscopes in the two and a half centuries after Leeuwenhoek sharpened the images of the external features of the spermatozoon but contributed little to our understanding of its internal structure. The introduction of the electron microscope, however, brought a dramatic acceleration of discovery; and in the past twenty years we have amassed far more information about this remarkable cell than was gathered in the previous 275 years.

**The Acrosome and Acrosome Reaction**
Among the more significant recent advances have been a clarification of the relations of the mammalian sperm acrosome to the nucleus and cell membrane, biochemical analyses of its complement of hydrolytic enzymes, and a description of the mechanism of their release in the acrosome reaction that is essential to fertilization. Originally described as a "granule" at the tip of the sperm head, the acrosome was shown in electron micrographs to be a membrane-bounded cap covering the anterior two-thirds of the nucleus (figure 5.13). The cell membrane is closely applied to its outer surface with little or no intervening cytoplasm. The apical portion of the acrosomal cap is greatly thickened in some mammals and extends some distance beyond the anterior margin of the nucleus. This portion of the acrosome often has a complex shape characteristic of the species (Fawcett 1970; Phillips 1972, 1974). The content of the cap is rich in carbohydrates including galactose, mannose, fucose, galactosamine, and sialic acids (Clermont, Clegg, and Leblond 1955). It also contains the enzymes hyaluronidase (Austin 1948; Zaneveld, Polakoski, and Schumadier 1973), acid phosphatase (Teichman and Bernstein 1972), $\beta$-glucuronidase, n-acetylglucosaminidase (Hartree and Srivastava 1963), and a trypsinlike protease called acrosin (Srivastava, Adams, and Hartree 1965; Stambaugh and Buckley 1970; Zaneveld, Srivastava, and Williams 1969; Zaneveld, Dragoja, and Schumacker 1972).

The recently ovulated egg is enclosed in a transparent proteinaceous envelope, the *zona pellucida*, and adhering to this are one or more layers of *cumulus cells* (figure 5.14). The penetration of these coverings of the ovum was originally attributed to vigorous motility of the spermatozoon providing the force necessary for a mechanical perforating action of the acrosome. Research of the past twenty years, however, has clearly established the importance of the hydrolytic acrosomal enzymes in this process. In the presence of a recently

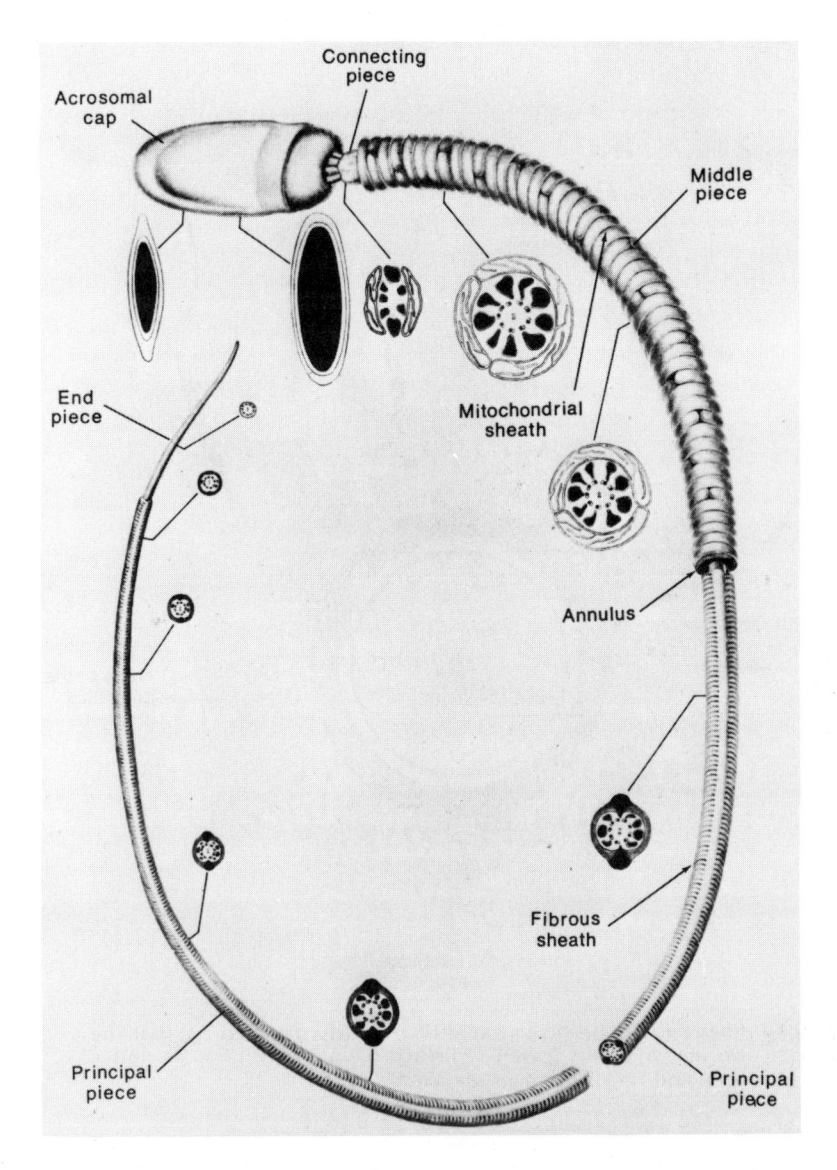

Figure 5.13 Drawing of a generalized mammalian spermatozoon with the cell membrane removed to show the arrangement of the underlying structural components. The appearance of cross sections as seen in electron micrographs at various levels is also depicted.
Source: D. W. Fawcett, *Developmental Biology* 44:394-436, 1975.

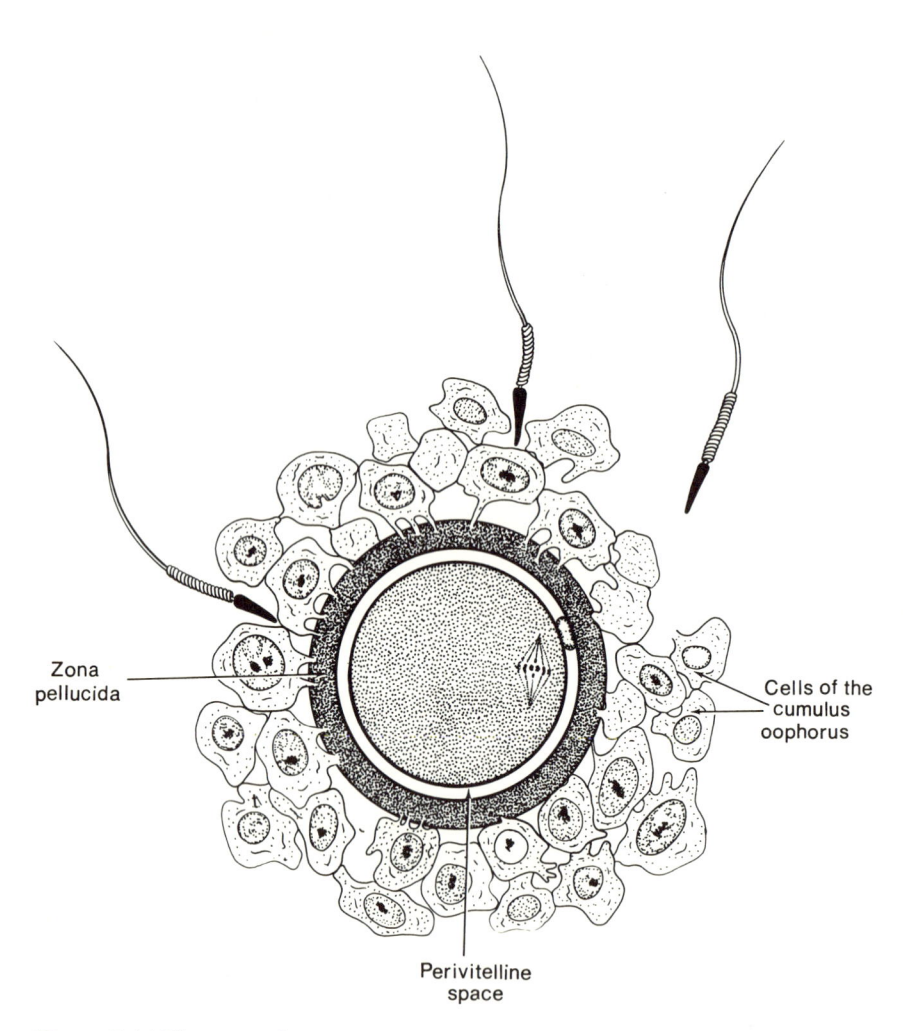

**Figure 5.14** Illustrates the barriers around the recently ovulated egg that the capacitated spermatozoa must traverse to reach the perivitelline space and achieve activation and fertilization of the ovum.

ovulated egg, the head of the spermatozoon undergoes a series of structural changes described as the *acrosome reaction*. The outer acrosomal membrane fuses at multiple points with the overlying cell membrane, creating openings through which the enzyme-rich contents of the acrosome are released (figure 5.15). This process of membrane fusion and vesiculation progresses until the major part of the outer acrosomal membrane and the overlying cell membrane are lost, leaving the anterior half of the sperm head invested only by the inner acrosomal membrane (Barros et al. 1968; Bedford 1968, 1971).

The ability of sperm to disperse the cells of the cumulus mass and to lyse a path through the zona pellucida has been repeatedly demonstrated and has led to the suggestion that the acrosome can be thought of as a highly specialized lysosome (Allison and Hartree 1970) or secretory granule that is activated in the presence of the ovum. One of the conditions necessary for this activation is the presence of calcium ions (Yanagimachi and Usui 1974), but what normally triggers the acrosome reaction in vivo remains an important gap in our knowledge. *Hyaluronidase* is believed to be involved in cumulus dispersion while *acrosin* is responsible for zona penetration (Srivastava, Adams, and Hartree 1965; Stambaugh and Buckley 1969, 1970). A paradoxical feature of this interpretation is the fact that the bulk of the acrosomal contents have been dispersed in traversing the cumulus and when the sperm head reaches the zona it is limited only by the inner acrosomal membrane (Bedford 1968, 1971). Thus, if acrosin were uniformly distributed in the acrosome, it would be largely lost in the early phases of the acrosomal reaction. Nonetheless, sperm are capable of digesting gelatin films (Gaddum-Rosse and Blandau 1972) and they still possess biochemically demonstrable protease activity after the plasma membrane and the contents of the acrosome have been lost (Brown and Hartree 1974). Labeled enzyme inhibitors (Stambaugh and Buckley 1970) and monospecific antibody to highly purified hyaluronidase and acrosin (Morton 1975) have been used to localize these enzymes. Hyaluronidase staining was more intense at the periphery of the intact acrosome; acrosin was concentrated near the inner acrosomal membrane and in the equato-

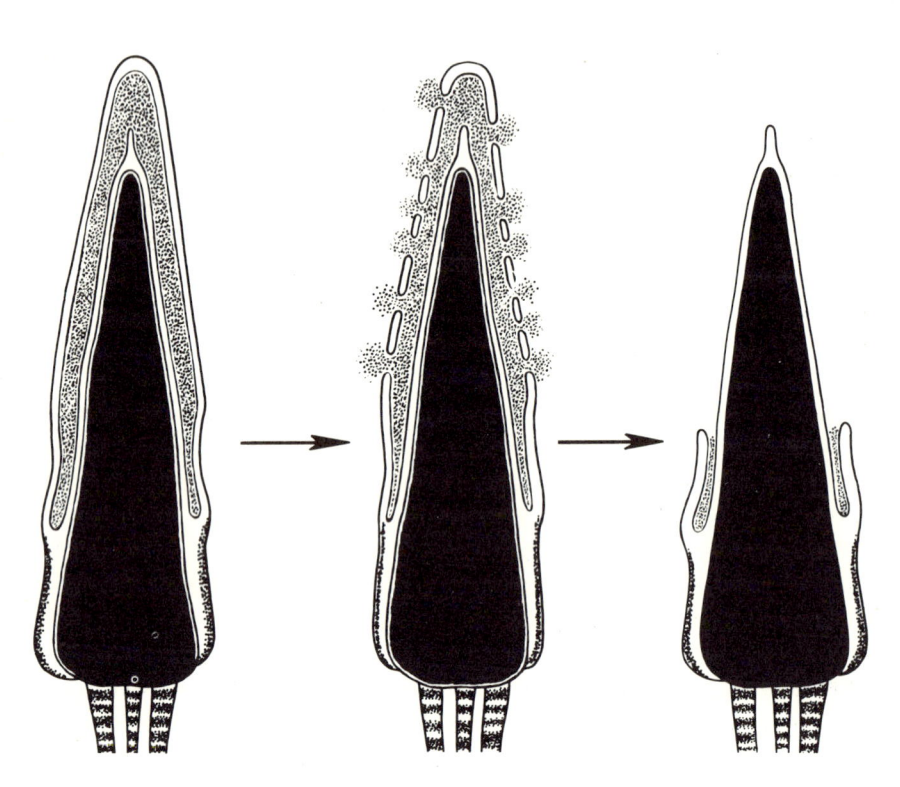

**Figure 5.15** Depicting successive stages of the acrosome reaction. The outer acrosomal membrane fuses with the cell membrane at multiple sites creating openings through which the enzyme-rich contents of the acrosome escape. This process leads ultimately to complete loss of the cell membrane over the anterior half of the head. Thereafter the inner acrosomal membrane is the limiting membrane of the sperm head over its anterior portion. The equatorial segment of the acrosome persists. Its function is poorly understood.
Source: D. W. Fawcett, *Developmental Biology* 44:394-436, 1975.

rial segment and persisted after experimental removal of all other components of the acrosome.

Realization that the enzymes liberated in the acrosome reaction are essential for fertilization has suggested new approaches to contraception involving selective stabilization of the acrosomal membrane to prevent enzyme release or development of specific inhibitors that would inactivate the enzymes (Stambaugh and Buckley 1970; Zaneveld et al. 1971). There are serious obstacles to success in these approaches to contraception, but no doubt much research will be devoted to them in the years ahead.

The cell membrane overlying the acrosome has been shown to differ from other regions of the cell surface in its charge (Nevo, Michaele, and Schindler 1961; Bedford 1963) and in its capacity to bind lectins (Edelman and Millette 1971). It has also been found in freeze-fracturing studies of guinea pig and rat spermatozoa that this region of the plasmalemma and the limiting membrane of the underlying acrosome both have a highly ordered internal organization not seen in membranes elsewhere (Friend and Fawcett 1974). It is speculated that the unique internal organization of these membranes may be related to their specific interaction in the acrosome reaction.

## The Motor Apparatus of the Spermatozoon

Because there is no fertility without sperm motility, the mechanism of sperm propulsion continues to be a subject of potential importance for population control. The structural analysis of the sperm tail with the electron microscope has been pursued in the past decade down to the molecular level and has provided the basis for a plausible theory of sperm motility.

The mammalian sperm tail, in common with the other motile cilia and other flagella, has as its principal motor component the *axoneme*—an axial bundle of microtubules arranged with two single microtubules in the center and nine doublet microtubules evenly spaced around them (figure 5.16). It is an intriguing example of the unity of nature that this nine plus two pattern of microtubules is constant in cilia and flagella throughout the biotic world from the

238                                  Major Advances and New
                                     Opportunities

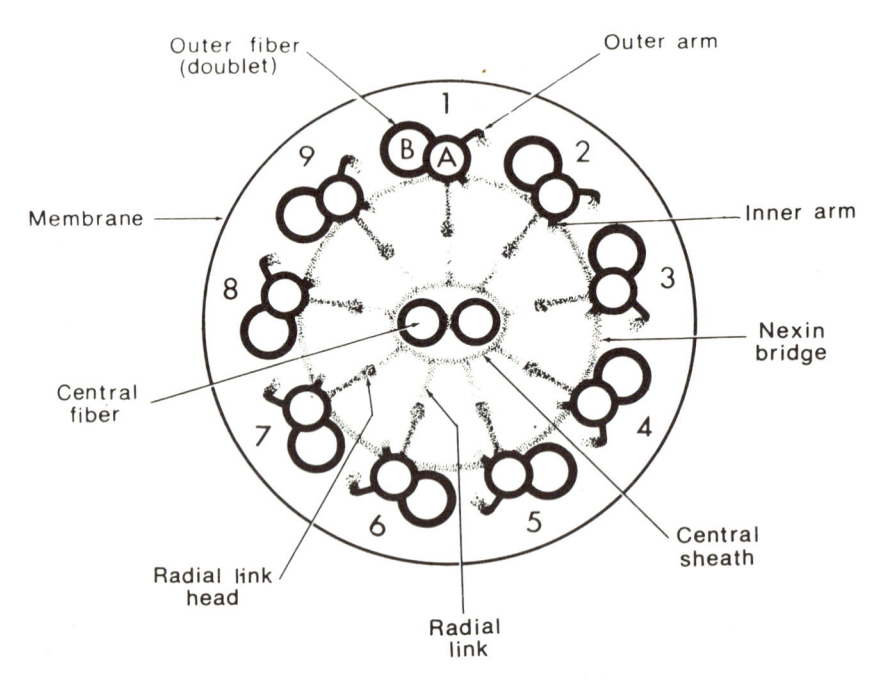

**Figure 5.16** Schematic representation of the current interpretation of the organization of the axoneme of cilia and sperm flagella.
Source: D. W. Fawcett, *Developmental Biology* 44:394-436, 1975.

protozoon in a drop of pond water to the cells lining the ventricles of the human brain. The functional advantage conferred by this particular arrangement and the reasons for its phylogenetic stability still elude us; but there has been rapid recent progress in our understanding of the chemical nature of the microtubules, their mode of assembly, and the way in which they may interact to produce the propagated waves of bending that pass along the tail of the swimming spermatozoon.

The doublet microtubules consist of *subfiber A*, which has a circular cross section, and *subfiber B*, possessing a C-shaped cross section attached at its ends to the wall of subfiber A. The cylindrical wall of subfiber A is made up of thirteen straight protofilaments and subfiber B of ten protofilaments, each 3.5 nm in diameter. The protofilaments, in turn, are linear polymers of 8 nm dimers of the protein *tubulin* (M.W. 55,000), associated end to end (Tilney et al. 1973; Warner 1970; Amos and Klug 1975). Subfiber A gives rise to two rows of diverging *arms* that project toward the next adjacent doublet (figure 5.16). The arms are spaced at regular intervals of 24 nm along the length of subfiber A and consist of *dynein* (M.W. 500,000), a protein with adenosine triphosphatase activity (Gibbons and Rowe 1965; Gibbons 1966). Also attached to subfiber A of each doublet are two slender *nexin links* that connect it to the adjacent doublets and a radial *spoke* that joins it to a delicate helical sheath wound around the central pair of microtubules (Stephens 1970).

Our understanding of the mechanism by which the axoneme produces propagated waves of bending is still incomplete, but it is known that the length of the doublets remains constant in different phases of the beat and therefore bending of the axoneme is not produced by active shortening or contraction of the microtubules (Satir 1968). The only plausible alternative is for local sliding to take place between neighboring doublets in a manner analogous to the sliding of interdigitating sets of filaments that is the basis for shortening of skeletal muscle. Strong experimental evidence for a sliding mechanism has recently been adduced by the demonstration that addition of adenosine triphosphate to segments of isolated axonemes caused doublet microtubules to move out opposite ends (Summers and Gibbons 1971). How the arms on one doublet interact with adjacent doublets to produce the sliding remains to be worked out, as does the mechanism for initiation and propagation of the wave of bending along the tail.

An interesting confirmation of the essential role of the dynein arms in flagellar movement has come from clinical studies of several infertile men who produced spermatozoa in normal numbers, none of which were motile. On electron microscopic examination the axonemes were found to lack the arms on the doublets (Pederson and Rebbe 1975; Afzelius et al. 1975). In addition to their infertility problem, these patients had situs inversus visceri and suffered from chronic sinusitis and bronchiectasis (Kartagener's syndrome) due to paralysis of the cilia of the respiratory tract associated with their genetic inability to synthesize dynein (Afzelius 1976).

Mammalian sperm tails differ from those of aquatic invertebrates in having a set of nine *outer dense fibers* in addition to the axoneme. Each of these courses longitudinally just peripheral to the corresponding doublet of the axoneme. Thus the cross-sectional pattern of the mammalian sperm tail is 9 + 9 + 2. The dense fibers have a thick medulla and a thin cortical layer that exhibits an oblique striation when viewed in surface replicas (Woolley 1969; Olson 1973). They are fixed to the nucleus anteriorly via the *connecting piece* and are attached to the wall of the corresponding doublet near their caudal termination. Therein lies one of the remaining difficulties in explaining mammalian sperm tail motility by a sliding mechanism, for these attachments would seem to impose serious restraint to sliding of axonemal components unless the dense fibers themselves are contractile or at least freely distensible.

When first observed in electron micrographs of sperm tails, the outer dense fibers were interpreted as accessory contractile elements that had evolved concurrently with development of internal fertilization to overcome the resistance to sperm locomotion in the female reproductive tract. The validity of this interpretation has recently been questioned as a result of chemical analyses of isolated outer dense fibers that show little or no similarity between them and any known contractile protein (Baccetti, Pallini, and Barrini 1973; Price

1973). The current view is that these fibers consist of a keratinlike scleroprotein rich in cysteine and that they are stiffening elements with passive elastic properties.

Also typical of the mammalian spermatozoon is a relatively long sheath of helically disposed mitochondria enveloping the axoneme and outer fibers in the initial segment of the tail called the middle piece. The close relationship of this *mitochondrial sheath* to the outer dense fibers suggested that it was essential to provide energy for the motor activities of the outer fibers. However, if these are not contractile, as recent biochemical evidence suggests, then the functional significance of the unusually long mitochondrial sheath of mammalian sperm is less clear, for the energy requirements of the axoneme are relatively small.

Caudal to the mitochondrial sheath, the axial components of the tail are enclosed by the *fibrous sheath*, a structure peculiar to mammalian spermatozoa. It consists of a series of hemicircumferential ribs on either side that pass halfway around the tail and terminate in two continuous longitudinal columns that run along the dorsal and ventral aspects of the tail for the greater part of its length. The thick longitudinal columns of the sheath would seem to impose considerable restraint to bending in the plane of the central pair of microtubules, but there would be little restraint to bending perpendicular to this plane, which would simply involve lateral flexion of the columns and a widening or narrowing of the interspaces between successive ribs of the fibrous sheath. The bending movements of the proximal part of the tail are, in fact, mainly perpendicular to the plane of the central pair of microtubules and only in the more caudal regions do the propagated waves become three dimensional.

The fibrous sheath has only recently been isolated in bulk for biochemical analysis (Olson, Hamilton, and Fawcett 1976) and at present its functional significance remains obscure. More research to determine the role of the outer dense fibers and the fibrous sheath may contribute to the control of fertility by suggesting means of suppressing sperm motility. The prospects of biochemical or immunological interference with the axoneme of the sperm tail are not encouraging because its principal protein, tubulin, is present in virtually all cells and is involved in cell division, intracellular transport, and other essential functions of organs throughout the body. On the other hand, if the proteins of the outer dense fibers and fibrous sheath prove to be unique to the spermatozoon, they might conceivably be used as antigens in the immunosuppression of sperm tail development or sperm motility.

Sperm Activation and Capacitation
Even after maturation in the epididymis, spermatozoa have still not attained their full capacity for fertilization. After they enter the female reproductive tract, they undergo further physiological changes that make them competent to penetrate the zona and fuse with the egg. This final preparation for fertilization is called *capacitation* (Austin 1952; Chang 1951). There is evidence that capacitation is a general prerequisite of conception in all mammals including man. For some time after discovery of this phenomenon, loss of the acrosome was thought to be equivalent to capacitation. This interpretation has been abandoned in the past decade and there is now a consensus that capacitation does not involve any visible morphological change in the spermatozoon. Instead it is believed to be a change at the molecular level, a change within the sperm membrane or possibly the removal of some component from its surface, and only after this has taken place does the acrosome reaction occur. There is immunocytochemical evidence for changes in the sperm membrane during capacitation (Oliphant and Brackett 1975), but components other than the surface membrane are also clearly affected. The pattern of energy metabolism changes upon the incubation of sperm in the uterus and, at least in some species, there is a noticeable change in the character and vigor of the tail movements. This latter component of capacitation is sometimes described as "activation." The time of residence in the female tract required for completion of this process also seems to vary with the species. In some species, both uterus and oviduct have the ability to capacitate, but a combination of the two appears to provide the optimal conditions. The ability of the uterus to accomplish capacitation depends upon the endocrine state of the female, whereas the oviduct in the rabbit retains its potential for capacitation

regardless of the state of the female with respect to circulating steroid hormones.

The interpretation of the nature of the change involved in capacitation has been complicated by the observation that sperm already prepared for fertilization by residence in the female tract can be *decapacitated* by exposure to seminal plasma (Chang 1957; Weinman and Williams 1964). An intensive search in the past fifteen years for an explanation of this secondary inhibition has only recently yielded promising results. Using antibodies to seminal plasma has demonstrated that antigenic plasma components that cannot be dislodged by simple washing are present on the surface of rabbit sperm. Upon incubation in utero however, there is a progressive decrease in these components with time, suggesting that capacitation involves removal of seminal plasma factors from the surface of the spermatozoon (Oliphant and Brackett, 1973; Aonuma et al. 1973). Those antigenic components can be removed from spermatozoa in vitro simply by exposure to hypertonic media (Brackett and Oliphant 1975), but it seems likely that their removal is enzymatic under the normal in vivo conditions of capacitation.

The prospects for control of capacitation are promising, but this will probably not be achieved in the next few years. The reversible inhibition of capacitated sperm clearly shows that a block to the capacitated state with nontoxic substances is possible, but the nature of the molecule involved must be established before approaching the more difficult problem of its delivery to the site of fertilization. It may also prove to be possible by systemic injection of drugs or hormones to alter the properties of the secretions of the female reproductive tract in such a way as to interfere with capacitation. Indeed, there is some reason to believe that the existing oral contraceptive methods for the female involving steroid hormones may, in part, owe their effect to an interference with capacitation.

## Zona Penetration and Fertilization

In penetration of the zona pellucida, the lytic action of enzymes bound to the inner acrosomal membrane or residing in the equatorial segment of the acrosome is believed to assist in creating a path for the vigorously motile spermatozoon. Having traversed this barrier, the spermatozoon lies free in the perivitelline space between the zona pellucida and the surface of the egg. The broad surface of the head then comes into contact with the egg and the membranes of the two gametes fuse (figure 5.17). Much investigative attention is now focused upon the portion of the sperm head immediately behind the acrosome because electron microscopic studies have shown that it is exclusively in this region that sperm attachment and membrane fusion take place (Barros and Franklin 1968; Stefanini, Oura, and Zambone 1969; Yanagimachi and Noda 1970). In freeze-fracture preparations, the membrane of this region is unusually rich in intramembrane protein particles, as are other metabolically active membranes. It is underlain by a thin dense layer, the *postacrosomal sheath*. The composition of this layer and its function in sperm-egg interaction are unknown. Clearly much more work is needed to identify the specific properties of this region of the sperm surface that are essential for gamete recognition and membrane fusion at fertilization.

Sperm-egg fusion induces the morphological and physiological changes described as *egg activation*. One of the visible manifestations of this process is the release of cortical granules from the peripheral cytoplasm into the perivitelline space. The chemical composition of these granules has not been determined, but there is indirect evidence suggesting that material released in their discharge alters the physicochemical properties of the zona pellucida in such a way as to prevent the penetration of other spermatozoa. The resulting *block to polyspermy* ensures penetration of a single spermatozoon. If we knew more about the biochemical mechanism of the block to supernumerary spermatozoa, it might be possible to simulate its action to prevent entry of the first spermatozoon.

Late in spermiogenesis the chromatin of the spermatid nucleus becomes condensed into a small, dense, metabolically inert mass that is highly resistant to factors in the environment and even to enzymatic digestion. Yet when the sperm enters the ooplasm at fertilization, the process of nuclear condensation is rapidly and completely reversed to give rise to the male pronucleus. This rapid transformation of the sperm nucleus in the ooplasm is one of the most remark-

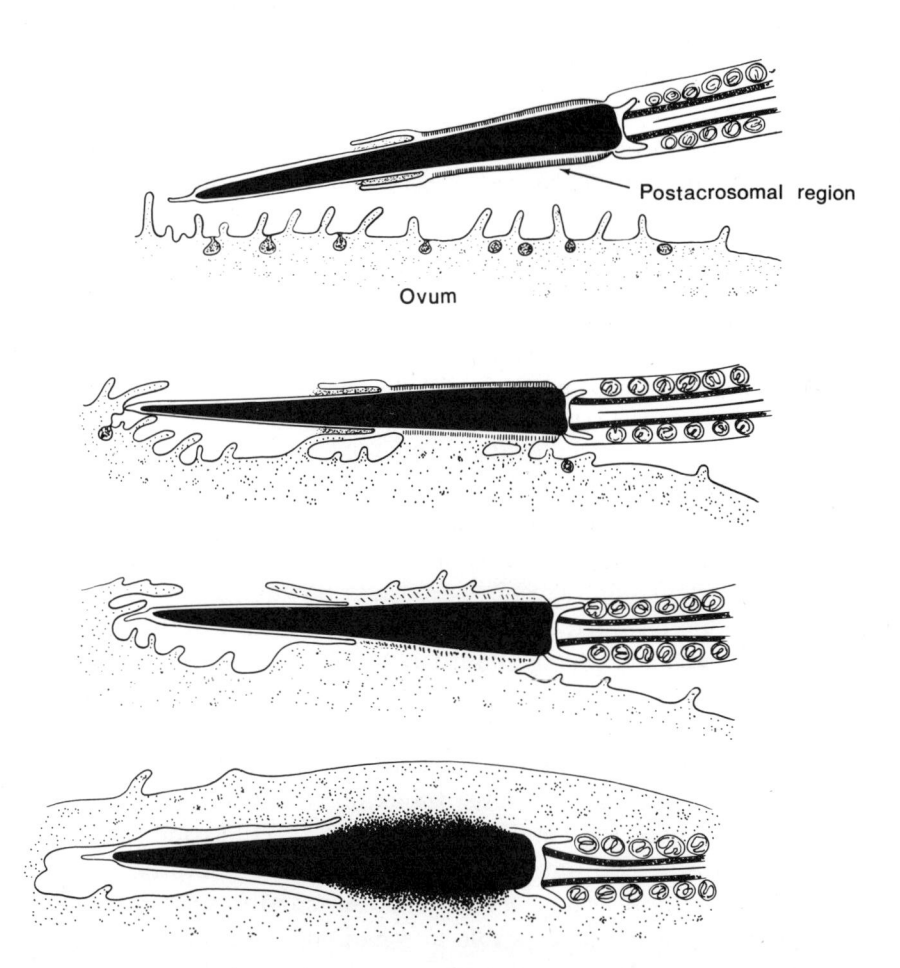

**Figure 5.17** The stages in fusion of the spermatozoon with the egg as seen in electron micrographs. The broad face of the sperm head contacts microvilli projecting from the surface of the ovum. The plasmalemma of the egg fuses with that of the postacrosomal region of the sperm. The sperm head sinks into the cytoplasm of the ovum and its chromatin begins to decondense to form the sperm pronucleus that is destined to fuse with the egg nucleus.
Source: Redrawn from R. Yanagimachi and Y. Noda, *Am. J. Anat.* 128:429, 1970.

able processes in reproductive biology. We still know very little about the process of chromatin condensation in the developing spermatozoon and even less about its decondensation in the fertilized egg. If the mechanism of either process were understood, it might prove vulnerable to interference.

Recognition of the need for capacitation stimulated efforts to achieve in vitro fertilization with spermatozoa capacitated in utero (Thibault, Dauzier, and Winterberger 1954), and by 1960 full-term young had been obtained from rabbit eggs fertilized in vitro and then transferred to recipient mothers (Chang 1959). It was later discovered that follicular or tubal fluid promoted capacitation in vitro (Yanagimachi and Chang 1963a,b; Yanagimachi 1969) and it was subsequently found that even these naturally occurring media were dispensable (Bavister 1969; Toyoda, Yokoyama, and Hosi 1971). The discovery of conditions for capacitation in chemically defined media was a significant advance permitting further studies of fertilization under controlled conditions (Miyamoto and Chang 1972, 1973; Iwamatsu and Chang 1969, 1970, 1971). There remain interesting differences between in vivo and in vitro fertilization that deserve further study. In the rat under normal conditions, there are only about forty-five sperm in each oviduct at the time of fertilization, whereas it requires three thousand to six thousand sperm for each egg under the conditions of in vitro fertilization. Polyspermy is rare in vivo but common in vitro. Thus the process of fertilization is more efficient under natural conditions and the female tract appears to play a role in selection, transport, and capacitation of spermatozoa that cannot yet be completely duplicated in vitro.

In 1960 our knowledge of fertilization was based mainly upon studies of the process in sea urchins with much less known about the mammal. Research since then has made in vitro fertilization of eggs from the common laboratory species a routine procedure. Extensive use of in vitro systems will now make it possible, in the next decade, to gain a better understanding of the basic molecular mechanisms involved in mammalian fertilization.

## Contraceptive Strategies Directed Against the Spermatozoon and Fertilization

Because the spermatozoon is the gamete that is transferred between the sexes, it is a logical target for conception control. The obvious approach of interposition of a mechanical barrier to sperm transfer has been fully exploited in the condom and diaphragm. The realization that successful fertilization in the mammal depends upon sperm motility made it reasonable to take advantage of their accessibility during transfer, to destroy motility by intravaginally administered surface active or cytotoxic spermicidal agents. These simple direct approaches remain the only widely used means of reversible fertility control acting upon the male gamete; but research of the past decade has now identified a number of vulnerable events in the complex processes of sperm transport, capacitation, and fertilization that may lend themselves to pharmacological or immunological interruption.

The reversible inhibition of capacitation by substances in epididymal and seminal plasma indicates that further effort should be devoted to establishing the nature of these substances. Their identification might well permit development of antifertility agents that would maintain spermatozoa in the decapacitated state. Because the release of lytic enzymes in the acrosome reaction is essential for fertilization, the possibility of blocking fertilization by inhibiting acrosin or hyaluronidase has appealed to several investigators. A cause of some concern in this approach is the fact that lysosomal enzymes are involved in normal physiological processes elsewhere in the body and a generalized inhibition of any of these enzymes might therefore have undesirable side effects. It has recently been shown, however, that *acrosomal hyaluronidase* is specific to sperm and this raises the hope that it may yet be possible to inhibit fertilization by raising antibodies to highly purified sperm hyaluronidase without affecting other hyaluronidases in the body (Metz 1973). Similarly the protease *acrosin* has been shown to be sperm specific (Zaneveld, Polakoski, and Schumadier 1973) and work is in progress to test its isoantigenicity and the feasibility of immunological inhibition of its activity to prevent sperm penetration of the zona pellucida. Another example of a sperm-specific isoenzyme is *lactic dehydrogenase-X*. Heterologous antibodies to this isozyme have been reported to suppress pregnancy in rabbits following postcoital passive immunization (Goldberg 1973). It is not yet clearly established whether the pregnancy suppression in these experiments was due to direct action of the antibodies on the fertilizing spermatozoa or to nonspecific interference with other biological processes. Nevertheless, these preliminary successes encourage further efforts at immunological inhibition of fertilization.

The existence of the zona reaction and the consequent block to polyspermy has drawn attention to the zona pellucida as a potential target of agents intended to prevent sperm penetration. Although the nature of the cortical granule factor that normally induces the zona reaction is not known, various treatments in vitro have been shown to mimic its action. Mild treatment of unfertilized eggs with trypsin-like enzymes is said to render the zona impenetrable to spermatozoa (Gwatkin, Anderson, and Hutchinson 1972). Nonenzymatic agents such as plant lectins that specifically bind to terminal saccharide residues of the zona substance are effective in blocking fertilization in vitro (Oikawa, Yanagimachi, and Nicolson 1973). It seems likely that a variety of other agents altering the molecular configuration of the zona pellucida may prove equally effective in preventing sperm penetration. Although such substances may have the desired effect on an in vitro test system, the probability that they would be tolerated in vivo or that they could be delivered to the site of fertilization in adequate concentration is not great.

There is abundant recent evidence of the importance of the sperm and egg membranes. Capacitation appears to involve changes in the sperm surface, which in turn affect sperm metabolism and motility patterns. The acrosome reaction depends upon a specific local response of the plasma membrane and the underlying acrosomal membrane. And the recognition and fusion of the gametes seems to depend upon special properties of the postacrosomal region of the sperm membrane. Further biochemical and immunological research on the regional specializations of the sperm membrane is clearly needed.

## Concluding Comments

The urgency of the population problem makes us impatient with what may seem to be slow progress toward practical solutions; but there is no doubt that in the more balanced perspective of future generations, the decades from 1960 to 1980 will be considered a golden era in the progress of basic understanding of reproductive biology of the male. It has seen the definition of the stages of the human spermatogenic cycle and description of the cytological events in germ cell differentiation, the discovery of the syncytial nature of the developing germ cells, and clarification of the mechanism of sperm release. Unique occluding junctions between Sertoli cells have been recognized as the structural basis of the blood-testis permeability barrier, and the physiological properties of the barrier have been thoroughly explored. The pathway of biosynthesis of testosterone has been worked out and its several steps have been localized to specific organelles of the Leydig cells. Receptors for the gonadotropins LH and FSH have been demonstrated on Leydig cells and Sertoli cells respectively. The mechanism of action of LH on its target cell has been elucidated and some of the biochemical effects of FSH on the Sertoli cell have been identified. The composition of testicular fluid has been determined and new information obtained on the control of epididymal function and its role in sperm maturation. The ultrastructure of the spermatozoon has been explored down to the macromolecular level. Most of its components have been isolated and chemically characterized. The secretory nature of the acrosome reaction has been established and its enzymes identified. The ultrastructural elements of the motor apparatus of the sperm tail have been described and a sliding mechanism of flagellar motility has been proposed. These represent but a few of the gains recorded in this fruitful period.

Many of these advances have depended upon morphological methods in a period when chemical approaches and molecular explanations are more highly valued. But this should not be surprising in the present stage of development of reproductive biology of the male. A reading of the history of medicine will reveal that morphologists have traditionally defined the problems and offered initial functional interpretations, often precariously based upon suggestive structural relationships alone. Physiological and biochemical validation of the functions has often come many years after the initial description of the structures involved.

The introduction of the electron microscope in the 1950s was probably the greatest single technological advance in the second half of this century. It would be difficult to exaggerate its ultimate impact on biomedical research. Before its advent, the morphologist was studying structures several orders of magnitude larger than those taking part in the reactions studied by the biochemist. The biochemist of that period naturally considered the findings of the morphologist largely irrelevant to his own interests and the morphologist, in turn, felt little need for chemical information. Communications between the basic science disciplines was poor because they were separated by a large informational gap, a no-man's land of biological structure in the dimensional range of 2 nanometers to 2 micrometers. The electron microscope opened up and rapidly obliterated that no-man's land by revealing the structural organization of cells and tissues down to the level of macromolecules. As a result, the boundaries between the disciplines have now become blurred. The morphologist and the biochemist are no longer investigating biological units orders of magnitude apart in size; they are often studying the same organelles or macromolecules and increasingly they speak a common language and use each other's methods. The morphological approach continues to play a germinal role by establishing a body of structural fact that continually redirects the thinking of the other basic sciences into productive new avenues of investigation and imposes realistic limits upon functional speculations. The whole pace of biomedical science has quickened. The crucial biochemical analysis or physiological experiment now follows quickly on the heels of new structural observations.

If the morphological and endocrinological contributions to male reproductive sciences in the past decade have been impressive, they are surely but a prelude to even more rewarding biochemical contributions. Biochemistry has permeated all branches of the life sciences and has increasingly become the common language of biology. Although it has long been an indispensable part of reproductive endocrinology, its impact upon our understanding of developmental processes has been seriously hampered by its inability to localize molecular events to specific cells in the heterogeneous populations that comprise the reproductive organs. There are indications that this limitation is now being overcome. The methods recently devised for dissociation of the seminiferous epithelium and separation of its cell types promise to make germ cell differentiation accessible to biochemical analysis. The maintenance of enriched fractions of certain cell types in vitro will make possible the identification of their products, the biosynthetic pathways involved, and their control mechanisms. The centrifugal segregation or selective solubilization of structural components of spermatozoa will lead rapidly to a better understanding of sperm locomotor mechanisms. The success achieved in capacitation and fertilization in vitro now opens the way for future investigations of the molecular reactions involved in gamete fusion and egg activation. The future looks bright.

The fact that the remarkably productive period of scientific discovery since 1960 has not produced an acceptable oral contraceptive for men should not be surprising or discouraging either to the scientists involved or to the agencies that support them. It need only be recalled that the existing oral contraceptives for women are based largely upon advances in our understanding of the hormonal control of the reproductive cycle that were made well before 1960. The past decade and a half has seen the exploitation of that knowledge to produce new and improved methods of conception control for women. Research on reproductive science of the male has lagged at least fifteen years behind that on the female. The number of basic and clinical investigators interested in andrology as opposed to gynecology is very small, and the investment in basic research on the male has been only a very small fraction of that devoted to the female. We are now in an exciting period of rapid advancement in our basic understanding of spermatogenesis and its hormonal control, sperm release, sperm maturation and capacitation, the physiology of the excurrent duct system, and the mechanisms of fertilization. A number of vulnerable steps in these complex processes have already been identified. The state of our knowledge of the male is now nearly comparable to that preceding the development of oral contraceptives for the female. If adequate research support can be sustained, there is every reason to expect that the next fifteen years will see the development of safe effective means of fertility control in the male.

Suppression of fertility is not the only objective of such research, nor will it be the only reward. The growth of our knowledge of male reproductive physiology will significantly improve our capacity to help those couples that earnestly desire a child but are unable to have one. Much of the fundamental research done to date on the male was motivated by a desire to *improve* animal fertility. The benefits have been great. Studies of spermatozoa and of methods for their long-term preservation and artificial insemination have been of inestimable value in animal breeding to produce food and fiber for an expanding population. Still more can be accomplished. These potential returns must also be taken into account in any assessment of the societal benefits of research in the reproductive sciences.

## References

Afzelius, B.A. 1976. A human syndrome caused by immotile cilia. *Science*, in press.

Afzelius, B.A.; Eliasson, R.; Johnsen, Ø.; and Lindholmer, C. 1975. Lack of dynein arms in immotile human spermatozoa. *J. Cell Biol.* 66:225-232.

Ahmad, N.; Haltmeyer, G.C.; and Eik-Nes, K.B. 1973. Maintenance of spermatogenesis in rats with intratesticular implants containing testosterone or dihydrotestosterone. *Biol. Reprod.* 8:411-419.

Allen, E. 1918. Studies on cell division in the albino rat. III. Spermatogenesis: The origin of the first spermatocytes and the organization of the chromosomes. *J. Morphol.* 31:133-185.

Allen, J.M. 1961. The histochemistry of glucose-6-phosphatase in the epididymis of the mouse. *J. Histochem. Cytochem.* 9:681-689.

Allen, J.M., and Slater, J.J. 1961. A cytochemical analysis of the lactic dehydrogenase diphosphopyridine nucleotide-diaphorase in the epididymis of the mouse. *J. Histochem. Cytochem.* 9:221-233.

Allison, A.C., and Hartree, E.F. 1970. Lysosomal enzymes in the acrosome and their possible role in fertilization. *J. Reprod. Fertil.* 21:501-515.

Amos, L., and Klug, A. 1975. Arrangements of subunits in flagellar microtubules. *J. Cell Sci.* 14:523-559.

Aoki, A. 1970. Hormonal control of Leydig cell differentiation. *Protoplasma* 71:209-223.

Aoki, A., and Fawcett, D.W. 1975. Impermeability of Sertoli cell junctions to prolonged exposure to peroxidase. *Andrologia* 7:63-76.

Aoki, A., and Massa, E.M. 1976. Subcellular compartment of free and esterified cholesterol in the interstitia cells of mouse testis. In press.

Aonuma, S.; Mayuma, T.; Suzuki, K.; Noguehi, T.; Iwai, M.; and Okabe, M. 1973. Studies on sperm capacitation. I. The relationship between a guinea pig sperm coating antigen and a sperm capacitation phenomenon. *J. Reprod. Fertil.* 35:425-432.

Austin, C.R. 1948. Function of hyaluronidase in fertilization. *Nature* 162:63.

_____. 1952. The capacitation of mammalian sperm. *Nature* 170:326.

Baccetti, B.; Pallini, V.; and Barrini, A.G. 1973. Accessory fibers of the sperm tail. I. Structure and chemical composition of the bull "coarse fibers." *J. Submicr. Cytol.* 5:237-256.

Baillie, A.H.; Ferguson, M.M.; and Hart, D.M. 1966. *Developments in steroid histochemistry*. New York: Academic Press.

Barros, C.; Bedford, J.M.; Franklin, L.E.; and Austin, C.R. 1967. Membrane vesiculation as a feature of the mammalian acrosome reaction. *J. Cell Biol.* 34:$C_1$.

Barros, C., and Franklin, L. 1968. Behavior of the gamete membranes during sperm entry into the mammalian egg. *J. Cell Biol.* 37:$C_{13}$.

Barry, M. 1843. Spermatozoa observed within the mammaliferous ovum. *Embryological Memoirs in Phil. Trans. Roy. Soc.* (London) 128.

Bavister, B.D. 1969. Environmental factors important for *in vitro* fertilization in the hamster. *J. Reprod. Fertil.* 18:544-545.

Becker, O. 1857. Uber Flimmerepithelium und Flimmergewegung in Geschlechtsapparat der saugetiere und das Menschen. *Molschotts Untersuch. Naturg. Menschen und Thiere* 2:71.

Bedford, J.M. 1963. Changes in the electrophoretic properties of rabbit spermatozoa during passage through the epididymis. *Nature* 200:1178-1180.

_____. 1966. Development of the fertilizing ability of spermatozoa in the epididymis of the rabbit. *J. Expl. Zool.* 163:319-329.

_____. 1968. Ultrastructural changes in the sperm head during fertilization in the rabbit. *Am. J. Anat.* 123:329-358.

_____. 1971. An electron microscopic study of sperm penetration into the rabbit egg after natural mating. *Am. J. Anat.* 133:213-259.

_____. 1973. The biology of primate spermatozoa. In *Advances in primatology reproductive biology of the primates*, vol. III, ed. W.P. Luckett. New York: Appleton-Century-Croft.

Bedford, J.M.; Calvin, H.I.; and Cooper, G.W. 1973. The maturation of spermatozoa in the human epididymis. *J. Reprod. Fertil.* (suppl. 18), 199-213.

Bellvé, A.R.; Anderson, E.; and Hanley-Bowdoin, L. 1976. Synthesis and amino acid composition of basic proteins in mammalian sperm nuclei. *Dev. Biol.* 471:349.

Bellvé, A.R., and Romrell, L.J. 1974. The transition of basic proteins during mammalian spermatogenesis. *J. Cell Biol.* 63:19a (abstr).

Benda, C. 1898. Uber die Spermatogenese der Vertebraten und hoherer Evertebraten. 2. Die Histogenese der Spermien. *Verh. d. Phys. Gesell.* Berlin.

Beneden van, E. 1875. La maturation de l'oeuf la fécondation, et les premieres phases du développement embryonnaire des mammifères. *Bull. Ac. Roy. de Belgique* 11.

Bennett, G.; Leblond, C.P.; and Haddad, A. 1974. Migration of glycoprotein from the Golgi apparatus to the surface of various cell types as shown by radioautography after labelled fucose injection into rats. *J. Cell Biol.* 60:258-271.

Benoit, M.J. 1926. Recherches anatomiques, cytologiques et histophysiologiques sur les voies excretrices du testicule, chez les mammiferes. *Arch. Anat. Histol. Embryol.* (Strasbourg) 5:173-412.

Bergstrom, S. 1949. Prostaglandinets kemi. *Nord. Med.* 42:1465.

Beyler, A.L.; Potts, G.O.; Coulston, F.; and Surrey, A.R. 1961. The selective testicular effects of certain bis-(di-chloro-acetyl)-diamines. *Endocrinology* 69:819-833.

Bhatnagar, Y.M.; Bellvé, A.R.; Romrell, L.J.; 1976. The integrity and characterization of histones in isolated spermatogenic cells. *J. Cell. Sci.* In press.

Blackshaw, A.W., and Samisoni, J.I. 1967. Histochemical localization of some dehydrogenase enzymes in the bull testes and epididymis. *J. Dairy Sci.* 50:747-752.

Blandau, R.J., and Rumery, R.E. 1964. The relationship of swimming movements of epididymal spermatozoa to their fertilizing capacity. *Fertil. Steril.* 15:571-579.

Blaquier, J.A. 1971. Selective uptake and metabolism of androgens by rat epididymis. The presence of a cytoplasmic receptor. *Biochem. Biophys. Res. Commun.* 45:1076-1082.

Blaquier, J.A., and Calandra, R.S. 1973. Intranuclear receptor for androgens in rat epididymis. *Endocrinology* 93:51-60.

Bohmer, T. 1974. Conversion of butyrobetaine to carnitine in the rat *in vivo*. *Biochim. Biophys. Acta* 343:551-557.

Bouin, B., and Ancel, P. 1903. Recherches sur les cellules interstitielles du testicule des mammiferes. *Arch. Zool. Exptl. Gen.* 1:437-526.

Bowen, R.H. 1920. Studies on insect spermatogenesis. History of the cytoplasmic components. *Biol. Bull.* XXIX.

Brackett, B.G., and Oliphant, G. 1975.

Capacitation of rabbit spermatozoa *in vitro. Biol. Reprod.* 12:260-274.

Branton, D. 1965. Fracture faces of frozen membranes. *Proc. Natl. Acad. Sci.* 55:1048-1056.

Bressler, R.S., and Ross, M.H. 1972. Differentiation of peritubular myoid cells of the testis: effects of intratesticular implantation of newborn mouse testes into normal and hypophysectomized adults. *Biol. Reprod.* 6:148-159.

Briggs, M., and Briggs, M. 1974. Oral contraceptive for men. *Nature* 252: 585-586.

Brokelmann, J. 1963. Fine structure of germ cells and Sertoli cells during the cycle of the seminiferous epithelium in the rat. *Z. Zellforsch. Mikroskop. Anat.* 59:820-850.

Brooks, D.E.; Hamilton, D.W.; and Mallek, A.H. 1974. Carnitine and glycerophosphorylcholine in the reproductive tract of the male rat. *J. Reprod. Fertil.* 36:141-160.

Brown, C.R., and Hartree, E.F. 1974. Distribution of a trypsin-like proteinase in the ram spermatozoon. *J. Reprod. Fertil.* 36:195-198.

Brown, P.D.C., and Waites, G.M.H. 1972. Regional blood-flow in the epididymis of the rat and rabbit: effect of efferent duct ligation and orchidectomy. *J. Reprod. Fertil.* 28: 221-233.

Brown-Woodman, H.D.C.; White, I.G.; and Salamon, S. 1975. Effect of $\alpha$-chlorohydrin on the fertility of rams and on the metabolism of spermatozoa *in vitro. J. Reprod. Fertil.* (abstr) 43:381.

Burgos, M.H., and Fawcett, D.W. 1955. Studies on the fine structure of the mammalian testis. I. Differentiation of the spermatids in the cat. *J. Biophys. Biochem. Cytol.* 1:287-300.

Burgos, M.H.; Sacerdote, F.L.; and Russo, J. 1973. Mechanism of sperm release. In *The regulation of mammalian reproduction*, ed. S.J. Segal, R. Crozier, P.A. Corfman, and P.G. Condliffe, pp. 166-182. Springfield: Charles Thomas.

Bustos-Obregon, E., and Holstein, A.F. 1973. On the structural patterns of the lamina propria of human seminiferous tubules. *Zeitschr. Zellforsch* 141:413-425.

Butenandt, A. 1931. Uber die chemische Untersuchungen der Sexualhormone. *Ztschr. f. angen. Chem.* 44: 905.

Butenandt, A., and Hanisch, G. 1935. Uber Testosteron. Unwandlung des Dehydroandrosterons in Androstendiol und Testosteron; ein Weg zur Darstellung des Testosteron aus Cholesterin. *Ztschr. Physiol. Chem.* 237:89.

Calvin, H., and Bedford, M.J. 1971. Formation of disulfide bonds in the nucleus and accessory structures of mammalian spermatozoa during maturation in the epididymis. *J. Reprod. Fertil.* 13:65-75.

Castro, A.E.; Alonso, A.; and Mancini, R.E. 1972. Localization of follicle stimulating and luteinizing hormones in the rat testis using immunohistological tests. *J. Endocrinol.* 52: 129-136.

Castro, A.E.; Seiguer, A.C.; and Mancini, R.E. 1970. Electron microscopic study of localization of labeled gonadotropins in Sertoli and Leydig cells of rat testis. *Proc. Soc. Expl. Biol. Med.* 133:582-586.

Cavazos, L. 1958. Effects of testosterone proprionate on histochemical reactions of the epithelium of the rat ductus epididymidis. *Anat. Rec.* 132: 209-228.

Chang, M.C. 1951. Fertilizing capacity of sperm deposited in the Fallopian tube. *Nature* 168:697.

_____. 1957. Some aspects of mammalian fertilization. In *The beginnings of embryonic development*, ed. A. Tyler, C.B. Metz, and R.C. von Borstel, pp. 109-134. Washington: American Association for the Advancement of Science.

_____. 1959. Fertilization of rabbit ova in vitro. *Nature* (Lond.) 184: 466-467.

Christensen, A.K., and Fawcett, D.W. 1961. The normal fine structure of opossum testicular interstitial cells. *J. Biophys. Biochem. Cytol.* 9:653-670.

Christensen, A.K., and Gillim, S.W. 1969. The correlation of fine structure and function in steroid-secreting cells, with emphasis on those of the gonads. In *The gonads*, ed. K.W. McKerns, pp. 415-488. New York: Appleton-Century-Crofts.

Christensen, A.K., and Mason, N.R. 1965. Comparative ability of seminiferous tubules and interstitial tissue of rat testis to synthesize androgens from progesterone-4-$^{14}$C in vitro. *Endocrinology.* 76:646-656.

Clermont, Y. 1958. Contractile elements in the limiting membrane of the seminiferous tubules of the rat. *Exp. Cell Res.* 15:438-440.

_____. 1960. The cycle of the seminiferous epithelium of the guinea pig. A method for identification of the stages. *Fertil. Steril.* 11:563-573.

_____. 1963. Cycle of the seminiferous epithelium in man. *Am. J. Anat.* 112:35-45.

Clermont, Y., and Bustos-Obregon, E. 1968. Reexamination of spermatogonial renewal in the rat by means of seminiferous tubules mounted "in toto." *Am. J. Anat.* 122:237-247.

Clermont, Y.; Clegg, R.C.; and Leblond, C.P. 1955. Presence of carbohydrates in the acrosome of the guinea pig spermatozoon. *Exp. Cell Res.* 8:453-458.

Clermont, Y., and Leblond, C.P. 1953. Renewal of spermatogonia in the rat. *Am. J. Anat.* 93:475-502.

Clermont, Y.; Leblond, C.P.; and Messier, B. 1959. Durie du cycle de l'epithelium seminal du rat. *Arch. Anat. Micr. Morph. Exp.* 48:37-56.

Cohen, M.P.; Hoffer, A.P.; and Rosen, S. 1976. Carbonic anhydrase localization in the epididymis and testis of the rat: histochemical and biochemical analysis. *Biol. Reprod.*, in press.

Cooper, G.W., and Bedford, J.M. 1971. Acquisition of surface charge by the plasma membrane of mammalian spermatozoa during epididymal maturation. *Anat. Rec.* 169:300-301.

Cooper, T.G., and Waites, G.M.H. 1974. Testosterone in rete testis fluid and blood of rams and rats. *J. Endocrinol.* 62:619-629.

Coppola, J.A. 1969. Extragonadal male antifertility agent. *Life Sci.* 8: 43-48.

Coppola, J.A., and Saldarini, R.J. 1974. A new orally active male antifertility agent. *Contraception* 9:459-470.

Coutinho, E.M., and Melo, J.F. 1973. Successful inhibition of spermatogenesis in man without loss of libido: A potential new approach to male contraception. *Contraception* 8:207-217.

Crabo, B. 1965. Studies on the composition of the epididymal content in boar and bull. *Acta Vet. Scand.* 6: (suppl 5)1-94.

Crabo, B., and Appelgren, L.E. 1972. Distribution of $^{14}C$ α-chlorohydrin in mice and rats. *J. Reprod. Fertil.* 30: 161-163.

Dawson, R.M.C.; Mann, T.; and White, I.G. 1957. Glycerophosphorylcholine and phosphorylcholine in semen and their relation to choline. *Biochem. J.* 65:627-633.

Dawson, R.M.C., and Rowlands, I.W. 1959. Glycerophosphorylcholine in the male reproductive organs of rats and guinea pigs. *Q. J. Exp. Physiol.* 44:26-34.

De Graaf, R. 1668. Tractatus de virorum organis generationi inservientibus, In *Opera Omnia*, Lugduni Batavarum: Hachiana.

de Kretser, D.M. 1967. Changes in the fine structure of the human testicular interstitial cells after treatment with human gonadotropins. *Z. Zellforsch.* 83:344-358.

_____. 1969. Ultrastructural features of human spermiogenesis. *Z. Zellforsch.* 98:477.

_____. 1974. The regulation of male fertility. The state of the art and future possibilities. *Contraception* 9:561-599.

de Kretser, D.M.; Kerr, J.B.; and Paulsen, C.A. 1975. Peritulular tissue in the normal and pathological human testis. An Ultrastructural study. *Biol. Reprod.* 12:317-324.

de Kretser, D.M.; Burger, H.G.; and Hudson, B. 1975. The pituitary testicular response to luteinizing hormone releasing hormone administration to normal men. *Austr. N.Z. J. Med.* 5:227.

Drobeck, H.P., and Coulston, F. 1962. Inhibition and recovery of spermatogenesis in rats monkeys and dogs medicated with bis(dichloroacetyl) diamines. *Exp. Mol. Pathol.* 1:251-274.

Dym, M., and Fawcett, D.W. 1970. Observations on the blood-testis barrier of the rat on the physiological compartmentation of the seminiferous epithelium. *Biol. Reprod.* 3:308-326.

---. 1971. Further observations on the numbers of spermatogonia, spermatocytes and spermatids joined by intercellular bridges in mammalian spermatogenesis. *Biol. Reprod.* 4:195-215.

Ebner, V. von. 1871. Untersuchungen uber den Bau der Samenkanalchen und die Entwicklung der Spermatozoiden bei den Saugethieran und beim Menachen. Leipzig.

---. 1888. Zur Spermatogeneses bei den Saugetieren. *Arch. Mikr. Anat.* 31:236-292.

Edelman, G.M., and Millette, C.F. 1971. Molecular probes of spermatozoon structure. *Proc. Natl. Acad. Sci.* 68:2436-2440.

Elkington, J.S.H.; Sanborn, B.M.; and Steinberger, E. 1975. The effect of testosterone proprionate on the concentration of testicular and epididymal androgen-binding activity in the hypophysectomized rat. *Mol. Cell. Endocrinol.* 2:157-170.

Erickson, R.P.; Friend, D.S.; and Tennenbaum, D. 1976. Localization of lactate dehydrogenase-X on the surfaces of mouse spermatozoa. *Exp. Cell Res.* 91:1-5.

Ericsson, R.J., and Baker, V.F. 1970. Male antifertility compounds: biological properties of U-5897 and U-15, 646. *J. Reprod. Fertil.* 21:267-274.

Ericsson, R.J., and Youngdale, G.A. 1970. Male antifertility compounds: structure and activity relationships of U-5897, U-15646 and related substances. *J. Reprod. Fertil.* 21:263-266.

Euler, U.S. von. 1936. On the specific vasodilating and plain muscle stimulating substances from accessory genital glands in man and certain animals (prostaglandin and vesiglandin). *J. Physiol.* 88:213-234.

Ewing, L.; Stratton, L.G.; and Desjardins, C. 1973. Effect of testosterone polydimethylsiloxane implants upon sperm production, libido, and accessory sex organ function in rabbits. *J. Reprod. Fertil.* 35:245-253.

Fawcett, D.W. 1956. The fine structure of chromosomes in meiotic prophase of vertebrate spermatocytes. *J. Biophys. Biochem. Cytol.* 2:403-406.

---. 1961. Intercellular bridges. *Exp. Cell Res.* (suppl.) 8:174-187.

---. 1970. A comparative view of sperm ultrastructure. *Biol. Reprod.* (suppl.) 2:90-127.

---. 1973. Observations on the organization of the interstitial tissue of the testis and on the occluding cell junctions in the seminiferous epithelium, pp. 83-100. *Advances in Biosciences* 10. Pergamon Press, Vieweg.

---. 1974. Interactions between Sertoli cells and germ cells. In *Male fertility and sterility*, ed. R.E. Mancini and L. Martini, pp. 13-36. New York: Academic Press.

_____. 1975a. Gametogenesis in the male. Prospects for its control. In *Developmental biology of reproduction*, pp. 25-54. New York: Academic Press.

_____. 1975b. The ultrastructure and functions of the Sertoli cells. In *Handbook of physiology*, sect. 7, Endocrinology, eds. D.W. Hamilton and R.O. Greep, pp. 21-55. Washington, D.C.: Physiological Society.

Fawcett, D.W.; Ito, S.; Slautterback, D.L. 1959. The occurrence of intercellular bridges in groups of cells exhibiting synchronous differentiation. *J. Biophys. & Biochem. Cytol.* 5:453-460.

Fawcett, D.W.; Anderson, W.A.; and Phillips, D.M. 1971. Morphogenetic factors influencing the shape of the sperm head. *Dev. Biol.* 26:220-251.

Fawcett, D.W., and Hoffer, A.P. 1976. Failure of androgen alone to prevent regression of the initial segment of the rat epididymis after castration or efferent duct ligation. *Anat. Rec.*, in press.

Fawcett, D.W., and Hollenberg, R.D. 1963. Changes in the acrosome of guinea pig spermatozoa during passage through the epididymis. *Z. Zellforsch.* 60:276-292.

Fawcett, D.W.; Ito, S.; and Slautterback, D.L. 1959. The occurrence of intercellular bridges in groups of cells exhibiting synchronous differentiation. *J. Biophys. Biochem. Cytol.* 5:453-460.

Fawcett, D.W.; Leak, L.V.; and

Heidger, P.M. 1970. Electron microscopic observations on the structural components of the blood-testis barrier. *J. Reprod. Fertil.* (suppl) 10: 105-122.

Fawcett, D.W.; Long, J.A.; and Jones, A.L. 1969. The ultrastructure of endocrine glands. *Recent Prog. Horm. Res.* 25:315-368.

Fawcett, D.W., and Phillips, D.M. 1969a. Observations on the release of spermatozoa and on changes in the head during passage through the epididymis. *J. Reprod. Fertil.* (suppl) 6:405-418.

_____. 1969b. The fine structure and development of the neck region of the mammalian spermatozoon. *Anat. Rec.* 165:153-184.

_____. 1970. Recent observations on the ultrastructure and development of the mammalian spermatozoon. *Proceedings of the 1st International Symposium on Comparative Spermatology, Quad. #137*, pp. 2-38. Accademia Nazionale Rome: dei Lincei.

Fawcett, D.W.; Neaves, W.B.; and Flores, M.N. 1973. Comparative observations on intertubular lymphatics and the organization of the interstitial tissue of the mammalian testis. *Biol. Reprod.* 9:500.

Felizet, G., and Branca, A. 1898. Histologie du testicule ectopique. *J. d'Anat. et Physiol.* 34:589-641.

Fellous, J.; Gachelin, G.; Buc-Caron, M.H.; Dubois, P.; and Jacob, F. 1976. Similar location of an early embryonic antigen on mouse and human spermatozoa. *Dev. Biol.*, in press.

Fevold, H.L.; Hisaw, F.L.; and Leonard, S.L. 1931. The gonad stimulating and luteinizing hormones of the anterior lobe of the hypophysis. *Am. J. Physiol.* 97:291-301.

Fleischer, B.; Fleischer, S.; and Ozawa, H. 1969. Isolation and characterization of Golgi membranes from bovine liver. *J. Cell Biol.* 43:59-79.

Flickinger, C.J., and Fawcett, D.W. 1967. The junctional specializations of Sertoli cells in the seminiferous epithelium. *Anat. Rec.* 158:207-222.

Flores, M.N., and Fawcett, D.W. 1972. Ultrastructural effects of the antispermatogenic compound WIN-18446 (bis dichloroacetyl diamine). *Anat. Rec.* 172:310.

Folman, Y.; Haltmeyer, G.C.; and Eik-Nes, K.B. 1972. The presence and formation of 5 α-dihydrotestosterone in rat testis *in vivo* and *in vitro*. *Endocrinology.* 91:702-710.

Franchimont, P.; Millet, D.; Vandrely, J.; Letawa, J.; Legros, J.J.; and Netter, A. 1972. Relationship between spermatogenesis and serum gonadotropin levels in ozoospermia and oligospermia. *J. Clin. Endocrinol. Metab.* 34:1003-1008.

Franchimont, P.D.; Chari, S.; and Demoulin, A. 1975. Hypothalamus, pituitary, testis interaction. *J. Reprod. Fertil.* 44:335-350.

Franchimont, P.; Chari, S.; Hagelstein, M.T. et al. 1975. Existence of a follicle stimulating hormone inhibiting factor 'inhibin' in bull seminal plasma. *Nature.* 257:402-404.

Frankel, A.I., and Eik-Nes, K.B. 1968. Steroidogenesis *in vitro* of the epididymis of the rabbit. *Red. Proc.* (abst.) 27:624.

―――. 1970. Metabolism of steroids in the rabbit epididymis. *Endocrinology.* 87:646-652.

Fray, C.S.; Hoffer, A.P.; and Fawcett, D.W. 1972. A reexamination of motility patterns of rat epididymal spermatozoa. *Anat. Rec.* 173:301-308.

French, F.S., and Ritzen, E.M. 1973. A high affinity androgen binding protein (ABP) in rat testis: Evidence for secretion into efferent duct fluid and absorption by epididymis. *Endocrinology.* 95:88-93.

Frick, J., and Bartsch, G. 1973. Inhibition of spermatogenesis. In *Physiology and genetics of reproduction*, part A., ed. E.M. Coutinho and F. Fuchs, pp. 259-274. New York: Plenum Press.

Friend, D.S., and Fawcett, D.W. 1974. Membrane differentiations in freeze-fractured mammalian sperm. *J. Cell Biol.* 63:641-664.

Fritz, I.B.; Kopec, B.; Lam, K.; and Vernon, R.G. 1974. Effects of FSH on levels of ABP in the testis. In *Hormone binding and activation in testis*, ed. M. Dufaye and A.R. Means, New York: Plenum Press.

Gaddum-Rosse, P. 1968. Sperm maturation in the male reproductive tract: Development of motility. *Anat. Rec.* 161:471-482.

Gaddum-Rosse, P., and Blandau, R.J.

1972. Comparative studies on the proteolysis of fixed gelatin membranes by mammalian sperm acrosomes. *Am. J. Anat.* 134:133-143.

Gatenby, J.B. 1918. The cytoplasmic inclusions of the germ cells. I. Lepidoptera. *Q. J. Micros. Sci.* 63:197-258.

Gerson, T.; Shortland, F.; and Dunckley, G. 1964. The effect of β-sitosterol on the metabolism of cholesterol and lipids in rats on a diet low in fat. *Biochem. J.* 93:385-390.

Gibbons, I.R. 1966. *Studies on the ATPase activity of 14S and 30S dynein from cilia of Tetrahymena.* J. Biol. Chem. 241:5590-5596.

Gibbons, I.R., and Rowe, A.J. 1965. Dynein: A protein with ATPase activity from cilia. *Science* 149:424-426.

Gilula, N.B.; Fawcett, D.W.; and Aoki, A. 1976. Ultrastructural and experimental observations on the Sertoli cell junctions of the mammalian testis. *Dev. Biol.* 50:142-168.

Glover, T.D., and Nicander, L. 1971. Some aspects of structure and function in the mammalian epididymis. *J. Reprod. Fertil.* (suppl) 13:39-50.

Goldberg, E. 1972. Amino acid composition and properties of crystalline lactate dehydrogenase X from mouse testis. *J. Biol. Chem.* 247:2044-2048.

_____. 1973. Infertility in female rabbits immunized with lactate dehydrogenase X. *Science* 181:458-459.

Gondos, B., and Zemjanis, R. 1970. Fine structure of spermatogonia and intercellular bridges in Maeaca nemestrina. *J. Morphol.* 131:431-446.

Greep, R.O. 1937. The dual nature of anterior pituitary influence on the testis. *Anat. Rec.* (suppl) 67:22.

Greep, R.O., and Fevold, H.L. 1937. The spermatogenic and secretory function of the gonads of hypophysectomized adult rats treated with pituitary FSH and LH. *Endocrinology* 21:611-618.

Greep, R.O.; Fevold, H.L.; and Hisaw, F.L. 1936. Effects of two hypophyseal gonadotropic hormones on the reproductive system of the male rat. *Anat. Rec.* 65:261-272.

Greep, R.O., and Jones, I.C. 1950. Steroid control of pituitary function. *Recent Prog. Horm. Res.* 5:197-261.

Greep, R.O.; Van Dyke, H.B.; and Chow, B.F. 1942. Gonadotropins of the sevine pituitary. 1. Various biological effects of purified Thylokentrin (FSH) and pure Metakentrin (ICSH). *Endocrinology* 30:635-649.

Gunn, S.A., and Gould, T.C. 1956. Differences between dorsal and lateral components of the dorsolateral prostate of rat in $^{65}$Zn uptake. *Proc. Soc. Exp. Biol. Med.* 92:17-26.

Gutman, E.B.; Sproul, E.E.; and Gutman, A.B. 1936. Significance of increased phosphatase activity at the site of osteoplastic metastases secondary to carcinoma of the prostate. *Am. J. Cancer* 28:485-495.

Gwatkin, R.B.L., and Anderson, O.F. 1969. Capacitation of hamster spermatozoa by bovine follicular fluid. *Nature* 224:1111-1112.

Gwatkin, R.B.; Anderson, O.F.; and Hutchinson, C.F. 1972. Capacitation of hamster spermatozoa in vitro: The role of cumulus components. *J. Reprod. Fertil.* 30:389-394.

Hagenas, L.; Ritzen, E.M.; and French, F.S. 1974. Testicular androgen binding protein (ABP): Localization of site of production. *J. Steroid Biochem.* (abstr) 5:382.

Hall, P.F. 1963. The effect of interstitial cell stimulating hormone on the biosynthesis of testicular cholesterol from acetate-1-$C^{14}$. *Biochemistry* 2:1232-1237.

_____. 1970. In *The androgens of the testis*, ed. K.B. Eik-Nes, pp. 73-115. New York: M. Dekker.

Hall, P.F., and Eik-Nes, K.B. 1964. The effect of interstitial cell-stimulating hormone on the production of pregnenolone by rabbit testis in the presence of an inhibitor of 17 α-hydroxylase. *Biochem. Biophys. Acta* 86:604-609.

Hamilton, D.W. 1971. Steroid function in the mammalian epididymis. *J. Reprod. Fertil.* (suppl) 13:89-97.

_____. 1972. The mammalian epididymis. In *Reproductive biology*, ed. H. Balin and S. Glasser, pp. 268-337. Amsterdam: Excerpta Medica Foundation.

Hamilton, D.W.; Jones, A.L.; and Fawcett, D.W. 1968. Sterol biosynthesis from (1-$^{14}$C) acetate in the epididymis and vas deferens of the mouse. *J. Reprod. Fertil.* (abstr) 18:156.

Hansson, V.; Ritzen, E.M.; and French, F.S. 1974. Androgen transport mechanisms in the testis and epididymis. *Acta endocr. Copnk* (suppl) 191:191-198.

Hansson, V.; Djoseland, O.; Reusch, E.; Attramadal, A.; and Torgerson, O. 1973a. An androgen binding protein in the testis cytosol fraction of adult rats: Comparison with the androgen binding protein in the epididymis. *Steroids* 21:457-474.

_____. 1973b. Intracellular receptors for 5-α-dihydrotestosterone in the epididymis of adult rats. Comparison with the androgenic receptor in the ventral prostate and the androgen binding protein (ABP) in the testicular and epididymal fluid. *Steroids* 22:19-33.

Harris, M.E. 1973. Concentration of testosterone in testis fluid of the rat. *Proc. Soc. Study of Reprod.* (abstr) 55:61.

Harrison, R.G. 1949. The comparative anatomy of the blood-supply of the mammalian testis. *Proc. Zool. Soc. Lond.* 119:325.

Hartree, E.F., and Srivastava, P.N. 1963. Chemical composition of acrosomes of ram spermatozoa. *J. Reprod. Fertil.* 5:225-232.

Heller, C.G.; Nelson, W.O.; Hill, I.C.; Henderson, E.; Maddock, W.O.; and Jungch, E.C. 1950. The effect of testosterone administration upon the human testis. *J. Clin. Endocrinol. Metab.* (abstr) 10:816.

Heller, C.G.; Moore, D.J.; Paulsen, C.A.; Nelson, W.; and Laudlow, W.M.; 1959. Effects of progesterone and synthetic progestins upon the reproductive physiology of normal men. *Fed. Proc.* 18:1057-1065.

Heller, C.G.; Moore, D.J.; and Paulsen, C.A. 1961. Suppression of spermatogenesis and chronic toxicity in men by a new series of bis-(dichloroacetyl)-diamines. *Toxicol. Appl. Pharmacol.* 3:1-11.

Heller, C.G.; Flageolle, B.Y.; and Matron, L.J. 1963. Histopathology of the human testis as affected by bis-(dichloroacetyl)-diamines. *Exp. Mol. Pathol.* (suppl) 2:107-114.

Hertwig, O. 1875. Beitrage zur Kemitniss der Beldung, Befruchtung und Teilung des tierischen Eiex. *Morphol. Jahrbuch.* (Liepzig), p. 347.

Hochereau, M.T. 1963. Etude comparee de la vague spermatogenetique chez le taureau et chez le rat. *Ann. Biol. Animale Biochim. Biophys.* 3:5-20.

Hoffer, A.P.; Hamilton, D.W.; and Fawcett, D.W. 1973. The ultrastructure of the principal cells and intraepithelial leucocytes in the initial segment of the rat epididymis. *Anat. Rec.* 175:169-202.

Holstein, A.F. 1969. Morphologische Studien am Nebenhoden des Menschen. *Zuanglose Abhandl. Gebiet Nom. Pathol. Anat.* 20:1-91.

Horan, A.H., and Bedford, J.M. 1972. Development of fertilizing ability of spermatozoa in the epididymis of the Syrian hamster. *J. Reprod. Fertil.* 30:417-423.

Hovatta, O. 1972. Effects of androgens and antiandrogens on the development of the myoid cells of the rat seminiferous tubules. *Zeitschr. Zellforsche.* 131:299-308.

Huckins, C. 1971. The spermatogonial stem cell population in adult rats. I. Their morphology, proliferation and maturation. *Anat. Rec.* 1969:533-548.

Huckins, C., and Oakberg, E.F. 1971. Cytoplasmic connections between spermatogonia seen in whole-mounted seminiferous tubules from normal and irradiated mouse testis. *Anat. Rec.* (abstr) 169:344.

Huggins, C. 1956. Control of cancer of man by endocrinologic methods. A review. *Cancer Res.* 16:825-830.

Huggins, C.; Masina, M.H.; Eichelberger, L.; and Wharton, J.D. 1939. Quantitative studies of prostatic secretion. I. Characteristics of the normal secretion; the influence of thyroid, suprarenal and testis extirpation and androgen substitution on the prostatic output. *J. Exp. Med.* 70:543-556.

Inano, H.; Machino, A.; and Tamaoki, B. 1969. *In vitro* metabolism of steroid hormones by cell free homogenates of epididymis of adult rats. *Endocrinology* 84:997-1003.

Iwamatsu, T., and Chang, M.C. 1969. *In vitro* fertilization of mouse eggs in the presence of bovine follicular fluid. *Nature* 224:919-920.

_____. 1970. Further investigation of capacitation of sperm and fertilization of mouse ova in vitro. *J. Exp. Zool.* 175:271-282.

_____. 1971. Factors involved in the fertilization of mouse eggs *in vitro*. *J. Reprod. Fertil.* 26:197-208.

Jackson, H. 1959. Antifertility substances. *Pharmacol. Rev.* 11:135-172.

_____. 1970. Antispermatogenic agents. *Br. Med. Bull.* 26:79-86.

Johansson, E.D.B., and Nygren, K.G. 1973. Depression of plasma testosterone levels in men with norethindrone. *Contraception* 8:219-226.

Johnson, M.H. 1970. An immunological barrier in guinea pig testis. *J. Pathol.* 101:129-139.

_____. 1973. Physiological mechanisms for immunological isolation of spermatozoa. *Adv. Reprod. Physiol.* 6:279-324.

Johnson, S.G. 1970. The stage of spermatogenesis involved in the testicular-hypophyseal feed-back mechanism in man. *Acta Endocrinol.* 64:193-210.

Kierszenbaum, A.L. 1974. RNA synthetic activities of Sertoli cells in mouse testis. *Biol. Reprod.* 11:865-876.

Kierszenbaum, A.L., and Tres, L.L. 1974. Nucleolar and perichromosomal RNA synthesis during meiotic prophase in the mouse testis. *J. Cell Biol.* 60:39-53.

Kirton, K.T.; Ericsson, R.J.; Ray, J.A.; and Forbes, A.D. 1970. Male anti-fertility compounds: Efficacy of U-5897 in primates (Macaca mulatta). *J. Reprod. Fertil.* 21:275-278.

Kolliker, R.A. von. 1841. Beitrage zur Kenntnis der Geschlechtwerhaltnisse und der Sämenflussigheit wirbelloser Thiere. Berlin.

Kormano, M. 1967. Dye permeability and alkaline phosphatase activity of testicular capillaries in the post natal rat. *Histochemie* 9:327-338.

_____. 1968. Penetration of intravenous trypan blue into the rat testis and epididymis. *Acta Histochem.* 30:133-136.

Kruegar, D.M.; Hodgen, G.D.; and Scherins, R.J. 1974. New evidence for the role of the Sertoli cell and spermatogonia in feedback control of FSH secretion in male rats. *Endocrinology* 95:955-962.

Krutscher, W., and Wolbergs, H. 1935. Prostataphosphatase. *Ztschr. Physiol. Chem.* 236:237.

Ladman, A.J. 1967. The fine structure of the ductuli efferentes of the opossum. *Anat. Rec.* 157:559-576.

Lam, D.M.K.; Furrer, R.; and Bruce, W.R. 1970. The separation, physical characterization, and differential kinetics of spermatogonial cells of the mouse. *Proc. Natl. Acad. Sci.* 65:192-199.

Landace, R.L., and Laughead, R. 1951. Seminal fructose concentration as an index of androgenic activity in man. *J. Clin. Endocrinol.* 11:1411-1424.

La Valette, St. George von. 1865.

Uber die Genese der Samenkorper I. *Arch. Mikr. Anat.* 1:403-414.

Leblond, C.P., and Clermont, Y. 1952a. Definition of the stages of the cycle of the seminiferous epithelium in the rat. *Ann. N.Y. Acad. Sci.* 55:548-573.

―――. 1952b. Spermatogenesis of rat, mouse, hamster and guinea pig as revealed by the "periodic-acid fuchsin sulfurous acid" technique. *Am. J. Anat.* 90:167-215.

Lee, P.A.; Jaffe, R.B.; Midgley, A.P.; Kohen, F.; and Niswender, D. 1972. Regulation of human gonadotropins. VIII. Suppression of serum LH and FSH in adult males following exogenous testosterone administration. *J. Clin. Endocrinol. Metab.* 35:636-664.

Lee, V.K.; Keogh, E.J.; de Kretser, D.M.; and Hudson, B. 1974a. FSH, LH and testosterone levels in the rat from birth to sexual maturity (Proc). *J. Reprod. Fertil.* 36:479-480.

―――. 1974b. Selective suppression of serum gonadotropins in castrated rats by testis homogenates. *Int. Res. Comm. Med. Sci.* 2:1406.

Leeuwenhoek, Anton van. 1677. *Phil. Trans. Roy. Soc.* (London).

―――. 1685. *Phil. Trans. Roy. Soc.* (London) vol. 15.

Leydig, F. 1850. Zur Anatomie der mämlicken Geschlechtsorgane und Analdrusen der Saugethiere *Z. wiss. Zool.* 2:1.

Long, J.A., and Jones, A.L. 1967. The fine structure of the zona glomerulosa and zona fasciculata of the adrenal cortex of the opossum. *Am. J. Anat.* 120:463-488.

Lubicz-Nawrocki, C.M., and Chang, M.C. 1973. The comparative efficacy of testosterone, progesterone and dehydroepiandrosterone for the maintenance of fertilizing capacity in castrated hamsters. *Biol. Reprod.* 9:279-295.

Lubicz-Nawrocki, C.M., and Glover, T.D. 1973. The influence of the testis on the survival of spermatozoa in the epididymis of the golden hamster. *J. Reprod. Fertil.* 34:315-329.

McCullagh, D.R. 1932. Dual endocrine activity of the testis. *Science* 76:19.

McCullagh, D.R., and Walsh, E.L. 1935. Experimental hypertrophy and atrophy of the prostate gland. *Endocrinology* 19:466-470.

McCullagh, D.R., and Schneider, I. 1940. The effect of a non-androgenic testis extract on the estrous cycle in rats. *Endocrinology* 27:899-902.

McGadey, J.; Baillie, A.H.; and Fergurson, M.M. 1966. Histochemical utilization of hydroxysteroids by the hamster epididymis. *Histochimie* 7:211-217.

McGee, L.C.; Jukn, M.; and Domm, L.V. 1928. The development of secondary sex characters in capons by injection of extracts of bull testis. *Am. J. Physiol.* 87:406-435.

McLean, W.S.; Tindall, D.J.; Nayfeh, S.N.; French, F.S.; and Hansson, V. 1974. Preliminary characterization of an androgen receptor in cytosol fraction from seminiferous tubules. Separation from testicular androgen binding protein (ABP). *J. Biol. Chem.*

MacLeod, J. 1965. Human cytology following the administration of certain antispermatogenic compounds. In *Agents affecting fertility*, ed. C.R. Austin and J.S. Perry, pp. 92-123. Boston: Little, Brown and Co.

Maneely, R.B. 1959. Epididymal structure and function: A historical and critical review. *Acta Zool.* 40:1-21.

Mann, T. 1945. Studies on the metabolism of semen. I. General aspects. Occurrence and distribution of cytochrome, certain enzymes and coenzymes. *Biochem. J.* 39:451-458.

_____. 1954. *The biochemistry of semen.* London: Methuen and Company, Ltd.

_____. 1968. *The biochemistry of semen and of the male reproductive tract.* London: Methuen and Company, Ltd.

Mann, T.; Davies, D.V.; and Humphrey, G.F. 1949. Fructose and citric acid assay in the secretions of the accessory glands of reproduction as indicators of male sex hormone activity. *J. Endocrinol.* 6:75-85.

Mann, T., and Parsons, U. 1950. Studies on the metabolism of semen. VI. Role of hormones. Effect of castration, hypophysectomy and diabetes. Relation between blood glucose and seminal fructose. *Biochem. J.* 46:440-450.

Marques, N.R., and Fritz, I.B. 1965. Effects of testosterone on the distribution of carnitine, acetylcarnitine and carnitine acetyltransferase in tissues of the reproductive system of the male rat. *J. Biol. Chem.* 240:2197-2200.

Mason, K.K., and Shaver, S.L. 1952. Some functions of the caput epididymides. *Ann. N.Y. Acad. Sci.* 55:585-593.

Mason, K.K.; Shaver, S.L.; Hodge, H.C.; and Maynard, E.A. 1952. Extensive intratubular and edema of the rat testis compatible with spermatogenic function. *Anat. Rec.* 109:402-403.

Mawson, C.A., and Fischer, M.I. 1957. Carbonic anhydrase and zinc in the prostate glands of rat and rabbit. *Arch. Biochem. Biophys.* 36:485-486.

Means, A.R. 1973. Specific interaction of $^3$H-FSH with rat testis binding sites. *Adv. Exp. Med. Biol.* 36:431-448.

Means, A.R., and Hall, P.F. 1967. Effect of FSH on protein biosynthesis in the testis of the immature rat. *Endocrinology* 81:1151.

_____. 1971. Protein biosynthesis in the testis. Action of FSH on polyribosomes in immature rats. *Cytobios.* 3:17-24.

Means, A.R., and Vaitukaitis, 1972. Peptide hormone receptors, specific binding of $^3$H-FSH to testis. *Endocrinology* 90:39-46.

Meistrich, M.L. 1972. Separation of mouse spermatogenic cells by velocity sedimentation. *J. Cell Physiol.* 80:299-312.

Merkow, L.; Acevedo, H.F.; Slifkin, M.; and Caito, B.J. 1968. Studies on the interstitial cells of the testis. 1. The ultrastructure in the immature guinea pig and the effect of stimulation with human chorionic gonadotropin. *Am. J. Pathol.* 53:47-61.

Metz, C.B. 1973. Role of specific sperm antigens in fertilization. *Fed. Proc.* 32:2057-2064.

Meves, F. 1899a. Uber Struktur und Histogenese der Samenfäden der Meerschweinchens. *Arch. Fur Mikr. Anat.* 54:329-402.

―――――. 1899b. Uber Struktur und Histogenese der Samenfaden des Meerschweinchens. *Arch. Fur. Mikr. Anat. LIV.*

―――――. 1901. Struktur und Histogenese der Spermien. *Ergebnisse der Anatomie und Entwinklungsgeschicte* 11:437-516.

Miller, R.G., and Phillips, R. 1969. Separation of cells by velocity sedimentation. *J. Cell Physiol.* 78:191-202.

Mills, S.C., and Scott, T.W. 1969. Metabolism of fatty acids by testicular and ejaculated spermatozoa. *J. Reprod. Fertil.* 18:367-369.

Miyamoto, H., and Chang, M.C. 1972. Development of mouse eggs fertilized *in vitro* by epididymal spermatozoa. *J. Reprod. Fertil.* 30:135-137.

―――――. 1973a. Effect of osmolality on fertilization of mouse ova *in vitro*. *J. Reprod. Fert.* 32:481-487.

―――――. 1973b. The importance of serum albumin and metabolic intermediates for capacitation and fertilization *in vitro*. *J. Reprod. Fert.* 32:193-205.

Moens, P.B. and Go, V.L.W. 1971. Intercellular bridges and division patterns of rat spermatogonia. *Z. Zellforsch.* 127:201-208.

Moens, P.B., and Hugenholtz, A.D. 1975. The arrangement of germ cells in the rat seminiferous tubule: An electron microscopic study. *J. Cell Sci.* 19:487-507.

Monesi, V. 1962. Autoradiographic study of DNA synthesis and the cell cycle in spermatogonia and spermatocytes of mouse testis using tritiated thymidine. *J. Cell. Biol.* 14:1-18.

―――――. 1965. Synthetic activities during spermatogenesis in the mouse RNA and protein. *Exp. Cell Res.* 39:197-224.

Monita, J. 1966. Some observations on the fine structure of the human ductule efferentes testis. *Arch. Histol. Jpn.* 26:341-365.

Moor, H. 1964. Die Gefrier - fixation lebenden Zellen und ikse annsudung in der Elektronen mikroskopie. *Z. Zellforsch.* 62:546-580.

Moor, H., and Muhlenthaler, K. 1963. Fine structure in frozen etched yeast cells. *J. Cell. Biol.* 17:609-628.

Moore, C.R. 1924a. VI. Testicular reactions to experimental cryptorchidism. *Am. J. Anat.* 34:269-276.

―――――. 1924b. VIII. Heat application and testicular degeneration; the function of the scrotum. *Am. J. Anat.* 34:337-358.

_____. 1926a. The biology of the mammalian testis and scrotum. *Q. Rev. Biol.* 1:4-50.

_____. 1926b. Scrotal replacement of experimental cryptorchid testes and the recovery of spermatogenic function. *Biol. Bull.* 51:112-128.

_____. 1928. Hormone production in the normal testis, cryptorchid testis and non-living testis grafts as indicated by the spermatozoon motility test. *Biol. Bull.* 55:339-357.

Moore, C.R., and Gallagher, T.F. 1930. Seminal vesicle and prostate function as a testis hormone indicator. *Am. J. Anat.* 45:39-69.

Moore, C.R.; Hughes, W.; and Gallagher, T.F. 1930. Rat seminal vesicle cytology as a testis hormone indicator and the prevention of castration changes by testis extract injection. *Am. J. Anat.* 45:109-135.

Moore, C.R., and McGee, L.C. 1928. On the effects of injecting lipid extracts of bull testes into castrated guinea pigs. *Am. J. Physiol.* 87:436-446.

Moore, C.R., and Price, D. 1932. Gonad hormone functions and the reciprocal influence between gonads and hypophysis, with its bearing upon sex hormone antagonism. *Am. J. Anat.* 50:13-71, 1932.

Morris, M.D., and Chaikoff, I.L. 1959. The origin of cholesterol in liver, small intestine, adrenal gland and testis of the rat: Dietary versus endogenous contributions. *J. Biol. Chem.* 234: 1095-1097.

Morton, D.B. 1975. Acrosomal enzymes: Immunochemical localization of acrosin and hyaluronidase in ram spermatozoa. *J. Reprod. Fertil.* 45: 375-378.

Moses, M. 1956. Chromosomal structures in crayfish spermatocytes. *J. Biophys. Biochem. Cytol.* 2:215-218.

_____. 1968. The synaptonemal complex. *Ann. Rev. Genet.* 2:263-412.

Moses, M.; Counce, S.J.; Paulsen, D.F. 1975. Synaptonemal complex complement of man in spreads of spermatocytes, with details of the sex chromosome pair. *Science* 187:363.

Moudgal, N.R.; MacDonald, G.J.; and Greep, R.O. 1971. Effect of hCG antiserum on ovulation and corpus luteum formation in the monkey. *J. Clin. Endocrinol. Metab.* 32:579-581.

Moyle, W.R. 1970. The action of luteinizing hormone on steroidogenesis in transplantable mouse testes tumors. Ph.D. Thesis, Harvard University, Cambridge, Massachusetts.

Moyle, W.R., and Armstrong, D.T. 1970. Stimulation of testosterone biosynthesis by luteinizing hormone in transplantable mouse Leydig cell tumors. *Steroids* 15:681-693.

Murad, F.; Strauck, B.S.; and Vaughn, M. 1969. The effect of gonadotropins on testicular adenyl cyclase. *Biochem. Biophys. Acta* 177:591-598.

Murota, S.; Shikita, M.; and Tamaoki, B. 1966. Androgen formation in the testicular tissue of patients with prostatic carcinoma. *Biochem. Biophys. Acta.* 117:241-246.

Murphy, H.D. 1965. Sertoli cell stimulation following intratesticular injections of FSH in the hypophysectomized rat. *Proc. Soc. Exp. Biol. Med.* 118:1202-1205.

Nagano, T., and Okumura, K. 1973. Fine structural changes of allergic spermatogenesis in the guinea pig. I. Similarity in the initial changes induced by passive transfer of anti-testis serum and by immunization with testicular tissue. *Virchow's Arch. Abt. B. Zellpath.* 14:223-236.

Nagano, T., and Suzuki, S. 1976a. Freeze-fracture observations on the intercellular junctions of Sertoli cells and of Leydig cells in the human testes. *Cell and Tissue Res.* 166:37-48.

———. 1976b. The postnatal development of the junctional complexes of the mouse Sertoli cells as revealed by freeze-fracture. *Anat. Rec.*, in press.

Neaves, W.B. 1973a. Permeability of Sertoli cell tight junctions to lanthanum after ligation of the ductus diferens and ductuli efferentes. *J. Cell Biol.* 59:559-572.

———. 1973b. Ultrastructural transformation of a murine Leydig cell tumor after gonadotropin administration. *J. Natl. Cancer Inst.* 50:1069-1073.

———. 1975. Biological aspects of vasectomy. In *Male reproductive system*, ed. R.O. Greep and D.W. Hamilton, pp. 383-403. Bethesda: American Physiological Society.

Nelson, W.O. 1936. The effect of various sex hormones on the testes of hypophysectomized rats. *Anat. Rec.* (abstr) 67:110.

Nelson, W.O., and Bunge, R.G. 1957. The effect of therapeutic doses of nitrofurantoin (Furadantin) upon spermatogenesis in man. *J. Urol.* 77:275-281.

Nelson, W.O., and Patanelli, D.J. 1965. Chemical control of spermatogenesis. In *Agents affecting fertility*, ed. C.R. Austin and J.S. Perry, pp. 78-92. Boston: Little, Brown and Co.

Neri, R.; Florance, K.; Koziol, P.; and Van Cleave, S. 1972. A biological profile of a non-steroidal antiandrogen 4'Nitro-3'trifluoromethylisobutyr anilide. *Endocrinology* 91:427-437.

Neumann, H.C.; Potts, G.O.; and Ryan, W.T. 1970. Steroidal heterocycles. 13. 4-alpha, 5-epoxy-5 alpha-androst-2 eno [2,3d] isoxazoles and related compounds. *J. Med. Chem.* 13:948-951.

Neumasin, F., and Berswordt-Wallrabe, R. 1966. Effects of the androgen antagonist cyproterone acetate on testicular structure, spermatogenesis, and accessory sexual glands of testosterone treated adult hypophysectomized rats. *J. Endocrinol.* 35:363-371.

Neutra, M., and Leblond, C.P. 1966. Radioautographic comparison of the uptake of galactose-$^3$H and glucose-$^3$H in the Golgi region of various cells secreting glycoproteins or mucopolysaccharides. *J. Cell Biol.* 30:137-150.

Nevo, A.C.; Michaele, I.; and Schindler, H. 1961. Electrophoretic properties of bull and of rabbit spermatozoa. *Exp. Cell Res.* 23:69-83.

Nicander, L. 1957a. On the regional histology and cytochemistry of the ductus epididymidis in rabbits. *Acta Morphol. Neerl Scand.* 1:99-118.

_____. 1957b. An electron microscopical study of absorbing cells in the posterior caput epididymidis of rabbits. *Z. Zellforsch.* 66:829-847.

_____. 1967. An electron microscopical study of cell contacts in the seminiferous tubules of some mammals. *Zeitschr. f. Zellforsch.* 83:375-397.

Niemi, M., and Kormano, M. 1965. Contractility of the seminiferous tubule of the postnatal rat testis and its response to oxytocin. *Ann. Med. Exp. Fenn.* 43:40-42.

Nishikawa, Y., and Waida, Y. 1952. Studies on the maturation of spermatozoa. 1. Mechanism and speed of transition of spermatozoa in the epididymis and their functional changes. *Bull. Nat. Inst. Agr. Sci.* series G, 3: 69-81.

Oikawa, T.; Yanagamachi R.; and Nicolson, G.L. 1973. Wheat germ agglutinin blocks mammalian fertilization. *Nature* 241:256-257.

Oliphant, G., and Brackett, B.G. 1973. Immunological assessment of surface changes of rabbit sperm undergoing capacitation. *Biol. Reprod.* 9:404-414.

_____. 1975. Capacitation of rabbit spermatozoa *in vitro. Biol. Reprod.* 12:260-274.

Olson, G. 1973. Surface structure revealed by replicas of critical-point dried spermatozoa. *J. Cell Biol.* 59: 254a.

Olson, G.; Hamilton, D.W.; Fawcett, D.W. 1976. Isolation and characterization of the fibrous sheath of rat epididymal spermatozoa. *Biol. Reprod.*, in press.

Orgebin-Crist, M.C. 1967. Sperm maturation in rabbit epididymis. *Nature* 216:816-818.

_____. 1968a. Maturation of spermatozoa in the rabbit epididymis. *J. Reprod. Fertil.* 16:29-33.

_____. 1968b. Studies on the function of the epididymis. *Biol.Reprod.* (suppl) 1:155-175.

Orgebin-Crist, M.C.; Danzo, B.J.; and Davies, J. 1975. Endocrine control of the development and maintenance of sperm fertilizing ability in the epididymis. In *Handbook of physiology*, section 7, Endocrinology vol. V, ed. D.W. Hamilton and R.O. Greep, pp. 319-338. Washington, D.C.: American Physiological Society.

Orgebin-Crist, M.C., and Davies, J. 1974. Functional and morphological effects of hypophysectomy and androgen replacement in the rabbit epididymis. *Cell Tissue Res.* 148:188-201.

Orgebin-Crist, M.C., and Tichenor, P. 1974. Effect of testosterone on sperm maturation *in vitro. Nature* 245:328-329.

Ortavant, R. 1956. Autoradiographie des cellules germinales du testicle de belier. Durce des phenomenes spermatogenetiques. *Arch. Anat. Micros. Morphol. Exp.* 45:1-10.

_____. 1958. La Cycle spermatogenetique chez le Belier. Thesis, Institut Nationale Recherche Agronomique, Paris.

Patanelli, D.J. 1975. Suppression of fertility in the male. In *Handbook of physiology*, section 7, Endocrinology, vol. V, ed. D.W. Hamilton and R.O. Greep, pp. 245-258. Washington, D.C.: American Physiological Society.

Patanelli, D.J., and Nelson, W.O. 1964. A quantitative study of inhibition and recovery of spermatogenesis. In *Recent progress in hormone research*, vol. 20, ed. G. Pincus, pp. 491-543. New York: Academic Press.

Paul, R.; Williams, R.O. and Cohen. E. 1974. Structure activity studies with chlorohydrins as orally active male antifertility agents. *Contraception* 9:451-457.

Paulos, A.; Voglmayr, J.K.; and White, I.G. 1973. Phospholipid changes in spermatozoa during passage through the genital tract of the bull. *Biochim. Biophys. Acta* 306:194-202.

Pearson, D.J., and Tubbs, P.K. 1967. Carnitine and derivatives in rat tissues. *Biochem. J.* 105:953-963.

Pederson, H., and Rebbe, H. 1975. Absence of arms in the axoneme of immobile human spermatozoa. *Biol. Reprod.* 12:541-544.

Perey, B.; Clermont, Y.; and Leblond, C.P., 1961. The wave of the seminiferous epithelium in the rat. *Am. J. Anat.* 108:47-77.

Peyre, A., and Laporte, P. 1966. Evolution des acides sialiques epididymaires chez le rat impubere et adulte. *C.R. Acad. Sci.* (Paris) 263:1872-1875.

Phillips, D.M. 1972. Substructure of the mammalian acrosome. *J. Ultrastruct. Res.* 38:591-604.

_____. 1974. *Spermiogenesis.* New York: Academic Press.

Ploen, L. 1973. An electron microscope study of the delayed effects on rabbit spermateleosis following experimental cryptorchidism for twenty-four hours. *Virchow's Arch. B. Abt. Zellpath.* 14:159-184.

Pokel, J.D., Moyle, W.R.; and Greep, R.O. 1972. Depletion of esterified cholesterol in mouse testis and Leydig cell tumors by luteinizing hormone. *Endocrinology* 9:323-325.

Prasad, M.R.N.; Rajalakshmi, M.; Gupta, G.; and Karkun, T. 1973. Control of epididymal function. *J. Reprod. Fertil.* (suppl) 18:215-222.

Prasad, M.R.N.; Singh, S.P.; and Rajalakshmi, M. 1970. Fertility control in male rats by continuous release of microquantities of cytoproterone acetate from subcutaneous silastic capsules. *Contraception* 2:165-178.

Price, J.M. 1973. Biochemical and morphological studies of outer dense fibers of rat spermatozoa. *J. Cell Biol.* 59:272a.

Raj, M., and Dym, M. 1976. The effects of selective withdrawal of FSH or LH on spermatogenesis in the immature rat. *Biol. Reprod.*, in press.

Rajalakshmi, M., and Prasad, M.R.N.

1968. Changes in the sialic acid content of the accessory glands of the male rat. *J. Endocrinol.* 41:471-476.

———. 1969. Changes in sialic acid in testis and epididymis of the rat during the onset of puberty. *J. Endocrinol.* 44:379-385.

Reddy, P.R.K., and Rao, J.M. 1972. Reversible antifertility action of testosterone proprionate in human males. *Contraception* 5:295-301.

Regaud, C. 1901. Etudes sur la structure des tubes seminifères et sur la spermatogénèse chez las mammifères. *Arch. Anat. Micr.* 4:101-155.

Reid, B.L., and Cleland, K.W. 1957. The structure and function of the epididymis. I. The histology of the rat epididymis. *Australian J. Zool.* 5:223.

Ritzen, E.M.; Dobbins, M.C.; Tindall, D.J.; French, F.S.; and Nayfeh, S.N. 1973. Characterization of an androgen-binding protein (ABP) in rat testis and epididymis. *Steroids* 21:593-607.

Rivarola, M.A.; Podesta, E.J.; and Chemes, H.E. 1972. In vitro testosterone-$^{14}$C metabolism by rat seminiferous tubules at different stages of development: Formation of 5 α-androstandiol at meiosis. *Endocrinology* 91:537-542.

Rolshoven, E. 1940. Die funktionelle Polymorphie des Sertoli-Syncytiums und ihr Zusammenhang mit der Spermatogenese. *Z. Zellforsch.* 31:156-164.

———. 1945. Spermatogenese und das Sertoli-Syncytium. *Z. Zellforsch.* 33:439-460.

Romrell, L.J.; Bellvé, A.R.; and Fawcett, D.W. 1976. Separation of mouse spermatogenic cells by sedimentation velocity: A morphological characterization. *Dev. Biol.* 49:119-131.

Roosen-Runge, E.C., and Giesel, L.O. 1950. Quantitative studies on spermatogenesis in the albino rat. *Am. J. Anat.* 87:1-30.

Ross, M.H. 1967. The fine structure and development of the peritubular contractile cell component in the seminiferous tubules of the mouse. *Am. J. Anat.* 121:523-534.

Ross, M.H., and Long, I.R. 1966. Contractile cells in human seminiferous tubules. *Science* 153:1271-1273.

Ruzicka, L.; Goldberg, M.W.; Meyer, J.; Brungger, H.; and Eichenberger, E. 1934. Uber die Synthese des Testikelhormons (Androsteron) und Stereoisomerer desselben durch Abban hydrierter Sterine. *Helv. Chim. Act.* 17:1395.

Ruzicka, L., and Wettstein, A. 1935. Uber die kunstliche Herstellung dis Testikelhormons Testosteron (Androsten-3-on-17-ol). *Helv. Chim. Act.* 18:1264.

Satir, P. 1968. Studies on cilia: III. Further studies on the cilium tip and a "sliding filament" model of ciliary motility. *J. Cell Biol.* 39:77-85.

Saxena, B.B.; Leyendecker, G.; W. Chin et al. 1969. Radioimmunoassay of follicle-stimulating (FSH) and luteinizing (LH) hormones by chromatoelectrophoresis. *Acta Endocrinol.* (suppl) 142:185-206.

Scott, T.W., and Dawson, R.M.C. 1968. Metabolism of phospholipids by spermatozoa and seminal plasma. *Biochem. J.* 108:457-463.

Scott, T.W.; Voglmeyer, J.K.; and Setchell, B.P. 1967. Lipid composition and metabolism in testicular and ejaculated ram spermatozoa. *Biochem. J.* 102:456-461.

Sertoli, E. 1865. Dell'esistenza di particolari cellule ramificati nei canalicoli seminiferi del testiculo humano. *Morgagni* 7:31.

Setchell, B.P. 1967. The blood-testicular fluid barrier in sheep. *J. Physiol.* (London) 139:64.

_____ 1970. Testicular blood supply, lymphatic drainage, and secretion of fluid. In *The testis*, ed. A.D. Johnson, W.R. Gomes, and N.L. Van Denmark, pp. 101-239. New York: Academic Press.

_____. 1974. Secretions of the testis and epididymis. *J. Reprod. and Fert.* 37:165-177.

Setchell, B.P.; Dawson, R.M.C.; and White, R.W. 1968. The high concentration of free myo-inositol in rete testis fluid from rams. *J. Reprod. Fertil.* 17:219-220.

Setchell, B.P., and Jacks, F. 1975. Inhibin-like activity in rete testis fluid. *J. Endocrinology* 62:675-676.

Setchell, B.P.; Voglmayr, J.K.; and Waites, G.M.H. 1969. A blood testis barrier restricting passage from blood into rete testis fluid but not into lymph. *J. Physiol.* (London) 200:73-85.

Setchell, B.P., and Waites, G.M.H. 1975. The blood testis barrier. In *Handbook of physiology*, section 7, Endocrinology, vol. V, ed. D.W. Hamilton and R.O. Greep, pp. 143-172. Washington, D.C.: American Physiological Society.

Sherins, R.J.; Gandy, H.M.; Thorsland, T.W.; and Paulsen, C.A. 1971. Pituitary and testicular function studies. I. Experience with a new gonadal inhibitor, 17-d-pregn-4-en-20-yno(2,3,-d)-isoxazol 17-ol (Danazol).*J. Clin. Endocrinol.* 32:522-531.

Sherins, R.J., and Loriaux, D.L. 1973. Studies in the role of sex steroids in the feedback control of FSH concentrations in man. *J. Clin. Endocrinol. Metab.* 36:886-893.

Shikita, M., and Hall, P.F. 1967. The action of human chorionic gonadotrophin *in vivo* upon microsomal enzymes of immature rat testis. *Biochem. Biophys. Acta.* 136:484-497.

Skoglund, R.D., and Paulsen, C.A. 1973. Danazol-testosterone combination. A potentially effective means for reversible male contraception, a preliminary report. *Contraception* 7:357-365.

Solari, A.J., and Tres, L. 1967. Ultrastructure of the human sex vesicle. *Chromosome* 22:16-31.

_____ 1970. The three dimensional reconstruction of the X-Y chromosomal pair in human spermatocytes. *J. Cell Biol.* 45:43-53.

Smith, P.E. 1930. Hypophysectomy and replacement therapy in the rat. *Am. J. Anat.* 45:205-273.

Smith, P.E., and Engle, E.T. 1927. Experimental evidence regarding the role of the anterior pituitary in the development and regulation of the genital system. *Am. J. Anat.* 40:159-217.

Spallanzani. 1780. Dissertazioni di fisica animale e vegetabile vol. II. Modena.

Srivastava, P.N.; Adams, C.E.; and Hartree, E.F. 1965. Enzymic action of acrosomal preparations on the rabbit ovum *in vitro. J. Reprod. Fertil.* 10: 61-67.

Stambaugh, R., and Buckley, J. 1969. Identification and subcellular localization of the enzymes effecting penetration of the zona pelucida by rabbit spermatozoa. *J. Reprod. Fertil.* 19: 423-432.

_____. 1970. Histochemical subcellular localization of the acrosomal proteinase effecting dissolution of the zona pellucida using fluorescein-labelled inhibitors. *Fertil. Steril.* 23: 348-352.

Stefanini, M.; Oura, C.; and Zambone, L. 1969. Ultrastructure of fertilization in the mouse. 2. Penetration of sperm into the ovum. *J. Submicr. Cytol.* 1:1-23.

Steinberger, A.; Heindel, J.J.; Lindsey, J.N.; Elkington, S.H.; Sanborn, B.M.; and Steinberger, E. 1975. Isolation and culture of FSH responsive Sertoli cells. *Endocr. Res. Commun.* 2:261-272.

Steinberger, E. 1971. Hormonal control of mammalian spermatogenesis. *Physiol. Rev.* 51:1-22.

Steinberger, E.; Steinberger, A.; and Sanborn, B. 1973. Endocrine control of spermatogenesis. In *Physiology and genetics of reproduction*, ed. E.M. Coutinno and F. Fuchs, pp. 163-182. New York: Plenum Press.

Stephens, R.E. 1970. Isolation of nexin—the linkage protein responsible for the maintenance of the 9-fold configuration of flagellar axonemes. *Biol. Bull.* (abstr) 139:438.

Stubblefield, E. 1973. The structure of mammalian chromosomes. *Int. Rev. Cytol.* 35:1-58.

Summers, K.E., and Gibbons, I.R. 1971. Adenosine triphosphate-induced sliding of tubules in trypsin-heated flagella of sea-urchin sperm. *Proc. Natl. Acad. Sci.* 68:3092-3096.

Suter, D.A.I.; Brown-Woodman, D.C.; Mohri, H.; and White, I.G. 1975. The molecular site of action of the antifertility agent α-chlorohydrin in ram spermatozoa. *J. Reprod. Fertil.* (abstr) 43:382.

Swierstra, E.E., and Foote, R.H. 1963. Cytology and kinetics of spermatogenesis in the rabbit. *J. Reprod. Fertil.* 5:309-322.

Tamaoki, B.I.; Inano, H.; and Nakano, H. 1969. In *The gonads*, ed. K.W. McKerns, pp. 547-613. New York: Appleton-Century-Crofts.

Teichman, R.J., and Bernstein, M.H. 1971. Fine structure localizations of acid phosphatase in rabbit and bull sperm heads. *J. Reprod. Fertil.* 27: 243-248.

Thibault, C.; Dauzier, L.; and Winterberger, S. 1954. Etude cytologique de la ficondation in vitro de l'ouf de la lapine. *C.R. Soc. Biol.* (Paris) 148: 789-790.

Tilney, L.G.; Bryan, J.; Bush, D.J.; Fujievara, K.; Mooseker, M.S.; Murphy, B.; and Snyder, D.H. 1973. Microtubules: Evidence for 13 protofilaments. *J. Cell Biol.* 59:267-275.

Toren, D.; Menon, K.M.; Forchielli, E.; and Dorfman, R.I. 1964. In vitro enzymatic cleavage of the cholesterol side chain in rat testis preparations. *Steroids* 3:381-390.

Toyama, Y. 1975. Actin-like filaments in peritubular myoid and Sertoli cells of swine and mouse testis. *Proc. 10th Int. Congr. Anat.* (Tokyo), (abstr) p. 443.

Toyoda, Y.; Yokoyama, M.; and Hosi, T. 1971. Studies on the fertilization of mouse ova *in vitro*. In vitro fertilization with fresh epididymal sperm. *Jap. J. Animal Reprod.* 16:147-151.

Tuck, R.R.; Setchell, B.P.; Waites, G.M.H.; and Young, J.A. 1970. The composition of fluid collected by micropuncture and catheterization from the seminiferous tubules and rete testis of rats. *European J. Physiol.* 318:225-243.

van der Molen, J.H.; de Bruijn, H.; Cooke, B.; and de Jong, F. 1973. In *The endocrine function of the human testis*, ed. V. James, M. Serio, and L. Martini. New York: Academic Press.

Van Thiel, D.H.; Sherins, R.J.; Meyers, G.H.; and DeVita, V.T. 1972. Evidence for a specific seminiferous tubular factor affecting follicle stimulating hormone secretion in man. *J. Clin. Invest.* 51:1009-1019.

Vernon, R.G.; Kopec, B.; and Fritz, I.B. 1973. Studies on the distribution of the high affinity testosterone binding protein in rat testis seminiferous tubules. *J. Endocrinol.* (abstr) 57:ii.

Vitale-Calpe, R.; Fawcett, D.W.; and Dym, M. 1973. The normal development of the blood-testis barrier and the effects of clomiphene and estrogen treatment. *Anat. Rec.* 176:333-344.

Voglmayr, J.K. 1975. Metabolic changes in spermatozoa during epididymal transit. In *Handbook of physiology*, sect. 7, Endocrinology, ed. D.W. Hamilton and R.O. Greep, pp. 437-451. Washington, D.C.: American Physiological Society.

Voglmayr, J.K.; Waites, G.M.H.; and Setchell, B.P. 1966. Studies on spermatozoa and fluid collected directly from the testis of the conscious ram. *Nature* (London) 210:861-863.

von Berswordt-Wallrabe, R., and Neumann, F. 1968. Successful reinitiation and restoration of spermatogenesis in hypophysectomized rats with pregnant mare's serum after a long regression period. *Experientia* 24:499-501.

Waites, G.M.H., and Einer-Jensen, N. 1975. Collection and analyses of rete testes fluid from macaque monkeys. *J. Reprod. Fertil.* 41:505-508.

Waites, G.M.H., and Setchell, B.P. 1969a. Some physiological aspects of the function of the testis. In *The gonads*, pp. 649-714, ed. K.W. Mc-

Kerns. New York: Appleton-Century-Crofts.

———. 1969b. Physiology of the testis, epididymis and scrotum. In *Advances in reproductive physiology*, pp. 1-63, ed. A. McLaren. London: Logos Press.

Walsh, E.L.; Cuyler, W.K.; and McCullagh, D.R. 1934. Physiological maintenance of male sex glands; effect of androitin on hypophysectomized rats. *Am. J. Physiol.* 107:508-512.

Warner, F.D. 1970. New observations on flagella fine structure: The relationship between matrix structure and the microtubule component of the axoneme. *J. Cell Biol.* 47:159-182.

Weinman, D.E., and Williams, W.L. 1964. Mechanism of capacitation of rabbit spermatozoa. *Nature* 203:423.

White, I.G., and Hudson, B. 1968. The testostcrone and dehydroepiandrosterone concentration in fluids of the mammalian male reproductive tract. *J. Endocrinol.* 41:291-292.

Williams-Ashman, H.G. 1954. Changes in the enzymatic constitution of the ventral prostate gland induced by androgenic hormones. *Endocrinology* 54:121-129.

Williams-Ashman, H.G., and Banks, J. 1954. The synthesis and degradation of citric acid by ventral prostate tissue. I. Enzymatic mechanisms. *J. Biol. Chem.* 208:337-349.

Wilson, J.D. 1975. Metabolism of testicular androgens. In *Handbook of physiology*, sect. 7, Endocrinology, vol. V, ed. D.W. Hamilton and R.O. Greep, pp. 491-508. Washington, D.C.: American Physiological Society.

Wislocki, G.B. 1949. Seasonal changes in the testis, epididymis and seminal vesicles of deer investigated by histochemical methods. *Endocrinology* 44:167-189.

Woolley, D.M. 1969. Striations in the peripheral fibers of rat and mouse spermatozoa. *J. Cell Biol.* 49:936-939.

Yanagimachi, R. 1969. In vitro capacitation of hamster spermatozoa by follicular fluid. *J. Reprod. Fertil.* 18:275-286.

———. 1972. Fertilization of guinea pig eggs *in vitro*. *Anat. Rec.* 174:9-20.

Yanagimachi, R., and Chang, M.C. 1963a. Fertilization of hamster eggs *in vitro*. *Nature* 200:281-282.

———. 1963b. In vitro fertilization of golden hamster ova. *J. Exp. Zool.* 156:361-375.

Yanagimachi, R., and Noda, Y. 1970. Electron microscopic studies of sperm incorporation into the golden hamster egg. *Am. J. Anat.* 128:429-462.

Yanagimachi, R., and Usui, N. 1974. Calcium dependence of the acrosome reaction and activation of guinea pig spermatozoa. *Exp. Cell Res.* 89:161-174.

Young, W.C. 1929a. A study of the function of the epididymis. II. The importance of an aging process in sperm for the length of the period during which fertilizing capacity is retained by sperm isolated in the epididymis. *J. Morphol.* 48:475-491.

_____. 1929b. A study of the function of the epididymis. I. Is the attainment of full spermatozoon maturity attributable to some specific action of the epididymal secretion? *J. Morphol. Physiol.* 49:479-495.

_____. 1929c. The influence of high temperature on the reproductive capacity of guinea pig spermatozoon determined by means of artificial insemination. *Physiol. Zool.* 2:1-8.

_____. 1931a. A study of the function of the epididymis. II. Functional changes undergone by spermatozoa during their passage through the epididymis and vas deferens of the guinea pig. *J. Exp. Biol.* 8:151-163.

Zaneveld, L.J.D.; Dragoja, B.M.; and Schumacher, G.F.B. 1972. Acrosomal proteinase and proteinase inhibitor of human spermatozoa. *Science* 177: 702-703.

Zaneveld, L.J.D.; Srivastava, P.N.; and Williams, W.L. 1969. Relationship of a trypsin-like enzyme in rabbit spermatozoa to capacitation. *J. Reprod. Fert.* 20:337.

Zaneveld, L.J.D., and Williams, W.L. 1970. A sperm enzyme that disperses the corona radiata and its inhibition by decapacitation factor. *Biol. Reprod.* 2:363-368.

Zaneveld, L.J.D.; Polakoski, K.L.; and Schumadier, C.F.B. 1973. Properties of acrosomal hyaluronidase from bull spermatozoa. *J. Biol. Chem.* 284:564-580.

Zaneveld, L.J.D.; Robertson, R.T.; Kessler, M.; and Williams, W.L. 1971. Inhibition of fertilization *in vivo* by pancreatic and seminal plasma trypsin inhibitors. *J. Reprod. Fertil.* 25:387-395.

Zondek, B., and Aschheim, S. 1927. Hypophysenvorder lappen und Ovarium. Beziehungen der endokrinen Drusen zur Ovarialfunktion. *Arch. f. Gynak.* 130:1-45.

# 6 New Methods for Fertility Regulation: Status Report

Contraceptive practice throughout the world was revolutionized in the late 1950s and early 1960s with the development of the oral contraceptive and the modern IUD. In subsequent years, scores of millions of couples have at some time adopted one of these methods voluntarily to limit fertility. Yet it is evident that these widely heralded technological advances, like the more traditional methods of fertility regulation, have significant limitations. As clinic or physician-oriented methods, they present programmatic difficulties in many situations. Both are associated with troublesome side effects that lead to frequent discontinuation of use. Both methods carry rare but serious risks to health and new safety issues with each emerge from time to time. In countries that have adopted programs to reduce fertility through the provision of family planning services, the limitations of the methods available, including the pill and IUD, have contributed to the limited results achieved.

Considerable attention has been given in the last decade to modifying existing fertility control methods in order to improve effectiveness, safety, or programmatic usefulness. Doses of oral contraceptives have been reduced, new forms of inert intrauterine devices have been designed, procedures for tubal ligation have been simplified, and there have even been changes made in the aesthetic appeal of the latex condom. In addition, the vacuum aspiration technique for first trimester termination of pregnancy, developed in China, has perhaps become the most widely used abortion procedure through the world.

There has also been some success in the development of new approaches to fertility regulation, and we may now be on the threshold of important technological changes. Pharmacologically active IUDs have been developed, tested under stringent drug regulations, approved by regulatory agencies, and adopted by millions of women around the world. A new chemical abortifacient, prostaglandins intended to replace intraamniotic saline for late abortions, has been developed. A method for postcoital contraception, oral estrogens, has been tested, shown to be effective, and given official approval for emergency use. In addition, tests are nearly completed with fertility

This chapter by Drs. Sheldon J. Segal and Egon Diczfalusy summarizes the essays commissioned for the review pertaining to contraceptive development.

regulation methods that would take the form of an intramuscular injection or a vaginal suppository to induce abortion as early as four weeks after a missed period; a vaccine against pregnancy is being tested in several countries; a contraceptive injection for men is being tested; and an entire new field of contraception through long-term drug delivery systems has captured the interest of both university and industrial scientists. Some of these approaches represent refinements of existing methods, but others are entirely new and derive directly from the findings of fundamental research since 1960.

## New Methods Recently Developed

### Copper-releasing IUDs

Studies in animals led scientists to investigate the effect of metal ions on the enzymes of the uterus. As a direct result of this work, copper-bearing IUDs were developed. Now between 5 and 6 million women in thirty-one countries have begun to use the Copper T or the closely related version marketed commercially as the Copper 7. The acceptance of this method, first made available in 1973, has thus been remarkably rapid. These devices are particularly useful for women who have not previously had a child for whom older forms of IUDs are not suitable. The Copper T is the first method of contraception classified as a drug that has been developed fully by the public sector (the Population Council and the International Committee for Contraception Research) for use by the public sector. As a result of the licensing arrangements entered into by the council to protect the public interest, the Copper T can be purchased by governmental or international agencies at less than market prices. The consequent savings are considerable: For its first purchase of 1,200,000 Copper Ts, the United Nations Fund for Population Activities paid $4 million less than the closest commercial price for a comparable competitive product. The savings on this purchase alone nearly offset the total public sector investment in the Copper T's development. The government of India has introduced this device as a major new method in its national family planning program and is setting up facilities for local manufacture to keep costs low and avoid the need for foreign exchange. Mexico is also seeking to manufacture the Copper T locally.

**Progesterone-releasing IUDs**
Progesterone-releasing IUDs have been approved for distribution in the United Kingdom, Canada, and Mexico and may soon be approved in several other countries. Their development culminates many years of work in polymer chemistry on drug release systems effective for long periods of time. Although the limit of one year effectiveness reduces their usefulness in public sector family planning programs, the devices are the first based on the principle of local action by a hormone released by an intrauterine device. Newer versions may overcome the limitation on period of use as well as reduce their cost. Like the copper-releasing IUDs, the progesterone-releasing IUD can be used by nulliparous women. It was developed under cooperative arrangements between industry and publicly supported programs. A private corporation obtained licenses from the Population Council for the use of the T-form intrauterine device and from a Ford Foundation grant recipient for the principle of hormone release from an IUD. The company's patented know-how was applied to develop the delivery system using the natural hormone progesterone, and WHO's network of clinical research centers participated in clinical evaluation. Price reductions are assured for public sector purchase of these devices as a condition of the nonroyalty bearing agreements between the company on the one hand and the Ford Foundation and Population Council on the other.

**Chemical Abortifacient**
Anthropological journals (and streetside vending stands in many countries) are replete with potions deemed to be abortifacients, but partial success in the long search for a scientifically proven chemical abortifacient was achieved only recently. Prostaglandins are ubiquitous naturally occurring substances that are under study for various possible applications to fertility regulation. These were first discovered in the 1930s and were the subject of considerable fundamental research attempting to establish their physiological role. This basic

work was hampered by the unavailability of an adequate quantity of pure material. A breakthrough occurred in 1967 with the isolation of the pure prostaglandins and the development of a method for total synthesis. One of the prostaglandins has consequently been shown to be useful for inducing midtrimester abortions and to offer some safety advantages over other methods available for these abortions. The intraamniotic instillation of prostaglandin can initiate uterine contractions and, therefore, cause evacuation of the uterus.

A prostaglandin analog (15-methyl prostaglandin $F_{2\alpha}$) has been tested sufficiently to be sponsored as a new drug by a U.S. pharmaceutical company for use as an intramuscular injection in place of curettage or aspiration during the first twelve weeks of pregnancy or during weeks twelve to fifteen, for which there is now no suitable methodology. The same compound is being tested in the form of a vaginal suppository. Work in this field, with respect to both synthesis of analogs and clinical testing, has been stimulated and expedited by industry and the public sector.

## Postcoital Estrogens

Hardly a new method in terms of unauthorized use throughout the world, estrogens were approved in 1975 by the U.S. Food and Drug Administration for use as a postcoital contraceptive in emergency situations. This action followed years of work in laboratory animals establishing the feasibility of the method and several prospective clinical studies that required several years to accumulate sufficient cases. The method consists of a high dose of synthetic or natural estrogen taken for five successive days. The procedure can be highly effective if started within seventy-two hours after possible fertilization. It cannot be used routinely both because of its long-term carcinogenic potential and because it frequently causes severe nausea and vomiting and disrupts the subsequent ovarian cycle. The U.S. regulatory agency approved postcoital use of estrogens under limited circumstances after the completion of extensive clinical studies. Studies of estrogenic substances are continuing, seeking compounds that provide the basic activity desired without the undesirable side effects.

## Methods Now under Investigation

The lengthy and complex process of contraceptive research and development includes fundamental studies, the development of animal models to study issues of effectiveness and safety, and applied product development work leading to trials in human subjects. Any new methods likely to emerge for use in the next three to five years must now be at some advanced stage and have some testing underway in human subjects. Less advanced projects may require an even longer period of development. Based on these criteria, the following new or modified contraceptives may phase in over the next decade, provided sufficient funds are available to assure that the work proceeds.

### Immunization of Women with Chorionic Gonadotropin

One of the most promising new approaches is *immunological*, by means of active immunization. Such a "vaccination" must be based on a completely specific antigen and rendered immunogenic without resorting to noxious adjuvants. It should result in the formation of specific antibodies directed exclusively to the biologically active principle in question. This objective has been pursued for many years and prompted the search for a highly specific antigen that could serve safely as a basis for an antipregnancy vaccine. Specific proteins from sperm, ovarian follicles, uterus, and various reproductive tract fluids have been studied. Major advances during the 1960s in basic protein chemistry rapidly unraveled the chemical structure of many large protein molecules, including the pregnancy-specific hormone human chorionic gonadotropin (hCG). This was the needed breakthrough that made possible the isolation of a chemically pure substance that does not occur normally in women except in pregnancy. The $\beta$ subunit of hCG is highly specific and its limited cross-reactivity with pituitary luteinizing hormone (LH) can be further reduced by using synthetic fragments of the molecule. The antibodies produced neutralize the activity of the entire hCG molecule without which pregnancy cannot become established and maintained.

An investigative program with a limited number of volunteer subjects was begun in India in 1974 and has been broadened to include clinical trials in Finland, Brazil, Chile, Sweden, and the Dominican Republic. This collaborative program between the All India Institute of Medical Sciences and ICCR employs an antigen in which a modified β subunit of hCG is linked to tetanus toxoid. The World Health Organization's Expanded Programme of Research, Development and Research Training in Human Reproduction also has sponsored a synthesis and testing program to develop an antigen with hCG fragments, and at least two pharmaceutical firms are working along similar lines.

**Pharmacologic Suppression of the Corpus Luteum**
The immunological approach described above is based on prevention of corpus luteum function, which during the first weeks of pregnancy is essential for pregnancy to become established and maintained. Along similar lines, drug-induced suppression of the corpus luteum (luteolysis) and prevention of progesterone production has been the subject of both laboratory and clinical studies for several years. The approach could serve as the basis for a once-a-month pill that would induce a menstrual flow even if fertilization had occurred. Indeed, the fertile cycle would pass unnoticed and there would be no need for the woman to take hormones in order to maintain menstrual regularity. In this approach could lie the solution to the problem of risks associated with continuous dosage forms of hormonal contraception, a solution that could yield near-perfect contraception.

Three programs are now in some stage of clinical testing using a natural plant product (*Montanoa tuberosa*, commonly called Zoapatle), an analog of the naturally occurring prostaglandins (15-methyl derivative of prostaglandin $F_{2\alpha}$), and a synthetic nonsteroidal hormone antagonist (Centchroman). Animal studies suggest that orally active extracts of the Mexican plant *Montanoa tuberosa* are capable of suppressing circulating progesterone levels. Infusions prepared from this plant are claimed to induce abortion in women in the second trimester of gestation by an oxytocic effect that starts uterine contractions. It is not known presently whether the luteolytic and oxytocic effects are due to the same or different principles. This research, supported by the Ford Foundation since 1971, is being carried out at the Medical Institute of the Social Security Service of Mexico and is now a component of a WHO collaborative effort.

In addition to the abortifacient activity noted above, 15-methyl, prostaglandin $F_{2\alpha}$ is also being studied for its possible luteolytic effects.

Luteolysis is also believed to be the basis for the antifertility activity of the synthetic compound Centchroman, which is being studied as a postcoital pill in India. The name is derived in part from its origin at the government's Central Drug Research Institute where a large series of similar compounds, related to diethylstilbestrol, have been synthesized. Animal studies suggest that this compound suppresses corpus luteum function. It is possible, however, that in women the principal mechanism of action is not luteolysis but interference with transport of the egg in the Fallopian tube.

Additional approaches to luteolysis are likely to emerge from increasing information on the enzymatic control of progesterone synthesis and the role of receptors in the action of hormones.

### Long-acting Forms of Contraceptive Steroids

Injectable steroid contraceptives are marketed in some countries. Each compound now in use has important issues still unresolved regarding safety, reversibility, or effectiveness. These uncertainties may be resolved satisfactorily with increasing level of use, but the distribution of these products is now limited. Meanwhile, scientists are seeking other compounds or new delivery systems that could be used safely as injectable contraceptives. The synthesis and testing of new progestins is proceeding under separate programs of WHO and National Institutes of Health's Center for Population Research (CPR), and ICCR has investigated the influence of different crystal sizes of steroid on the release rate from the injection site. Several programs have turned attention to preparations involving steroids coated with biodegradable substances such as those used for absorbable surgical sutures (polylactic acid). Suitable technology to coat steroids with such biodegradable materials has undergone considerable development and now appears to be available.

A variant of the contraceptive injection that has become feasible as a result of this new technology is the placement of a biodegradable coated steroid preparation directly into the cervix where very small amounts of drug can be released over a long period of time. The steroid could exert a local action in the cervix or uterus, thus avoiding the need for significant blood levels. If so, steroid-related side effects might be virtually eliminated.

Among the new methods based on the administration of contraceptive steroids through a sustained drug delivery system, the development of subdermal contraceptive implants is the most advanced. These implants consist of small capsules or rods placed under the skin, where they continuously release drugs for from one to eight years. The main work on contraceptive implants is being carried out by ICCR.

The first studies were based on the principle of continuous low dose progestin therapy (similar to the minipill), but the incidence of menstrual bleeding disturbances was high. To overcome this problem and improve effectiveness, most of the recent work has shifted to capsules that deliver sufficient quantities of steroid to suppress ovulation. Over a dozen steroids have been tested clinically, and at least two have shown promise. The ICCR has begun a comparative study of two different subdermal implant regimens. Each regimen involves the use of six Silastic capsules 30 mm in length; one system utilizes the new progestin, norgestrienone, which would provide an effective lifetime of three years; the second utilizes the widely used steroid d-norgestrel, which could last for as long as eight years. Laboratory studies are proceeding with advanced technology for achieving the same release rates and lifespans with smaller and fewer implants. To eliminate the need for removal of implants at the end of their effective lifetime, biodegradable polymers that could be utilized as contraceptive rods are under development.

Many drugs are rapidly absorbed through the vaginal mucosa into the bloodstream. Silicone or plastic rings containing contraceptive drugs have been designed to take advantage of this site of absorption. The rings are placed in the same position as a vaginal diaphragm, where they release constant amounts of drug. The most advanced version of this method was investigated by a U.S. company and then discontinued; it is now under development by ICCR. In this program, rings have been designed to release sufficient quantitites of progestational steroids to block ovulation. These rings are removed after three weeks of continuous placement in the vagina; several days later withdrawal bleeding similar to normal menses occurs. The ring is reinserted for the next cycle. As an alternative schedule the ring is left in place continuously to produce long periods without menstrual bleeding. Replacement of the steroid-depleted ring with a new one is required every three months. A third approach, sponsored by WHO, is to develop vaginal rings that will provide constant delivery of low doses of contraceptive steroids, thus achieving contraceptive efficacy through a minipill mechanism. A difficulty in all vaginal ring programs has been menstrual bleeding disturbances. New progestational compounds as well as the addition of estrogen are presently under investigation in an effort to reduce this problem.

Absorption of steroids through the skin, a principle long employed for cosmetic purposes, is now being tested as a basis for contraception. As with subdermal implants, a silicone polymer is used as the carrier of progestin. It is molded into the form of a snugly fitting bracelet on the upper arm. The clinical tests are being carried out by ICCR. The work needed to develop a contraceptive bracelet is mainly product design and subsequent clinical testing; the principle of steroid administration by this route is sufficiently established. Feasibility for contraception depends on a bracelet design that will provide a sufficiently constant rate of absoprtion.

The concept of controlled drug release from a device has been applied to the development of intracervical devices in the WHO and CPR programs. Inert devices can be placed and retained within the cervical canal. The devices are currently being used as drug delivery systems for contraceptive steroids, and clinical studies are being conducted to evaluate their performance. The objective is to achieve a local antifertility action in the cervix without significant uptake of the steroid into the bloodstream.

The common feature of injectable contraceptives, subdermal contraceptive implants, vaginal rings, contraceptive bracelets, and intracervical devices is that they are long-acting forms of steroidal contraception. Whether they can improve upon the effectiveness of oral contraceptive steroids or eliminate concerns over the safety of continuous use of steroids is questionable; indeed, they raise new issues while resolving others. What may be achieved with these new methods is the advantage of long-term protection after a single act of contraception.

Studies aimed at methods based on identification of the time of ovulation so that the couple may avoid intercourse at this time continue. Although use of these techniques has decreased to low levels in recent years, there is interest in developing more effective methods for use by couples who for a variety of reasons wish to avoid medicines and devices. Several agencies are now investigating the efficacy and acceptability of new approaches based on self-observation by the woman of the condition of her cervical mucus. One such study is underway in the United States and several others are contemplated in other countries.

**Improvement in Intrauterine Devices**
Many new varieties of conventional IUDs are undergoing clinical testing. Work in this field is included in the programs of CPR, WHO, and the International Fertility Research Program (IFRP) of the U.S. Agency for International Development (AID). In addition, the Indian Council of Medical Research maintains an evaluation program for new devices. The objective of all these programs is to seek changes in design that will improve effectiveness and reduce the incidence of bleeding or of spontaneous expulsion associated with the IUDs now being used.

The search has focused on devices possessing an optimal combination of small size (tending to reduce bleeding and pain, but increase expulsions) and rigidity (tending to decrease expulsions, but increase bleeding and pain). To reduce expulsion, some devices have also been designed with projections on the sides that anchor the IUD to the uterine wall near the tubal ostia. Another approach is to establish a barrier between the two walls of the uterus. About a dozen variations based on these principles are being tested, but they do not appear to show more than incremental change, if any, compared to the best devices already available.

A special need exists for an IUD suitable for insertion in the immediate postpartum period. Available IUDs have high expulsion and failure rates when inserted immediately after parturition, largely because during this period the uterus is relatively large and soft and has not yet involuted following delivery. A recent approach is the design of a special copper-bearing T-shaped device for postpartum use. It was developed by ICCR and is being tested by WHO's Clinical Research Centers.

**Simplified Techniques of Female Sterilization**

Simplified techniques for female sterilization are being sought, particularly to make it possible to perform the procedure on an outpatient basis in situations in which hospital bed space is limited. Most of the new approaches require improved methods for occluding the Fallopian tubes. Techniques under development include tubal clips or bands, new types of electrocoagulation, caustic chemicals, and plugs or tissue adhesives of various kinds. This work has been spurred by the interest in laparoscopic and culdoscopic sterilization procedures. Most laparoscopic sterilizations to date have been achieved through the use of electric current to coagulate the Fallopian tubes. However, electrocoagulation has been associated with infrequent but serious complications; furthermore the tubes sometimes heal sufficiently following such cauterization to restore fertility. To simplify tubal occlusion and facilitate reversibility procedures a variety of clips have been investigated, but neither the rigid nor the spring-loaded designs thus far tested appears to close the tubes with sufficient reliability or permanency. A different approach has centered on the development of a Silicone rubber band that is slipped over a loop of the Fallopian tube, constricting around and occluding it.

Female sterilization might be radically simplified by avoiding surgery altogether. This could be achieved by reaching the Fallopian tubes through the body's natural passageways: the vagina, cervical canal, and uterus. By this transcervical approach the tubes could be occluded either by using caustic chemicals to cause the formation of scar tissue or through the placement of plugs within the tubes. However, it has been impossible thus far to deliver liquids solely to the lumen of the Fallopian tubes without some spillover into the uterus or abdominal cavity. Methods for the blind delivery of polymerizing solutions or materials, such as plugs or meshes, have not yet been successfully developed; but further work on such delivery systems is continuing. Current research to avoid spillage is focusing on a polymer that changes from a liquid to a gel when injected from a syringe into the Fallopian tubes.

## Pharmacologic Male Contraception

Ever since the development of oral contraceptives for women, efforts have been underway to identify suitable drugs for inhibition of fertility in the male. In principle, this can be achieved using drugs that suppress the formation of sperm by the testis; drugs that block proper maturation of the sperm in the epididymis; or drugs that impair the transport of sperm from the testis through the epididymis, vas deferens, and urethra. Chemical substances that interfere with each of these processes have been identified and studied clinically.

Among the drugs known to suppress the production of sperm by the testis are a variety of antimetabolic agents and steroid hormones. The antimetabolic agents cannot be considered serious contraceptive possibilities due to their general systemic toxicity. At appropriate dose levels, estrogens, progestins, androgens, and antiandrogens block the production of sperm, probably by interfering with the availability of testosterone in the testis.

Estrogens are among the most potent agents for this purpose; but it is known that long-term administration of estrogens to men can cause breast enlargement, loss of libido, and an increase in thromboembolic disease. Nonetheless, limited clinical investigations are being conducted in which these problems are minimized by using very low doses of estrogens in conjunction with testosterone. The preliminary results of these clinical studies have been promising; suppression of sperm production has been achieved without evident side effects. However, it is likely that the toxicity concerns with the administration of long-term estrogens will discourage vigorous development of this approach.

Progestins are less potent inhibitors of spermatogenesis than estrogens; but because there is no evidence linking them to an increased incidence of cardiovascular disease, they have been much more widely tested in men. So far it has been established that sperm production can be suppressed with progestins supplemented by androgens while maintaining normal levels of plasma testosterone and that this procedure is readily reversible. Current research seeks to identify effective dosages of suitable combinations that can be administered as monthly injections or daily pills without an unacceptable incidence of side effects. One regimen utilizing monthly injections of Depo-Provera and testosterone enanthate has yielded encouraging results. A male method based on combined administration of a progestin and an androgen may possibly become available within a few years.

Testosterone itself is known to inhibit sperm production in men if given in sufficiently high doses. Frequent intramuscular injections are required, however, and there is concern about health hazards (cardiovascular problems in particular) over the long term with the high doses required. As a potential means of avoiding some of these toxicity concerns, some work has been done with orally active steroids that are less androgenic than testosterone. A common finding to date has been only partial effectiveness at the relatively high doses of the anabolic agents studied.

A steroid that exhibits antiandrogenic activity has also been tested for its ability to block sperm production. High doses of this drug, cyproterone acetate, will impair but not abolish spermatogenesis; and it is uncertain to what extent even this partial effect is due to the antiandrogenic activity of the drug and to what extent it results from its known progestational activity. However, cyproterone acetate at low doses may be effective as a male contraceptive agent because of a second pharmacologic action of the drug. Sperm produced during treatment with the drug appear to be unable to penetrate the cervical mucus of women as do normal sperm. This effect may be a consequence of the reported inhibitory action of cyproterone acetate on sperm maturation in the epididymis, a finding that remains controversial and must be confirmed. Clinical studies of cyproterone acetate are underway in several centers, and other antiandrogenic substances are being synthesized and tested in order to expand efforts based on this approach. This work, which has high priority in the WHO program, is based on fundamental discoveries in laboratory animals done by a Ford Foundation grantee at Delhi University in India. The project for development of a male injectable contraceptive is sponsored by ICCR, and a similar program is proceeding in the NIH Contraceptive Development Branch.

## Reversible Vas Occlusion

The major drawback of vasectomy is its uncertain reversibility. Efforts are underway to overcome this limitation by developing methods for reversibly occluding the vas deferens.

One approach centers on improvements in surgical techniques for rejoining the cut ends of the vas. It is possible to restore sperm to the ejaculate of previously vasectomized men by means of surgical reanastomosis of the vas deferens, but this does not always restore fertility. Whether the subsequent low fertility of some men is due to an immunologically induced irreversible impairment associated with the vasectomy or the imperfections in the surgical reversal is still unknown. Recent reports suggest that very much higher rates of recovery of fertility can be achieved using microsurgical techniques for reanastomosis. These findings will undoubtedly stimulate current efforts to improve the reversibility of the conventional vasectomy procedure.

An alternate approach that has been intensively pursued is the development of valves, clips, or plugs that can be used to occlude the vas deferens reversibly. Reversal would be accomplished through removal of the clip or plug or, for a valve device, switching into the "open" mode. A wide variety of devices has been developed and tested, mainly in animals; but thus far none has proved suitable. Active investigations are continuing except in one major agency, which has decided after several years of study that this approach is not biologically feasible.

## New Opportunities in Fertility Regulation

Chapters 4 and 5 summarized the remarkable progress made in recent years in fundamental reproductive research. (For a more detailed account, see the companion volume.) Many of these significant basic discoveries create important new opportunities for applications in fertility regulation. These discoveries have resulted partly from the development of highly sensitive assays for determining minute amounts of hormones in blood and other body tissues, partly from the application of methodological advances (such as amino acid sequencing and polypeptide synthesis) to problems relating to reproductive processes, and mainly from the fact that the process of reproduction is now being probed at the subcellular molecular levels.

The isolation and synthesis of the gonadotropin releasing factor from the brain has stimulated a number of studies aimed at using the decapeptide to regulate ovulation, with only limited results thus far. However, improved understanding of structure-activity relationships of the decapeptide is now leading to the synthesis of analogs with higher levels of activity than the natural releasing factor and of analogs that are potent antagonists. Both agonists and antagonists have important implications for fertility control in either the female or the male.

Another neuroendocrine advance relates to the role of the pineal gland in regulating pituitary-ovarian function. An ovulation inhibiting substance produced by the pineal in lower mammals has been identified as the biologic amine, melatonin. Melatonin is now known to be produced by human subjects, both male and female, and excreted in the urine in a diurnal pattern. Whether melatonin has a role in regulating human fertility remains to be established, but the identification of a pattern of melatonin excretion, when correlated with animal data, provides the basis for renewed interest in the pineal and its role in human reproduction. Other significant work indicates a possible role of polypeptides in pineal control of pituitary-ovarian function.

The chemistry of the gonadotropic glycoproteins has attracted the attention of protein chemists, and structural analysis is virtually complete for the three principal human gonadotropins. Beyond the immunologic approach, based on the findings noted previously, there are other potential applications of new knowledge of gonadotropin chemistry for fertility regulation. The enzymatic removal of the carbohydrate moieties drastically reduces the molecule's biological activity without reducing its capacity to occupy binding sites on target cells. There are no evident pharmacologic barriers to the testing of this potential chemical block to hCG action in women for menses induction. Although it is likely that more work will be needed to establish the proper molecular modifications, dosage, and schedule of use, the approach is amenable to rapid testing of feasibility and could open a major new area of clinical study.

Also advancing is work aimed at identifying the structural features of the gonadotropin receptors on the membranes of target cells. Analysis of the location of gonadotropin receptors is already developing rapidly. In the male, follicle stimulating hormone (FSH) receptors are localized specifically to the seminiferous tubules on the Sertoli cells. LH receptors are localized to the Leydig cells and, perhaps, associated with some epididymal epithelial cells. In the male, as in the female, extragonadal receptor locations are still under investigation (particularly in neural tissue). The ovary has hCG receptors associated with granulosa cells of medium-sized follicles, but not small follicles. The role of FSH may be to stimulate the formation of LH receptors on those cells that will luteinize after ovulation. The most advanced work on the chemistry of these receptors is with the luteal cell receptors of hCG that cross-react with LH. There are sugar moieties extending from the surface that may be subject to interference through pharmacologic intervention.

The receptor chemistry and intracellular molecular biology of steroid hormones has been broadly elucidated in a remarkably short period of time. Specific cytoplasmic receptors in target cells carry the steroid (estrogen or progesterone, and probably all biologically active steroids) to the nucleus where a transcriptional activity brings about quantitative and qualitative changes in messenger RNA synthesis and alters the pattern of structural and secretory products of the cell, converting it from nonstimulated to stimulated state. Many steps in this sequence are now clearly worked out. Factors also have been described that play a role in the generation of cytoplasmic receptors, in the rate of receptor degeneration, in the binding of the steroid to the receptor, and in the binding of the steroid-receptor complex to chromatin. The synthesis of chemical substances that can block the normal steroid-receptor interaction has been reported, and clinical application of this principle or others mentioned may soon become feasible.

Whenever attention has been given to cell membranes in the reproductive process, significant new information has emerged. The sperm surface has been found to have gradients of electrical charge that change during the maturation process. With the use of chemical mapping agents, such as conconavalin A, it has been possible to analyze the surface distribution pattern of specific carbohydrate moieties. The significance of this new information is not yet sufficiently clear to reveal implications for fertility regulation. Membrane biology has produced new understanding of the fertilization process itself. Aspects of the structure of sperm receptors on the egg membrane have been revealed for some marine invertebrates, and it has been postulated that polyspermy is blocked after the sperm enters the egg by sperm-induced chemical alteration of the egg membrane, which inactivates the sperm receptors. Greater understanding of these phenomena could make possible experiments to discover means of inactivating the sperm receptors independent of sperm-egg interaction. Alternatively, prevention of the block to polyspermy, which would be equally effective in preventing the development of viable embryos, may prove more feasible.

Several findings regarding mammalian gametes have been announced in a relatively short period of time; and with these advances, new possible applications to fertility regulation are coming into focus. The concept of sperm "capacitation" is now more or less classical, even though the biochemical characterization of this final maturation event has not been achieved. Recent work reveals that the epididymal maturation phase of mammalian spermatozoa (including man) is androgen dependent and may be controlled by an extrinsic factor produced by the epididymal epithelium. The logical approaches for interfering with epididymal maturation of sperm appear to include those employing antiandrogens, those aiming at blocking production of androgen-binding protein by the seminiferous tubules, and those based on inhibition of the androgen-dependent extrinsic maturation factor.

Follicular maturation factors, long recognized as normal constituents of the ovaries of marine organisms, appear to play a role in the mammalian ovary as well. Studies in the rat reveal that oocyte arrest during the first meiotic division may be the result of a specific maturation inhibitor produced by a component of the primary ovarian follicle. As the follicle grows, reinitiation of oocyte maturation may result from withdrawal of this extrinsic inhibitor and the appearance of a maturation factor not present in small follicles. These substances, along with those that inhibit or stimulate luteinization, represent a previously unrecognized group of chemical agents in the reproductive process. They may provide new prospects for immunologic approaches to fertility regulation, as well as for pharmacologic interventions, once their chemistry and normal physiological roles are better understood.

Basic research of past decades made possible the new fertility regulation techniques that changed the contraceptive practices of millions throughout the world and provided the foundation for approaches that promise even greater advances in the next ten years. The fundamental discoveries of the recent past will serve as the basis for the contraceptive technology of the future if support levels are adequate so that the new leads and opportunities summarized here can be intensively pursued. Both the fundamental and the goal-oriented work are coequal parts of the same continuous process leading to technological innovation: Greater understanding of the fundamental processes of reproduction, the identification of new principles, and the flow of new ideas provide the essential basis for the applied research needed to develop an improved technology of fertility regulation.

# III Institutional Setting and Constraints

# 7 The Institutional Base for Research and Training in the Reproductive Sciences

A diverse mix of professional and institutional resources is involved in the spectrum that ranges from fundamental research and training in the reproductive sciences at one end to contraceptive development and testing at the other. Fundamental research and training is primarily a university function, augmented by important work at a few free-standing institutes. At the applied end of the spectrum the pharmaceutical industry is a major factor, but in recent years several public sector organizations supported by government and philanthropic funds have emerged that are also heavily involved in contraceptive development.

The review assembled considerable information on these resources, both directly through queries to experts who prepared country reports and indirectly through analysis of the principal investigators and institutions receiving grant and contract awards from the major funding agencies. Unfortunately, however, the organizational patterns of scientific work vary widely from country to country, so that much of the information gathered is not strictly comparable. As a result, we cannot attempt a systematic analysis of the research and training resources of the field worldwide. But we can present material that describes the major resources and suggests a partial answer to whether or not the field has capacity in being that can contribute to the expanded effort needed, as chapter 6 makes clear, if an improved, safer, and more acceptable contraceptive technology is to be developed.

In this chapter, therefore, we focus on the resources available for fundamental research. The institutional base in the industrialized nations is large and well known. On the other hand, the fundamental research capacity in the developing nations is little known and to some, unexpected. As a result, there is special interest in the research resources of the developing world and comparatively more attention is paid here to these nations. We also summarize briefly the facilities available for the training of new research personnel. Finally we present a concise summary of the views expressed by active scientists in various nations of the needs and challenges facing the field. Comparable information on industrial and public sector organizations concentrating on contraceptive development is presented in chapter 8.

## Basic Research Institutions

Based on the national and regional reports and the records of the principal funding agencies in the field,[1] we could identify 419 institutions throughout the world that undertake at least modest levels of activity in the reproductive sciences and contraceptive development. These include 176 institutions in the United States, 162 in the fifteen other industrial nations represented in the review, and 61 in the nine developing nations and regions. (Although funding data was not available from Latin America, it has been estimated that approximately twenty institutes are involved in reproductive research in that region [Perkin 1975].) From the grant listings, at least 1,370 scientists of varying levels of experience are currently studying aspects of the reproductive process. It appears, however, that this figure represents only a partial accounting of the experienced scientific cadre available for work in the field; examination of the membership lists of the principal professional societies relevant to the reproductive sciences suggests that the number of reproductive researchers worldwide is about three thousand.[2] (For

---

[1] *Government and Multilateral Agencies:* National Institute of Child Health and Human Development-NICHD,
National Institutes of Health-NIH (excluding NICHD),
National Institute of Mental Health-NIMH,
National Science Foundation-NSF,
Atomic Energy Commission-AEC,
Food and Drug Administration-FDA,
Agency for International Development-AID,
World Health Organization-WHO,
United Nations Fund for Population Activities-UNFPA;
*Philanthropic Agencies:* Ford Foundation-FF,
Rockefeller Foundation-RF,
Population Council-PC,
Scaife Family Charitable Trust;
*Countries responding:* United States; *Other Industrialized Countries:* Australia, Belgium, Britain, Canada, France, Germany, Israel, Japan, Italy, Netherlands, New Zealand, Scandinavian countries (geographic area)-Denmark, Finland, Norway, Sweden. *Developing countries*: Egypt, Hong Kong, India, Philippines, Thailand, South Korea, Equatorial Africa (geographic area), Iran, and Turkey.

[2] Based on several professional society lists (Society for the Study of Reproduction, Endocrine Society, and Society for the Study of Fertility) there are at least sixteen hundred researchers in the United States and Canada and four hundred in Britain. Adding the number of researchers in other nations listed in the funding data brings the total number of scientists engaged in reproductive research to at least twenty-five hundred but probably nearer to three thousand.

comparative purposes, in the United States alone there were twenty-nine thousand doctorate life scientists engaged in research and development in all sectors as of 1972 [National Science Foundation 1974].)

As a measure of activity, the number of papers in the reproductive sciences and contraceptive development published in internationally recognized journals from 1970 through 1975 is presented in table 7.1. This tabulation, compiled by the Reproduction Research Information Service,[3] includes an estimated 70 percent of published articles and omits articles in journals of local circulation. The totals further understate relevant scientific activity because results of clinical studies, relatively more frequent in the developing world, are less likely to be published in international journals than are findings of laboratory research.

As would be expected, the bulk of the published papers were prepared by investigators from the industrial nations of North America, Western Europe, Japan, Israel, and Oceania. More than 8 percent of the papers, however, were contributed by scientists in the developing nations of Asia, Latin America, and Africa. Moreover, the total num-

Table 7.1 Published Research Papers on Reproduction and Contraceptive Development, by Geographic Area of Authors, 1970-1975

| Geographic Area | Papers Number | Percent |
| --- | --- | --- |
| Mexico, Caribbean Area, Central and South America | 1,141 | 2.5 |
| Africa and the Near East | 864 | 1.9 |
| Asia | 1,765 | 3.9 |
| North America | 19,961 | 44.2 |
| Western Europe | 17,468 | 38.8 |
| Japan, Israel, and Oceania | 3,861 | 8.7 |
| Total | 45,060 | 100.0 |

Source: Reproduction Research Information Service, Ltd. 1976. Excludes 3701 papers by researchers in the USSR and Eastern Europe.

[3] The report was compiled from the *Bibliography of Reproduction*, which it publishes monthly. The *Bibliography* is prepared from direct review of about 250 journals and from listings in a number of abstracting services.

ber of articles published increased from 6,557 in 1970 to 9,150 in 1975 (not shown), and the proportion contributed by developing country investigators increased slightly over the six-year period covered.

### Research in the Industrialized Nations

The research institutions working in the field vary both in administrative setting and size. Most are in university or medical school departments, of course, but a number are attached to hospitals, government laboratories, or free-standing institutions. Some detailed information is available about U.S. institutions that have thus far undertaken the bulk of the work in the field. Of the 176 research groups in the United States, 106 include the full-time equivalent of one senior scientist supported by junior scientists and can be considered to have only a "minor" degree of involvement in the field; 49 have full-time equivalent of at least two senior scientists, as well as supporting staff, and can be considered to have a "major" commitment; the remaining 21, each with five to ten senior investigators and substantial supporting staffs, are essentially at the level of "institutes" devoting their principal attention to the field. Although this classification does not provide conclusive information, it does suggest that in relatively few U.S. universities and research laboratories is the reproductive process a major full-time focus of work. Moreover, the funding data presented in chapter 11 imply strongly that the involvement of many of these institutions and scientists in reproductive research has been a relatively recent phenomenon, spurred by the greater availability during the last decade of funds to support this branch of inquiry.

To encourage more institutions to develop systematic and concentrated programs in this field, the Center for Population Research (CPR) of the National Institute of Child Health and Human Development began in 1971 a program of support for Population Research Centers. In the biomedical sciences, there are now twelve centers—eleven at universities and the twelfth at the Population Council—that have up to thirty-five scientists on their staffs. CPR also awards program project grants for the support of a group studying three or more interrelated research projects. Such awards supported reproductive research in 1976 in eleven different medical schools and universities, as well as in the Mayo Foundation, the Salk Institute for Biological Studies, and the Oregon Regional Primate Research Center.

In other industrial nations, the review identified important clusters of research institutions working in the field. Activities were reported at five universities in Belgium, six in Italy, nine in the Netherlands, eight in Australia, five in Israel, and seven in Japan. In Great Britain, twelve research units financed by the Medical Research Council, usually located within a university, are engaged in reproductive studies to some extent. Only one, the Reproductive Biology Unit in Edinburgh, directs its research specifically to the field of inquiry. A number of senior reproductive scientists in other institutions are also funded by the MRC, and several institutes working in reproductive research are supported by the Agricultural Research Council. In Sweden, Denmark, Finland, and Canada, Medical Research Council funds also support considerable work by scientists based primarily in medical schools (twenty-two research groups are located in the Scandinavian countries and thirteen in Canada). Reproductive research in sixteen German universities is supported by the Deutsche Forschungsgemeinschaft. And in France seventeen research units, primarily in hospitals and medical schools, carry out reproductive research. Their funds stem from several different ministries of the government—Health (INSERM), Education (CNRS), Agriculture (INRA), and Research and Industrial Development.

The review was unable to obtain comparable information on the research resources in the USSR and other nations of Eastern Europe. Five years ago, however, the WHO Feasibility Study documented that there are at least three important reproductive research institutes in Moscow and Leningrad, three each in Poland and Hungary, and two in Yugoslavia (WHO 1971). WHO established a Research and Training Centre at the All-Union Scientific Institute of Obstetrics and Gynaecology in Moscow in 1972.

Research in the Developing World

The industrialized nations are well known for their capabilities in many fields of science, but the fact that many developing nations have mounted serious programs in the reproductive sciences is little known. A number of important recent advances in the field are attributable to these researchers. The addition of trace elements to intrauterine devices and the use of quinocrine to produce tubal occlusion were developments in Chile. The oral contraceptive minipill was first studied in Mexico. Various types of steroid implants and steroid contraceptives for men are under development in Brazil. A variety of new intrauterine devices and other approaches to fertility regulation are originating in Peru. Several of the newer approaches to tubal ligation were first studied by Mexican scientists, who are also exploring a natural plant product that may have luteolytic as well as oxytocic activity and is orally active. A promising effort to develop a contraceptive vaccine linking the $\beta$ subunit of human chorionic gonadotropin to a tetanus toxoid carrier is under investigation at the All-India Institute of Medical Sciences. According to one observer (Segal 1974, p. 601), Indian scientists have done significant work on "the role of trace metals in reproductive systems of both male and female, the effect of intrauterine foreign bodies on protein constitution of the intrauterine fluid and enzymatic patterns in the endometrium, the immunology of gonadotropic hormones, the hormonal control of sperm maturation in the epididymis, the metabolism of contraceptive steroids, and mechanism of hormone action."

These advances in fundamental research are in addition to the role in clinical trials that developing world researchers have also carried out. Clinical testing of fertility regulation methods must be carried out in nations with varying conditions because, as a recent WHO workshop (Kessler and Standley 1974, p. 577) concluded,

For established methods, dosages appropriate for women of a high level of nutrition may not be applicable [in some developing nations]. Side effects of concern in the West may not manifest themselves while new side effects may appear arising from disease and deficiency states in the developing countries. For new methods developing countries should be involved at an early stage of the research to determine acceptability. They must also be involved in testing at all stages of development so that safety and effectiveness data can be acquired rapidly for different populations.

While acknowledging the need for clinical testing in developing nations, some officials of governments and international assistance agencies question whether scarce financial and manpower resources should be used to equip and staff laboratories in developing nations for fundamental research studies. However, as noted above, well-established research groups in Latin America, Asia, and Africa already are contributing to the field's knowledge base in fundamental as well as applied areas and could make effective use of additional resources for continued and expanded research activity.

These contributions are the products of a sizable cadre of experienced scientists engaged in reproductive research in developing countries. At least 172 scientists who held reproductive research fellowships provided by the Population Council or the Ford Foundation between 1963 and 1973 are now engaged in reproductive studies in developing nations. This grossly understates the number of qualified scientists in these countries because it does not include senior scientists who received their training prior to 1963 and others who were trained locally. Many scientists in developing nations resent the suggestion that they should limit their work to clinical trials and not have the opportunity to conduct research on topics to which they give highest priority and find most intellectually challenging.

The problems associated with capacity building for reproductive research in developing nations are similar to those encountered in connection with institution building for research and training efforts in a variety of fields. University structures and the priorities given to research vary from country to country and often are not conducive to a sustained research effort. Free-standing research institutes or research sections of government departments may be more appropriate institutional bases. Adequate full-time research positions and salaries are lacking in most developing countries, requiring medically qualified research scientists to enter private practice or hold several positions simultaneously in order to make a living.

Because the encouragement of reproductive research in developing nations must deal with a number of unique problems, particularly when outside assistance agencies are involved, the regional surveys undertaken as part of this review were conducted by scientists who are citizens of the regions surveyed. In addition to gathering information on institutional strengths or weaknesses and levels of funding, they were asked to sound out the opinions of their colleagues with respect to the most appropriate institutional arrangements, the kinds of support, and the mechanisms needed in order to achieve a more effective research effort in accordance with priorities as they see them.

## Latin America

Research in the reproductive sciences is more widespread in Latin America than in other developing regions (Perkin 1975). More than twenty research groups in seven countries are actively engaged in research of direct relevance to contraceptive development, and at least three hundred scientists are conducting research on either a part- or full-time basis.

The history of reproductive research in Latin America is unusual in several ways. Unlike in other regions, in Latin America the development of an institutional capacity to undertake research in the reproductive sciences preceded the development of official government-supported family planning programs. In Chile, Mexico, Brazil, and Colombia this existing research capacity almost certainly contributed to accurate local information on the advantages and disadvantages of specific contraceptive methods.

Argentina and Mexico are the countries with the greatest number of research groups specializing in reproduction; and Chile, Brazil, Uruguay, Peru, and Colombia each has one or more well-established groups. In most other countries occasional research projects, usually on clinical subjects, may be undertaken by individual physicians; but institutions devoted to such research have not yet been established.

An impressive level of research productivity in Latin America has been achieved despite obstacles similar to those encountered in other developing regions. Most medical schools give low priority to biomedical research. Throughout much of the region reproductive research receives even lower priority, and as a consequence investigators must look beyond their universities and often beyond their national boundaries for the support they require. Salary supplements are generally needed to augment the part-time salaries many investigators receive from their universities. Funds for major pieces of equipment can often not be found locally. Because contraceptive research is frequently viewed as a sensitive issue, researchers may be unable to obtain support from national research funds. Latin scientists have therefore depended to a large degree on international sources for research support.

Of special interest is the organization in 1974 of the Latin American Program of Research in Human Reproduction (PLAMIRH) by a group of seven Latin American scientists from six countries, with support of the Ford Foundation and the International Development Research Centre of Canada. PLAMIRH awards research grants on a competitive basis. Scientific review is conducted by an independent panel of Latin American scientists in accordance with Latin American priorities. During the first round of the competition in May 1975 nearly fifty research applications were received from scientists in the region.

## South Asia

Research in reproductive sciences in South Asia is dominated by the work at Indian laboratories and clinics. Between 1970 and 1975, more than 40 percent of the articles on reproductive topics published in internationally recognized journals by developing world scientists were produced by India (Reproduction Research 1976). The All-India Institute of Medical Sciences has been designated one of four Research and Training Centres by the World Health Organization's Expanded Programme of Research, Development and Research Training in Human Reproduction and there is substantial participation by Indian scientists in most of the WHO task forces on specific subjects in reproductive research. At least sixteen university-based departments and free-standing institutes are carrying out significant programs, and a review of reproductive research in India observes (Segal 1974, p. 601),

A generation of Indian scientists has been developed who do not have to go abroad for their postdoctoral training and who are doing work of a calibre that compares favorably with that done anywhere else in the world. There used to be a concentration of such work in the ten or twelve institutions originally supported in the first round of Ford Foundation grants in this field in 1962-63. But dissemination of activity is becoming evident. Now considerable potential for scientific contribution exists in a second level of Indian institutions, not clearly identified in the past.

A major program in reproductive research operated collaboratively by the University of Delhi Department of Zoology, the All-India Institute of Medical Sciences, the Central Family Planning Institute, and the Indian Council of Medical Research is training high quality Ph.D.s in the reproductive sciences. To develop suitable career opportunities for these younger investigators, an eminent Indian researcher urges that a system of career awards, visiting fellowships, and other incentives be established to capitalize on the availability of scientific manpower for research in reproduction (Prasad 1974). Support for reproductive research has been provided by the Indian government through the Indian Council of Medical Research, the Council of Scientific and Industrial Research, and the University Grants Commission.

Southeast Asia

The research effort in Korea, Malaysia, the Philippines, Singapore, and Thailand is not well developed, although interest has increased as the nations of this region have begun to devote greater resources to national family planning programs. Major emphasis is placed on study of the effects of modern contraception on the health status of women in the region.

Thailand has a core of scientists in university medical schools, but low salaries and the relatively low status given to research has militated against development of a major research capacity. Some support for research in the applied sciences comes from the National Research Council of Thailand, which makes small grants for supplies, equipment, and miscellaneous expenses. Of the universities only Chulalongkorn has endowment funds earmarked for research. These are given to staff in income supplements up to $50 a month to encourage research, but none of these funds to date has been used for support of the reproductive sciences. At the university, however, an

Institute for Health Research has recently been established with an active Human Reproduction Unit. This institute will establish career positions and modern laboratories to attract long-term funding from national and international sources. The unit's research program was developed and stimulated with the aid of a WHO research team engaged in a series of studies of the effects of contraceptive hormones on glucose tolerance, liver function, lactation, and other clinical studies relating to the effects of contraception on the Thai population (Dusitsin 1974).

**People's Republic of China**
Although it is a less developed country as measured by the usual indices, reports of knowledgeable visitors indicate a high level of sophistication in research in the reproductive sciences in the People's Republic of China, particularly as directed toward improved contraceptive methodology (Chang 1974; Djerassi 1974).[4]

Several of China's senior reproductive scientists were trained in Western universities and worked in Europe and the United States. For example, Dr. Huang Minlon, who worked at Schering in Berlin during the 1930s and at Harvard and Berkeley during the 1940s, organized an active program of steroid chemistry at the Institute of Organic Chemistry of the Chinese Academy of Sciences in Shanghai (Djerassi 1974). However, younger scientists now are receiving their training in China (Chang 1974). Laboratories are not air conditioned and laboratory instruments are adequate but not elaborate. According to Chang, (p. 155) the scientists he visited "were very proud of their China-made instruments because self-reliance is one of the guiding principles. The synthetic steroids and non-steroidal compounds were all manufactured in China." Djerassi notes that major equipment dates to the late 1950s or early 1960s and is typically of Japanese or European origin.

Because of his primary interest in the manufacture of steroidal contraceptives by the PRC pharmaceutical industry and his own preeminence in the field, Djerassi focused on this area of reproductive

---

[4] For abstracts from Chinese medical journals indicating extensive research activity in the early 1960s, see Orleans 1973.

research. He reports advanced chemical research groups of the Institute of Materia Medica in Peking and the Institute of Organic Chemistry in Shanghai. He notes (p. 25),

The Peking group under Dr. Huang Liang has completed the total synthesis of norgestrel . . . this complicated process is now presumably in the hands of a pharmaceutical factory, since norgestrel appears to be the active component in the Chinese once-a-week and once-a-month pill [currently under clinical scrutiny] . . . and a daily pill. Similarly, in ten months the Peking group completed the total synthesis of prostaglandin $F_{2\alpha}$ by minor modifications of [a published procedure]. . . . These are impressive achievements especially if one considers the short time required to repeat these difficult multi-step syntheses and the fact that all reagents and starting materials are produced in the PRC.

At the Institute of Physiology of the Chinese Academy of Sciences in Shanghai, a special section on antifertility compounds employs twenty scientists and technicians under T. D. Feng. This group uses radioimmunoassays of estrogens, progesterone, and luteinizing hormones to study the mechanism of action of once-a-month pills developed in PRC. This group is also involved in testing local herbs for contraceptive effectiveness and the use of IUDs as a carrier for steroids to increase their effectiveness (Chang 1974).

**Subsaharan Africa**
While subsaharan Africa has a number of excellent university centers at which some research in reproductive sciences is carried out, it is nonetheless the region with the fewest institutional resources for training and research. Relatively little activity in the reproductive sciences and contraceptive development is currently underway, although the review of this region revealed that several groups of experienced scientists express a strong interest in expanding their work (Adadevoh, Olatunbosun, and Dada 1974). There is interest in both fundamental and applied research, with an emphasis on development of contraception appropriate to African women, reinforced by a concern that existing methods are offered in Africa without attention to special problems of acceptability and side effects specific to the region. Because infertility is recognized as a major problem

throughout broad areas of East and West Africa, the treatment of this condition is a major stimulus for research in reproduction. Throughout this region there are several departments of animal physiology or husbandry concerned with improving the reproductive performance of cattle. In a number of instances opportunities for collaborative research with departments concerned with human physiology have been identified.

The general status of reproductive research in Equatorial Africa is summarized by the head of the Reproductive Biomedicine Research Laboratory at the University of Ibadan, Nigeria (Adadevoh 1974, p. 330):

In Africa reproductive biology research and contraceptive development has generally received a low priority to date, although interest has definitely been stimulated amongst scientists and institutions all over. The understanding and projections for the future in the area, however, varies—emphasis being placed on clinically applied research by some whilst some favor this and basic research both in man and animals (including animal husbandry). . . . Except for a few scientists and university institutions to which funds were made available by grants from outside agencies, research in the area has been minimal and the marginal funds directed to research have derived either from university department funds or governmental departmental funds given for research in general. National research councils have in general not identified reproductive biology projects for funding.

## Middle East

The most extensive research program in reproduction in the Middle East is in Egypt, and efforts are underway to increase research activity in Iran and Turkey. There is thus far little significant activity elsewhere.

Although clinical studies in reproduction were carried on in Egypt in the early 1960s, in 1966 the departments of obstetrics and gynecology at Cairo and Alexandria universities began major research programs in reproduction with assistance from the Ford Foundation. The principal stimulus to expand research in this field was the decision of the Egyptian government to initiate a national family planning program with two thousand clinics providing contraceptive services emphasizing the pill. Those responsible for the family planning program were concerned with the side effects of oral contraceptives and the need to find optimal contraceptive methods for Egyptian women (Hefnawi 1974). Research in reproduction is presently carried out in ten institutions. The major contributions of these research groups, numbering five to ten scientists and technicians, about half at the doctoral level, lie in clinical studies on safety, acceptability, and effectiveness of current and potential contraceptive techniques. Although receiving foreign exchange support from international agencies, particularly UNFPA (via WHO) and the Ford Foundation, none of these institutions is adequately equipped with modern instrumentation, spare parts, and consumable supplies; nor are there adequate reference library facilities containing up-to-date periodicals, texts, and monographs. Nor is their funding adequate to provide financial and professional encouragement to the increasing number of young scientists and physicians who have received advanced training abroad.

Training

As in other areas of science, the traditional process of developing research skills in reproductive research follows the route of a general university education, professional or graduate school work usually to the doctoral level, and frequently special postdoctoral training in a subspecialty of the biomedical area in which the degree has been issued. Postdoctoral training has become increasingly common in recent years, both for new Ph.D.s and more experienced researchers, as a result of the great expansion of knowledge in the biological sciences. Although postdoctoral training has developed in part to permit scientists to master the complex instrumentation that has been introduced in biology, its more important objective is to provide indepth exposure to the concepts, theories, and objectives of the host laboratory and to learn what questions need to be asked to advance knowledge in the special subject under investigation. These general principles apply to training in the reproductive sciences and contraceptive development.

There has been a major international effort over the past twenty years to strengthen the professional capacity for biomedical research in reproduction in both industrialized and developing countries. For scientists in developing nations the principal sources of support for training and fellowship awards until 1974 have been the Ford Foundation and Population Council. A total of 339 study awards or fellowships were awarded to scientists from thirty-nine countries by these two agencies alone between 1955 and 1974. With the establishment of the NIH Center for Population Research in 1968, special attention was given to strengthening professional capacity through training grants, fellowships, and research career awards. Between 1970 and 1974 some six hundred U.S. scientists were trained in various aspects of the reproductive sciences. In addition these U.S. based agencies have supported training efforts in Japan, Canada, Europe, Israel, and Oceania, thus supplementing the support provided by medical research councils and other governmental and private agencies in these countries. All told, several hundred scientists from these countries have received special postdoctoral training in the reproductive sciences from CPR, the Ford Foundation, and the Population Council. Additional fellowships awarded by the Rockefeller and Lalor Foundations and the Scaife Family Charitable Trusts in the U.S. and the Nuffield Foundation and Oliver Byrd Fund in the United Kingdom bring to fifteen hundred the estimated number of scientists who have received training for research in reproduction over the past two decades. In addition the World Health Organization has awarded two hundred forty research training grants and has given assistance to approximately four hundred individuals for short-term travel and study awards.

Research training was not goal-oriented and didn't include a major clinical component until relatively recently. It aimed at increasing research manpower in the field of reproduction in industrial and developing countries, without reference to the unique needs of either. In the last five years there has been an effort, in which WHO has taken the lead, to relate research training more closely to research problems in the control of fertility. This has been particularly true for research training for developing countries and is due to several factors. One of them has been the difficulty of assessing the impact of broad research training on fertility regulation. Another has been the increasing demand by administrators of national family planning programs for research manpower to study problems arising in the programs (such as clinical trials of new methods or assessment in particular populations of the side effects of existing methods). The administrators believe that this research is best conducted by local scientists trained in a more applied approach than in a broad academic orientation.

Efforts have also been initiated to develop manpower in disciplines essential to contraceptive development, such as epidemiology, clinical pharmacology, and the biosocial sciences. Manpower in these disciplines is in short supply everywhere.

The effectiveness of training programs is often questioned primarily due to the lack of clear-cut evaluation criteria and impact measures. For developing nation scientists trained in industrialized nations, one simple measure that has been used is the number of trainees who return to their home countries and remain active in research on reproduction. Precise information on this matter is not available. Some trainees return home immediately on completion of their training, others delay their return for a few years waiting for a suitable career opportunity at home, and others settle permanently in a developed country. Reasonable estimates can be made, however, because most training centers are able to show where their trainees are located and what positions they hold. Their records show that approximately two-thirds of the trainees return to their homelands and continue research in reproduction. The percentage was somewhat lower in the midsixties when employment in industrialized nations was fairly easy to obtain. The attraction has all but vanished due to the severe fiscal constraints on research activity in those countries. Moreover, many training centers in industrial nations now do not accept candidates from developing countries unless they have written assurance of a position at home to which they can return after completing training. However, it is sometimes advantageous to all concerned for the trainee to remain in an industrialized country, especially those trainees who have gained expertise in particular lines of work that could not be pursued effectively, if at all, were they to return to their home base.

An effective strategy for generating professional manpower for the reproductive sciences and contraceptive development, suitable for the industrial as well as the developing world, must thus involve a multiplicity of considerations. There is no doubt, however, that training is an indispensable element in maintaining the viability of any field of science. Essentially it is the continuous infusion of young minds and new ideas that maintains dynamism and vigor. Thus, although it is difficult to establish specific standards to evaluate the effectiveness of training programs, the status of the field itself gives an important insight into the success or failure of previous training efforts. The reproductive sciences and contraceptive development are now experiencing the same explosion of seminal discoveries that characterized the physical sciences a few decades ago. Many of these discoveries emanate from laboratories of scientists trained in the last two decades for advanced work in this field.

## The Views of Reproductive Scientists

Laboratories, institutes, and training mechanisms are as important in the reproductive sciences as in any other area of scientific inquiry, but the field's principal resources ultimately are the investigators actively working on problems related to human reproduction. The policies of public and private sector funding agencies and of the scientific community in general will undoubtedly be critical in determining the evolution of the field during the next several decades. But its future course will also be influenced by the perceptions and preferences of the existing cadre of reproductive researchers. Policy makers accordingly will want to give serious consideration to the viewpoints expressed by these men and women who carry the day-to-day responsibilities for the research effort.

This review conducted two surveys to ascertain the range of opinions held by reproductive scientists. The first was a mail questionnaire sent to two thousand identified researchers in approximately twenty-five nations (Knobil and Pullum 1974). For the second, approximately one hundred scientists of recognized distinction were either interviewed orally or asked to communicate their comments and suggestions in writing. Responses to the mail questionnaire were obtained from 654 investigators, representing a response rate of about one-third. No claim can be made that the responses to either survey necessarily are representative of the views of all reproductive scientists, but we believe that the viewpoints summarized here provide a sense of the principal concerns of investigators in the field.

Respondents to the mail survey covered the spectrum of career status from trainee to senior investigator, with a median age of forty-two. Two-thirds were located in universities and the rest were in government, industry, or independent research institutes. Among the university group, half were in medical schools; and of these, three-fifths were in clinical departments and the remainder in basic medical science departments. More than one-third of the respondents were located in English-speaking countries outside of North America.

The responses document the fact that the great majority of workers currently look to government as the principal source of support for their research. Seventy percent have government support and 40 percent have *only* government support. In contrast, very few are solely dependent for support on either foundations (6 percent) or industry (5 percent).

Although respondents worked on problems ranging from the most theoretical and fundamental to the most practical, more than three-quarters expressed the view that both basic *and* applied research need to be increased in order to yield improved means of fertility regulation. Not surprisingly, university scientists showed a strong preference for basic research and those working in industry preferred applied studies. In estimating the probable impact of basic research on contraceptive development over the next ten years, nearly all respondents felt that it would be major or at least moderate; only 4 percent felt that it would have a minor impact.

Although respondents expressed a strong commitment to research aimed at improved contraceptive methods, many indicated their belief that biomedical research needs to be complemented with social research before the regulation of fertility can actually be achieved. More than a third of the respondents, for example, said that existing methods are sufficiently effective to enable individuals to regulate their fertility more efficiently than at present. Many of these respondents urged that greater emphasis be placed on social research to determine how current methods could be made more acceptable to individuals living in different nations. The survey unfortunately did not elicit opinions on the potential of technological developments, in the form of safer, more effective, and easier-to-use methods, to make contraceptives more acceptable in the areas of greatest need.

When asked to specify the best mechanisms for advancing reproductive research, respondents favored increasing both the number of research grants and opportunities for training. There was a sharp division between North Americans and those from other countries on support for research centers or institutes. Support of these institutions found little favor among North Americans but was especially attractive to European respondents. It seems possible that in Europe the conditions for research and career development in research institutes or centers are better than in the universities, where monolithic departmental structures remain traditional.

When asked the relative importance of training grants, fellowships, and job opportunities for increasing the number of investigators in the reproductive sciences and contraceptive development, respondents overwhelmingly choose in favor of job opportunities. This response was inversely related to age (which may be taken as a proxy for job security). Training grants ranked second. Young investigators seem more concerned with the demand for their services than with instruments for increasing the supply of scientists.

Not unexpectedly, the scientists expressed a preference for long-term committed support: 27 percent favored three years, 47 percent favored five years, and 7 percent favored six or more years. Nearly four out of five respondents favored recognizing the reproductive sciences as an academic discipline. More than nine out of ten preferred research grants for the support of basic studies; 61 percent favored contracts for contraceptive development studies. The two types of research are evidently viewed as distinct, requiring different methods of funding.

An overwhelming majority (70 percent) stated that foundations should support areas, such as innovative research, not covered by government funding. Many stressed the importance of small "venture grants" for bold new ideas and start-up grants for beginners. Interestingly, twice as great a proportion of Europeans as North Americans believe that foundations should supplement government funding, probably reflecting the low levels of government support available in many European countries. Six out of ten felt it appropriate for foundations to collaborate with and support research in industry. Workers in industry, however, were almost uniformly opposed to such support.

These viewpoints of the mail survey respondents were complemented and amplified by the statements obtained from many senior investigators who have contributed significantly to the advancement of the field. Because their highly individualistic views are not easily summarized, we present extensive excerpts from their statements. In general, we judge the following to be representative of their views.

Like scientists in other fields, reproductive researchers are primarily concerned with two overriding considerations: One is to obtain support for research in which they have an abiding interest and the other is to ensure opportunities for career development for younger investigators. These primary needs must be met to ensure maximal progress in basic, applied, or clinical research. Investigators today find themselves fraught with uncertainties. They must rely on outside support, and their careers often depend upon their ability to obtain all or a large part of their salaries from external sources. Most granting agencies, however, favor the support of projects rather than people.

There is a great need to facilitate the transition of young investigators from training status to staff positions on the academic ladder. The NIH Senior Research Fellowships, Markle Scholar and American Cancer Society Scholar awards, and Rockefeller Foundation Fellowships, with salary support for up to five years, have been stepping stones for many top ranking young scientists, including reproductive biologists. There is need to train more young persons in the reproductive sciences, but they must also be given opportunities to exercise their skills and develop professionally. These opportunities are diminishing in number at a time when they should be increasing.

The scientist's operational needs are approximately the same for basic, applied, or clinical research: adequate laboratory facilities, supplies, technical assistants, equipment, and opportunities for communication and interaction with other investigators. Inadequacies in any of these areas can be detrimental to the effective pursuit of research. The modern research enterprise is no longer a single-handed effort by a lone investigator working with simple tools, single-minded dedication, and private means. Research today utilizes an array of complex instruments and techniques and encompasses the sophisticated coordination of facilities, supplies, and personnel. In a very real sense the limiting factor in the operation of a modern laboratory is the availability of technical personnel with specialized skills. For many it is an earning and learning experience before taking up work toward an advanced degree. There is an all-too-common tendency among granting agencies to regard technical assistants as dispensable, whereas they are often critically necessary to the success of the research program.

These basic themes are reflected in the specific comments of reproductive scientists. These comments were obtained from the mail survey and are presented anonymously as stipulated in the questionnaire. The opinions cover a wide range of topics with much, but not universal, agreement.

### Need for Contraceptive Research and Funding Strategies

"We do not have in hand acceptable means of family planning suitable for use in all circumstances. . . . When it comes to reproductive research, we seem to express our needs only in demographic terms. When we evaluate kidney, heart, or cancer research, we look to the comfort and well-being of the individual. Over and above the fact that population pressures are still a major economic and social issue is the consideration that the individual couple is often victimized by inability to use properly the methods presently available. This results in a substantial incidence of unwanted pregnancy. At this point the solution is often abortion."

"Contraception is not a take-it-or-leave-it matter. *The future of mankind depends not on the conquest of cancer but on the control of human reproduction.*" [emphasis added]

"An explosion in basic and applied knowledge has occurred in the area of reproductive biology and endocrinology in the past fifteen years. This information is of immense importance in understanding human disease, population control, treatment of infertility, animal husbandry, and in general biology and psychology. . . . We are struck by the importance of diverse approaches to funding both basic and applied research and to the training of professional research personnel."

"It would be best to concentrate the bulk of research support on the project grant system, with particular attention to providing adequate support for the most talented young researchers just beginning their careers. The support of a limited number of large program-directed laboratories is justified to the extent that they provide a place for young workers to demonstrate their talents and for projects that require particularly expensive instrumentation."

"Granting agencies must assume greater responsibility for the achievement of balanced progress in all areas of reproductive biology while maintaining the flexibility to focus effort where practical application seems most promising."

"Research in this area should be goal-oriented . . . strategic research with an application horizon of perhaps ten to twenty years, rather than tactical research with a three to five year horizon. Project grants would ensure flexibility and the potential of switching resources from one part of the research front to another as circumstances changed."

"Support should be given to individual research workers still actively engaged in research, rather than to a department or to an institute [which] . . . is most likely to be used for administrative purposes."

"There is a great need for what may be termed 'risk' or venture capital—the availability of relatively modest sums of money to test new ideas and approaches which are currently difficult to get funded."

"As the depth of research increases and as the data required become more complex, it is harder for the individual scientist to have access to enough techniques to answer a research question satisfactorily. Thus collaboration becomes ever more important. While obviously a great deal of collaboration already takes place in the field, it is often somewhat ephemeral and haphazard. If more coordination and transfer of information in an atmosphere of complete trust could be achieved, one could envisage complex permanent teams made up of individual workers or laboratories specifically encouraged by means of contract funds to carry out research in depth. Under the leadership of a senior scientist, biophysicists, biochemists, physiologists, surgeons, clinicians, cytologists, etc., could be coordinated to make maximum use of material and equipment."

"Collaboration between reproductive biologists and protein chemists, molecular biologists or immunochemists is in my mind likely to breed fruitful synergism leading to fresh thinking and insights at a molecular level."

### Support of Basic and Applied Research

Although all of the commentators were either basic scientists or clinical investigators, many urged a program of balanced support of fundamental and applied research as complementary activities:

"No hard and fast rules should be applied in planning research into reproduction, nor should emphasis necessarily be put on applied research to the detriment of basic research, or vice versa. Both are equally important and are also entirely complementary. Few advances in the applied aspects of reproductive knowledge could be made in the absence of considerable progress in our basic understanding of the process. On the other hand, without the stimulus of a social and economic need to use knowledge of reproductive processes to control our world, basic research would be inefficient."

"Basic science and applied science are two aspects of the same endeavor. . . . Basic research can yield information for applied science and vice versa."

"The past ten years has demonstrated beyond question that many of the significant advances made in the biological and clinical areas of reproduction have been dependent on basic biochemical knowledge previously developed."

"The possibilities for contraception based upon present knowledge depend either upon accelerating or inhibiting known processes. The concept is a simple one and is only broadly conceived because our knowledge of physiological processes is not profound enough to be more specific. It becomes clear that research into the fundamentals of these processes must proceed not only on a broad front but also in detailed depth."

"Both basic and applied research may be valueless unless they have a direct relationship to better contraception. . . . In view of the crisis presented by nature of the population and environment problem, the 'so what?' test may be applied to work done."

**Stability of Support**
"Research needs stability of support. The current dismantling of research and funding difficulties are leading to disenchantment of young scientists."

"The undependability of support makes long-term studies difficult and pushes potential young investigators into other fields."

"The whole field is still at a fairly early state of development and would probably be best developed by long-term support."

"Uncertainty as to future funding will eventually reduce productivity very nearly to zero. There is, however, a regrettable tendency to divert large long-term support to departmental or other activities. The responsible investigator, rather than the department or administrative official, should be given control of the grant or option of grant renewal."

"The trend toward restricted short-term funding is definitely counterproductive. Reproductive research requires a long-term investment and commitment. . . . The uncertainties of the economics of research in reproduction has led to a growing disenchantment, particularly of young investigators. Generally especially for the investigator with clinical training, either medical or veterinary, the opportunities for productive and lucrative nonresearch careers are increasingly attractive."

"Personal finance is nearly always a perpetual worry to younger scientists. Permanent security . . . is obviously an impossible expectation [but] . . . to expect scientists to tackle awkward research problems satisfactorily on annually or biennially renewable fellowships, contracts, or grants is to ignore the adverse physiological strain imposed by such a Damocletian situation. It seems to me a realistic possibility for research funding bodies to be able to ensure a scientist's livelihood in units of five years."

Training
"Many of the substantial advances made in the past ten years have resulted from the active participation of younger investigators who have worked in the laboratories of more senior scientists under training and fellowship programs. The availability of opportunities in these programs, especially the latter, has been sharply curtailed in the past few years primarily because of federal cutbacks. In addition to fewer scientists being trained, this also results in a corresponding decrease in the amount of research which can be pursued. We would strongly urge augmentation of training and fellowship programs. . . . Efforts should be made to attract younger scientists to this field if we are to expect future progress equal to that of the past decade."

"There should perhaps be at least two types of basic training—training courses of short duration (one month) for workers from different countries who need a better understanding of the known mechanisms of conception to help them in fostering practical or field programs in their respective countries. Secondly, there is the need for training in the laboratory for those who seem likely to be able to advance basic knowledge in the field."

"Training programs should be continued in order to provide continuity in the supply of young research workers and to maintain options flexible enough to accommodate unanticipated future needs. However, the number of students accepted for training in Ph.D. programs should be severely limited, with talent and motivation being the criteria for acceptance."

"More research support in the realm of male reproduction is clearly needed. However, more money for research grants alone may not be sufficient to correct the deficits of the last decade. There is also a shortage of qualified research workers in male reproductive biology. Beyond the availability of increased research support, funds for training, especially at the postdoctoral level, will be necessary. In the meantime, investigation of particularly urgent matters may require recruitment of developed talents in other areas."

"An important 'amplifier' effect on the growth of this field [would result from] training more neuropharmacologists."

## References

Adadevoh, B.K. 1974. Review of biomedical research in reproduction in Equatorial Africa. In *A review of biomedical research in reproduction biology in countries of the developing world*, ed. J.T. Lanman, pp. 300-303. Mimeographed. New York: Population Council.

Adadevoh, B.K.; Olatunbosun, D.A.; and Dada, O.A. 1974. Geographic report—East and West Africa. Report commissioned for the reproduction review by the Ford Foundation.

Chang, M.S. 1974. Family planning services in the People's Republic of China. In *China medicine as we saw it*, ed. J.R. Quinn, pp. 147-162. DHEW Publication #(NIH) 75-684, Bethesda: Fogarty International Center.

Djerassi, C. 1974. Fertility limitation through contraceptive steroids in the People's Republic of China. *Studies in Family Planning* 5(1):19-20.

Dusitsin, N. 1974. A review of research and support in reproductive biology and contraceptive development in Thailand. Report commissioned for the reproduction review by the Ford Foundation.

Hefnawi, F. 1974. Report on Egyptian system of funding research in reproductive biology and contraceptive development. Report commissioned for the reproduction review by the Ford Foundation.

Kessler, A., and Standley, C.C. 1974. Human reproduction and family planning: research strategies in developing countries. *Nature* 251(5476):577-579.

Knobil, E., and Pullum, T. 1974. Attitude survey of scientists in reproductive research. Draft. Report commissioned for the reproduction review by the Ford Foundation.

Lanman, J.T., ed. *A review of biomedical research in reproduction biology in countries of the developing world.* Mimeographed. New York: Population Council.

National Science Foundation. 1974. *Federal support to universities, colleges, and selected nonprofit institutions, fiscal year 1972.* Washington, D.C.: National Science Foundation NSF 74-305.

Orleans, L.A. 1973. Family planning developments in China, 1960-66: Abstracts from medical journals. *Studies in Family Planning* 4:197-215.

Perkin, G. (Ford Foundation, Brazil) 1975. Personal communication.

Prasad, M.R.N. 1974. A report on reproductive biology in India. Report commissioned for the reproduction review by the Ford Foundation.

Reproduction Research, Information Service Limited. 1976. Output of research reports for the developed and developing worlds, in the period

1970-1975. Mimeographed. Cambridge, England.

Segal, S.J. 1974. Review of biomedical research in reproduction: Bangladesh, India, Pakistan, Sri Lanka. *A review of biomedical research in reproduction biology in countries of the developing world*, ed. J.T. Lanman, pp. 600-612. Mimeographed. New York: Population Council.

World Health Organization (WHO). 1971. Expanded Programme of Research, Development and Research Training in Human Reproduction. *Report of a Feasibility Study*. Appendix A. Mimeographed. Geneva.

# 8  Contraceptive Development in Industry and the Public Sector

There has been a traditional division of labor in the evolution of new technologies: Universities and scientific institutes carry out research leading to new findings, which are then exploited by industry and developed into new products for manufacture and distribution. Although it is doubtful that the functions have actually ever been so clearly delineated, many observers in the early 1960s expected the pharmaceutical industry to take the major responsibility for the development of new contraceptive products, leaving government and private philanthropy free to concentrate their resources on support of the underlying fundamental investigations. Instead, industry's share of the funds supporting contraceptive development research has actually declined during the last decade (as the funding data in chapter 11 demonstrate). This reality has given rise to a reappraisal of the traditional division of labor, and some actual shifting of functions has already occurred with the formation since 1971 of active "public sector" programs supported by governmental and philanthropic funds to undertake functions in the development process that industry carried out exclusively a decade ago.

Because both the pharmaceutical industry and the new public sector programs constitute important parts of the institutional resources available in the field, the review sought to assemble information on the scope of their contraceptive development activities. This chapter summarizes these findings, emphasizing particularly the patterns of collaboration that are emerging between industry and the public sector.

The traditional view of the central position of the pharmaceutical industry in development of modern drugs is set forth by Djerassi in these terms (1970, p. 943):

Except for certain biologicals (special vaccines) essentially all modern prescription drugs were developed by pharmaceutical companies. I know of no case in which all of the work (chemistry, biology, toxicology, formulation, analytical studies, and clinical studies through phase III) leading to governmental approval of a drug (for example, by the FDA in the United States) was performed by a government laboratory, medical, or a non-profit research institute. This does not mean that many of the basic discoveries leading to the development of a drug ultimately used by the public are not discovered in such nonindustrial laboratories, or that certain important steps (for example, much of the clinical work) are not performed outside of industry. Nevertheless, it is a simple fact that in modern industrial nations pharmaceutical firms play an indispensable role in the development of any drug.

Applying this viewpoint to the prospects for development of new contraceptives, he asserts (1970, p. 949):

The organizational abilities of the pharmaceutical industry are a *sine qua non* for the development of practical birth control agents. . . . Given that major advances in birth control will be based on chemical methods, the access to the large and highly productive organic chemical research groups in the pharmaceutical industry is an indispensable prerequisite.

Much of the early developmental work on current modern contraceptives, however, was done by nonprofit research institutions. The Worcester Foundation for Experimental Biology carried out the early work on the pill in the 1950s to the point where it could be turned over to the industry for further development and distribution (Battelle 1973). In the 1960s, the Population Council in collaboration with university researchers did the comparable early studies of the modern IUDs; pharmaceutical firms became involved in their testing, packaging, and distribution after acquisition of proprietary rights from their inventors. Industry has also played a secondary role in the development of improved methods of sterilization and pregnancy termination, which are largely the products of university-based medical research. Much of the industry's effort since 1960 has been devoted to refinements of the pill, but pharmaceutical firms also have been actively involved in the development of medicated IUDs and research on prostaglandins.

To trace the industry's past efforts in contraceptive development and determine its future plans, this review commissioned a survey that collected information from forty-four U.S. industrial and nonprofit research organizations (Duncan 1974). Comparable information was also sought, with only limited success, from pharmaceutical firms outside the United States. Much of the information sought is regarded as proprietary and can be published only without identifying individual firms.

Fifteen U.S. pharmaceutical firms reported some expenditures for contraceptive research and development.[1] Of these fifteen, seven firms reported expenditures of $1 million or more in at least one year since 1965, but only five reported expenditures at this level in 1974 and only four predicted increased expenditures by 1980. The aggregate expenditures of the respondent companies have fluctuated between $13 and $15 million (in current dollars) since 1970;[2] in constant dollars, however, the expenditures of U.S. firms peaked in 1971 and by 1974 had declined to a level lower than in 1965 (see table 11.2). The aggregate level of predicted 1980 expenditures is less than that in 1974 when anticipated inflation is taken into account.

Industry tends to rely almost exclusively upon its own funds to support work in contraceptive research and development. As of the time of the survey, however, two firms had accepted NIH and WHO contracts accounting for approximately 5 percent of their expenditures. Industry spends about three-quarters of its contraceptive research and development funds within its own organization, with the remainder supporting clinical research by outside physicians, work in unaffiliated commercial laboratories, and to a limited extent, studies in universities and nonprofit institutions.

Contraception is only a minor part of industry's total research and development effort; in fact, even those firms spending in excess of $1 million annually for contraceptive development devote 10 percent or less of their total research budgets to the field.

As would be expected, the dominant factor causing industry to undertake or enlarge contraceptive research and development programs is the expectation of a potentially profitable market. Although a number of firms cited the "social need to control the population expansion" as their reason for entering the contraceptive development field, the more frequently cited motivation was "a combination

[1] Defined as "including all basic and applied research required to identify, characterize, formulate, fabricate, test, and evaluate contraceptives up to the point of marketing. It excludes sales and marketing costs, but includes the proportion and share of overhead required to administer and support the research and development" (Duncan 1974).

[2] This finding is compatible with Djerassi's "incomplete personal survey among five pharmaceutical companies (Lilly, Ortho, Searle, Syntex, and Upjohn) [that] has shown that their cumulative 5-year expenditure (1965-1969) in this field amounted to $68 million" (1970, p. 945).

of a burgeoning market place and (what appeared at the time to be) leads that wanted exploitation." Conversely, declining market potential was given as the main reason for discontinuing or not expanding research efforts. The high costs and protracted length of time required to identify and develop a new drug to the stage of marketability was cited as the principal deterrent. Though in some cases a large market might exist, the expected return on investment was not considered adequate to defray the costs of developing new concepts or even to undertake refinement or improvement on existing drugs, especially for those types of contraceptives where safe, effective, and relatively well accepted product lines were readily available. A failure to identify acceptable product candidates after reasonable expenditures of resources was another reason given for discontinuing research efforts.

Although only limited information was obtained on the activities of pharmaceutical firms outside the United States, they express similar motivations for initiating contraceptive development programs (as well as for reducing or abandoning them) as their U.S. counterparts. The questionnaire used to survey U.S. firms was sent to one hundred pharmaceutical firms and their subsidiaries in eleven nations outside the United States. Thirty-seven did not respond; forty indicated that they were not involved in the field or were unwilling to provide information; twenty-three completed the questionnaire in whole or in part. Unfortunately, the estimates of total expenditures by non-U.S. firms is understated because two of the most active firms in contraceptive development would not provide their expenditures (although they did respond to other questions). The figures submitted by responding firms in eight European countries, Canada, and India totaled between $2 and $4 million in the years from 1970 to 1974.

As is the practice of U.S. firms, these companies provide most of their own financing for contraceptive research and development. In 1975 a major break with this tradition came with acceptance by Schering AG of a $1.2 million cost-sharing contract from the NIH Center for Population Research for development of a long-acting injectable female contraceptive agent, norethindrone enanthate, for which Schering holds the patent.

Contraceptive development also represents only a minor part of the total research activities of non-U.S. firms, ranging from less than 1 percent to 20 percent of their research and development budgets. Two Canadian firms are exceptions. One was organized to develop contraceptive products and between 1960 and 1971 devoted all of its research funds to contraceptive development; in recent years it has explored other areas with 20 percent of its research budget. Another spent half its research and development budget on contraceptive development in 1971, but plans to reduce this proportion to 25 percent by 1980.

## How Much Can Be Expected of Industry?

A major policy issue turns on an assessment of the extent to which the pharmaceutical industry effort can be expected to meet the world's needs for an adequate array of methods of fertility control. Some observers argue that the more closely a contraceptive product approaches the ideal for mass use in many nations of the world (the ideal product being one that is inexpensive and long lasting), the less profitable it becomes to industry. Industry spokesmen assert, furthermore, that the contraceptive development process has become so costly, because of increasingly stringent regulation and the potential of product liability, as to discourage corporate investment. On the other hand, some knowledgeable scientists believe that firms would commit greater resources if a really exciting lead to a potential contraceptive method were to emerge from the research laboratory.

Respondents to the survey who indicated their discouragement with the prospects for the development of new contraceptive methods agree with the arguments advanced by Djerassi (1970). For example, a Syntex official notes (Bennett 1974, p. 491),

The elaborate regulatory requirements governing the development of oral contraceptives are among the chief reasons why so many large pharmaceutical companies have either completely ceased all research toward new contraceptives or markedly reduced their research expenditures in this area. A unique oral chemical contraceptive with a new mechanism of action now costs about $10 million to develop over a time space of about ten to fourteen years.

A Warner-Lambert executive asserts that to realize adequate return on investment, a firm has to achieve $21 million in annual sales of a new contraceptive with a product life of ten years (after a development period of ten years at a total cost of $20 million) (Edgren 1972). "Few contraceptives currently marketed can approximate this figure and one would be optimistic in the extreme to project such sales for any new development," he states, concluding (Edgren, 1972, p. 145),

These estimates do not augur well for the future of contraceptive research, or for that matter for drug research in general. They may explain the fact that over the past several years in the U.S. industrial pharmaceutical research in general has not expanded. One company has completely terminated drug research, and none appears to be expanding its efforts in this critical area.

Not all firms responding to the survey are as negative about the prospects of future corporate investment in contraceptive development. Two firms expect to commit about $5 million in 1980, one, $3.4 million, and two others, about $2 million each.

Much of the industrial research effort is considered confidential, making it difficult to assess the extent to which industry expenditure is aimed at refinement of existing methods (for example, lower dosages of the compounds contained in current oral contraceptives) and the extent to which it represents research on genuinely new methods. According to Bennett (1974), industry is concentrating on improving existing methods. In response to a follow-up questionnaire sent to the firms responding to the survey asking them to specify their current and projected research emphasis, eleven firms stated that they are indeed pursuing research and development activities in most avenues of contraceptive development, as well as some kinds of fundamental research.

Respondents to the survey provided a broad list of suggestions for encouraging greater participation by the pharmaceutical industry in contraceptive development, including (Duncan 1974):

• nonrestrictive government support of metabolic, toxicologic, and clinical evaluation of drug candidates;

• identification of animal models that have predictive value for human studies;

- increased emphasis on use of subhuman primates for laboratory investigations;
- relaxation of restrictive or cumbersome federal drug regulation procedures;
- provision of a means for rapidly and effectively assessing the ultimate product potential of laboratory discoveries;
- objective cross cultural assessment of the deficiencies of present contraceptive products;
- reassessment, by the medical and scientific community, of the benefit:risk ratio of existing or potential contraceptive agents and dissemination of this information to the lay public as well as to government health officials;
- a greater degree of collaboration between programs in both industrial and public sectors to minimize duplication of effort and to assign priorities to project areas;
- development of a more favorable posture by the courts and the media to manufacturers regarding the publicity and damage claims associated with contraceptive usage.

Djerassi (1970) also suggests modifying the patent life of contraceptive drugs because of the extremely long-term testing required before a product may be marketed. Much of the seventeen-year life of the patent, he argues, may be consumed in the ten- or fifteen-year period of testing. He proposes that use-patent protection be offered for, say, ten years after the Food and Drug Administration (FDA) approves a product for general distribution.

Survey respondents also indicated receptivity to collaboration with such public sector programs as the World Health Organization's Expanded Programme of Research, Development and Research Training in Human Reproduction, the Contraceptive Development Branch of the National Institute of Health's Center for Population Research (CPR), and the contraceptive development programs supported by the Agency for International Development (AID) and the Population Council. For the most part, these arrangements subsidize the costs of clinical trials and/or provide access to clinical facilities for testing of drug candidates not readily available to industry. They also provide a system for exchange of information on new laboratory findings and areas for needed product development or innovation, as well as, to some extent, direct financial support.

Djerassi (1970) also proposes much more extensive government subsidy of contraceptive development by pharmaceutical firms. Pointing to the large expense of long-term toxicity studies and clinical trials, he suggests that firms be enabled to apply to a government agency for full funding of these studies, repaying the government on an annual royalty basis if a commercial product is developed and sharing profits with the government after the initial funding has been repaid. This would follow precedents of government-industry collaboration in other major technological efforts such as the space program.

Not-for-Profit Research Organizations

Government funding in the United States supports contraceptive development activity by several not-for-profit research organizations.[3] Leading examples of these institutions are the Worcester Foundation, Population Council (not included in the Duncan survey), the Battelle Memorial Institute, and the Stanford Research Institute. (Research institutes operating on a for-profit basis are not active in contraceptive research and development.) Seven of these institutions in the United States carry on significant programs in this field; their combined expenditures rose from $0.5 million in 1960 to a peak of $6.1 million in 1973 and then declined in subsequent years. The projected expenditures of these institutions for 1980 is only $4.8 million. Because they are dependent on government sources for 70 to 80 percent of their funding, these projections reflect the uncertain availability of such funding in the future.

Contraceptive Development in the Public Sector

During the past five years a number of "public sector" mechanisms dedicated to the development of new contraceptives have been established with governmental and/or philanthropic funds. These pro-

[3] Survey of not-for-profit organizations included in Duncan 1974.

grams not only provide funds to others but actually carry out most of the functions generally performed by private industry, sometimes alone but often in collaboration with one or more pharmaceutical firms. Four programs are identified for detailed description: the Population Council's International Committee for Contraception Research (ICCR), the WHO Expanded Programme of Research, Development and Research Training in Human Reproduction, the CPR's Contraceptive Development Branch, and the Research Division of the Office of Population, U.S. Agency for International Development.

**International Committee for Contraception Research**

In 1971, the Population Council launched ICCR to provide an international, noncommercial mechanism for identifying promising new contraceptive leads and for rapidly developing and testing these leads to determine their feasibility. ICCR is composed of prominent clinical investigators from Austria, Brazil, Finland, India, Sweden, and the United States. Its budget was $450,000 in 1971, increasing to $1.2 million in 1974, $1.4 million in 1975, and an expected $2.7 million in 1976. ICCR has received its support from the Ford and Rockefeller Foundations, the International Development Research Centre of Canada, and the General Service Foundation.

The ICCR initially focused its work on eight promising approaches to new contraceptive methodology: the Copper T intrauterine device, a weekly pill, a menses-inducing pill, a precoital pill, improved methods of female sterilization, a subdermal implant for women, an injectable, and a vaginal ring device. These leads were chosen on the basis of their technological promise, their potential utility in developing countries, and a determination that ICCR could hasten their development. As several passed through the stages of selection, rapid evaluation, and preliminary product development, certain leads were dropped and others added. Among those dropped were a loop/clip technique of vas occlusion, a postcoital pill, and spermatogenesis suppression.

Clinical research is first undertaken at the institutions of committee members; more advanced clinical investigations may be conducted by other investigators under ICCR grants or contracts. ICCR also sponsors research by committee members, other investigators, and at the Population Council. To gain access to compounds and technology held by industry, formal agreements are negotiated. A fundamental ICCR objective in entering such agreements is to assure a continuing supply of the final product at low cost to public sector family planning programs in return for granting to industry distribution rights to sales in the private sector.

Of the leads, the postcoital pill, the precoital pill, and the loop/clip technique of reversible vas occlusion have been eliminated from further consideration because of demonstrated lack of effectiveness. Work on transcervical sterilization and the weekly pill has been set aside for the present. Further research on two other leads—the injectable contraceptive and menses induction—is continuing at a modest level pending the development of promising new approaches. The Copper T 200 device is considered a finished product, now available both commercially and through the public sector in over two dozen countries. A new lead, a contraceptive vaccine, has recently been added, making use of the $\beta$ subunit of human chorionic gonadotropin linked to tetanus toxoid vaccine; and the remaining three leads continue under intensive investigation.

Since 1971, two hundred dosage regimens have been tested clinically; excluding tests of the Copper T, over seventy-five hundred subjects in thirty-four countries have participated in these trials. Of these dosage regimens, approximately twenty are still under investigation. Twelve Investigational New Drug Applications have been approved by the FDA and two others are awaiting approval as of this writing (January 1976).

**WHO Expanded Programme of Research, Development and Research Training in Human Reproduction**

Between 1965 and 1969, the member states of the World Health Organization urged the WHO, through resolutions passed in the World Health Assembly, to develop a program in family planning including research in human reproduction, particularly fertility regulation. Special contributions from the United States, Sweden, and the United Nations Fund for Population Activities (UNFPA) helped to lay the basis for the extensive program of contraceptive research and development that began in 1970 with a feasibility study.

Over the next year, consultations with officials from governments and medical research councils; researchers from academic institutions, industry, and other agencies; and site visits to research centers in all parts of the world helped in determining the program strategy. Funds have been made available since 1972 for the Expanded Programme of Research, Development and Research Training in Human Reproduction through special contributions to WHO by Canada, Denmark, Finland, Mexico, Norway, Sweden, the United Kingdom, UNFPA, the International Development Research Centre of Canada, and the Ford Foundation. The budget grew from $4.5 million in 1972 to $12 million in 1975. The program now involves more than six hundred scientists from sixty countries.

The Expanded Programme aims at assessing in different populations the safety and effectiveness of existing methods of fertility regulation, improving these methods, and developing a variety of new ones suited in particular to the requirements of developing countries. Research is conducted on presently available hormonal contraceptives and intrauterine devices in populations exposed to such environmental hazards as malnutrition and endemic diseases. At the same time, the design of IUDs is being improved and medication included to reduce the side effects of pain and bleeding. Research on injectable contraceptives ranges from indepth assessment of the few currently available preparations to development of new polymers for constant release and synthesis of new compounds.

Vaginal rings and intracervical devices developed in the program are now being clinically tested, as are a variety of prostaglandin analogs for the termination of first and second trimester pregnancy. Different current methods of abortion and sterilization are also being assessed and new methods of tubal occlusion are being explored. Simpler and more reliable methods for the prediction and detection of ovulation are sought for couples wishing to use natural family planning methods.

The Expanded Programme has also included from its beginning research on methods with a longer development time: pills and injectables for men, postcoital agents for women, and birth control vaccines for both men and women.

Social science research is simultaneously being carried out on the characteristics of fertility regulation methods that relate to their acceptability. This should provide guidance for the strategy of the program and assist administrators who wish to use the methods at the national level. Research on the service implications of the methods constitutes another aspect of the WHO Programme; it focuses on the suitability of different service settings and different types of personnel for the delivery of specific methods.

Most of the research in the WHO Programme is conducted through task forces within each area of effort; they determine the detailed strategy, identify scientists, and monitor progress. They are essentially multinational and multidisciplinary, bringing together whichever skills are required for the development of a method, from the investigation of reproductive processes through chemical synthesis, pharmacology, toxicity to clinical trials, product development, and field trials. A network of twenty WHO designated Collaborating Centres for Clinical Research in Human Reproduction in nineteen countries provides the facilities for the clinical testing of methods based on common protocols. Four WHO Research and Training Centres in Human Reproduction also take part in these clinical studies as well as acting as major multidisciplinary foci for research in the priority areas of the program.

Another major aim is to strengthen resources for research in this field in developing countries. This is done not only by building up the staff and equipment of WHO collaborating institutions but also through research training, long-term consultant assistance, the provision of supplies, and the organization of meetings and workshops.

## Center for Population Research

The Center for Population Research of the National Institutes of Health (specifically the National Institute of Child Health and Human Development) is the world's single largest source of funding for the reproductive sciences and contraceptive development. CPR carries out its role as donor through its Population and Reproduction Grants Branch, in which it supports fundamental biomedical research, multidisciplinary programs of population research, and research training. CPR also directs three contract branches that have more targeted mandates: Contraceptive Development, Contraceptive Evaluation, and Behavioral Sciences. These three are somewhat similar in mission, but not in operation, to the WHO Expanded Programme and ICCR.

The Contraceptive Development Branch carries out a contract program for directed fundamental research as well as research in product development. In the first category emphasis is given to basic studies in several well-defined areas of the reproductive sciences in order to expand the scientific base for the product development area. This includes investigations on the detection of ovulation, the oviduct, the cervix, the ovarian corpus luteum, studies of the male reproductive process, and mechanisms of hormone action. These contracts for fundamental research differ from CPR's general support of basic research through grants in that the contract programs are initiated and coordinated by CPR, with staff input as necessary, and are geared to the rapid expansion of the science base in those areas that appear promising for the development of new contraceptive technology.

Contract research by the Contraceptive Development Branch is now focused on product development, aimed at the discovery and testing of new contraceptive modalities. These extend from laboratory synthesis to testing of new chemicals and screening for contraceptive activity leading eventually to clinical trials involving both drugs and devices. The drug development program includes syntheses and biological evaluations of potential antifertility agents. Chemically the program encompasses steroids, polypeptides, prostaglandins, and a variety of other substances broadly classified as nonsteroids and includes a drug testing facility to evaluate compounds prepared under the drug development program. Compounds are selected by the staff from a variety of sources and are screened for hormonal and antifertility activity. Over five hundred compounds and five drug delivery systems have been evaluated in more than a thousand assays. The Contraceptive Development Branch is also concerned with studies of drug delivery systems, development of male and female sterilization techniques, and support of clinical studies on potentially useful contraceptive methods. For example, recently completed studies demonstrated that sperm count could be severely reduced in men who take Danazol daily and receive a monthly injection of testosterone enanthate. Additional studies have been initiated to determine whether effective inhibition of spermatogenesis can be maintained if the drug burden upon the body is lessened by spacing Danazol administration. The branch is also contemplating additional clinical trials aimed at the evaluation of testosterone enanthate alone.

Complementing the Contraceptive Development Branch of the CPR is the work of the Contraceptive Evaluation Branch, which assesses the safety of contraceptive methods currently in use. The scope of the program includes the long-term surveillance of contraceptive users for cardiovascular, carcinogenic, diabetogenic, and possible other effects; case-control studies of specific disease entities known or thought to be associated with oral contraceptive exposure; interactions of contraceptive steroids with essential dietary nutrients; the use of animal models in evaluating the effects of contraceptive steroids in an effort to identify more suitable animal models to predict effects in people; and the evaluation of possible adverse effects on subsequent progeny related to previously used methods of contraception. Investigations have also focused on the natural history of the hypertensive response seen in some women on oral contraceptives, the risk factors associated with its occurrence, and its reversibility. In addition, a collaborative program was recently developed to evaluate the occurrence of serious complications associated with intrauterine devices currently used in the United States, and studies of the effects of induced abortion on subsequent pregnancy performance are under development. Possible immunological, endocrinological, and other systemic effects subsequent to vasectomy both in animal models and in man are also under evaluation. In addition to the surveillance of users of various methods of contraception, program emphasis is aimed at identifying risk factors for particular adverse effects known to be associated with oral contraceptives.

### Agency for International Development

Between the year of its inception, 1967, and 1975, AID's Office of Population has obligated a total of $39.4 million to support applied biomedical research with the objective of developing new and improved means of fertility control for use in developing countries. The appropriation of these funds peaked in 1972 with $7.6 million and has subsequently declined to $2.2 million in 1974. Since then, appropriations have remained at the lower level. Funds have been applied in the following three major areas: research on a once-a-month self-administered method, improvement of current means, and field studies.

Investigations on the once-a-month regimen have included over forty studies seeking new ways to control corpus luteum function and block progestational activity and the development of inhibitors of gonadotropin releasing factors. A wide range of studies on prostaglandins have also been supported for use in this same regimen, including research on new means of measurement, new approaches to synthesis, studies of formulation and delivery methodology, compound screening, studies of the effects on mammalian primate reproductive physiology, human clinical trials, and a prostaglandin information service.

Improvement of the current means of fertility control has encompassed work on several techniques. Steroidal contraceptives have been studied primarily for their health effects in the populations of the developing countries. Research on the intrauterine device has led to the development of a pleated plastic membrane form. Additional IUD studies focusing on causes of early discontinuation (such as pain and bleeding) were funded in 1975.

AID has sponsored studies to simplify and improve the safety of postconceptive control of fertility by surgical termination of pregnancy. These studies are joining engineering input to a system of field trials with the objective of developing improved models of both electrical and nonelectrical equipment. The most advanced of these research activities has developed new, simple equipment, the "minisuction," for menstrual regulation.

A number of experimental approaches to new techniques of female sterilization are under investigation using cornual trauma, cryosurgery, tissue glues, and plugs and transcervical sterilization methods that would eliminate the need for an operation. A recent development of a single aperture laparoscopic sterilization with tubal (Hulka-Clemens) clips and (Yoon) Fallope bands now permits female sterilization to be done under local anesthesia and on an outpatient basis avoiding the two main hazards of laparoscopic female sterilization—general anesthesia and electrocautery.

Studies for a more reliably reversible means of male sterilization and simplified means of male sterilization for field use are also being pursued. This work includes studies of nonsurgical male sterilization by injection of the vas and implantation of reversible vas valves.

In fiscal year 1972 Northwestern University received $3.4 million to coordinate a contract program of awards of up to $60,000 annually to a number of investigators to carry out applied research on improved means of fertility control. The Program for Applied Research on Fertility Regulation (PARFR) is currently managing twenty active subcontracts.

To improve currently used means of fertility control and to evaluate fertility control methods that may have differing efficacy and risks associated with them when used in the developing countries, the AID research program is collaborative and runs comparative clinical trials of new methods. The focus of this effort is the epidemiological evaluation of the success and performance characteristics of these methods under use conditions in the field by a network of collaborating investigators. These field studies have also made it possible to carry out double blind trials of new methods in the same clinic setting.

Beginning in fiscal year 1967, AID supported the development of the International IUD Program of the Pathfinder Fund. This field study of IUD performance characteristics has provided comparative data from forty countries. Uniform records and centralized data processing have allowed the determination of which performance patterns are related to IUD's user and clinical characteristics.

To extend the availability of a clinical network for field trials, an International Fertility Research Program (IFRP) was initiated in fiscal year 1971. Up to fiscal year 1975 a total of $9 million has been provided to the IFRP to support conduct of collaborative field trials of new IUDs, sterilization techniques, pregnancy termination techniques, prostaglandins, and pharmacologic contraceptives in many countries.

In addition to the $39.4 million obligated for research to develop new means of fertility control, $3.2 million has been provided to the Population Information Program to make up-to-date research findings available to research scientists, policy makers, and family planning program administrators in developing countries through a series of population reports.

**Public Sector Agency Relationships**

The contraceptive development process in the public sector thus involves annual expenditures of some $26 million, an amount greater than the pharmaceutical industry's investment in the field. Each agency has its own priorities, operating style, and constituency, although there is in fact considerable overlap among their advisory bodies. Several agencies may support contraceptive development projects at the same institution. There is both formalized collaboration and competition among the different agencies, and there is a mix of relationships between these public sector programs and private industry.

In the absence of the profit motive the public sector agencies are less intent on withholding information on research progress from other agencies than is the pharmaceutical industry, generally regarded as one of the most secretive of the high technology industries. However, the terms of agreements between public sector agencies and industry, as in the case of testing proprietary compounds, often require confidentiality.

Because of their comparative advantages in different aspects of the contraceptive development process, a lead developed by one organization on occasion has been "given up" for further clinical testing by another. Optimal allocation of resources in the public sector contraceptive development field does not necessarily require the absence of redundancy and competition. There are well-known examples in which a race among leading scientific groups has led to important developments; early development of the oral contraceptive was characterized by a "technology race."

Collaboration with Industry

Public sector contraceptive development intersects with industrial research at many stages in the contraceptive development process, and the major public sector programs have let contracts to industry for a variety of research and development undertakings. Many of the products tested for their potential as fertility regulating methods by public sector agencies are products on which patents are held by private industry. In this process private interests may be greatly enhanced by public funds, and arrangements are required between the public sector agencies and industry to preserve the interests of the public in the potential fruits of collaboration. At the same time such arrangements must maintain private industry's incentive to turn over promising products for testing by public sector agencies or to invest in further development of an invention growing out of research supported by government or philanthropy.

Patent policy is an essential element of these collaborative arrangements. The questions at issue are by no means unique to the contraceptive field; they pertain broadly to the whole gamut of scientific research and the transfer of the resulting technology to industry. In a sense, the public interest in widespread application of the output of science that justifies the initial investment of public funds in research appears to conflict with industry's interest in profitability, which patent laws are designed to protect by assuring the company a monopoly for a limited period. But universities and research laboratories are not industrial or marketing entities, and the distribution of the technology that results from research is facilitated by industry's capabilities. As a result, both parties have an interest in attempting to reconcile, at least to some degree, the underlying divergence of interest.

Governmental and international agencies and private foundations are experimenting with varying approaches to patent questions in the contraceptive field (McLean 1975). The Ford Foundation and the Population Council have, for example, entered into patent agreements with grantee institutions and principal investigators whereby the institution is responsible for obtaining patents on inventions and is required to grant nonexclusive licenses of any patentable invention resulting from the sponsored research. The agreement requires the foundation's or council's consent before the institution may permit an exclusive license of the patent. Drug companies interested in further development and marketing of the invention usually request the consent of the foundation or council to obtain exclusive licenses. At present, the foundation and council have made an initial decision not to demand royalties in return for their consent to an exclusive license, although another approach would be to negotiate standard royalty arrangements with pharmaceutical firms and feed these royalties back into further research. Instead, these institutions, in their agreements with the drug companies, seek to assure that the "public sector" (that is, national and voluntary family planning programs) will be able to purchase the new contraceptive devices at a price lower than what the drug company would charge the "private sector" (that is, commercial suppliers to private physicians).[4] The key issues forming the basis of these agreements are (1) definition of the "public sector," (2) pricing formulae, and (3) guarantee of supply to the public sector. The details vary with circumstances, such as the sums of money the various parties have contributed, or will have to contribute, to the research undertaking.

The International Development Research Centre of Canada has adopted a different approach toward patent policy or research grants. IDRC retains complete right, title, and interest, in all coun-

[4] As noted in chapter 6, the savings from UNFPA's first purchase of the Copper T nearly offset the total research and development costs as a result of ICCR's arrangements regarding public sector pricing.

tries of the world, to any new devices or processes or improvements on inventions made as a result of a research grant. IDRC also denies the right of the institution or inventor to publish information relating to the patentable invention without IDRC's permission. (Publication too soon may in some jurisdictions bar the right to get a patent.) IDRC's theory is that patent laws vary and worldwide patent administration is so complex that it cannot rely on most inventors or grantee institutions to perfect and adequately maintain patents.

IDRC's agreement with its grantee institutions provides that unless it regards its support as having been substantially responsible for the invention it will not ask to receive a large share of royalties. The agreement also provides that it is IDRC's *intention* to grant back and release to the institution rights to patent the invention, subject to safeguards concerning availability of the invention for use by developing nations.

WHO reserves the public sector rights on any patents taken out on products or processes derived from projects supported by the organization. WHO may also take out patents. In collaborating with industry, WHO and the company assess the relative participation of each in the development of the product to arrive at mutually agreeable terms for the cost to the public sector of the product and its manufacture in adequate quantities.

If a company has a major new product and the public agency contribution is limited to a few clinical studies, the agency is in a weak position to demand special low prices for the public sector. In cases where the public agency has taken a compound "off the shelf" and contributed substantially to its development (toxicology, pharmacology, clinical studies), the agency is in a much stronger position to negotiate special concessions.

Industry traditionally has been reluctant to offer its proprietary compounds for testing under U.S. government auspices for fear of losing patent protection for these compounds. NIH patent policy gives to the Assistant Secretary for Health and Scientific Affairs "sole and exclusive power to determine whether or not and where a patent application shall be filed, and to determine the disposition of all rights in such invention, including title to and rights under any patent application or patent which may issue theron" (U.S. Department of Health, Education and Welfare 1972). But in order to facilitate industry-government collaboration, it is now NICHD policy that if the pharmaceutical firm holds the patent on a compound or its uses that NICHD wishes to test for a new use in fertility control, the government will continue to recognize these rights in funding further development of the proprietary compounds.

These experimental approaches are continuing to evolve with experience. Patent policies thus promise to become an important means of facilitating public sector collaboration with private industry in contraceptive development, despite the apparently differing interests of the parties involved.

### References

Battelle. 1973. Oral Contraceptives. In *Interactions of science and technology in the innovative process; some case studies.* Columbus, Ohio: Battelle.

Bennett, J.B. 1974. Hindrances to new contraceptive development. *J. Reprod. Fertil.* 37:488-498.

Djerassi, C. 1970. Birth control after 1984. *Science* 169:941-951.

Duncan, G.W. 1974. A survey of contraceptive research and development. Report commissioned for the reproduction review by the Ford Foundation.

Edgren, R.A. 1972. A viewpoint of industry. *Clinical Proceedings*, pp. 144-154. IPPF Southeast Asia and Oceania Congress.

McLean, S.A. 1975. Patents and inventions financed by the public sector: A case study in contraceptive research. Report commissioned for the reproduction review by the Ford Foundation.

U.S. Department of Health, Education and Welfare. 1972. General Provisions #316, Patents Rights Clause #20. Washington, D.C.: Government Printing Office.

# 9 The New Climate for Research and Development

As we enter the last quarter of the twentieth century, public attitudes toward the scientific enterprise are changing rapidly in many nations; and these changes are redefining the rules governing research and development activities. If the years following World War II in the United States and other industrial nations were characterized by a "wonder drug" mentality that looked to the laboratories for immediate solutions to urgent problems, the 1960s were marked by the emergence of a countervailing skepticism and suspicion of science, in many ways no less simplistic. The "magic cures" of the 1940s and 1950s gave way to the "overmedicated society" of the 1960s.

The factors underlying this shift will doubtless be explored and debated by historians and students of social development. That there has been a change—and that it has substantially altered the climate in which research and development efforts are conducted—is, however, beyond dispute. Pervasive questions are now raised concerning the necessity for, and desirability of, many scientifically determined interventions in the natural order, including medical interventions designed (at least by their discoverers and practitioners) to preserve life and maintain health. These developments have led to increasing regulation by governments of scientific research in general and medical research in particular, as concerns over the safety and efficacy of new medications and treatment modalities have increasingly been expressed by consumers and policy makers. They have also led to even more fundamental questioning of the ways in which research involving human subjects is conducted and of the ethical limits of science-based decisions and interventions.

In the United States and other nations, these new issues are generic to all branches of scientific research. But they have particularly affected research in the reproductive sciences and contraceptive development for at least three reasons: First, safety is a special concern because contraceptive methods are for the most part used by healthy persons. Second, they are employed for long periods of time by the vast majority of individuals of childbearing age in many nations. Third, fertility regulation commands widespread public attention because it is directly linked with sexual, cultural, and religious practices. In the United States since 1970, there have been congressional hearings or regulatory actions involving the safety and/or efficacy of most current contraceptives—the oral pills, the intrauterine devices, injectables, and a "morning-after" pill.

Issues of contraceptive safety, regulation, and the ethics of human experimentation thus are emerging as powerful factors shaping research policy in the United States and other nations. This review accordingly commissioned a number of papers from varying perspectives (Barber 1974; Braun 1975; Hutt 1975; Kordon and Gaudin 1975; McLean 1975), gathered information in interviews with researchers and industry and government officials (Djerassi 1975; Schering AG 1975; Igarishi 1975), and examined relevant published material. Each of these subjects, of course, is complex enough to warrant a detailed study of its own, and considerable work along these lines is underway in government agencies, universities, and research institutions. Our purpose, therefore, was not to attempt to resolve these issues but rather to gain a sense of how the new climate is affecting reproductive research and contraceptive development and to assess the implications of these effects for the field's future. In this chapter we present an overview, inevitably oversimplified, of these processes.

**Regulation of Fertility Control Agents**

Nations have established a variety of mechanisms to regulate the distribution and sale of drugs; the principal purpose of these regulatory processes is presumably to assure the maximum degree of safety and efficacy possible. Despite this universality of purpose, there is considerable variation from country to country in the requirements imposed by regulatory agencies before a drug may be marketed. As a result, there is also much variation in how the regulatory process is viewed in different parts of the world.

The development of new approaches from the status of a laboratory curiosity to that of an approved, safe, effective marketed product in the United States is now a complex and lengthy process. Assuming that effects on fertility are demonstrated in a suitable model system, selected toxicology studies must be planned. The major objective of these initial animal studies is to identify potential sites of target organ toxicity produced under dosages similar or exaggerated to the projected human experience. The testing protocol, including dose levels, duration, and selection of the most relevant species, must be carefully designed because the effects of the hormonal agents so often forming the basis of contraceptive methodology may produce profound species-specific effects on neuroendocrine regulatory mechanisms that may not be clearly extrapolated to man.

The design of toxicology studies of compounds with known pharmacological attributes need only include sufficient information about metabolic pathways to assure that the results of model systems using previous other versions are applicable. In the case of novel compounds, however, a full metabolic profile should be developed in order to assist in species selection and projecting human dose level and frequency. Other considerations of design include selection of species with appropriate reproductive and endocrine function and application of the drug or device in a manner closely resembling that to be used clinically. In addition, teratologic and general studies of fertility and reproductive performance in female animals must be completed prior to the initiation of clinical trials in women.

When these studies are completed, initial clinical pharmacology studies in patients may begin. The purpose of phase I clinical studies is to make preliminary observations concerning systemic and local tolerance and to determine whether the metabolic profile of the drug is that expected from the animal models. These are followed by phase II studies designed to yield evaluations of clinical efficacy; such studies might involve one to two thousand women studied closely for six months in order to estimate product effectiveness. Once a likely effective dose regimen is established and the occurrence of relatively common side effects observed, phase III is initiated. A typical objective of a phase III contraceptive study would be to collect information on ten to twenty thousand cycles of use.

The rate at which drug development moves through these phases is conditioned by the nature of the product, the results achieved, and regulatory criteria. For example, the progression of the combinational estrogenic and progestational agents used chronically for contraception requires completion of a carefully orchestrated series of toxicology studies as follows: initial segments of teratologic studies prior to phase I, one year of general toxicology studies prior to phase II, studies on all aspects of reproduction, two years of general toxicology studies, and initiation of seven to ten year oncogenic studies in dogs and monkeys prior to entering phase III. It should be noted that detailed study of reproductive performance in animals receiving fertility control agents has been particularly challenging. Evaluation of the multisystemic effects of these drugs has made appraisal of carbohydrate and lipid metabolism and blood coagulation virtually routine if not mandatory, although no parameters have been identified as yet that are directly predictive of clinical disease in humans. Because of the profound effects of these agents on the endocrine system, regular study of such effects and their reversibility has received continuing attention.

The theoretical effectiveness of combination hormonal contraceptive products taken under ideal conditions is virtually 100 percent, so it is difficult to demonstrate the superiority of a new product. Although studies encompassing a hundred thousand cycles are considered to yield adequate information to document use-effectiveness in the field, studies twenty to one hundred times larger would be required to demonstrate a comparative twofold improvement in efficacy.

The elaborate animal and clinical protocols that have been designed for the study of the oral contraceptives do not necessarily apply either from a scientific or regulatory view to nonsteroidal or mechanical fertility control agents, but inevitably they have come to serve as a basis for comparison. Thus, as a legacy of the development of hormonal contraceptives, a new standard of needed information with respect to pharmacology, metabolism, side effects and acceptability, theoretical and use-effectiveness, reversibility in long-term use, and potential carcinogenic effects of chronic exposure to drugs and/or devices has been adopted.

Something of a worldwide regulatory picture was provided by key officials of a European-based pharmaceutical company that is one of the largest firms operating in the international contraceptive market.

In their experience, no premarketing registration is required at all in 36 (mostly smaller) countries, and some kind of registration is required in 102. The latter group, however, is very diverse. It includes 64 nations in which the registration process is more or less confined to filing a statement that the product is already licensed and sold in the country of origin. At the other end of the spectrum are such nations as the United States, Australia, Canada, Great Britain, Sweden, and Germany, where the licensing requirements are most stringent.[1] It has been the experience of this company that the licensing requirements for contraceptive products in many nations are more demanding than for any other class of drugs.

As might be expected, the specific standards and practices in different nations also vary considerably. Even among nations with generally comparable requirements, there are important differences. In Great Britain, Sweden, Germany, and a number of other countries, for example, six-month studies are accepted to meet the long-term toxicology requirements; in the United States two-year studies are required (and the United States, in addition, requires that seven- and ten-year toxicology studies must be initiated prior to the beginning of phase III clinical trials). In Japan, it is reported that official permission is not required to begin testing a new drug and that the data requirements with respect to efficacy, toxicology, and safety are generally less stringent (Igarashi 1975). As a result of these and other factors, specific contraceptive products (and probably other drugs as well) are freely marketed in some countries but are restricted or not available in others. The United States, for example, withheld approval of Depo-Provera for contraceptive purposes in 1974 (largely as a result of congressional concern over its carcinogenic potential), although it has been approved as a contraceptive in sixty-two nations of Europe, Africa, the Middle East, Asia, and Latin America (Upjohn 1976).

Our interest in the regulatory process is primarily to assess its influ-

---

[1] Practices also vary among different companies within a single nation; in some countries, national companies face fewer regulatory requirements than foreign companies. It has, for example, been impossible for any foreign company to obtain a license to market oral contraceptives in Japan (Schering AG 1975).

ence on decisions by pharmaceutical firms and public sector agencies to engage in research related to reproduction and fertility regulation. From this perspective, the regulatory policies of some countries clearly have more impact than the regulatory policies of others. The policies carried out by the U.S. Food and Drug Administration (FDA) are thought to be the most influential in the world for several reasons: The size of the U.S. market (the United States accounts for a large portion of all drugs sold in the world, Eastern Europe and China excluded) makes it inevitable that a company would expect a major share of its sales of new products to come from the United States and thus would seek approval from the FDA. The greater availability of scientific institutions and personnel working on contraceptive research also makes it likely that much of the research and testing of new products will be done in the United States, subject to FDA requirements.

Regulatory requirements have become more stringent in many nations in the last decade, and industry officials tend to date this trend to the adoption in the United States of the Food and Drug Amendments of 1962, following the thalidomide tragedy. These amendments increased the FDA's power not only to assess safety but also to evaluate the efficacy and appropriateness of proposed research designs to investigate new drugs. As a consequence, the FDA has formulated extensive requirements that must be met before a company is permitted to test or market a new drug. The requirements for contraceptive drugs are, for the most part, more stringent than for other classes of medications, as is illustrated in table 9.1. The FDA's power is not limited to approval or disapproval of New Drug Applications; it also extends to monitoring marketed drugs. This has led to requirements that new labeling be prepared for oral contraceptives to inform users of potential hazards and to the withdrawal of two intrauterine devices from the market.

Consequences of the Regulatory Requirements

Almost no group *directly* questions the value of this sort of regulation for keeping dangerous drugs off the market; some consumer

Table 9.1 Principal FDA Requirements for Testing and Approval of New Drugs and Contraceptives

| Human Studies | Drugs in General | Contraceptive Drugs[a] |
|---|---|---|
| Phase I (less than 10 subjects) (clinical pharmacology) | 3-4 species; Acute studies<br>2 species; 2-4 weeks<br>2 species; 3 months | 3 species; 3 months |
| Phase II (50 subjects) (clinical pharmacology) | No additional studies | 3 species; 1 year |
| Phase III (clinical trials on larger number of subjects) | 2 species; 6 months or longer<br>1 species; Fertility and general reproductive performance<br>2 species; Teratology<br>1 species; Perinatal-postnatal | 3 species; 2 years<br>1 species; Fertility and general reproductive performance<br>2 species; Teratology<br>1 species; Perinatal-postnatal<br>2 species; Start 7-year dog and 10-year monkey studies |
| New Drug Application | 2 species; 1 year (nonrodent)<br>1½-2 years (rodent)[b] | No additional studies; Update 7- and 10-year studies |

[a]The preclinical testing of contraceptives, estrogens, and progesterones, are presently defined as regulatory *requirements* that differ from the guidelines (D'Aguanno 1973) recommended for other new drugs. A major difference in the contraceptive preclinical requirements is that specific animal species are defined. The guidelines for noncontraceptive drugs leave the choice of the experimental animal to the discretion of the scientist who is ultimately responsible for establishing the risk potential of the new drug.

[b]Second rodent species may be required for carcinogenic evaluation.

Source: Braun 1975.

groups, in fact, contend that the practices of even the most rigorous regulatory agencies are not strict enough adequately to protect the health of persons using drugs. *Indirectly*, however, significant questions have been raised in some quarters: What is the appropriate calculus within which to weigh the benefits of new drugs against their risks? How is one to understand and justify the fact that medications are available and used clinically in some scientifically advanced nations with comparable regulatory mechanisms but are not approved for use in other countries, specifically in the United States? Are U.S. citizens suffering from a "drug lag" caused by supercaution on the FDA's part that deprives them of beneficial medication used elsewhere? On the other hand, are citizens of other countries less well protected than those of the United States? Echoes of these issues are also heard on more technical questions: Can the findings of tests on dogs and monkeys (required for the seven- and ten-year toxicology studies of new contraceptives) be extrapolated to human beings? Are there animal models more appropriate for use in research applicable to humans (Briggs and Diczfalusy 1974)?

None of these questions, of course, has a simple answer and the debates continue, involving scientists, industry and regulatory officials, consumer spokesmen, and policy makers. Almost all parties agree, however, on several points that are pertinent to the subject of this review. There seems to be no prospect that regulation of drug research and development will diminish in any country; rather, as the former FDA general counsel puts it, "governmental regulation and control will gradually increase for the foreseeable future" (Hutt 1975, p. 19). Almost everyone would also agree that the regulatory process has lengthened the time required for approval of new products, although there are varying estimates of the magnitude of the difference. Schering AG officials (1975) state that they could register a new drug for testing in Germany in less than two weeks in 1961, a process that now takes an average of two and a half years. The vice-president for research of the Ortho Pharmaceutical Corporation believes that the development time for new drugs generally is now eight to ten years and possibly longer for new contraceptives: "The lengthening interval that now separates the present from the next generation of contraceptive agents has now become longer" (Braun 1975, p. ii). The Stanford chemist instrumental in the synthesis of steroids at the Syntex Corporation in the 1940s and 1950s states that the time frame for the emergence of a new contraceptive product has increased to about fifteen years (Djerassi 1970).

There is also considerable agreement that the development of a new product has now become more costly, although the cited figures vary. Braun estimates that the cost of a new contraceptive has now risen to about $11 million per product, three to four times the amount required before 1962. Djerassi puts the figure at $18 million. Schering AG believes that the cost of expensive long-term toxicology studies, which Djerassi estimated at $4 million in 1970, would be considerably higher today.

There are other possible consequences on which there is less agreement. As a result of increased regulatory requirements, some believe that smaller pharmaceutical companies have been driven out of the development of new drug products, leaving the field to the larger firms (Wardell and Lasagna 1975). Pharmaceutical officials cite the regulatory requirements in the contraceptive field to explain why the industry invests less today than a decade ago in development of new contraceptives. "That is the reason why so much less money is being spent by industry now," Djerassi (1975) contends. Braun (1975, p. ii) sees the consequences of increased regulation as including "diminished levels of spending for innovative fertility research within industry, and—perhaps most alarming—the clearly waning interest of the investment community in contraceptive manufacturing as a growth industry, signaling possibly an eventual drying up of capital funds for the discovery and development of innovative agents for birth control."

Because of lack of comparable data, it is difficult to be certain that the interest in contraceptive manufacturing has indeed waned. It is clear, however, that advances in contraceptive development will be greatly facilitated by collegial cooperation among industry, the public sector, and regulatory agencies. A systematic effort is needed to develop mechanisms to ensure that regulatory and policy decisions are based on scientifically valid information.

## Ethical Considerations

These increased regulatory requirements, spurred in large part by consumer concerns over drug safety, thus have established a new set of parameters for the conduct of biomedical research. Concerns over the ethical values implicitly or explicitly embodied in scientific investigation have prompted another set of parameters governing the research enterprise. The root justification of all scientific effort is that additional knowledge will be beneficial to human society. In recent years, however, this global justification has been subject to challenges that raise disturbing questions, such as, Who benefits from whose risks? Who decides these issues and on what criteria? What are the rights and responsibilities of the respective parties—donors, researchers, human subjects—to the research effort?

The testing of new drugs or devices is inevitably associated with some risk to the experimental subject. Ever since Hippocrates admonished physicians to abstain from intentional injury and wrongdoing, medical scientists have been aware of the tension between advances in medicine and medical techniques and safe treatment. In the past decade social scientists, physicians, philosophers, biologists, and lawyers have begun to develop a new discipline of bioethics concerned explicitly with the immediate welfare of the human subjects of medical research and experimentation; and formal mechanisms, such as the National Commission for the Protection of Human Subjects of Biomedical and Behavioral Research in the United States, have been established to formulate guidelines for the ethical conduct of research. The Department of Health, Education and Welfare took an important first step in 1971 by formulating its policy on protection of human subjects, and in subsequent years it issued a series of regulations focusing on such special situations as research on minors and prisoners. Comparable efforts are beginning in other countries, such as Mexico and India. In recent years a number of private foundations have adopted a policy on human research.[2]

---

[2] The Ford Foundation policy, adopted in 1972, states, "The Ford Foundation recognizes that safeguarding the rights and welfare of human subjects or patients involved in biomedical research supported by the Foundation is the responsibility of the grantee. At the same time, the Foundation is concerned that this responsibility is adequately discharged. It is, therefore, the policy of the Ford Foundation that no work shall be initiated under Foundation grants which en-

Like other areas of biomedical research, some reproductive studies are not intended to benefit the immediate subject at all (for example, experiments on hysterectomy specimens to study early implantation or experiments on spermatozoa to develop inhibitors of motility). Many others, although beneficial to the subject, are principally directed toward gathering information of long-term benefit to society (for example, experiments with new techniques of surgical sterilization or with new forms of intrauterine contraception). These ethical dilemmas are particularly apparent in contraceptive development that is not entirely typical of biomedical research because it attempts to regulate a normal biological process in healthy individuals, rather than cure a disease or malfunction.

Two major means of reconciling the benefits of biomedical research with the risks to its subjects are now in use. The first offers a subject who is mentally competent sufficient information about the research to make an informed and voluntary decision to participate or not ("informed voluntary consent of subjects"). The second provides review of the ethical implications of research protocols by the investigator's peers ("peer review"). Informed voluntary consent and peer review have developed in tandem, but each provides its own distinct safeguards to the experimental situation.

To require informed voluntary consent is obviously complex because how informed a person can be depends on many variables: the subject's general educational level and actual working capacity to understand a researcher's explanation of what will be done in the study and what are the attendant risks and benefits. (A subject who is a minor or mentally incompetent may never be able to give "informed" consent in a legal sense because the presumption would be that such a person is unable to comprehend the explanation.)

tails research involving human subjects or patients unless assurances, satisfactory to the Foundation's professional staff and expert reviewers, are obtained that (a) the rights and welfare of the subjects or patients involved in the research are adequately protected, (b) methods used to obtain freely given, informed consent of subjects or patients are adequate and appropriate, (c) the risks and potential medical benefits to the subjects or patients are deemed acceptable by the research investigator and the subjects or patients, and (d) the research will be carried out in accordance with local and national policies and regulations pertaining to research involving human subjects or patients. The Board will from time to time review the effectiveness of this policy statement in the light of experience in its application and of new developments in this rapidly evolving area."

It is frequently even more difficult to determine when consent is voluntary. Many physicians project an image of omniscience, making it difficult for patients to evaluate critically the physician's suggestion of participation in research. Research subjects are more often clinic and ward patients than private patients; they tend to be poorer and less well educated than private patients and in the United States many are nonwhites. For these reasons, some ethicists believe that research subjects may be particularly vulnerable to pressure by physicians-authority figures. Moreover, payments for participation in research—whether in dollars or in provision of otherwise unattainable medical services for themselves or their families—have an entirely different meaning for poor persons than for more affluent persons. For similar reasons, some ethicists and lawyers also doubt that prisoners and other captive populations, such as the institutionalized mentally retarded, are ever able to consent voluntarily. Others feel that under certain protected conditions, prisoners should be permitted to participate in medical research so they can make a contribution to society and enhance their self-esteem.

Peer review procedures, developed in the last three decades to evaluate the scientific merit of research proposals, have been extended to assess their ethical implications as well; and most, if not all, major institutions conducting clinical research on humans in the United States now have such mechanisms. In brief, peer review committees are based on the premise that the investigators may be so deeply involved in the research and its expected results that responsibility for reviewing matters of merit, propriety, and ethics should be shared with less-involved but well-informed associates.[3]

Peer review committees have only recently begun to be established outside the United States, often as a result of an NIH grant. For example, Mexico has no regulations on the ethical aspects of research on humans corresponding to those promulgated by DHEW. But the attitudes of the Mexican scientific community have begun to change, and a few years ago peer review committees were formed in some of the more prestigious centers. These committees are voluntary and

[3] For excellent discussions of the motivation of scientific research, see Ratnoff and Ratnoff 1967; Barber et al. 1973; Moore 1969.

thus reflect an important institutional commitment to focus on ethical aspects of human research, independent of regulation by government or professional societies.

Peer review is not a panacea. Often the investigator is the most expert person in the field being studied, and others may have difficulty understanding the project and assessing adequately its ethical implications. The investigator frequently has developed relationships with other researchers in the same institution. If, for example, review committee members are subordinates of the researcher or expect subsequently to have their own work reviewed by the project's sponsor, they may consciously or unconsciously avoid raising hard ethical or scientific issues. In addition, protocols vary widely in the care with which they are prepared and reviewers approach evaluations with different standards.

Despite these limitations, peer review is viewed as a helpful mechanism for protecting human subjects of research. Peers can identify risks not apparent to the investigator, and collegial discussion with scientists from several disciplines—as well as with informed nonscientists—may promote a broader consideration of ethical issues than limiting review to individuals with the same areas of interests, training, and background. Indeed, peer review is frequently the only assurance against overreaching by the researcher where the available subjects are less likely themselves to give informed voluntary consent.

Most of the discussion and regulation of ethical aspects of human experimentation has thus far occurred in the industrialized world, especially in the United States; but awareness of these issues is increasing in the developing world. Indeed, there are world medical standards, such as the Nuremberg Code and the Declaration of Helsinki, that provide general ethical benchmarks for researchers in all countries. But these codes do not define standards at the detailed level of the DHEW guidelines. Most universities, hospitals, and medical schools in the developing world do not yet have rules, review committees, or procedures to assure informed voluntary consent. But researchers in these countries are developing their own ways of dealing with ethical issues. For example, one Latin American researcher will not permit a subject to participate in his projects until she has spoken with another woman who has already participated in the research.

As a whole, however, the marked difference in practices between industrialized and developing nations raises difficult questions for which satisfactory answers are yet to emerge. It would be unwise, and probably unethical, for grantors to seek to impose their standards on researchers in other nations. Moreover, withholding medical research and aid out of ethical considerations would itself pose an ethical dilemma because it may deprive members of a society of the benefits of medical research. Partial answers may be found in the accumulation of greater experience by researchers in different nations with the principles of peer review and voluntary informed consent and the emergence of international norms for protection of human subjects of research. Because conditions in each nation vary, the benefit-to-risk ratios may also vary, justifying some differences in norms and standards. Progress toward internationally acceptable standards will depend in part also on the development of more rational means of assessing differential benefits and risks.

In the contraceptive field, the issue of research on "captive" populations takes an additional form in the assertion that a disproportionate share of the clinical testing of new contraceptive products is conducted on Third World women (Warwick 1975). Numerous clinical tests have been done in developing countries (although industry officials contend that because of the regulatory requirements in industrialized countries, which constitute the bulk of the market for new contraceptive products, most of the clinical research, especially the early studies, for new contraceptives is in fact done in industrialized nations (Schering AG 1975). The typical form of the debate obscures important medical considerations that are also ethical considerations. For example, if a drug proved to be safe and effective when tested on women in industrialized nations, would it be appropriate to conclude that it will also be safe and effective when used by women in developing countries who are exposed to different endemic diseases and have varying nutritional and health levels? As a result of this concern, officials in many developing countries have insisted that new products must be tested in their nations before they can be licensed for distribution and sale.

It seems evident that the ethically appropriate criteria for the conduct of medical research generally and fertility research in particular will be subject to increasing discussion and debate among scientists, policy makers, consumers, and donor agencies that support research. In the long term, this should be a positive development, leading to emergence of a closer perceived community of interest in the research enterprise by all parties. In the short run, however, because these questions necessitate changes in the way research has been conducted, it is apparent that they will also impose additional constraints on research activities, including projects aimed at development of new contraceptive techniques.

## Conclusion

In this chapter we have summarized two major new factors, regulatory and ethical requirements, that have changed the climate in which medical and contraceptive research takes place. Our intent was not to deal extensively with these factors, but rather to delineate in a general way changes believed to have important effects on decisions by industrial and academic researchers and institutions to engage in specific lines of scientific inquiry.

Several conclusions are suggested that are pertinent to the principal subject of this study:

1. The new regulatory and ethical requirements have made contraceptive research less, rather than more, attractive to the pharmaceutical industry. As a result, the strategies developed in the early 1960s that depended on the major involvement of the pharmaceutical industry in the development of new contraceptives now have to be reappraised. Unless current trends can be modified significantly, it is unlikely that the industry will play the substantial role originally envisioned. There appear to be three remedies available that are not mutually exclusive: either (1) reduce the regulatory and ethical requirements (which is neither desirable nor feasible) or (2) develop alternative institutions and mechanisms to undertake the kind of work origi-

nally assigned to industry; or (3) develop incentives for industry participation in the form of subsidies.

2. Regardless of which means, or which combination, are chosen, the development of new contraceptive methods will surely take longer in the 1970s and 1980s than it did in the 1950s and 1960s. Some industry spokesmen assert that if the current regulatory rules had been applied two decades ago, it is doubtful that the oral contraceptive or the IUD would have been approved for marketing and distribution until perhaps the late 1960s or early 1970s. Whether or not one shares this conclusion, it seems clear that the current rules will inevitably mean that more time will elapse between the conceptualization of a new contraceptive and its emergence as an approved product.

3. Finally, it also seems evident that the costs of product development will be higher. This seems clear whether or not government and private philanthropies subsidize some of the development costs, as urged by some industry officials. In any case, the total costs will be greater than they were a decade ago; both the additional technical requirements of longer studies in both animals and humans and the new ethical requirements mean that more time and effort must go into each study. The point here is not to suggest that these new requirements should be waived, but rather to make explicit what should be obvious: The cost of research related to reproduction and contraceptive development has increased. As the *New York Times* (1976) editorialized (in reference to the ethical requirements of research), "Though experimentation only upon subjects who can give voluntary consent and who may have recourse in the courts if they are not adequately protected may cost more in dollar terms, that is surely a price society should pay for such knowledge. It is nothing more than the price necessary to maintain the civilized balance between the society's need for greater medical knowledge and its equally urgent need to respect the rights of all individuals in the community and to erect effective safeguards for the very weakest."

The same factors suggest that the pharmaceutical industry is unlikely to shoulder this additional burden (and may even be unwilling to continue to bear the share of the burden it assumed in the 1960s and early 1970s). The principal consequence of these varied processes, therefore, is that the search for new contraceptives and their subsequent distribution now depends, even more than in previous years, on decisions made and activities undertaken by the noncommercial sector—government and international agencies and philanthropic foundations.

## References

Barber, B. 1974. Ethical aspects of clinical research in the field of human reproduction. Report commissioned for the reproduction review by the Ford Foundation.

Barber, B.; Lally, J.J.; Makarushka, J.L.; and Sullivan, D. 1973. *Research on human subjects: Problems of social control in medical experimentation.* New York: Russell Sage Foundation.

Braun, G. 1975. Regulatory requirements seen from the viewpoint of the pharmaceutical firms. Report commissioned for the reproduction review by the Ford Foundation.

Briggs, M., and Diczfalusy, E. eds. 1974. *Pharmacological models in contraceptive development: Animal toxicity and side-effects in man.* Proceedings of a meeting organized by the World Health Organization in Geneva, 1973. Stockholm: Karolinska Institute.

D'Aguanno, W. 1973. Guidelines for reproductive studies for safety evaluation of drugs for human use. In *Introduction to total drug quality*, F.D.A., pp. 41-46. Washington, D.C.: Government Printing Office.

Djerassi, C. 1970. Birth control after 1984. *Science* 169:941-951.

_____. (Stanford University) 1975. Personal communication with Roy O. Greep.

Hutt, P.B. 1975. Regulatory requirements seen from the viewpoint of the FDA. Report commissioned for the reproduction review by the Ford Foundation.

Igarashi, M. (University of Gunma, Japan). 1975. Personal communication with Richard T. Mahoney.

Kordon, C., and Gaudin, T. 1975. Ethics and morality of human testing. Report commissioned for the reproduction review by the Ford Foundation.

McLean, S. 1975. Ethical implications of research on human subjects in reproductive biology. Report commissioned for the reproduction review by the Ford Foundation.

Moore, F.D. 1969. Therapeutic innovation—ethical boundaries in the initial clinical trials of new drugs and surgical procedures. *Daedalus* 98:502-522.

*New York Times.* January 23, 1976. "Experiments in People," p. 15.

Ratnoff, O.D., and Ratnoff, M.F. 1967. Ethical responsibilities in clinical investigation. *Perspectives in Biology and Medicine* 11:82-90.

Schering, AG 1975. Personal communication with Gordon Perkin and Richard Mahoney.

Upjohn Co. 1976. Personal communication.

Wardell, W., and Lasagna, L. 1975. *Regulation and drug development.* Washington, D.C.: American Enterprise Institute for Public Policy Research.

Warwick, D. 1975. Contraceptives in the third world. *Hastings Center Report* 5(4):9-12.

# IV

**Financing: Past, Present, and Future**

World War II marked a major turning point in the way in which Western nations have provided financial support for scientific research. Before the war, support of scientific research was relatively haphazard: There were few efforts by government to channel financial support to researchers (except in the agricultural and military fields) and the bulk of financial support for science came from industry, private philanthropy, or university budgets. Federal management of weaponry and other military research during both world wars laid the groundwork for subsequent governmental support of basic science. The three decades since World War II have witnessed the emergence of what has been called "big science"—the formalization of institutional arrangements for support of science and of procedures for obtaining such support—on a scale unparalleled in history. This process has significantly expanded the level of activity not only in the physical sciences most closely related to military and industrial needs, but in the biological and social sciences as well.

Not all branches of science shared equally in the post-World War II expansion. The relative funding levels have depended on social and political perceptions of priority and on the ability of existing personnel and facilities in the different branches to exploit new findings that offer the promise of solving urgent problems. Paradoxically, although research in the reproductive sciences related to contraceptive development is critical to the solution of some of the most urgent problems facing nations and individuals, these fields were virtually excluded from the rapid expansion of biomedical research in the two decades following World War II and only began to receive support from established medical research agencies around 1965. There are doubtless several reasons for this paradox. One factor is that these branches of biomedical research deal with sex and human reproduction, and there have been—and to a lesser degree, still are—strong social taboos against the systematic study of these subjects. To initiate and sustain serious research in the reproductive sciences has required for more than a half century concerted effort by interested individuals and private organizations, mainly from outside the mainstream of the biomedical research community.

In this chapter we draw primarily on information related to these developments in the United States because it has not been possible to assemble systematic information on comparable developments in other nations. We feel confident, however, that the basic picture that emerges would not be substantially altered if data from other nations were available.

**The 1920s and 1930s**

Organized support of reproductive research had its inception in the United States in 1921 with the organization of the Committee for Research in Problems of Sex by the National Academy of Sciences' National Research Council. By the time the committee ceased to function forty-two years later, it had provided slightly more than $2 million for research in the field. The committee was supported with funds provided principally by the Rockefeller family, through Social Hygiene, Inc., until 1931 when the Rockefeller Foundation assumed financial responsibility for the program. Its avowed purpose was "to establish and defend the importance and the dignity" of studies related to sex and human reproduction and thus to overcome the "relative disrepute" associated with the field (Aberle and Corner 1953, p. 1).

In its first year of operation (1922-1923) the committee made awards ranging from $250 to $4,000 to seven investigators. In the second year two new grantees were added; one was Edgar Allen, who was to determine whether the hormonal activity newly discovered in the follicular fluid from cow ovaries would substitute for the endocrine function of the ovaries. This turned out to be a breakthrough discovery that opened the field of reproductive endocrinology to much fruitful research. The awards made in the first year totaled $6,096 and in the second, $20,324. The most that was expended by the committee in any year was $74,415 in 1931 (Aberle and Corner 1953).

The quality of the workers selected for grant aid was very high, as is evident retrospectively from the fact that the roster of grantees is replete with the names of distinguished pioneering investigators in reproductive biology and sex behavior. They are the group that produced the first (1932) and second (1939) editions of *Sex and Internal Secretions*, viewed today as classic texts for both the fundamental importance of the accomplished research and their perceptions of problems yet to be resolved. The committee was determined to demonstrate that sex and the reproductive system could be studied with the same scientific detachment and standards of quality as any other natural phenomenon. They were also anxious to foster public acceptance of the need for enlightenment in this area. Their concern was justified. When the plan for organizing this committee was first presented to the Division of Medical Sciences of the National Academy of Sciences, it was turned down on the basis that sex and reproduction were not suitable subjects for dignified scientific study and discourse. In the late forties, the committee devoted a large share of its resources to Alfred Kinsey's investigations of human sexual behavior. This study was a major factor in the gradual demise of neo-Victorian prudery regarding the natural function of reproduction in the human species.

The NAS-NRC entry into basic reproductive research was complemented in 1923 with an initiative focused on the clinical and social aspects of reproduction when a group of obstetricians, demographers, and reproductive biologists established the National Committee on Maternal Health to encourage ". . . the scientific investigation of the medical aspects of maternal health and human fertility" (NCMH 1941, p. 3). A survey carried out by the committee to ascertain the state of the art concluded that "it is easier to find rapidly the newest information about fertility patterns of vegetables and farm animals than it is to obtain such information about human beings!" (NCMH 1941, p. 8). Much of the committee's efforts were devoted to assembling such information, and it sponsored the publication of several classic treatises on reproductive topics: Robert Latou Dickinson's *Human Sex Anatomy*; Norman Himes's *History of Contraception*; and Carl Hartman's *The Time of Ovulation in Women*. It was not until 1937, however, that the committee itself undertook to support more fundamental research; over the next four years it expended $123,000 on a variety of studies mostly concerned with the prevention of conception. For several years thereafter,

NCMH funds were used to support the organization of family planning efforts in areas of the world. In 1957, the committee accepted a grant from the Population Council to evaluate the effectiveness, acceptability, safety, and cost of methods of fertility control; and ten years later its staff and activities became a part of the council's Biomedical Division (Tietze 1975). During these latter years, NCMH was headed by Christopher Tietze.

In 1936, another NAS-NRC mechanism for support of basic reproductive research was started with the formation of the Committee on Research in Endocrinology supported by the John and Mary R. Markle Foundation. To maximize their joint effectiveness, a liaison was established with the Committee on Research in Problems of Sex. Although the intention was to "consider research on sex and reproduction as outside its domain [nevertheless, the endocrine orientation was such that] . . . the operation of gonadal factors on bodily processes outside the sex sphere might properly be its concern" (National Research Council 1945, p. 3). Adherence to the separation of fields of overlapping interest proved difficult. During the fourteen years (1936 to 1950) that the endocrine committee functioned, it spent $561,000 on grants-in-aid for research, $71,000 of which was allocated to fourteen investigators for research that contributed directly or indirectly to an understanding of reproductive phenomena (NRC 1945).

These efforts initiated by private philanthropy constituted the principal organized efforts in human reproductive research in the United States prior to World War II. During that period, government involvement had been indirect at best. The U.S. Department of Agriculture did, however, sponsor work in animal husbandry. Fractionation of steroids was used by researchers in animal husbandry to develop an index of specific hormonal activity. Methods of collection and preservation of sperm were devised. Physiologically buffered solutions were used to facilitate improved study of spermatozoa. Insemination was successfully carried out with frozen sperm. Although this research was government sponsored, it was left to the philanthropists to recognize its potential for the understanding and control of human fertility (NCMH 1941).

Prior to World War II, a few pharmaceutical firms also conducted research that would eventually lead to the better understanding of the human reproductive processes, although their goal at the time was not specified as fertility control. Schering AG of Berlin, Germany, spent approximately $5 million on endocrine research in the 1930s; and CIBA of Basel, Switzerland, committed a total of $1 million (Gibian 1975).

The data of the pre-World War II period are fragmentary and can be used only to suggest the meager level of support for the work in the field. Information is lacking on activities in the universities of Europe and other centers of scientific activity. Table 10.1 shows the amounts the U.S. groups directly involved in reproductive research were able to spend on research grants from the inception of the Committee on Research in Problems of Sex in 1921 to 1945; the average amount over the twenty-four-year period was $69,450 a year. Reports from one representative year, 1941, reveal that the $124,000 spent by these private donors to support reproductive research was less than 1 percent of the total funds contributed by U.S. philanthropic agencies to medical research and less than 0.3 percent of the total expenditure on medical research in the United States (The American Foundation 1955).

Table 10.1 Private Agency Reproductive Research Expenditures in the U.S., 1922-1945 (current dollars in 000s)

| Year | Amount |
| --- | --- |
| 1922-1925 | 99.6 |
| 1926-1930 | 323.5 |
| 1931-1935 | 352.7 |
| 1936-1940 | 520.1 |
| 1941-1945 | 370.9 |
| Total | 1,666.8 |

Source: NAS-NRC Committee on Research in Problems of Sex; NAS-NRC Committee on Research in Endocrinology; and the National Committee on Maternal Health.

## Post-World War II, the 1940s and 1950s

The end of World War II ushered in an era of expansion of research in many branches of U.S. science as the federal government took the lead in support of fundamental research. Medical research was particularly favored during this period with the expansion of the National Institutes of Health. The National Institute of Health had been created in 1930, superseding the Hygienic Laboratory and orienting medical research toward the cure of chronic diseases. A trend toward basic research paralleled the increased emphasis on chronic diseases. Subsequently, through the Social Security Act of 1935, funds were finally authorized for the "investigation of disease and problems of sanitation" (U.S. 1935) but government-sponsored medical research remained essentially an intramural task and the actual appropriations remained small—a total of $375,000 in fiscal 1936 (Swain 1962). The National Cancer Institute, whose grants-in-aid and fellowship program set the stage for today's large extramural research organization, was created in 1937. The legal groundwork, however, was not laid until the enactment of the Public Health Service Act of 1944. Among other things, it empowered the Surgeon General to "make grants-in-aid to universities, hospitals, laboratories, and other public or private institutions, and to individuals" (U.S. 1944). With the precedent having been set by the programs of the Cancer Institute, the number of disease-oriented institutes increased, each authorized to carry out a large grant and fellowship program. To represent the new status, NIH pluralized its name in 1948; and within the next decade it burgeoned into the nation's principal source of support for medical and health-related investigations. Federal support for medical research increased from $27 million in 1947 to $471 million in 1960, and the federal share of total U.S. support for medical research grew from 31 percent to 59 percent (NIH 1966).

Reproduction was not a chronic disease, however, and was not looked upon with favor during this period of influx of government funds into medical research. As a matter of fact, the NIH was forbidden to support research having any explicit relationship with birth control prior to 1959. However, with the establishment of the

National Institute of Arthritis and Metabolic Diseases (NIAMD) in 1950 and formation of the Endocrine Study Section (initially called the Endocrinology and Metabolism Study Section) in 1951, a trickle of support was made available, especially for research on the gonadal steroids. The field received another slight boost in 1955 with the creation of the Human Embryology and Development Study Section. The applications considered for support by this study section did not deal with fertility regulation but rather with the study of events beyond the time of implantation (Hertz 1975; Graff 1975; Hill 1975).

The National Science Foundation was established by the government in 1950 as an independent agency devoted to the support of science and science education without regard to specific practical missions. By the time this authorization was made, the extramural program of NIH was sufficiently well established that the new agency did not undertake the support of disease-oriented research. Instead it subsidized basic research in the chemical and biological sciences, contributing only indirectly to medical research. Reproductive research again did not fit the perceived mandate of this new agency and received only scant attention: In 1957, NSF awarded only 7 grants for reproductive studies, totaling $41,000, out of 997 grants in all fields of science and a total of $21.5 million. Two years later it supported 38 reproductive studies totaling $308,000, out of 1,809 grants amounting to $64.5 million (NSF 1957; 1959). Most of these grants were in the area of neuroendocrine function and clearly were not designated as research in the direction of fertility regulating agents. Again, as with NIH, the modicum of support for the study of reproductive processes derived not from a conscious effort on the part of the U.S. government but was due to the very nature of fundamental research (knowledge accrued by such research is often applicable to several fields of inquiry).

During the 1945 to 1960 period the major source of government funds for the support of reproduction research continued to be the Department of Agriculture. With the outbreak of World War II, production goals for meat and dairy products were sizably increased, giving incentive to the study of breeding efficiency. Investigators studied the use of hormones in stimulating milk and egg production and the readying of livestock and fowl for market. The New Jersey Agricultural Experiment Station examined the possibility of synchronizing fertility periods in sheep by using injections of pregnant mare serum. Artificial insemination was the center of much attention at the stations in Ohio, Wisconsin, Missouri, and at Cornell University. Research of this period stimulated more work in the direction of quick and sensitive assays for hormones, techniques that were applicable to humans as well as animals. The operating budget of these Agricultural Experiment Stations was approximately $11 million in 1942 and rose to $71 million by 1951 (U.S. Department of Agriculture 1942-1951). It is estimated that 2 to 4 percent of these funds was spent annually on animal reproduction.[1] Although not a large portion of this budget, these funds constituted the major governmental contribution to the field of reproduction during the 1945-1960 period.

Philanthropic support of the reproductive sciences and contraceptive development also grew during this period but at a rate considerably slower than that of support for medical research in general (which implies that the taboos were still strong). Moreover, expansion of the field required the initiation of several new efforts, again outside the mainstream. The Committee on Human Reproduction was established by the National Research Council in 1947, pursuant to a contract between the National Academy of Sciences and the National Committee of Maternal Health (with the Planned Parenthood Federation of America concurring). Little came of this contract because the sponsor failed to raise the funds needed for a major program. When the NRC Committee was disbanded in 1951, it had funded only nine research projects and held four annual conferences on reproduction, at a total cost of $112,000 (Goldblum 1975).

The Planned Parenthood Federation of America initiated a limited grants program in reproductive research during the fifties with funds ranging from $40,000 to $100,000 a year. Clinical investigations on conventional contraceptive techniques, the diaphragm and spermi-

[1] Percent calculated from funds designated for Animal Research-Husbandry under the Bureau of Animal Industry (Agricultural Research Administration 1948, 1950, 1952, 1954).

cidal jellies, were supported with these funds (Horowitz 1975).

Support for the NAS-NRC Committee on Research in Problems of Sex was continued by the Rockefeller Foundation during this postwar period. However, several university research programs originally supported through this committee outgrew the committee's resources and were aided directly by the foundation. The universities of California, Chicago, and Rochester and Columbia and Cornell universities received a total of $1,066,000 during the 1950s through this program (RF 1965). The University of Lund and McGill University also received Rockefeller Foundation funds during the fifties (RF 1948-1961).

Some funds not specifically earmarked for the reproductive sciences filtered in from the Milbank Memorial Fund and the NAS-NRC Committee on Growth, established in 1945 with support from the American Cancer Society. The relationship of nutrition to pregnancy was a topic of interest to Milbank; support was limited, however, and a total of $150,000 was distributed during the fourteen-year period 1947-1961 (Kasius 1975).

Between 1945 and 1955, the Committee on Growth approved recommendations totaling $20 million for support of grants-in-aid, fellowships, and scholarships in cancer research. During this period, the committee served as a significant source of funds for studies relating to reproductive hormones and tissues. Although a few of the grants dealt directly with cancer of the male and female reproductive organs, specifically the prostate, cervix, and uterus, the majority were concerned with the relationship of steroid hormone metabolism to normal and abnormal growth. Some measure of what this research program meant to the reproductive sciences during this period can be gauged from the fact that among the 163 grants awarded by the committee in 1947-1948 for a total of $1,508,207, three fellowships and twenty-four awards totaling $164,895 were specifically for research related to the reproductive system. The support continued at a slightly higher level through 1956, when the committee was disbanded (NRC 1946-1954).

Despite the modest growth of support for reproductive research in the decade following World War II, it is of interest that the funds needed to develop the first important innovation in contraceptive technology in the twentieth century—the oral contraceptive—did not come from these institutional sources. Rather, the funds necessary to transform Makepeace's 1937 demonstration that progesterone suppresses ovulation in rabbits into a contraceptive suitable for use by humans were provided by an individual philanthropist concerned with the urgency of the population problem and by the pharmaceutical industry. Gregory Pincus's work that led to the marketing of the pill in 1960 began in 1951 when Margaret Sanger, the birth control pioneer, introduced Pincus to Mrs. Stanley McCormick, the widow of the son of the manufacturer of farm equipment.[2] Mrs. McCormick made a small grant through the Planned Parenthood Federation of America that enabled Pincus and his coworker, M.C. Chang, to explore the potential of artificially synthesized progesterone for conception control. In the next decade Mrs. McCormick contributed more than $2 million to the work at the Worcester Foundation for Experimental Biology, using PPFA as the mechanism through which her contributions could remain anonymous (Hoagland 1975).[3] Pincus's research was also aided by support from the G. D. Searle Company for his work in steroid chemistry.

In early 1951, the Syntex Corporation, under the leadership of Carl Djerassi, also initiated an intensive research program aimed at the synthesis of 19-norsteroids, primarily to find analogs of cortisone that would lack that compound's undesirable side effects. One result was the synthesis of norethisterone, which was found to be an extremely strong progestational material. Similar work at about the same time at the Searle Company, under the direction of F. B. Colton, led to the synthesis of norethynodrel, which also showed strong progestational activity. Both norethisterone and norethynodrel were made available in 1953 to Pincus and Chang for evaluation, although neither was originally intended for contraceptive use. It is estimated that Syntex invested nearly $2.5 million in this program between

[2] See Battelle 1973 and Rock 1963 for a review of the development of the oral contraceptive.

[3] Another woman contributed $25,000 a year to the research program for several years, but she remains anonymous.

1950 and 1959 (Djerassi 1975); the amounts allocated by Searle are unavailable.

The most significant factor in the growth of institutional support for reproductive research during the immediate postwar period was the establishment of the Population Council by John D. Rockefeller III, in 1952 for the purpose of stimulating and supporting significant activities in the broad field of population. The council's Biomedical Division was largely a granting agency for research and training in the reproductive sciences during its initial years. In 1957, however, it established its own research laboratories in the Rockefeller Institute (now Rockefeller University) and later built expanded laboratory facilities on the university's campus. The Biomedical Division, funded principally by the Ford and Rockefeller foundations and the Scaife philanthropies, became a significant source of support for research and training in reproductive research and contraceptive development throughout the world. The biomedical grant program grew from an estimated $13,000 in its first year of operation to approximately $270,000 in 1959. In the same year, expenditures for biomedical fellowships amounted to almost $50,000, having started at $17,000 in 1954; by 1964, ninety-nine fellows from all parts of the world had received one to three years of training (PC 1952-1960; 1965).

Another significant event for the field's development took place in the late 1950s when the Ford Foundation initiated sponsorship of research in the reproductive sciences (Harkavy 1975). From a modest beginning with a grant to Pincus for a training program in human reproduction at the Worcester Foundation, the Ford program was to emerge as the principal private source of support for the field throughout the world during the 1960s, outstripping for much of the decade the amounts made available even by government agencies.

The surviving data on support levels in the United States for the period from 1945 to 1960 (shown in table 10.2) are again fragmentary but nonetheless instructive in demonstrating the meager level of support for the reproductive sciences and contraceptive development as recently as two decades ago.

In the fifteen years following World War II, the average annual

Table 10.2 Private Agency Reproductive Research Expenditures in the U.S., 1946-1960 (current dollars in 000s)

| Year | Amount |
| --- | --- |
| 1946-1950 | 1,415.3 |
| 1951-1955 | 855.5 |
| 1956-1960 | 2,603.0 |
| Total | $4,873.8 |

Source: NAS-NRC Committees on Research in Problems of Sex, in Endocrinology, on Growth and on Human Reproduction; Rockefeller Foundation; Milbank Memorial Fund; National Committee on Maternal Health; Planned Parenthood Federation of America; Sunnen Foundation; Ford Foundation; and Population Council.

amount of support for reproductive research had increased to about $325,000, about four times the prewar average. By contrast, the amount the federal government spent on medical research in 1960 was seventeen times greater than in 1947, and the amount spent on medical research by all sources (government, industry, and philanthropy) grew nearly tenfold in the same period (NIH 1966). Moreover, more than 40 percent of the funds made available for the reproductive sciences during these fifteen years came through the newly formed Population Council and another 15 percent was provided by the Rockefeller Foundation. A detailed study for the year 1958[4] showed that private philanthropic funds for reproductive research represented a little more than 0.5 percent of total contributions by

[4] The survey, carried out by Sheehan and Weil-Fisher for the Ford Foundation, estimated that "... approximately $900,000 was expended on medical research related to reproductive physiology in the United States [in 1958]. Of this amount $215,000 was provided by the Population Council. The United States Government, through the National Institutes of Health, ... and the National Science Foundation, contributed about $300,000. More than half of the latter amount, however, was spent on projects directly related to pregnancy and maternal health. The National Research Council's Committee for Research in Problems of Sex has an annual budget of $100,000, only $35,000 of which goes into projects on reproductive physiology. $75,000 a year is spent by the Planned Parenthood Federation of America on research, while the Sunnen Foundation of St. Louis has spent $100,000 a year in Puerto Rico for improvement and testing of spermicidal compounds. In addition, a substantial amount of undetermined magnitude is spent by commercial firms on research and development of devices and drugs 'that approach practicability and saleability' " (Sheehan and Weil-Fisher 1958, p. 1).

U.S. private donors for medical research, approximately half the proportion that had been received by reproductive research in 1941; the governmental contribution was only 0.1 percent of its total medical research budget. The 1960 expenditures of U.S. pharmaceutical firms for contraceptive development, estimated for this study at $3.4 million (Duncan 1974), constituted 1.6 percent of expenditures by pharmaceutical firms for the development of all ethical drugs (Pharmaceutical Manufacturers Association 1975).

It is apparent that these fragments of information do not present a complete picture of the sources of support for work in the field prior to 1960. That there *were* other sources of support may be inferred from the work going on in laboratories in other nations, as well as in the United States. It seems doubtful, however, that the field received any substantial amount of support in any nation or that it was accorded degrees of attention by the governmental and scientific communities of other nations significantly higher than in the United States.

Despite these limitations, the available data demonstrate the low level of support for the reproductive sciences and contraceptive development prior to 1960. The decades of underfinancing had taken their toll. Regarding the methods of fertility regulation then in existence, Sir Alan Parkes, Professor Emeritus of Physiology at Cambridge University, stated at that time that "established methods . . . are so crude as to be a disgrace to science in this age of spectacular technical achievement" (1961, p. 570). Regarding our knowledge of the basic processes of reproduction, Sir Solly Zuckerman (1959, p. 1263), the distinguished British anatomist, summed up the 1959 Conference on Mechanisms Concerned with Conception sponsored by the Population Council and the Planned Parenthood Federation of America in these words: "Vast areas of the subject are still cloaked in an ignorance which prevents a rational and scientific approach to the problem of population control. . . . The subject . . . is still littered with legends."

**The Transition Period, 1961 to 1965**

The early 1960s witnessed increased advocacy of support for research related to population growth, including studies aimed at development of more effective fertility regulating agents, and in the United States, the inauguration of a new President more favorably disposed to research in the field. Public pressure led the National Institutes of Health to initiate in 1961 a survey of research on reproduction in any way related to fertility control. The initial version of the Albritton report (named after its principal author, Errett Albritton) listed U.S. research on fertility at $5.7 million in 1961, of which NIH contributed $1.7 million (Broder 1962, 1963). Furthermore, it recommended that NIH provide nearly $17 million annually for fertility research, plus $4 million in capital costs for eight population research institutes at various university and medical centers. Such strong advocacy of reproductive research caused alarm in the higher echelons of the Department of Health, Education and Welfare, and the report became known to the public only through news leaks. Under public and congressional pressure, a report altered to omit the funding recommendations was finally released at the end of 1962. By NIH's reckoning, the report cited some 198 relevant projects supported by $3.4 million of NIH funds, and an additional 560 projects receiving $4.7 million from other federal agencies, the private foundations, and industry (U.S., DHEW 1963). Although the criteria for relevance utilized in the report are subject to question, the prevailing attitude at NIH is clearly revealed in the text, which characterized the situation in terms quite different from those of Parkes: "The pressures and objectives for research on birth and population control are somewhat different from those involved in other types of biomedical research. *The problem requiring solution is not one of physical disease.* There would seem to be little doubt that effective contraceptive techniques are available. A research program to find new techniques ... *is not an objective toward which the NIH has a planned effort*" (U.S. DHEW 1963, p. 27) (emphasis added). However, in 1963 the National Institute of Child Health and Human Development was established with a broad mandate to study the processes of human development, emphasizing such adverse outcomes of pregnancy as mental retardation. NICHD's mission, in the view of the Kennedy administration, included the reproductive sciences and contraceptive development and, as stated explicitly several years later,

"... a responsibility for improving family planning methods as a partial solution to the problem ... and more importantly—to investigate the health aspects of various family planning techniques" (U.S. Congress 1967, p. 47). However, NICHD funds for reproductive research and contraceptive development were slow in coming. That the NICHD's initial level of commitment for the study of fertility regulating agents was less than adequate is implicit in the recommendations adopted unanimously in 1965 by NICHD's Advisory Council.

The role of NIH in general, and of the NICHD in particular ... is clear. There is a compelling need for research on the basic phenomena of reproductive biology and of human reproduction on which can be based safe, acceptable and efficient means of family planning. ... The NIH must *openly and willingly* express its readiness to explore all the significant scientific questions relating to human reproduction. ... It is the responsibility of the Reproductive Biology Program of NICHD to *initiate, expand* and *improve* research in reproduction. ... It is most *urgent* that NICHD take steps to *stimulate and support increased research* on problems of human fertility, sterility and family planning. This effort should be appropriate to the magnitude of the problem. (emphasis added)

The admonition, couched in unusually strong language for such a committee, apparently went unheeded for at least two more years.

While the U.S. government hesitated, private philanthropies moved forward. In 1963, the Rockefeller Foundation formally declared population to be one of its five areas of prime interest, indicating the intention to enlarge its earlier programs in this field. In the same year, the Ford Foundation raised its work in population to the status of a full program. Within its laboratories at the Rockefeller Institute, the Population Council continued to expand its program focused on fertility control; and it was this program, rather than that of NIH, that was primarily responsible for an extensive international effort to reevaluate the intrauterine device as a method of contraception and for the development of several current IUDs. Interest in IUDs was revived in 1959 with the publication of reports on their use and effectiveness by Israeli and Japanese investigators (Oppenheimer 1959; Ishihama 1959) and the council began to support both developmental studies (particularly those utilizing the new inert plastics) and clinical tests of effectiveness and safety. In the first half of the 1960s, the council invested almost $950,000 in this effort and sponsored two international conferences on intrauterine contraception (Tietze and Lewit 1962; Segal, Southam, and Shafer 1964).

The United Nations added its support in 1962 in a resolution entitled "Population Growth and Economic Development." Developing countries were offered assistance, on request, in the form of demographic and statistical studies. Although no explicit mention was made of research for improved fertility control agents, the World Health Organization took its first step in the field of human reproduction in 1963 by convening a Scientific Group on the Biology of Human Reproduction to review the state of knowledge, identify major needs, and recommend ways and means of stimulating research. The cautious debut of WHO into this still sensitive area was reflected in the Scientific Group's limited recommendations that WHO should assist in the development of fundamental knowledge of human reproductive biology and convene meetings of appropriate groups of specialists to consider methods of servicing and promoting specific research projects. Within two years, the Health Assembly noted that the 1965 United Nations Population Conference attached high priority to research and other activities in the field of reproduction and fertility. In that year, WHO convened a group to consider the clinical aspects of oral gestagens and in the following year another scientific group met to review the basic and clinical aspects of intrauterine devices. This early program was initiated by WHO's Maternal and Child Health Unit. In 1965 a new Unit in Human Reproduction was established to handle these activities and initiate a grants-in-aid program for the support of studies in reproduction with emphasis on the epidemiological and health aspects of human fertility.

The first half of the 1960s thus witnessed conflicting trends: On the one hand, burgeoning concern over population growth in both governmental and philanthropic circles led to increased funding levels in all sectors. At the same time, however, the entry of governmental and international agencies into the field, under constant public and political pressure, was marked by hesitation and timidity. These trends are reflected in the surviving funding data assembled in table 10.3. Significantly higher levels of funding became available to the field during this period; governmental funds began to support reproductive research for the first time, and the level of philanthropic support increased nearly seven times over that made available from 1956 to 1960. Compared to earlier periods, support for reproductive research thus mushroomed from 1961 to 1965. Compared to investment in other fields of medical research, however, the increase for reproductive research was considerably less spectacular. The U.S. government's contribution of $9.3 million in 1965 still represented less than 1 percent of the federal billion dollar medical research budget (NIH 1975). In that same year, contributions from U.S. private philanthropy equaled 4 percent of total private outlays for medical research and U.S. pharmaceutical firms expended approximately $10 million, nearly 3 percent of their ethical drug research budget (PMA 1975).

### The Past Decade—Government Becomes Active

The year 1965 marked a turning point in the U.S. government's policy. In his State of the Union Message, President Johnson proposed to

Table 10.3 Reproductive Research Expenditures in the U.S., 1961-1965 (current dollars in millions)

|      | U.S. Government | Philanthropy | Total |
|------|-----------------|--------------|-------|
| 1961 | 1.7             | 4.0          | 5.7   |
| 1962 | 2.0             | 1.0[a]       | 3.0   |
| 1963 | 3.4             | 4.7          | 8.1   |
| 1964 | 2.7[b]          | 2.3[a]       | 5.0   |
| 1965 | 9.3             | 6.6          | 15.9  |
| Total| 19.1            | 18.6         | 37.7  |

[a] Includes only the Rockefeller and Ford Foundations and the Population Council.
[b] Includes NICHD only.
Source: U.S. government agencies, Rockefeller and Ford Foundations, Population Council, and other philanthropic agencies in the United States. In addition, WHO reported spending $19,000 in 1964 and $55,000 in 1965.

"... seek new ways to use our knowledge to help deal with the explosion in world population and the growing scarcity in world resources" (Johnson 1965, p. 16). This was the first of scores of presidential references to the gravity of the situation. The Food for Freedom Act was amended a year later to "... make available resources to promote voluntary activities in other countries dealing with the problem of population growth and family planning" (U.S. Congress 1966, p. 4). In 1967, the Office of the War on Hunger, including a population program, was established within the Agency for International Development (AID) and Congress took three actions on both the international and domestic fronts: It earmarked $35 million for AID programs relating to population growth, including research, information services, and provision of medical assistance and supplies; it accorded family planning priority within the "war on poverty"; and it earmarked no less than 6 percent of maternal and child health funds for domestic family planning programs.

The advent of an explicit AID population program was accompanied by pressure from outside AID for funding of basic research in reproduction. As a result, AID was permitted to finance research projects that "are demonstrably relevant to LDC (less developed country) needs" (Gaud 1968). AID's first contract in 1967 awarded $194,000 to the Pathfinder Fund for field trials of IUDs. In the subsequent years most of its research funds were dispensed for support of leads AID considered promising, such as prostaglandins, analogs of releasing factors, and antiprogestational agents. With approximately 5.6 percent of the AID population budget allocated for biomedical research, this amounted to over $6.3 million in 1974 (AID 1975).

In terms of the U.S. government's normal mechanisms for support of biomedical research (NIH), AID funding was atypical. The demands on AID to finance reproductive research reflected the feeling of many that the government's effort still was not commensurate with the size of the population problem. In 1967 DHEW's Assistant Secretary for Health and Scientific Affairs asked a group of outside consultants to review the status of the department's overall programs in research and delivery of services. The ensuing report concluded,

The volume of research and training [in the reproductive sciences and contraceptive development] supported through the NICHD programs and through . . . other sections of NIH does not adequately meet the world's need for speedy development of radically improved contraception. . . . If NIH is to meet its obligation to encourage actively the expansion of research in reproductive biology directed towards the development of superior methods of contraception, it must act much more aggressively, with larger staff and budget. Throughout its history, NIH has made intensive efforts to encourage the development of one special field or another . . . but it has yet to mount a special effort [in reproductive research] comparable, for example, with its concern with mental retardation. (Harkavy, Jaffe, and Wishik 1967, pp. 178-179)

The report, coupled with continuing congressional pressure and the conclusions of President Johnson's Committee on Population and Family Planning, led in 1968 to the establishment within NICHD of a Center for Population Research. CPR has since become the principal U.S. agency in the field and indeed, the single largest source of support for reproductive research and contraceptive development in the world. Originally the center was assigned responsibility for organizing a program of contract research and providing a focus for information exchange and coordination of the efforts of the various federal agencies. In July 1969, CPR's responsibilities were increased to include administration of the grants program in basic reproductive research, thus centralizing in the center responsibility for most of the institute's extramural programs in the population field.

The recommendations of President Johnson's committee were reinforced by President Nixon in his 1969 Population Message when he called for "additional research in birth control methods of all types and the sociology of population growth (Nixon 1969, p. 17). DHEW, through the Center for Population Research, was urged to "take the lead in developing, with other federal agencies, an expanded research effort, which is carefully related to those of private organizations, university research centers, international organizations and other countries" (Nixon, 1969, p. 17). These varied initiatives led to enactment of the Family Planning Services and Population Research Act of 1970, the only population-related statute adopted by the U.S.

Congress. This act, which was renewed in 1975, authorized increased levels of funding for both the research and service programs.

In 1965 another federal agency, the Food and Drug Administration, had become increasingly concerned about reports on the adverse effects of oral contraceptives. An Advisory Committee on Obstetrics and Gynecology was formed to look into this and other matters related to drugs and devices used to prevent pregnancy. An initial report by that committee was published in 1966, followed by a final version in 1969. The initial report resulted in the 1967 allocation of $1 million by Congress to NICHD to award contracts to evaluate the medical effects of the oral contraceptives. The FDA has since sponsored and conducted research on the drug safety of compounds used, or likely to be used, for contraceptives.

By the late 1960s, the U.S. government was thus committed to research leading to new birth control agents. This policy shift, however, coincided with diminishing public support of science as a major means for solution of social problems. As a consequence perhaps, funding levels for most fields of medical research sponsored through NIH have decreased even though an overview of NIH appropriations shows a 28 percent increase in real dollars over the last six years (1969 to 1975). This upward trend, however, is reflected primarily in the funding for the Cancer and Heart and Lung institutes. The National Institute for Child Health and Human Development has maintained an approximately equal level since 1972 in real dollars. This is due in part to the establishment of CPR within the institute and the 1970 legislation assuring more funds for research and service in family planning. Appropriations for other institutes have actually decreased by 11 percent since 1969.

The historic taboos also delayed the emergence of active programs in population research on the international level until the late 1960s. Food shortages in the mid-1960s precipitated a United Nations General Assembly resolution of 1966 calling upon the entire UN system to give assistance when requested "in further developing and strengthening national and regional facilities for training, research, information and advisory services in the field of population ... " (UNFPA 1972, p. 55). In response to this and continuing pressure from concerned activists, the Secretary-General of the United Nations established a trust fund in 1967 to finance work in the population field. Over the next two years $5 million in voluntary government contributions supported U.N. projects, primarily to strengthen and improve statistical and demographic capabilities in developing nations. Early in 1969, the fund was renamed the United Nations Fund for Population Activities (UNFPA) and its mandate was broadened to include support of population activities carried out by all agencies within the U.N. system and by appropriate nongovernmental bodies. By 1972 the UNFPA had received the endorsement of the General Assembly. Pledges from donor governments increased to a total of $70.5 million in 1974. Statistical and demographic studies, family planning, population dynamics, and policy are among the areas supported; but research for improved fertility regulating agents has received little emphasis, never exceeding 0.5 percent of UNFPA's budget.

The World Health Organization, on the other hand, has played an increasing role in support of research to assess existing methods of fertility regulation, improve them, and develop a variety of new methods particularly for use in the developing countries. WHO launched its Expanded Programme of Research, Development and Research Training in Human Reproduction with a budget of $4.5 million contributed by governmental agencies and the Ford Foundation (1972). The Expanded Programme is now a major sponsor of research and training in human reproduction and family planning with a budget that reached $12 million in 1975. Contributions to the WHO Expanded Programme have come from Canada, Denmark, Finland, Mexico, Norway, Sweden, the United Kingdom, as well as from UNFPA, the IDRC of Canada, and the Ford Foundation. This program is mission-oriented and places emphasis on the development and application of new and improved methods of fertility regulation. In addition it supports fundamental studies in specified areas deemed to have particular relevance to the control of fertility (WHO, 1973, 1974, 1975).

Three other European countries are known to have government programs specifically directed to funding reproductive research. As with the U.S. government, their commitment to the field came late in the 1960s. In 1968 the Medical Research Council of Britain determined that obstetrics and gynecology should be regarded as one of the disciplines "requiring special encouragement" in view of the inadequacy of the existing level and standards of investigation (MRC 1970). As a result, the MRC Reproductive Biology Unit was established in Edinburgh as a research institute devoted to the reproductive sciences. But it was not until 1972 that the MRC concluded that research on population control requires special attention and support. The present policy (1974) is to continue support for fundamental research and to encourage further applied research of immediate relevance to population control (MRC 1974). However, since the economic downturn of 1969-1970, government support of medical research in Great Britain has either declined or had minimal year-to-year increases. Correcting for inflation in Britain, total MRC expenditures were actually 10 percent less in 1974 than in the previous year; reproductive research funds from MRC received the same cut (Hoult 1975).

The Deutsche Forschungsgemeinschaft (DFG) initiated a priority program in the physiology and pathology of reproduction in 1968, granting somewhat more than $1 million to the effort for five years. The purpose of the program is not considered to be fertility control but rather a better understanding of infertility. Another West German federal agency, the Ministry for Research and Technology, sponsored a pilot study, begun in 1974, on the side effects of oral contraceptives. Slightly more than half a million dollars has been allocated to this project (Voigt and Wollmer 1975).

A department of the French Ministry of Industry and Research, the Délégation générale à la Recherche Scientifique et Technique (DGRST), established a committee in 1973 to stimulate new programs in the field of reproduction and development. The total budget for the past five years, including 1975, was $4.5 million. If the committee is retained for five more years, it intends to upgrade research, particularly on the use of the oral contraceptive pills by very young girls at the time when hypothalamic cycling begins to operate and on gametogenesis and fertilization (Thibault 1974).

In the other industrialized countries surveyed, reproductive research is carried out with funds intended for general biomedical research. These governments generally fund the universities, which in turn allocate funds to various departments. Research in reproduction is usually carried out in departments of obstetrics and gynecology by investigators whose own interests determine the direction of their work.

Of those canvassed, India is the only developing country that places special emphasis on the study of the reproductive sciences. In the early 1960s, the government provided increased funds for such research, primarily through the Indian Council of Medical Research. Budget restrictions during 1973-1974 caused the tempo of research to slow down considerably; this was partially reversed by the end of 1974 (Prasad 1974).

In the other developing countries surveyed (Iran, South Korea, Egypt, Thailand, the Philippines, and Turkey), UNFPA, WHO, or the Ford Foundation have been instrumental in stimulating research by helping to fund laboratories specializing in the study of the reproductive sciences. Governments of these countries finance maintenance of laboratory buildings and salaries of the researchers, but the research usually requires outside subsidy by international assistance agencies. The research carried out is, for the most part, clinically oriented, directed to the testing of new contraceptives.

Thus, the historic taboos surrounding research in the reproductive sciences and contraceptive development *began* to be lifted only during the first half of the 1960s—and only as a result of determined and sustained pressure from the public and the political community. Developments varied in different nations, but it seems clear that only during the last half of the 1960s were the taboos lifted sufficiently to enable the medical research establishments in the United States and other industrialized nations to enter the field. As a result, important population research programs emerged in three U.S. foundations, a U.S. government agency, six foreign assistance agencies, a multilateral health agency, and the medical research arms of five other nations.

# References

Aberle, S.D., and Corner, G.W. 1953. *Twenty-five years of sex research.* Philadelphia: W.B. Saunders Co.

Agency for International Development (AID), Office of Population. 1975. *Population program assistance: United States aid to developing countries.* Washington, D.C.: Government Printing Office.

Agriculture Research Administration, Bureau of Animal Industry. 1948, 1950, 1952, 1954. *Budget of the U.S.* Washington, D.C.: Government Printing Office.

The American Foundation. 1955. American medical research: In principle and practice. In *Medical research: A midcentury survey*, vol. I. Boston: Little, Brown and Co.

Battelle. 1973. Oral Contraceptives. In *Interactions of science and technology in the innovative process: Some case studies.* Contract NSF-C667. Columbus, Ohio: Battelle.

Broder, D. *Washington Sunday Star.* 30, December 1962.

―――. *Washington Evening Star.* 25, April 1963.

Djerassi, C. (Stanford University). 1975. Personal communication.

Duncan, G.W. 1974. A survey of contraceptive research and development. Report commissioned for the reproduction review by the Ford Foundation.

Gaud, W.S. (Agency for International Development). 1968. Personal communication with David Bell.

Gibian, H. (Schering AG). 1975. Personal communication.

Goldblum, J.F. (National Academy of Sciences). 1975. Personal communication.

Graff, M. (National Institutes of Health). 1975. Personal communication.

Harkavy, O. (Ford Foundation). 1975. Personal communication.

Harkavy, O.; Jaffe, F.S.; and Wishik, S.M. 1967. *Implementing DHEW policy on family planning and population.* In *Hearings before the Subcommittee on Foreign Aid Expenditures of the Committee on Government Operations*, U.S. Senate, Ninetieth Congress, S. 1676, Part 1, pp. 163-206. Washington, D.C.: Government Printing Office.

Hertz, R. (George Washington University Medical Center). 1975. Personal communication.

Hill, R. (National Institute of Child Health and Human Development, Reproductive Physiology Section). 1975. Personal communication.

Hoagland, H. (Worcester Foundation for Experimental Biology). 1975. Personal communication.

Horowitz, B. (Planned Parenthood Federation of America). 1975. Personal communication.

Hoult, J.R. 1975. Reproductive biology research and contraceptive development in the United Kingdom. Report commissioned for the reproduction review by the Ford Foundation.

Ishihama, A. 1959. Clinical studies on intrauterine rings, especially the present state of contraception in Japan and the experience in the use of intrauterine rings. *Yokohama Medical Journal* 10:89-105.

Johnson, L. 1965. Transcript of the President's message to Congress on the state of the union. *New York Times*, 5 January 1965, p. 16.

Kasius, R.V. (Milbank Memorial Fund). 1975. Personal communication.

Medical Research Council. 1970. *Report—Council Subcommittee on Obstetrics and Gynecology*. (Peart Report to MRC). Mimeographed.

———. 1974. *Annual report, 1973-74*. London: Her Majesty's Stationery Office.

National Committee on Maternal Health, Inc. (NCMH). 1941. *The undeveloped medical aspects of human reproduction*. New York: NCMH.

National Institute of Child Health and Human Development (NICHD). 1965. *Recommendations of the National Advisory Child Health and Human Development Council on research in fertility, sterility, and population dynamics*. Mimeographed.

National Institutes of Health (NIH). 1975. *Basic data relating to the NIH, 1975*. Washington, D.C.: Government Printing Office.

National Institutes of Health, Office of Program Planning. 1966. *Resources for medical research*. No. 10.

National Research Council (NRC). 1945. *Summary report of the Committee on Research in Endocrinology 1936-1945*. Mimeographed. Washington, D.C.

———. 1946-1954. *Annual reports to the American Cancer Society*. Washington, D.C.: National Academy of Sciences.

National Science Foundation (NSF). 1957 and 1959. *Annual reports*. Washington, D.C.: Government Printing Office.

Nixon, R. 1969. *President's message to Congress on population*. Washington, D.C.: Population Crisis Committee.

Oppenheimer, W. 1959. Prevention of pregnancy by the Grafenberg ring method. *Am. J. Obstet. Gynecol.* 78: 446-454.

Parkes, A. 1961. The menace of overpopulation. *The New Scientist* 10:570-574.

Pharmaceutical Manufacturers Association (PMA) 1975. *Annual survey report, 1973-74*. Washington, D.C.

Population Council (PC). 1952-1960. *Annual reports*. New York: Population Council.

———. 1965. *The Population Council, 1952-64*. New York: Population Council.

Prasad, M.R.N. 1974. Research in reproductive sciences and contraceptive development in India. Report commissioned for the reproduction review by the Ford Foundation.

Rock, J. 1963. *The time has come*. New York: Knopf.

Rockefeller Foundation (RF). 1948-1961. *Annual reports*. New York: Rockefeller Foundation.

_____. 1965. *The population program of the Rockefeller Foundation.* Mimeographed.

Segal, S.J.; Southam, A.L.; and Shafer, K.D., eds. 1964. *Intra-uterine contraception: Proceedings of the second international conference, October 1964.* Amsterdam: Excerpta Medica Foundation.

Sheehan, R., and Weil-Fisher, E. 1958. Memo to the Ford Foundation. Adapted from their research for "The birth control pill." *Fortune* April 1958.

Swain, D.C. 1962. The rise of a research empire: NIH, 1930-1950. *Science* 138(3546):1233-1237.

Thibault, C. 1974. Research in reproductive sciences and contraceptive development in France. Report commissioned for the reproduction review by the Ford Foundation.

Tietze, C. (Population Council). 1975. Personal communication.

Tietze, C., and Lewit, S., eds. 1962. *Intra-uterine contraceptive devices: Proceedings of the conference, April 30-May 1, 1962.* Amsterdam: Excerpta Medica Foundation.

United Nations Fund for Population Activities (UNFPA). 1972. *Report, 1969-1972.* New York: United Nations.

U.S. Congress, House of Representatives, Committee on Agriculture. 1966. *The food for freedom act of 1966.* Report No. 1588, 89th Congress, 2nd Session, Washington, D.C.

_____, Senate, Committee on Appropriations. 1967. *Report on labor-health, education and welfare appropriations.* 90th Congress, 1st Session, Washington, D.C.

U.S. Department of Agriculture, Office of Experiment Stations. 1942-1951. *Reports on the agriculture experiment stations.* Washington, D.C.: Government Printing Office.

U.S. Department of Health, Education and Welfare. 1963. *A survey of research on reproduction related to birth and population control (as of January 1, 1963).* PHS Publication No. 1066.

U.S. *Statutes at Large.* 1935. No. XLIX.

_____. 1944. No. LVIII.

Voigt, K.D., and Wollmer, W. 1975. Research on reproductive sciences and contraceptive development in Germany. Report commissioned for the reproduction review by the Ford Foundation.

World Health Organization Expanded Programme of Research, Development and Research Training in Human Reproduction. 1972. *Report on programme implementation during the first year.* Geneva: WHO.

_____. 1973, 1974, 1975. *Annual Report.* Geneva: WHO.

Zuckerman, S. 1959. Mechanisms involved in contraception. *Science* 130:1260-1264.

# 11 The Decade of Growth: 1965-1974

The study of the reproductive sciences related to the regulation of fertility, as we have seen, was for many decades considered outside the scope of legitimate scientific inquiry. This fundamental bias limited the funds allocated to this area of investigation and made it difficult, if not impossible, for the field to develop in the same way as have other branches of biomedical research during this century. It was only with the emergence of active concern over rapid population growth in the 1960s that these historic limiting factors *began* to be reversed. As a consequence, the level of funds made available for the reproductive sciences increased beginning in the mid-1960s. The resulting expansion of work in laboratories in many countries is responsible for the improved knowledge of the human reproductive process that has been summarized in chapters 4 and 5. Although the recent increase in funds has enabled a larger number of researchers to work on reproductive problems and to train for careers in the field, it is questionable if it has redressed the imbalances caused by decades of taboo, stigmatization, and underfinancing. It is also doubtful if the current level of support is commensurate with the importance of the field to individuals and nations or with the exponential growth of world population or if it is adequate compared to support of other fields of medical and scientific inquiry.

To attempt to answer these questions requires various kinds of information. The state of knowledge in the field is an important consideration and much of this review is devoted to such an assessment. Another factor is the field's capacity to expand significantly its current level of activity to take advantage of new findings. Finally, accurate information on the level of funding currently available to the field is critical.

In the last decade and a half, there have been several efforts to assemble information on the amounts available to support research in reproductive sciences and contraceptive development. In 1961 and 1962, as noted in chapter 10, the U.S. National Institutes of Health carried out a study of the funds available for this research from sources in the United States and abroad (U.S. DHEW 1962, 1963). In subsequent years, inventories of research grants thought to be related to the field became the yearly task of the Interagency Committee on

Population Research (U.S. DHEW 1968, 1969a, 1970, 1971a, 1971b, 1972, 1973a, 1973b, 1974); and other investigators made attempts to gather pertinent funding data (Harkavy and Maier 1970, 1971; U.S. Congress 1967; U.S. DHEW 1969b). These studies added useful information, but each had acknowledged limitations. All were carried out by U.S. investigators and concentrated to a great extent on expenditures by U.S. governmental agencies and foundations. Support from other industrialized and developing nations was omitted from these analyses except where data were readily available. Expenditures by pharmaceutical companies—thought by some to be considerable—proved to be largely inaccessible to the previous investigators and thus also had to be omitted or crudely estimated.

A systematic survey was conducted for this review to determine past and present levels of funding for the reproductive sciences and contraceptive development and estimates of expected levels in the next several years. Information on the purpose and annual amounts of grants, and names and locations of grantees, was assembled from each of the nine U.S. government and U.N. agencies that allocate $1 million or more annually to the field, as well as from four U.S. private foundations with expenditures in this range.[1] Consultants in thirty-eight industrialized and developing nations and geographic areas where the level of reproductive research is known to be more than minimal were asked to obtain similar data from their respective governmental and private agencies. The support of contraceptive development provided by U.S. pharmaceutical firms was obtained through a special survey conducted by Dr. Gordon Duncan of the Battelle Institute, and this questionnaire was also distributed among firms outside the United States.

Despite the difficulties associated with international studies of this sort, the responses from government and private agencies and firms

[1] *Government and Multilateral Agencies:* National Institute of Child Health and Human Development - NICHD, National Institutes of Health - NIH (excluding NICHD), National Institute of Mental Health - NIMH, National Science Foundation - NSF, Atomic Energy Commission - AEC, Food and Drug Administration - FDA, Agency for International Development - AID, World Health Organization - WHO, United Nations Fund for Population Activities - UNFPA; *Philanthropic Agencies:* Ford Foundation - FF, Rockefeller Foundation - RF, Population Council - PC, Scaife Family Charitable Trust.

were unusually complete. Information was obtained from all the major agencies in the field, as well as from sixteen industrialized and nine developing nations and geographic areas.[2] The principal omissions relate to the levels of relevant research activity in the Union of Soviet Socialist Republics, the nations of Eastern Europe, and the People's Republic of China.[3] The Western nations for which data are lacking, with the exception of Mexico, are known to have relatively low levels of activity in reproductive research. In Mexico it is estimated that various government agencies and pharmaceutical firms support the reproductive sciences at the level of $2 million to $3 million annually.[4]

A more significant bias in the data derives from definitional difficulties and differences in methods of accounting for research expenditures. In an effort to minimize these errors, respondents were provided with definitions of laboratory and clinical research subjects in the reproductive sciences and contraceptive development that had been identified as relevant to this study (see appendix C). The thirteen major funding agencies supplied lists of grants, including each project's title, major investigator, location, and annual funding level. Based on these data, review staff coded each grant for computer tabulation (see appendix D for expenditures of each funding agency). The consultants located outside the United States were asked to ana-

[2] *Countries responding:* United States; *Other Industrialized Countries:* Belgium, Britain, Canada, France, Germany, Israel, Japan, Italy, Netherlands, Australia, New Zealand, Scandinavian countries—Denmark, Finland, Norway, and Sweden; *Developing Countries:* Egypt, Hong Kong, India, Philippines, Thailand, South Korea, Equatorial Africa, Iran, and Turkey.

[3] *Countries not responding:* Seven Eastern European countries, Mexico, Central and South America, the People's Republic of China, USSR, Indonesia, Malaysia, Singapore, Sri Lanka. In regard to South Africa, Dennis A. Davey, Ph.D., Professor of Obstetrics and Gynecology of the University of Cape Town, responded that "A number of clinical trials of different contraceptives (particularly oral contraceptives and intra-uterine devices) have been carried out but it is my impression that no real research has been carried out in reproductive biology and contraceptive development except possibly in my own department in Cape Town."

[4] It is also estimated that approximately $500,000 per year is spent on reproductive research in Argentina; $250,000 in Colombia, Peru, Uruguay, and Chile combined; and about $115,000 in Brazil. These amounts exclude outside support (Perkin 1975).

lyze the grant lists of their respective governmental and private agencies in terms of the review's definitions of relevant topics and to report only on funding originating from sources internal to each country (see appendix C for expenditures of each country represented). Nevertheless, it is likely that biomedical studies only tangentially related to the reproductive sciences and contraceptive development as conceptualized in this review have been included in the totals. These factors tend to inflate the amounts reported here as made available for support of work aimed at improved regulation of fertility.

Also tending to inflate the amounts reported here are changes in university accounting practices during the last decade. In 1971, for example, CPR awarded $10.5 million for research grants (Projected Program Projects and Center Grants) and training activities; of this total, $2.3 million, or 22 percent, constituted "indirect costs" of management administration. In 1974, however, "indirect costs" absorbed $7 million, or 32 percent, of CPR's total expenditures of $22.1 million (Douglas 1975). Over the decade, therefore, the proportion of total dollars awarded that actually financed research activities in the laboratory declined significantly and the proportion supporting other university costs increased. To determine how much was actually available to finance research would require information for each agency, nation, and year on the amounts spent on indirect costs (as well as on the accounting bases employed in their computation). These data are not available and such an analysis could not be done.

An offsetting difficulty is the fact that the level of support of reproductive research (and other medical research) derived from university budgets in the form of salaries and laboratory maintenance could not be specified in seven of the twenty-five countries and geographic regions on which data have been received. The nature of this indirect support varies from country to country. Failure to include contributions from university budgets thus tends to understate the amounts reported here as directed to research in reproduction. However, there have also been changes in university practices here that make it difficult to determine accurately the funds actually available to pay for research work. In the United States, for example, the typical practice in the past was for a professor's salary to be paid by the university. Some observers point out that in recent years, universities have increasingly demanded that research grants include faculty salaries previously borne by university budgets. As a result, the burden of faculty support has increasingly been shifted from the universities to the funding agencies and from less visible to more visible sources. The net effect is that the amounts actually available to finance research have increased less than the totals reported here indicate.

On balance, it is our judgment that the figures reported here tend to overstate the amounts that have actually been made available to support the reproductive sciences and contraceptive development and should be read as indicating the maximum level of support available in Western nations for this field. Our main interest, however, will be on the overall trends of support over the last decade and the distribution of emphasis within the field. We feel confident that these patterns, as reported here, are accurate.

The amounts allocated to research support will be presented in two ways in this chapter: First, in terms of current U.S. dollars at exchange rates prevailing in the specified year; and second, in order to provide a more reliable estimate of the level of research activity the dollar amounts purchased in a period of acute inflation, in terms of constant dollars, using 1970 as the base year.[5]

Overview

Table 11.1 shows the total amounts spent on research in the reproductive sciences and contraceptive development by the United States, fifteen other industrialized nations, and nine nations and regions in the developing world that support studies in this field. In 1965, government and international agencies, private foundations, and the pharmaceutical industry allocated slightly more than $31 million for these areas of inquiry. Ten years later, the worldwide aggregate had increased nearly fourfold to about $119 million. These

[5] For all countries, the consumer price indices published in the *U.N. Statistical Handbook* were used to make the adjustments. Values for 1974 were communicated by Mr. M. Pardo of the U.N. Statistics Office.

Table 11.1 Total Expenditures for Research in the Reproductive Sciences and Contraceptive Development by Country of Origin, 1965, 1969-1974 (current and constant U.S. dollars in 000s, 1970 = 100)

| Country of Origin[a] | 1965 | 1969 | 1970 |
|---|---|---|---|
| **Total All Countries** | | | |
| Current Dollars | 31,022 | 54,546 | 71,462 |
| Constant Dollars | 38,270 | 57,893 | 71,462 |
| **United States** | | | |
| Current Dollars | 25,928 | 41,537 | 55,009 |
| Constant Dollars | 32,010 | 44,188 | 55,009 |
| **Other Industrialized Countries (15)** | | | |
| Current Dollars | 4,886 | 11,992 | 15,267 |
| Constant Dollars | 5,992 | 12,634 | 15,267 |
| **Developing Countries (9)** | | | |
| Current Dollars | 208 | 1,017 | 1,186 |
| Constant Dollars | 268 | 1,070 | 1,186 |
| **Percent Distribution** (based on constant U.S. dollars) | | | |
| All Countries | 100.0 | 100.0 | 100.0 |
| United States | 83.6 | 76.4 | 77.0 |
| Other Industrialized Countries | 15.7 | 21.8 | 21.3 |
| Developing Countries | 0.7 | 1.8 | 1.7 |

[a]*Countries:* United States; *Other Industrialized Countries:* Belgium, Britain, Canada, France, Germany, Israel, Japan, Italy, Netherlands, Australia, New Zealand, Scandinavian countries—Denmark, Finland, Norway, and Sweden; *Developing Countries:* Egypt, Hong Kong, India, Philippines, Thailand, South Korea, Africa, Iran, Turkey. (Mexico, not included above, supports reproductive research and contraceptive development at the estimated level of $2-3 million annually [Perkin 1975].)

figures are in current dollars. When the impact of inflation is taken into account, the increase in actual spending levels is considerably less; in 1974, the purchasing power of funds for the reproductive sciences and contraceptive development was only a little more than twice that in 1965, a year marked by low funding levels following a half-century of underfinancing. The increase in the last decade has thus been a real one, but much less than the raw figures would indicate. The annual rate of increase in funding from all sources was on the order of 25 percent from 1969 through 1971, but the 1972 total was only 12 percent greater than the previous year's. In 1973, there was a slight increase and in 1974, there was a sharp decline of nearly

| 1971 | 1972 | 1973 | 1974 |
|---|---|---|---|
| 94,483 | 110,224 | 117,430 | 118,723 |
| 90,390 | 100,825 | 100,957 | 91,146 |
| 71,663 | 79,724 | 82,070 | 79,104 |
| 68,906 | 73,819 | 71,991 | 62,287 |
| 21,126 | 28,811 | 33,039 | 38,029 |
| 19,908 | 25,505 | 27,093 | 27,771 |
| 1,694 | 1,689 | 2,321 | 1,590 |
| 1,577 | 1,501 | 1,873 | 1,088 |
| 100.0 | 100.0 | 100.0 | 100.0 |
| 76.3 | 72.4 | 69.9 | 68.3 |
| 22.0 | 26.1 | 28.1 | 30.5 |
| 1.7 | 1.5 | 2.0 | 1.2 |

10 percent in total support, based on constant dollars.

The bulk of the funds for this field has come from U.S. sources. Throughout the last decade, U.S. government, philanthropic agencies, and pharmaceutical firms have provided from two-thirds to five-sixths of the support available worldwide. Within this general pattern, however, the proportion provided by U.S. agencies has declined from 84 to 68 percent, and the share of the fifteen other industrialized nations has increased from 16 to 31 percent. As would be expected, the proportion provided by the developing nations themselves has been small; but it has also increased over the ten-year period. In constant dollars, the amount from developing countries has in fact quadrupled, that from the fifteen other industrialized nations has more than quadrupled, and the U.S. total has not quite doubled. U.S. funds peaked in 1972 and declined 3 percent in 1973 and 14 percent in 1974. In the other industrialized nations, expenditures in the field doubled between 1969 and 1972, but increased little in 1973 and 1974. Since the inception of the WHO Expanded Programme of Research, Development and Research Training in Human Reproduction in 1972, an increasing proportion of the total funds spent by the other industrialized nations for this type of research, particularly by Sweden, Norway, Finland, Denmark, and Canada, has been contributed through this program.

Sources of Support

The respective levels of governmental, philanthropic, and industrial funding for reproductive research over the decade are presented in Table 11.2. In constant dollars, governmental funds increased from $16.9 million in 1965 to $70.9 million in 1973 and then declined to $63.2 million in 1974. Philanthropic funds rose from $8.3 million in 1965 to $23 million in 1971 and declined to $13 million in 1974. The amounts provided by pharmaceutical firms over the decade have been more uniform—$13.1 million in 1965, increasing to $17.4 million in 1972, and declining to $14.9 million in 1974. In each year of the decade, governmental agencies have provided the largest share of funds for the field, but the pattern has changed markedly. The governmental share was a little less than half of the total in 1965 but increased to more than two-thirds in 1974. This change is a result of two other trends that have great significance for future funding patterns: The share borne by a handful of private foundations, which essentially were the main source of support for the field until the mid-1960s, has dropped from 22 percent in 1965 to 14 percent in 1974. An even greater proportional decline has been registered by pharmaceutical firms. A decade ago, when contraceptive development appeared to offer substantial profits, pharmaceutical companies accounted for more than one-third of the funds entering the field; in 1974, their share had dropped to 16 percent.

Because the United States accounts for such a large proportion of worldwide support for the field, the patterns of U.S. expenditures are similar to the worldwide trends. Different patterns prevail in the other industrialized nations and the developing world. Private foundations and pharmaceutical firms account for a much smaller share of funding in the other industrialized nations than in the United States; in these countries, governmental agencies have provided between 83 and 90 percent of funds in the last decade. Virtually all of the available funds in the developing nations have come from government sources.

Government agencies, therefore, through direct grants to investigators or by contributing to such multilateral agencies as the World Health Organization and the U.N. Fund for Population Activities, have become the principal source of support for the field in all parts of the world. From the information assembled in this review, it seems clear that this will continue to be the case, not simply because government has become the principal support for medical research of all kinds. The few private foundations that provided most of the seed money to launch this area of scientific inquiry have had to work with increasingly limited funds as a consequence of inflation and the vagaries of the equity markets in industrial nations. In 1974, for example, the Ford Foundation was forced to announce that it plans to reduce by half its grants in all fields in 1978. At the same time, the level of spending by pharmaceutical firms has declined at least partly as a result of revised anticipations of profits resulting from the increasingly high cost of bringing new contraceptives to the market.[6] These factors make evident that the future of the field depends largely on the policies adopted by governmental research establishments, primarily in the United States and other developed nations.

A closer look at these patterns, nation by nation, is therefore in order. At least eight U.S. government agencies have provided some degree of support for the field, but the bulk of U.S. government funds comes from three: the Center for Population Research of the National Institute of Child Health and Human Development, other branches of the National Institutes of Health, and the U.S. Agency

[6] For a report on current thinking of U.S. pharmaceutical firms, see Karp 1975.

Table 11.2 Total Expenditures and Percent Distribution for Research in the Reproductive Sciences and Contraceptive Development by Country of Origin and Sector, 1965, 1969-1974 (constant U.S. dollars in 000s; 1970=100)

| Country of Origin and Sector | 1965 | 1969 | 1970 |
|---|---|---|---|
| Totals[a] | 38,270 | 57,893 | 71,462 |
| Government | 16,865 | 31,769 | 39,099 |
| Philanthropy | 8,336 | 11,317 | 17,345 |
| Industry | 13,068 | 14,808 | 15,018 |
| United States (Total) | 32,010 | 44,188 | 55,009 |
| Government | 11,423 | 19,928 | 25,307 |
| Philanthropy | 8,153 | 11,059 | 16,796 |
| Industry | 12,433 | 13,201 | 12,906 |
| Other Industrialized Countries (15) | | | |
| (Total) | 5,992 | 12,634 | 15,267 |
| Government | 5,183 | 10,787 | 12,620 |
| Philanthropy | 183 | 252 | 549 |
| Industry | 626 | 1,596 | 2,098 |
| Developing Countries (9) | | | |
| (Total) | 268 | 1,070 | 1,186 |
| Government | 259 | 1,054 | 1,172 |
| Philanthropy | 0 | 6 | 0 |
| Industry | 8 | 11 | 14 |

[a]Columns may not add up to totals due to rounding.

| Percent Distribution | 1965 | 1969 | 1970 |
|---|---|---|---|
| All Countries (Total) | 100.0 | 100.0 | 100.0 |
| Government | 44.1 | 54.9 | 54.7 |
| Philanthropy | 21.8 | 19.5 | 24.3 |
| Industry | 34.1 | 25.6 | 21.0 |
| United States (Total) | 100.0 | 100.0 | 100.0 |
| Government | 35.7 | 45.1 | 46.0 |
| Philanthropy | 25.5 | 25.0 | 30.5 |
| Industry | 38.8 | 29.9 | 23.5 |
| Other Industrialized Countries (15) | | | |
| (Total) | 100.0 | 100.0 | 100.0 |
| Government | 86.5 | 85.4 | 82.7 |
| Philanthropy | 3.1 | 2.0 | 3.6 |
| Industry | 10.4 | 12.6 | 13.7 |
| Developing Countries (9) | | | |
| (Total) | 100.0 | 100.0 | 100.0 |
| Government | 97.0 | 98.5 | 98.8 |
| Philanthropy | 0.0 | 0.5 | 0.0 |
| Industry | 3.0 | 1.0 | 1.2 |

| 1971 | 1972 | 1973 | 1974 |
|---|---|---|---|
| 90,390 | 100,825 | 100,957 | 91,145 |
| 50,293 | 66,167 | 70,924 | 63,247 |
| 23,024 | 17,271 | 14,562 | 12,962 |
| 17,074 | 17,387 | 15,471 | 14,936 |
| 68,906 | 73,819 | 71,991 | 62,287 |
| 32,034 | 43,250 | 44,578 | 37,980 |
| 22,585 | 16,876 | 14,161 | 12,662 |
| 14,288 | 13,694 | 13,252 | 11,645 |
| 19,907 | 25,505 | 27,093 | 27,771 |
| 16,701 | 21,430 | 24,494 | 24,179 |
| 435 | 395 | 393 | 300 |
| 2,772 | 3,680 | 2,206 | 3,291 |
| 1,577 | 1,501 | 1,873 | 1,088 |
| 1,558 | 1,487 | 1,852 | 1,075 |
| 4 | 0 | 8 | 0 |
| 14 | 14 | 13 | 13 |

| 1971 | 1972 | 1973 | 1974 |
|---|---|---|---|
| 100.0 | 100.0 | 100.0 | 100.0 |
| 55.6 | 65.7 | 70.3 | 69.4 |
| 25.5 | 17.1 | 14.4 | 14.2 |
| 18.9 | 17.2 | 15.3 | 16.4 |
| 100.0 | 100.0 | 100.0 | 100.0 |
| 46.5 | 58.5 | 61.9 | 61.0 |
| 32.8 | 22.9 | 19.7 | 20.3 |
| 20.7 | 18.6 | 18.4 | 18.7 |
| 100.0 | 100.0 | 100.0 | 100.0 |
| 83.9 | 84.0 | 90.4 | 87.1 |
| 2.2 | 1.6 | 1.5 | 1.1 |
| 13.9 | 14.4 | 8.1 | 11.8 |
| 100.0 | 100.0 | 100.0 | 100.0 |
| 98.8 | 99.1 | 98.9 | 98.8 |
| 0.3 | 0.0 | 0.4 | 0.0 |
| 0.9 | 0.9 | 0.7 | 1.2 |

for International Development. Mounting public support for efforts to deal with population growth in the 1960s brought about increased levels of funding from these agencies, but a principal stimulus was the adoption of the Family Planning Services and Population Research Act of 1970. AID spending peaked in 1972; CPR expenditures peaked in 1973 and have been declining since as a result of budgetary constraints on spending for social programs imposed by the Nixon and Ford administrations. The allocations to this field by other U.S. government agencies, including the National Science Foundation, have been minimal. Expenditures by the U.S. Food and Drug Administration, primarily to support studies of the safety of existing contraceptives, have declined since 1972, although concern over safety among consumers has continued to mount.

Table 11.3 summarizes total expenditures by government and philanthropic agencies in the United States and fifteen other industrialized nations over the decade. Governmental expenditures peaked in Britain and Italy in 1971 and in Belgium, Canada, Australia, and Israel in 1973; in Denmark, Finland, and the Netherlands the increases from 1973 to 1974 were minimal. In eight of the fifteen industrialized nations other than the United States, consultants report that support leveled off in 1974 and no increases are anticipated in the foreseeable future. Taken together, this means that of the eight countries that have supported research in the reproductive sciences and contraceptive development at the level of $1 million or more annually, six plateaued as of 1974 or are actually decreasing their funds. Data on private foundation support for research in six industrialized countries indicate the severely limited interest of private philanthropy in industrialized nations other than the U.S.;[7] in no country has its contribution exceeded $400,000 in any year. In the United States, where the bulk of private funds has come from the Ford, Rockefeller, and Scaife Foundations, allocations peaked at $22.6 million in 1971 and have since dropped by almost half. Tables 11.1, 11.2, and 11.3 summarize data presented in detailed tables for each country in appendix C.

[7] Large-scale philanthropic support of scientific research is primarily a U.S. phenomenon.

## The Components of Reproductive Research and Contraceptive Development

The data presented thus far document the overall trends for support for the field from all sources during the last decade. Within the field, and in science policy deliberations generally, questions are frequently raised regarding the appropriate allocation of resources to ensure orderly growth and development. Some observers, for example, have contended that increased priority for mission-oriented research and development programs distorts the growth of knowledge by reducing public support for the fundamental studies that cannot promise immediate solutions to urgent problems but are essential in the long run if such solutions are to be devised. Others have questioned the appropriate division of labor within the international scientific community.

The data gathered for this review permit us to evaluate some of these questions in a more systematic fashion than has heretofore been possible. In this section we will examine changes in the allocation of research resources in the field during the last decade.

### The Effects of a Mission-Oriented Effort

Table 11.4 shows how the funds expended in the field have been divided during the last decade among the four main program components with which this review is concerned. *Fundamental studies in the reproductive sciences* includes grants and contracts supporting basic research; core support to research centers, conferences, and symposia; and information dissemination. *Strengthening professional capacity* includes the training of researchers. Because training funds could not always be disentangled from monies supporting fundamental research, fundamental studies may be slightly overestimated and strengthening professional capacity underestimated. *Contraceptive development* is defined as all studies concerning agents being clinically tested for their fertility regulating effects. Pharmaceutical funds and core support to clinical research centers were categorically assigned to this research area. *Studies on safety of current fertility*

Table 11.3 Expenditures in the Reproductive Sciences and Contraceptive Development of Major Funding Agencies in Industrialized Nations, by Sector, 1965, 1969-1974 (constant U.S. dollars in 000s, 1970=100)

| Sector | No. of Agencies | 1965 | 1969 | 1970 |
|---|---|---|---|---|
| **Government Agencies** | | | | |
| United States | 8+ | 11,423 | 19,928 | 25,307 |
| Australia | 5+ | 469 | 515 | 697 |
| Belgium | 2 | 28 | 23 | 86 |
| Canada | 6 | 462 | 976 | 1,380 |
| Denmark | 2 | — | 17 | 48 |
| Finland | 2 | 10 | 12 | 22 |
| France | 6+ | 2,986 | 4,144 | 4,507 |
| Germany | 2 | — | 411 | 650 |
| Great Britain | 3 | 1,161 | 2,883 | 3,086 |
| Israel | 6 | 37 | 69 | 83 |
| Italy | 2 | 14 | 1,340 | 1,348 |
| Japan | 2 | — | — | — |
| Netherlands | 3 | — | — | — |
| New Zealand | 1 | 17 | 45 | 56 |
| Norway | 3 | — | 30 | 29 |
| Sweden | 2 | — | 320 | 628 |
| Totals | | 16,607 | 30,713 | 37,927 |
| **Philanthropic Agencies** | | | | |
| United States | 5+ | 8,153 | 11,059 | 16,796 |
| Australia | 2+ | — | 23 | 31 |
| Belgium | 1+ | 9 | 0 | 113 |
| Canada | 1 | — | 1 | 20 |
| Finland | 1 | 17 | 16 | 20 |
| Great Britain | 2 | 154 | 209 | 366 |
| Israel | 1+ | 2 | 3 | 0 |
| Totals | | 8,335 | 11,311 | 17,346 |

— = data not available.
(e) = estimated.

| 1971 | 1972 | 1973 | 1974 |
|---|---|---|---|
| 32,034 | 43,250 | 44,578 | 37,980 |
| 791 | 831 | 860 | 693(e) |
| 224 | 227 | 306 | 267 |
| 1,608 | 1,580 | 2,632 | 2,508(e) |
| 75 | 256 | 388 | 403 |
| 45 | 34 | 31 | 80 |
| 5,684 | 6,695 | 7,261 | 6,797 |
| 1,107 | 1,267 | 1,716 | 1,953 |
| 3,550 | 3,254 | 3,501 | 3,083 |
| 110 | 141 | 157 | 107 |
| 1,352 | 1,311 | 1,162 | 915 |
| — | 373 | — | — |
| 1,549 | 1,730 | 1,970 | 2,071 |
| 78 | 76 | 75 | 75(e) |
| 45 | 618 | 803 | 1,182 |
| 484 | 3,040 | 3,631 | 4,046 |
| 48,736 | 64,683 | 69,071 | 62,160 |
| 22,585 | 16,876 | 14,161 | 12,662 |
| 53 | 54 | 109 | 90(e) |
| 157 | 130 | 150 | 135 |
| 27 | 29 | 25 | 25 |
| 16 | 14 | 18 | 13 |
| 181 | 164 | 87 | 35 |
| 0 | 5 | 4 | 2 |
| 23,019 | 17,272 | 14,544 | 12,962 |

Table 11.4 Percent Distribution of Expenditures in all Reporting Countries for the Reproductive Sciences and Contraceptive Development by Purpose, 1965, 1969-1974 (based on constant U.S. dollars, 1970=100)

| Purpose | 1965 | 1969 | 1970 |
|---|---|---|---|
| Fundamental Studies in the Reproductive Sciences | 55.4 | 57.6 | 61.6 |
| Contraceptive Development | 35.3 | 30.0 | 24.7 |
| Studies on Safety of Current Fertility Control Methods | 2.7 | 5.0 | 7.3 |
| Strengthening Professional Capacity | 6.6 | 7.4 | 6.4 |
|  | 100.0 | 100.0 | 100.0 |

*control methods* includes evaluation of agents presently available to the public. The base data for the table are the expenditures in constant dollars reported to this review by agencies in the United States and other industrialized and developing nations, classified by the purpose of the investigation into the four principal program components. The table thus provides an overview of the worldwide emphasis within the field and of year-to-year variations in emphasis.

Throughout the decade, the lion's share of funds have supported basic investigations (57 percent in 1974) and more focused studies to develop new contraceptive methods have received a smaller percentage (30 percent in 1974). In fact, the share devoted to basic studies increased over the decade from 55 percent in 1965 to 62 percent in 1970, and has since declined slightly. The proportion allocated to contraceptive development declined from 35 percent in 1965 to 25 percent in 1970, and has increased to 30 percent since; this increase since 1970 reflects the emergence of the public sector programs described in chapter 10. The overall decline is, however, unexpected because the results of basic studies should have opened up new leads

| 1971 | 1972 | 1973 | 1974 |
|---|---|---|---|
| 60.6 | 57.3 | 61.7 | 56.6 |
| 27.1 | 28.3 | 26.2 | 29.5 |
| 7.5 | 9.1 | 7.2 | 8.7 |
| 4.8 | 5.3 | 4.9 | 5.2 |
| 100.0 | 100.0 | 100.0 | 100.0 |

for development and led to an increasing share for this component. Training expenditures stood at 7 percent in 1965, but have hovered around 5 percent since 1971. Studies related to safety began the decade as the smallest component at 3 percent and have received 8 to 9 percent since 1971; this is perhaps not unexpected in view of the increasing concerns expressed throughout the world by users of existing methods.

These overall allocations represent the sum of the varying priorities and interests of governmental and philanthropic agencies and pharmaceutical firms in the nations represented in this review. Some indications of this diversity are provided in table 11.5, which shows the share of support for each of the four main program components provided by the principal funding sources. About half of the funds allocated to fundamental studies has come from U.S. government agencies in most years of the decade; the share borne by U.S. private foundations has declined from one-fourth to one-sixth. There has been a corresponding increase over the decade in support from the World Health Organization and the residual group of other funding

Table 11.5 Percent Distribution of Expenditures in the Reproductive Sciences and Contraceptive Development by Purpose and Source of Funds, 1965, 1969-1974 (based on constant U.S. dollars, 1970=100)

| Purpose and Source of Funds | 1965 | 1969 | 1970 |
|---|---|---|---|
| **Fundamental Studies** | | | |
| U.S. government | 47.9 | 44.5 | 41.5 |
| WHO | .3 | .1 | .1 |
| U.S. private foundations | 26.8 | 21.4 | 27.7 |
| Other | 25.0 | 34.0 | 30.7 |
| Total | 100.0 | 100.0 | 100.0 |
| **Contraceptive Development** | | | |
| U.S. government | .6 | 8.8 | 8.4 |
| WHO | 0.0 | .1 | 0.0 |
| U.S. private foundations | 3.0 | 5.3 | 7.1 |
| U.S. pharmaceutical firms | 91.8 | 76.9 | 73.0 |
| Other | 4.6 | 8.9 | 11.5 |
| Total | 100.0 | 100.0 | 100.0 |
| **Monitoring Safety** | | | |
| U.S. government | 12.5 | 51.4 | 62.7 |
| WHO | 0.0 | 0.0 | .4 |
| U.S. private foundations | 76.5 | 27.6 | 25.6 |
| Other | 11.0 | 21.0 | 11.3 |
| Total | 100.0 | 100.0 | 100.0 |
| **Strengthening Professional Capacity** | | | |
| U.S. government | 39.9 | 40.1 | 36.6 |
| WHO | 0.0 | 0.0 | 0.0 |
| U.S. private foundations | 52.4 | 53.2 | 57.4 |
| Other | 7.7 | 6.7 | 6.0 |
| Total | 100.0 | 100.0 | 100.0 |

| 1971 | 1972 | 1973 | 1974 |
|---|---|---|---|
| 35.8 | 44.6 | 44.4 | 48.7 |
| .2 | 1.9 | 2.5 | 2.9 |
| 31.2 | 19.4 | 15.8 | 16.0 |
| 32.8 | 34.1 | 37.3 | 32.4 |
| 100.0 | 100.0 | 100.0 | 100.0 |
| 20.3 | 26.3 | 26.3 | 26.2 |
| .6 | 3.6 | 6.7 | 8.8 |
| 10.3 | 9.1 | 7.9 | 6.8 |
| 58.3 | 48.5 | 51.1 | 43.9 |
| 10.5 | 12.5 | 8.0 | 14.3 |
| 100.0 | 100.0 | 100.0 | 100.0 |
| 78.8 | 87.4 | 76.8 | 84.9 |
| .2 | 1.6 | 4.2 | 4.9 |
| 13.5 | 5.8 | 6.1 | 4.7 |
| 7.5 | 5.2 | 12.9 | 5.5 |
| 100.0 | 100.0 | 100.0 | 100.0 |
| 35.3 | 31.2 | 23.7 | 35.4 |
| — | 16.9 | 20.4 | 21.0 |
| 57.1 | 45.7 | 50.8 | 38.9 |
| 7.6 | 6.2 | 5.1 | 4.7 |
| 100.0 | 100.0 | 100.0 | 100.0 |

sources—medical research councils, private foundations, and pharmaceutical firms in other industrialized nations.

Support for targeted projects to develop new contraceptive methods came almost entirely from U.S. pharmaceutical firms at the beginning of the decade; the U.S. government had a minimal program and WHO had none. The pharmaceutical share had declined to less than half by 1974; the U.S. government's share had increased to more than one-fourth as a result of policy changes; and WHO's program accounted for 9 percent of all expenditures. It is of interest that the proportion provided by U.S. foundations reached as much as 10 percent in only one year.

A reverse pattern is shown in allocations for studies of contraceptive safety. Three-quarters of the funds for these studies in 1965 came from U.S. private foundations and one-eighth from U.S. government sources. By 1974, the U.S. government accounted for 85 percent of the funds, with the remainder provided in equal shares by WHO, U.S. private foundations, and other sources.

The U.S. government's share of total expenditures for training has remained fairly constant over the decade and the share borne by U.S. private foundations has declined. This decrease has been more than offset by the entry of WHO, which now accounts for one-fifth of training expenditures.

**Involvement of the International Scientific Community**

In chapters 1, 2, and 3 we showed the stakes that all nations, industrialized and developing, have in improved knowledge of reproductive physiology and fertility regulation. It would be desirable for the international scientific community to participate in studying a problem such as human reproduction, which is worldwide in scope, both because of the diverse conditions under which human reproduction takes place in different parts of the world and out of equity considerations. Biomedical research resources, however, are not evenly distributed throughout the world. The United States, Canada, Japan, Israel, Australia, and many nations of Western Europe have advanced medical research establishments, but the developing nations (and a number of industrialized nations as well) have far fewer personnel and institutions available with which to conduct research. One would expect, therefore, that this underlying maldistribution would be reflected in the geographical distribution of the worldwide reproductive and contraceptive research effort. Our data permit some of these patterns to be identified.

Table 11.6 shows the percent distribution of reproductive and contraceptive research expenditures by the geographic area in which the research was conducted during the last decade. As might be expected, the bulk of the work has been done in U.S. laboratories, which in 1965 accounted for 78 percent of total worldwide expenditures. By 1974, however, the United States share had declined to 68 percent and an increasingly larger share of the work was being done by investigators and institutions located in Western Europe, Canada, Japan, Israel, and Australia. These other industrialized nations accounted for only 19 percent of the effort in 1965, compared to 27 percent in 1974. There was also an increase in developing nations from less than 4 percent in 1965 to more than 5 percent in 1974.

The data suggest that the decade has been marked by an increasing involvement of the scientific communities of other industrialized and developing nations in the reproductive and contraceptive research enterprise. Comparison of table 11.6 with the percent distribution of research funds by country of origin in table 11.1 shows that this trend has paralleled the involvement of industrialized nations other than the United States in financial support of the field. The proportion of research funds expended in the laboratories of the fifteen other industrialized nations over the decade has increased in direct relationship to the increased share of funds originating in these nations.

Another way of illuminating these patterns is shown in table 11.7, which charts the growth of reproductive and contraceptive research expenditures (in constant U.S. dollars) in each of the three groups of nations, using 1965 as the base year. During the decade, expenditures for work in U.S. laboratories has more than doubled, the level in the other industrialized nations has increased three and a half times, and it has almost quadrupled in developing nations.

Table 11.6  Percent Distribution of Expenditures for Research in the Reproductive Sciences and Contraceptive Development by Geographic Area in Which Research Is Conducted, 1965, 1969-1974 (based on constant U.S. dollars, 1970=100)

| Geographic Area | 1965 | 1969 | 1970 | 1971 | 1972 | 1973 | 1974 |
|---|---|---|---|---|---|---|---|
| United States | 77.8 | 69.8 | 70.3 | 70.3 | 66.9 | 67.2 | 67.7 |
| Other industrialized countries | 18.7 | 25.9 | 25.0 | 25.7 | 26.5 | 27.3 | 26.9 |
| Developing countries | 3.5 | 4.3 | 4.7 | 4.0 | 6.6 | 5.5 | 5.4 |
| | 100.0 | 100.0 | 100.0 | 100.0 | 100.0 | 100.0 | 100.0 |

Table 11.7  Relative Growth of Expenditures for Research in the Reproductive Sciences and Contraceptive Development in Institutions Located in the United States, Other Industrialized Countries and Developing Countries, 1965, 1969-1974 (1965=100)

| Geographic Area | 1965 | 1969 | 1970 | 1971 | 1972 | 1973 | 1974 |
|---|---|---|---|---|---|---|---|
| United States | 100 | 134 | 168 | 213 | 223 | 210 | 215 |
| Other industrialized countries | 100 | 207 | 250 | 323 | 369 | 356 | 356 |
| Developing countries | 100 | 181 | 247 | 268 | 486 | 379 | 381 |

## Perspectives for Assessing the Adequacy of the Worldwide Effort

The field has evidently received increased support from many sources during the last decade. There are several ways in which the adequacy of these levels may be evaluated.

Table 11.8 relates the aggregate level of expenditures to the number of women of childbearing age in the nations represented in this study. The measure is crude, but nevertheless suggests a useful perspective because women of reproductive age and their partners comprise the principal direct consumers of the results of this effort. On this measure, the level of expenditure is obviously miniscule—37 cents per woman of childbearing age in 1973 in the nineteen countries in which funding and census data are available, and less, of course, per woman worldwide. The U.S. expenditure for each American woman was higher, $1.59 in 1973, but still far from a sizable amount. In the eleven other industrialized nations for which expenditure and population data are available, the expenditure per woman of childbearing age was considerably lower—52 cents in 1973. In the developing countries, with their very large populations, the expenditure averaged only 1 cent per woman.

A more refined measure is the relationship between expenditures on the reproductive sciences and a nation's allocation for all medical research. There is no single proportion of the medical research budget that unambiguously could be regarded as adequate for reproductive research, but the measure is useful to suggest the priority a nation's scientific and political communities assign to this field. Patterns of support for medical research vary greatly from country to country. In some, the philanthropic sector is an important source of support; in others, governmental allocations account for the lion's share. It was not possible in this survey to obtain adequate comparable data on both philanthropic and governmental expenditures for medical research in all countries. To avoid errors that would result from incomparabilities, therefore, our analysis will be limited to governmental expenditures for reproductive and medical research in the United States and thirteen other industrialized and developing nations for which adequate data could be obtained.

Table 11.8 Expenditures in the Reproductive Sciences and Contraceptive Development Per Woman of Reproductive Age, 1965, 1969-1973 (constant U.S. dollars in 000s, 1970=100; Women in millions)

|  | 1965 | 1969 |
|---|---|---|
| **All Countries** |  |  |
| Reproductive research expenditures | — | — |
| Women aged 15-44 in nations represented |  |  |
| $ Reproductive research per woman |  |  |
| **United States** |  |  |
| Reproductive research expenditures | 32,010 | 44,188 |
| Women aged 15-44 in U.S. | 38.9 | 41.8 |
| $ Reproductive research per woman | 0.82 | 1.06 |
| **Other Industrialized Countries** |  |  |
| Reproductive research expenditures | 5,992 | 10,678 |
| Women aged 15-44 | 43.1(9) | 35.0(12) |
| $ Reproductive research per woman | 0.14 | 0.31 |
| **Developing Countries** |  |  |
| Reproductive research expenditures | — | — |
| Women aged 15-44 |  |  |
| $ Reproductive research per woman |  |  |

[a]The number of countries for which both reproductive research funding and census data were available in that year.
[b]1972 census data used for Israel, Sweden, and Norway.
[c]1972 population estimates used for Denmark, Finland, and France.
Source: Number of women aged 15-44 for United States and other industrialized countries—U.N. Statistical Office Files; for developing countries—Nortman 1970, 1972, 1973, 1974.

Table 11.9 presents these data, expressing governmental expenditures for the reproductive sciences and contraceptive development as a proportion of total governmental medical research allocations. In most countries for which data were available the proportion has increased over the decade, but the increase is less than the dollar amounts suggest. Less than 1 percent of governmental medical research expenditures in the fourteen nations represented was devoted to reproductive research in 1965; the proportion had risen to only 1.6 percent by 1974, and in no year did it exceed 2.2 percent. Reproductive research's *share* of governmental medical research budgets has increased less than the *absolute* dollar amounts because *total* governmental investment in medical research was increasing generally during the last decade.

| 1970 | 1971 | 1972 | 1973 |
|---|---|---|---|
| — | — | — | 98,187<br>262.6(19)[a]<br>0.37 |
| 55,009<br>42.4<br>1.30 | 68,906<br>44.3<br>1.56 | 73,819<br>44.4<br>1.66 | 71,991<br>45.3<br>1.59 |
| — | 19,505<br>61.0[b](14)<br>0.32 | — | 24,323<br>46.6[c](11)<br>0.52 |
| 1,186<br>137.4(5)<br>0.01 | 1,577<br>142.5(5)<br>0.01 | 1,501<br>154.0(6)<br>0.01 | 1,873<br>170.7(7)<br>0.01 |

Because virtually no governmental funds were spent for reproductive research before 1960, the data imply that the interest in population growth in the early 1960s succeeded in obtaining enough attention for reproductive research so that the field attained a very modest share of governmental medical research expenditures by 1965. But the escalation of worldwide concern over rapid growth in the last decade, symbolized by the convening of the first intergovernmental population conference in Bucharest in 1974, has not succeeded in obtaining a significant level of additional priority for the field.

Another finding is implicit in table 11.9. The fourteen nations represented account for about four-fifths of worldwide governmental spending on reproductive research but an unknown proportion of worldwide governmental spending on general medical research. There is reason to believe that other nations, which do not support reproductive research, do support medical research. Thus the proportions shown tend to overstate the worldwide share of governmental medical research spending devoted to reproductive research. It seems fair to conclude that this field of inquiry currently receives perhaps slightly more than 1 percent of governmental medical research expenditures throughout the world, which in the United States and a number of other industrialized nations now constitute the principal source of support for medical research. Thus it seems clear that reproductive research is receiving only a very small share of all funds going into medical research.

There is some variation in these patterns in the different nations.[8] In the United States, the share devoted to reproductive research rose from 0.8 percent in 1965 to 2.2 percent in 1973 and then dropped back to 1.7 percent in 1974. In the other industrialized nations as a group, the proportion peaked at 2 percent in 1971 and 1972 and declined to 1.5 percent in 1974. For most of the decade, the proportions allocated by the Scandinavian nations to reproductive research have fairly consistently fluctuated around 5 percent of total governmental medical research expenditures; in Great Britain and West Germany, it has fluctuated around 3 percent. In Canada, on the other hand, it has in most years been less than 1 percent. It is noteworthy

---

[8] Sources for government medical research budgets in table 11.9 (obtained via country representatives—see appendix D):
*United States*— U.S. DHEW 1975. Limitations: Coverage limited to research and development; support of activities such as research training or capital outlays for research facilities are not included.
*Other Industrialized Countries* —
Australia — National Health and Medical Research Council (NH & MRC)
Canada — Medical Research Council (MRC); National Research Council (NRC) (research budgets from 1965-1973); the International Development Research Centre (IDRC) (research budget 1970-1973).
Denmark — Medical Research Council (MRC).
Finland — Medical Research Council (MRC).
France — Ministries of Research & Industrial Development (Research Institutes-research funds only), Agriculture (National Institute of Agriculture Research [INRA]), Education (National Center for Scientific Research [CNRS] and university budgets [estimates]), and Health (National Institute of Health [INSERM]).
Germany — Deutsche Forschungsgemeinschaft (DFG) (Life Sciences Division).

that India has consistently devoted between one-fifth and one-sixth of its governmental medical research budget to the field.

The meager attention accorded to reproductive sciences and contraceptive development by governmental medical research agencies is demonstrated graphically by the National Planning Association in the United States. The National Planning Association relates NIH outlays to incidence measures of cases associated with each category of research (1974). These NPA measures are presented in table 11.10. The range is from 32 cents for allergy research per day of bed disability from allergies and 88 cents for family planning research per woman of childbearing age to more than $1,000 for cancer research per death from cancer. There are legitimate differences of viewpoint concerning the appropriateness of the denominators used in these measures, and it is difficult to conceptualize a set of denominators that would be so clearly objective as to command unanimous support. Yet the range is so enormous as to make clear that almost regardless of what denominator is substituted, reproductive research (included in the family planning total) continues to receive very low priority from the U.S. government's scientific establishment. The data obtained in this survey suggest that the same fundamental patterns would be shown in comparable analyses in other developed nations.

Comparable data are not available with which to do a systematic analysis of the place of reproductive research in the development assistance effort. Fragmentary information for the 1974-1975 period

Great Britain — Office of Health Economics (estimates including budgets of the Medical Research Council [MRC], University Grants Committee [UGC], and the Department of Health and Social Security [DHSS]).
Netherlands — Total expenditure on scientific research in medical facilities of the universities on life sciences via the Netherlands Organization for the Advancement of Pure Research (Z.W.O.) and on health research via The Netherlands Organization for Applied Research (T.N.O.).
Norway — Medical Research Council (MRC).
Sweden — Medical Research Council (MRC).
*Developing Countries* — India — Total research budget including M.C.H. and Organization.
South Korea — Government medical research budget.
Thailand — National Research Council (funding of all areas of scientific research).

Table 11.9 Government Expenditures for Medical Research and Research in the Reproductive Sciences and Contraceptive Development, U.S. and Selected Countries, 1965, 1969-1974 (constant U.S. dollars in millions, 1970=100)

| | Government Expenditures 1965 Medical Research | Reproductive Research Amount | % | Government Expenditures 1969 Medical Research | Reproductive Research Amount | % | Government Expenditures 1970 Medical Research | Reproductive Research Amount | % |
|---|---|---|---|---|---|---|---|---|---|
| Total-Selected Countries | 1,890.1 | 16.1 | .9 | 2,328.1 | 29.1 | 1.3 | 2,302.6 | 37.0 | 1.6 |
| United States | 1,518.0 | 11.4 | .8 | 1,857.0 | 19.9 | 1.1 | 1,743.0 | 25.3 | 1.6 |
| Other Developed Countries[a] (Total) | 370.7 | 4.4 | 1.2 | 466.3 | 8.4 | 1.8 | 553.8 | 9.4 | 1.7 |
| Australia | — | — | | 1.55 | .06 | 3.6 | — | — | |
| Canada | 10.32 | .23 | 2.2 | 29.76 | .49 | 1.6 | 100.49 | .68 | .7 |
| Denmark | — | — | | .25 | .02 | 6.9 | 1.12 | .05 | 4.3 |
| Finland | .20 | .01 | 4.8 | .21 | .01 | 5.6 | .24 | .02 | 9.0 |
| France | 265.60 | 2.99 | 1.1 | 318.31 | 4.14 | 1.3 | 317.19 | 4.51 | 1.4 |
| Germany | — | — | | 15.37 | .41 | 2.7 | 21.92 | .65 | 3.0 |
| Great Britain | 94.54 | 1.16 | 1.2 | 94.39 | 2.88 | 3.1 | 105.26 | 3.09 | 2.9 |
| Netherlands | — | — | | — | — | | — | — | |
| Norway | — | — | | .87 | .03 | 3.4 | .76 | .03 | 3.9 |
| Sweden | — | — | | 5.63 | .32 | 5.7 | 6.83 | .37 | 5.4 |
| Developing Countries[b] (Total) | 1.51 | .26 | 17.6 | 4.78 | .84 | 17.5 | 5.83 | 1.06 | 18.2 |
| India | 1.51 | .26 | 17.6 | 4.28 | .83 | 19.4 | 4.65 | 1.06 | 22.7 |
| South Korea | — | — | | .39 | + | 1.6 | 1.09 | + | .4 |
| Thailand | — | — | | .11 | + | .4 | .09 | + | .1 |

Note: Columns may not add up to totals because of rounding.
[a]Information on Government Medical Research budgets unavailable for: Belgium, Italy, Israel, Japan, and New Zealand.
[b]Information on Government Medical Research budgets unavailable for: Africa, Egypt, Hong Kong, Iran, Philippines, and Turkey.
+less than $10,000
— = data not available.
Source: Government Medical Research data (see footnote 8 in text).

Table 11.9 (cont.)

| | 1971 Medical Research | 1971 Reproductive Research Amount | 1971 % | 1972 Medical Research | 1972 Reproductive Research Amount | 1972 % | 1973 Medical Research | 1973 Reproductive Research Amount | 1973 % | 1974 Medical Research | 1974 Reproductive Research Amount | 1974 % |
|---|---|---|---|---|---|---|---|---|---|---|---|---|
| Total-Selected Countries | 2,547.6 | 46.2 | 1.8 | 2,070.9 | 60.0 | 2.2 | 2,823.7 | 59.8 | 2.1 | 2,743.2 | 45.2 | 1.6 |
| United States | 1,880.0 | 32.0 | 1.7 | 2,061.0 | 43.3 | 2.1 | 2,028.0 | 44.6 | 2.2 | 2,273.0 | 38.0 | 1.7 |
| Other Developed Countries[a] (Total) | 662.0 | 13.4 | 2.0 | 704.0 | 14.3 | 2.0 | 790.5 | 14.5 | 1.8 | 469.2 | 7.3 | 1.5 |
| Australia | 2.20 | .11 | 5.2 | — | — | — | 2.86 | .23 | 8.1 | — | — | — |
| Canada | 104.05 | .82 | .8 | 102.34 | .80 | .8 | 103.64 | .90 | .9 | — | — | — |
| Denmark | 1.41 | .08 | 5.3 | 1.72 | .08 | 4.6 | 1.96 | .09 | 4.7 | 1.99 | .05 | 2.5 |
| Finland | .26 | .04 | 17.5 | .27 | .03 | 12.6 | .26 | .03 | 12.0 | .38 | .03 | 6.7 |
| France | 359.23 | 5.68 | 1.6 | 402.44 | 6.69 | 1.7 | 495.00 | 7.65 | 1.5 | 456.91 | 6.80 | 1.5 |
| Germany | 32.04 | 1.11 | 3.5 | 33.78 | 1.27 | 3.8 | 50.30 | 1.72 | 3.4 | — | — | — |
| Great Britain | 117.32 | 3.55 | 3.0 | 110.35 | 3.25 | 2.9 | 127.63 | 3.50 | 2.8 | — | — | — |
| Netherlands | 36.91 | 1.55 | 4.2 | 44.22 | 1.73 | 3.9 | — | — | — | — | — | — |
| Norway | .87 | .05 | 5.2 | .94 | .05 | 4.9 | 1.25 | .05 | 4.2 | 1.28 | .06 | 4.9 |
| Sweden | 7.69 | .37 | 4.8 | 7.94 | .37 | 4.7 | 8.61 | .35 | 4.1 | 8.62 | .31 | 3.7 |
| Developing Countries[b] (Total) | 5.63 | .81 | 14.3 | 5.94 | .75 | 12.5 | 5.21 | .67 | 12.2 | — | — | — |
| India | 4.34 | .80 | 18.5 | 4.41 | .74 | 16.7 | 4.07 | .66 | 16.1 | — | — | — |
| South Korea | 1.19 | + | .4 | 1.43 | + | 0.4 | 1.03 | .02 | 1.2 | 0.95 | + | 0.6 |
| Thailand | .10 | 0 | 0 | .10 | + | 3.1 | .11 | + | 1.4 | — | — | — |

Table 11.10 National Planning Association Measures of U.S. Government Outlays for Specific Categories of Medical Research, 1974 (expressed in 1969 dollars)

| Category | Unit of Measure | Expenditure |
| --- | --- | --- |
| Family planning | per woman aged 15-44 in U.S. | $ 0.88 |
| Cancer | per cancer death in U.S. | $1,019.00 |
| Cardiovascular and pulmonary | per heart and lung death in U.S. | $ 293.00 |
| Allergy and infectious diseases | per day of hospitalization in U.S. | $ 0.32 |
| Arthritis, metabolic and digestive diseases | per incident causing limited activity in U.S. | $ 5.97 |
| Neurological and visual | per incident causing limited activity in U.S. | $ 19.00 |
| Mental health | per incident of chronic mental disorder causing limited activity in U.S. | $ 65.00 |

Source: All measures derived from National Planning Association 1974.

suggests that this field also has not been perceived as a high priority by governmental and international agencies concerned with providing assistance to developing nations. The nine agencies,[9] multilateral, governmental, and private, that were the principal institutions in the development assistance community supporting population or family planning work allocated about $26 million to reproductive research, or 9 percent of their total expenditures for population programs (Saunders and Leonard 1975). This total, however, constituted only about 0.2 percent of the $11.3 billion in official development assistance funds provided by member nations of the Organization for Economic Cooperation and Development's Development Assistance Committee (OECD 1975, table 17). Even if some support for reproductive research was provided by other agencies than the nine represented in this comparison, it could not significantly alter the resulting proportion.

[9] World Bank, UNFPA, USAID, Swedish International Development Authority, IDRC, International Planned Parenthood Federation, Ford Foundation, Rockefeller Foundation, and the Population Council.

## Prospects for the Future

Thus the reproductive sciences, which were historically tabooed and underfinanced, began to receive a modicum of support in the last decade but still are accorded little priority. This analysis has been presented in terms of money amounts, which are of course useful to make diverse types of information comparable. Yet they tend to obscure the underlying reality that important research projects are *not* being carried out because of lack of funding. This has been particularly true as a consequence of the cutbacks of the last several years. NICHD's Center for Population Research, the largest single source of funds for the field, has had to turn down increasing proportions of applications that have been approved as meritorious by scientific review committees. CPR was able to fund about half of approved applications from 1969 to 1971; the proportion increased to 62 percent in 1972, then dropped to half again in 1973, 37 percent in 1974, and 47 percent in 1975 (Douglas 1975).

In other words, increased investment during the 1960s, by training a cadre of qualified investigators and attracting others to the scientific and social importance of the field, laid the groundwork for a substantial increase in research activity in the mid-1970s. This is now reflected in a larger number of meritorious applications, which increasingly must be turned down because of lack of funds.

The prospect for the future is not much brighter. In the survey, respondents were asked to estimate the levels of funds expected in the next several years; but these estimates were not forthcoming for all countries. The data available for sixteen nations are presented in constant dollars in table 11.11, which shows that the totals expected to be available in 1975 were 17 percent less than those available in 1973. Governmental funds were expected to decline 11 percent, private foundation expenditures were expected to be cut in half, and the pharmaceutical industry was expected to cut its allocations by 16 percent. The decline is attributable almost entirely to sharp cutbacks in U.S. expenditures. Government expenditures in the United States, which account for almost half of total worldwide expenditures, were

Table 11.11 Estimated 1975 Reproductive Research Expenditures Compared to 1973 in U.S. and Fifteen Other Industrialized Countries, by Sector (constant U.S. dollars in 000s, 1970=100)

| | 1973 | 1975 Estimate[a] | Percent Change 1973-1975 |
|---|---|---|---|
| Total | 99,084 | 81,850 | −17.4 |
| Government | 69,072 | 61,235 | −11.4 |
| Philanthropy | 14,554 | 7,550 | −48.1 |
| Industry | 15,458 | 13,065 | −15.5 |
| United States (Total) | 71,991 | 55,083 | −23.5 |
| Government | 44,578 | 36,142 | −18.9 |
| Philanthropy | 14,161 | 7,441 | −47.5 |
| Industry | 13,252 | 11,500 | −13.2 |
| Other Industrialized Countries (Total) | 27,093 | 26,767 | −1.2 |
| Government | 24,494 | 25,093 | +0.6 |
| Philanthropy | 393 | 109 | −72.3 |
| Industry | 2,206 | 1,565 | −29.1 |

[a] 1974 Consumer Price Index used in each case; for other industrial nations the support from each country was handled individually with the appropriate CPI.

expected to be cut by nearly one-fifth; U.S. private agencies were expected to reduce their funding levels by almost half; and pharmaceutical companies were expected to cut 13 percent.

Unless these trends are reversed, it seems clear that there will be considerably less research activity in the reproductive sciences and contraceptive development in the next several years.

## References

Douglas, C.D. (National Institutes of Health, Division of Research Grants). 1975. Personal communication with S. Lieberman.

Harkavy, O., and Maier, J. 1970. Research in reproductive biology and contraceptive technology: Present status and needs for the future. *Family Planning Perspectives* 2(3):3-18.

_____. 1971. Research in contraception and reproduction: A status report, 1971. *Family Planning Perspectives* 3(3):15-17.

Karp, R. 1975. Drugs on the market: Oral contraceptives are selling ex-glamour. *Barrons's*, February 24.

National Planning Association (NPA), Center for Health Policy Studies. 1974. *Chartbook of federal health spending 1969-74.* Washington, D.C.: NPA.

Nortman, D. 1970, 1972, 1973, 1974. Population and family planning programs: A factbook. *Reports on Population/Family Planning* No. 2.

Organization for Economic Cooperation and Development (OECD). 1975. *Development cooperation 1975 review.* Paris: OECD.

Perkin, G. (Ford Foundation, Brazil). 1975. Personal communication.

Saunders, L., and Leonard, A. 1975. *Population policies and programs post-Bucharest.* Mimeographed. New York: Ford Foundation.

U.S. Congress, Senate. 1967. *Population crisis: Hearings before the subcommittee on foreign aid expenditures of the committee on government operation.* 90th Congress.

U.S. Department of Health, Education and Welfare. 1962. *A survey of research on reproduction related to birth and population control (as of December 1, 1962).* Washington, D.C.: Government Printing Office.

_____. 1963. *A survey of research on reproduction related to birth and population control (as of January 1, 1963).* Washington, D.C.: Government Printing Office.

_____. 1968. *Current population research: 1966.* Report No. 1. Washington, D.C.: Government Printing Office.

_____. 1969a. *Population research: 1968.* Report No. 2. Washington, D.C.: Government Printing Office.

_____. 1969b. *Population research: A prospectus. Committee report to the Assistant Secretary of Health and Scientific Affairs.* Mimeographed. Washington, D.C.

_____. 1970. *The federal program in population research—inventory of population research supported by federal agencies during fiscal year 1970.* Washington, D.C.: Government Printing Office.

_____. 1971a. *Population research: 1969.* Report No. 3. Washington, D.C.: Government Printing Office.

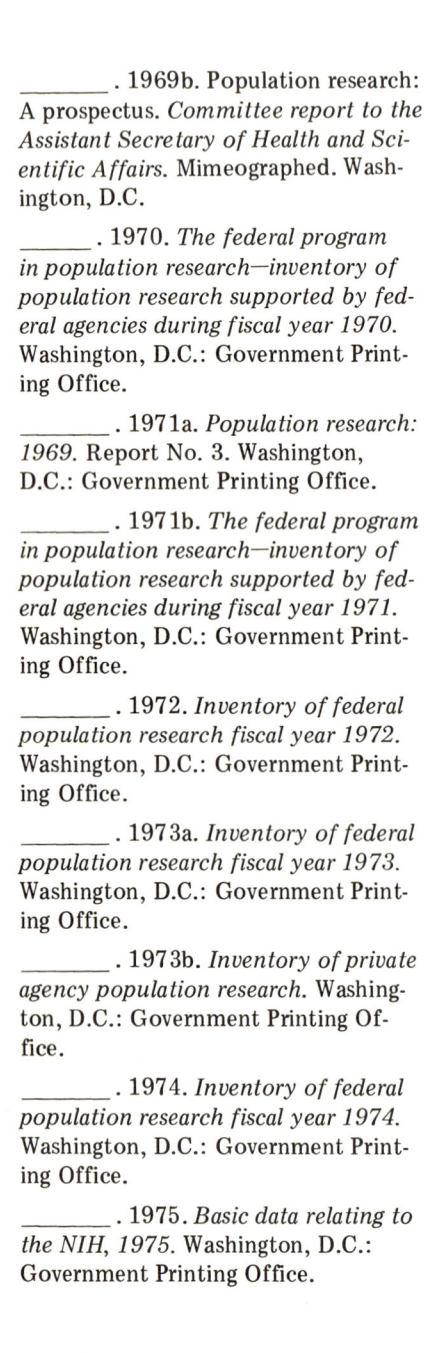

_____. 1971b. *The federal program in population research—inventory of population research supported by federal agencies during fiscal year 1971.* Washington, D.C.: Government Printing Office.

_____. 1972. *Inventory of federal population research fiscal year 1972.* Washington, D.C.: Government Printing Office.

_____. 1973a. *Inventory of federal population research fiscal year 1973.* Washington, D.C.: Government Printing Office.

_____. 1973b. *Inventory of private agency population research.* Washington, D.C.: Government Printing Office.

_____. 1974. *Inventory of federal population research fiscal year 1974.* Washington, D.C.: Government Printing Office.

_____. 1975. *Basic data relating to the NIH, 1975.* Washington, D.C.: Government Printing Office.

# 12 Toward an Adequate Worldwide Research Effort

Support for research in the reproductive sciences and contraceptive development, as we have seen, increased during the 1960s and reached a peak in 1973-1974. These gains, however, were insufficient to compensate for the historic underfinancing of this branch of scientific inquiry. Despite the burgeoning of worldwide concern during the last decade over both rapid population growth and improved pregnancy outcome, research in this field has never received a significant share of allocations for medical research or development assistance. The reproductive sciences received a maximum of 2.2 percent of governmental medical research expenditures in the United States in 1973 and in all nations represented in this study in 1972; by 1974, the proportion had declined to 1.6 percent. In 1974-1975, the field received about 0.2 percent of total funds for development assistance.

There is, of course, no simple way to determine what the proportion devoted to the reproductive sciences and contraceptive development *should* be. Despite the increased attention in recent years to science policy analysis, criteria have not yet emerged that would indicate whether the proportion allocated to any particular branch of inquiry is appropriate (or whether the share of gross national product devoted to medical research in general is adequate).[1] In the absence of objective criteria, each branch of science asserts its claims with reference to such considerations as the gravity of the problems that might be solved by additional knowledge or the readiness of the field to transform prior discoveries into usable technologies.

It is not possible for us to transcend these constraints in our analysis of the level of activity now underway throughout the world in the reproductive sciences and contraceptive development. As this review has shown, the problems addressed by this branch of science are urgent and affect individuals, families, and nations in both the industrialized and developing world. In addition, many breakthroughs that are susceptible to rapid exploitation and development have been made in recent years, and significant institutional and human resources are available to exploit these findings.

Although the worldwide perspective is useful, a better sense of the place of this field within the broader arena of scientific inquiry gen-

---

[1] For an interesting discussion of these questions, see Burger 1975.

erally may be gained by closer examination of patterns within the U.S. National Institutes of Health, which, as we have seen, has become the largest single source of support for this field in the world. NIH expenditures for the reproductive sciences and contraceptive development have in recent years constituted approximately 2 percent of total NIH research expenditures, despite the fact that a priority effort was called for verbally by three U.S. presidents and enacted into law by Congress. Two percent does not seem to be a proportion that can be accurately characterized as constituting a "priority program."

Some notion of what a "priority program" looks like is provided by the expenditure data for two other fields of inquiry—research on cancer and on heart and lung disease. Both of these fields have been the subject of special legislation adopted in order to assign them higher priority. Table 12.1 shows the proportions of total NIH research expenditures allocated to cancer and to heart and lung studies from 1972 through 1975 contrasted with the proportion allocated to biomedical studies of reproduction through the Center for Population Research. It is evident that cancer research has been especially favored since the adoption of the National Cancer Act of 1971; between 1972 and 1975, cancer's share of total NIH expenditures increased from one-quarter to one-third. Heart and lung did not fare as

Table 12.1 Percent of Total NIH Research Expenditures Devoted to Cancer, Heart and Lung, and Reproductive Sciences and Contraceptive Development, FY 1972-1975 (Based on current U.S. dollars)

|  | Fiscal Years | | | |
| --- | --- | --- | --- | --- |
|  | 1972 | 1973 | 1974 | 1975 |
| National Institutes of Health | 100.0 | 100.0 | 100.0 | 100.0 |
| National Cancer Institute | 24.6 | 27.7 | 30.1 | 33.4 |
| National Heart and Lung Institute | 14.8 | 16.6 | 16.3 | 15.7 |
| Center for Population Research (biomedical studies) | 2.0 | 1.7 | 2.4 | 2.0 |

Source: For NIH, NCI, and NHLI, DeWald 1975; for CPR, 1972-1974 grant and contract listing tabulated by computer (see table D.3, appendix D); for 1975; Center for Population Research 1975.

well, receiving between 15 and 17 percent of total NIH expenditures.

It seems clear that programs receiving actual priority, such as cancer and heart and lung, are treated differently from those where the "priority" assignment is more verbal than real. We would not argue that a program must receive as much as cancer or heart and lung before it can be deemed a priority effort; too many other factors are present to permit such automatic conclusions. But given the fact that the share devoted to these preferred programs has ranged from 15 to 33 percent, we find it difficult to believe that any program could be deemed a "priority" effort if it receives much less than 10 percent of total NIH expenditures. If CPR had received 10 percent of NIH expenditures from 1972 through 1975, its appropriation (in current dollars) would have been $152, $175, $176, and $207 million respectively—roughly five times the amounts actually appropriated. If the field had received 10 percent of total medical research expenditures in the nations represented in this study, its funding level in 1974 would have been $473 million.

The principal purpose of priority setting, of course, is to guide research institutions and personnel into fields of inquiry regarded as socially urgent or scientifically promising. There can be little doubt that more scientists today than five years ago are devoting their time and energies to studies related to cancer and to heart and lung disease. The effect on other branches of inquiry is complementary: relatively fewer researchers—or at least fewer than would have been the case given equivalent funding—devote their efforts to fields that do not receive priority allocations. The effects are seen not only on the present generation of established researchers but also on future generations because a greater number of fellowships and other mechanisms for the training and development of scientific manpower are available in the preferred fields. Both of these processes have operated in the last several years to the detriment of the reproductive sciences and contraceptive development.

A common measure of these processes is the proportion of grant applications approved as scientifically meritorious after surviving the peer review process at NIH and subsequently funded. The measure is limited because the number of applications submitted in the first place is a function, in part, of the funds available to a given institute; scientists are usually reluctant to prepare time-consuming applications unless there seems a reasonable chance they will be funded. Table 12.2 shows these proportions for the National Cancer Institute, the National Heart and Lung Institute, and the Center for Population Research. In 1974, an approved NCI application had about twice the probability of being funded as an approved application to CPR; the probability of funding of an approved NHLI project was 72 percent greater. These data suggest some of the very real impact of the scarcity of funds on the level of scientific activity in this field. They also suggest that existing institutions and personnel in reproduction research have the capacity to expand the current level of activity in the field, if sufficient support were made available.

The priorities underlying medical research policy in the United States have increasingly been questioned and a Presidential Commission on Biomedical Research was appointed in 1975 to examine these issues (Wade 1976). Witnesses testifying before the commission in July 1975 pointed out that current policies emphasize research into diseases that affect the end of the life cycle and can have only limited impact on the quality of human existence. One of the most pointed statements was made by Andre Hellegers, professor of obstetrics and gynecology and director of the Kennedy Institute at Georgetown University, who called for giving greater priority to research into the beginnings of life (1976): "We know the life span is genetically limited. What one is doing [by emphasizing cancer and heart research] is to trade one dying method for another dying method without any assurance that one set of dying methods is preferable

Table 12.2 Percent of Council Approved Grants Awarded by NCI, NHLI, and CPR, FY 1969, 1971, 1973 and 1974

| | Fiscal Years | | | |
|---|---|---|---|---|
| Institute | 1969 | 1971 | 1973 | 1974 |
| National Cancer Institute | 55.9 | 68.2 | 59.6 | 71.7 |
| National Heart and Lung Institute | 78.4 | 62.1 | 46.7 | 63.0 |
| Center for Population Research | 49.5 | 56.2 | 51.3 | 36.7 |

Source: NCI and NHLI data from Burger 1975; CPR data from Douglas 1975.

to another.... I personally wouldn't hesitate to say let [the increased budget for the reproductive sciences] come from heart and cancer right now into reproduction...."

A final consideration that must be taken into account before we project the needs of the field for an adequate program in the next period of time is inflation, which in the last decade has affected all nations in which reproductive research is conducted. Studies conducted for NIH have shown that the costs of medical research have increased more rapidly than consumer prices in general (Woolley 1975), in part because of the higher costs of the increasingly sophisticated methodology now utilized. Barnes (1975) has pointed out that a milliunit of hCG cost $30 in 1970 and $180 in 1975; a scanning electron microscope now costs about $100,000; guinea pigs have gone from $6 in 1970 to $11 in 1975, small monkeys from $55 to $125, and board for monkeys increased 350 percent from 1972 to 1975. As we have seen (table 11.1), the failure of reproductive research appropriations to keep pace with these cost increases resulted in an actual decline in 1974 in the real dollars available for work in the field, even though the number of current dollars allocated was slightly larger than in 1973. Thus, the first step in projecting a future funding program for the field is to estimate the funding required simply to maintain current levels of activity. The difficulty in such a projection, of course, is that no one can be certain what the rate of inflation will be in the future; the best that can be done is to take the recent secular trends and project them forward. Table 12.3 attempts to do this by projecting forward to 1980 the funding levels (in current dollars) allocated to the field during 1974. For the United States, the estimated rate of inflation is based upon a special biomedical research price deflator developed for NIH (Woolley 1975); for other nations, the rate is based on the average annual rate of increase in their consumer price indices between 1969 and 1974.[2]

The results are that in 1980, nearly $200 million will be needed

---

[2] The projection assumes that the proportion of total funds for the field originating in the United States, the other industrialized nations, and the developing nations will be the same in 1980 as it was in 1974. If the share coming from the other industrialized nations were to increase, the annual totals would increase also because the inflation rates in those countries have been higher than in the United States.

Table 12.3 Projected 1980 Funding Levels for Reproductive Research and Contraceptive Development Required to Match 1974 Expenditures (Current U.S. Dollars in 000s)

|  | 1974 Expenditures | Projected 1980 Expenditures[a] |
|---|---|---|
| Total | $118,723 | $197,473 |
| United States | 79,104 | 125,921 |
| Other industrialized nations | 38,029 | 68,135 |
| Developing nations | 1,590 | 3,417 |

[a]Inflation factors for U.S. derived from a biomedical research price index (Woolley 1975); for other nations, they are based on an extrapolation to 1980 of the average annual rate of increase in their consumer price indices for the years 1969-1974.

simply to keep the level of research activity in the reproductive sciences and contraceptive development constant with the level achieved during 1974. Expenditures from U.S. governmental, philanthropic, and industrial sources would have to increase from $79 million in 1974 to $126 million. In the fifteen other industrialized nations, the amounts would have to increase from $38 to $68 million; in developing nations, an increase from $1.6 million to $3.4 million would be needed. The difficulty with this approach, of course, is that it would perpetuate current patterns. It is hoped these projections would compensate for inflation; but if the field is currently underfinanced, it would continue to be so in 1980.

Our data document the fact that in 1974 total expenditures for the reproductive sciences and contraceptive development fell far short of all funding targets for the field set by expert groups in the last decade.[3] Perhaps the most systematic earlier assessment of the field's needs was done in 1969 for the Assistant Secretary of Health of the U.S. Department of Health, Education and Welfare (Family Planning Services 1970). This report set forth a detailed five-year plan that called for expenditures on biomedical research by U.S. agencies in

[3] In the United States, various targets for funding of biomedical research were set in Harkavy, Jaffe, and Wishik 1967; President's Committee on Population and Family Planning 1968; Family Planning Services 1970; and the U.S. Commission on Population Growth in 1972. None of these levels had been achieved by 1974.

1974 of about $228 million. Because the actual expenditures by government, industry, and private philanthropy in the United States in 1974 was only about $79 million, the achieved funding level fell short by almost two-thirds of the amount targeted in 1969. To reach the targeted funding level in 1980, taking anticipated inflation into account, expenditures by U.S. agencies would have to increase to $330 million.

A different approach to the estimation of future needs was adopted for the review. From the comprehensive evaluations prepared by recognized experts of the current state of knowledge in the various subjects comprising the overall field, it was possible to identify 230 key questions that are ready for further intensive investigation. The subjects and the state-of-the-art papers from which they are drawn are listed in summary form in appendix E, summary E.1. It is important to note that this research agenda is not composed of unmanageably global subjects, but rather of specific issues delineated by past findings and capable of being studied using existing knowledge, techniques, and instrumentation. Each subject, in our view, is important enough to our understanding of the human reproductive process to warrant being studied by an average of five research units led by two Ph.D.-level scientists over a five-year period.[4] The current average annual cost of operating a research unit of this size in both industrialized and developing nations is about $120,000. Accordingly, we estimate the worldwide cost of an adequate research effort in fundamental studies in the reproductive sciences at $165 million in 1976 (see table E.1, appendix E).

Our estimates of the costs of contraceptive development were prepared in a similar manner, based on what it would take to carry out the work on those leads that are in relatively advanced stages of development, as well as those that are already known but are less advanced. We placed immunologic and luteolytic approaches, sustained release forms of contraceptive steroids, a postpartum IUD, protaglandins, and a steroid to suppress spermatogenesis in the former cate-

[4] These are, of course, average figures. The investigation of some subjects would be completed more quickly; others would undoubtedly take longer than five years.

gory. In the latter category are such possibilities as antagonists to releasing hormones, sperm antigens, receptor regulation, and means of detecting ovulation and of prolonging lactation. Estimates were prepared of the costs of the studies required to pursue each of these leads through the complex process of development—animal pharmacology to determine mechanism of action, animal toxicology, human pharmacology and clinical effectiveness, evaluation of effects on patients, and long-term monitoring of safety. We concluded that the development of these leads could absorb $101 million in 1976 (see tables E.2a, b, c, d, appendix E).

These estimates reflect our judgment that many of the experienced investigators needed to carry out the research are already in the field or can be attracted to join in this effort. Additional personnel and institutions would be needed, however, to mount the expanded level of activity proposed here. The training of new personnel and the creation of new institutes or centers committed to work in this field would also require attention and funding. In addition, of course, the field's cadres will be subject to attrition and replacement personnel will be needed. Based on these considerations, we estimate that $22.5 million would be needed in 1976 to strengthen the field's professional capacity (see table E.4, appendix E).

Finally, we have noted at various points in this volume the concerns over safety of existing contraceptive methods that are increasingly being expressed by consumers and regulatory agencies. We identified the key questions of safety requiring investigation and estimated the costs of such studies for each of the principal current methods. Some issues could be resolved with relatively short-term and inexpensive studies; but others, particularly prospective studies on large populations, are inherently longer in duration and considerably more costly. In our judgment, an adequate program to evaluate the safety of existing contraceptives would cost $72 million in 1976 (see table E.3, appendix E).

The determinations underlying these summary estimates are presented in detail in appendix E. We emphasize that the building blocks for these estimates are the specific assessments of the state of the field—what is known, what remains at issue, what could be resolved with current techniques and instrumentation. In essence, therefore, the projected 1976 program is one for which the prerequisite knowledge already exists, a significant professional and institutional capacity is in place, and the additional capacity required could be developed.

In table 12.4, we compare these projected funding levels for each branch of the field with the levels actually achieved in 1974. The total amount needed in 1976, $360.5 million, is three times the amount available in 1974; put another way, the field had available in 1974 perhaps one-third of the funds needed to exploit adequately existing knowledge. The funds required for fundamental studies are about two and a half times the 1974 amounts; the totals needed for contraceptive development are about three times greater and for training three and a half times greater. The required funding level of safety studies, however, is about seven times the amounts available, reflecting our conviction that rapid expansion of this branch of the field is urgently necessary.

In this chapter, we have suggested three ways of projecting the field's future funding needs: a *maintenance* program simply to keep constant the level of activity achieved in 1974; a *high priority* program receiving a fixed proportion (10 percent) of medical research

Table 12.4 1974 Expenditures for Reproductive Research Compared to Projected 1976 Program Needs (Current U.S. Dollars in Millions)

| Purpose | 1974 Expenditures (Actual) | 1976 Program Needs (Projected) |
| --- | --- | --- |
| Fundamental studies in the reproductive sciences | 67.1 | 165.0 |
| Contraceptive development | 34.4 | 101.0 |
| Studies on safety of current fertility control methods | 10.9 | 72.0 |
| Strengthening professional capacity | 6.3 | 22.5 |
| Total | 118.7 | 360.5 |

Source: Actual expenditures (1974)—detailed account in appendixes C and D. Projected needs (1976)—detailed account in appendix E.

expenditures; and an *adequate-exploitation-of-existing-knowledge* program developed inductively from the findings of this review about the state of the field and its potential. Table 12.5 brings the three approaches together and adjusts the estimate for 1976 to take account of anticipated inflation through 1980. In summary:
• To maintain the 1974 level of activity, the field would need this year $142.7 million. Our data indicate that even this minimum level is not being achieved.
• To achieve a high priority effort, the field would need more than half a billion dollars this year and the figure would increase to $766 million in 1980 as a result of inflation.
• To adequately exploit existing knowledge, the field would need at least $360.5 million this year, increasing to $498 million in 1980.

The 1976 figure will obviously not be achieved, but the 1980 target could be reached if officials who decide on research allocations can be persuaded to face up to the urgency of the problems of population growth, fertility regulation, and reproductive outcome that this field addresses. Based on current sources of funds for the field, expenditures by government, industry, and private philanthropy in the United States would have to increase to $328 million by 1980 to achieve this target of $498 million. The total in other industrialized nations would have to increase to $162 million and in developing nations to $7 million. This distribution of funding responsibility, of course, could very well change, particularly if a greater number of scientists and research institutions in industrial nations other than the United States are stimulated to participate in this worldwide effort. Such a development would increase the overall totals slightly because the inflation rates in the other industrialized nations are higher than in the United States.

Table 12.5 Three Levels of Projected Funding for the Reproductive Sciences and Contraceptive Development, 1974-1980 (Millions of current U.S. dollars)[a]

| | 1974 | 1976 | 1977 | 1978 | 1979 | 1980 |
|---|---|---|---|---|---|---|
| 1. Maintenance of 1974 level of activity | 118.7 | 142.7 | 156.5 | 170.5 | 184.3 | 197.5 |
| 2. High priority, assuming 10% of medical research expenditures | 472.6 | 566.0 | 619.7 | 671.8 | 721.8 | 766.3 |
| 3. Adequate exploitation of existing knowledge and opportunities[b] | .... | 360.5 | 395.2 | 430.3 | 464.8 | 497.6 |
| United States | .... | 246.2 | 269.2 | 291.1 | 311.3 | 328.1 |
| Other industrialized nations | .... | 110.0 | 121.1 | 133.6 | 147.2 | 162.3 |
| Developing nations | .... | 4.3 | 4.9 | 5.6 | 6.3 | 7.2 |

[a] Inflation factors for 1974-1980 derived as in table 12.3.
[b] Distribution assumes that the proportion of total funds for the field in 1974 from the U.S. (68.3%), other industrial nations (30.5%) and developing nations (1.5%) will be the same in 1976 and will change in subsequent years only as a result of differential rates of inflation.
.... = not applicable

## References

Barnes, A. 1975. Follow-up: Dial B for banana. *RF Illustrated* 2(3):9.

Burger, E.J. Jr. 1975. Science for medicine—time for another appraisal. *Fed. Proc.* 34:2106-2114.

Center for Population Research. 1975. Progress report of the population research program of the National Institute of Child Health and Human Development: 1975. *Population Research*. September.

DeWald, D.D. (National Institute of Child Health and Human Development). 1975. Personal communication.

Douglas, C.D. (National Institutes of Health, Division of Research Grants). 1975. Personal communication with S. Lieberman.

Family Planning Services. 1970. Hearings before the Subcommittee on Public Health and Welfare, pp. 162-180. Committee on Interstate and Foreign Commerce, House of Representatives, 91st Congress, Second Session. Washington, D.C.: Government Printing Office.

Harkavy, O.; Jaffe, F.S.; and Wishik, S.M. 1967. *Implementing DHEW policy on family planning and population—a consultant's report*. In *Hearings before the Subcommittee on Foreign Aid Expenditures of the Committee on Government Operations*, U.S. Senate, Ninetieth Congress S. 1676, Part 1, pp. 163-206. Washington, D.C.: Government Printing Office.

Hellegers, A. 1976. In President's

Biomedical Research Panel, *Written Statements Supplementing Verbal Testimony of Witnesses*, (DHEW Pub. No. [internal O.S.] 76-508). Washington, D.C.: DHEW.

President's Committee on Population and Family Planning. 1968. *Population and family planning: The transition from concern to action*. Washington, D.C.: DHEW.

U.S. Commission on Population Growth and the American Future. 1972. *Population and the American future*. Washington, D.C.: Government Printing Office.

Wade, N. 1976. Congress looks harder at cancer. *Science* 191:364.

Woolley, H.B. 1975. Biomedical research price assumptions for the FY 1977-FY 1981 NIH Plan. Mimeographed. National Institutes of Health.

# Appendixes

Appendix A         Bibliography of Reports

## Scientific Essays

Bahl, O.
Chemistry and Biology of Human Chorionic Gonadotropin and its Subunits.

Barber, B.
Ethical Aspects of Clinical Research in the Field of Human Reproduction.

Blandau, R.J.; Boling, J.L.; Broderson, S.; and Brenner, R.M.
The Oviduct. I. Subsection on Transport Mechanisms.

Cavazos. L.F.
The Mammalian Accessory Sex Glands—A Morphological and Functional Analysis.

Chang, M.C.; Austin, C.R.; Brackett, J.M.; Hunter, R.H.F.; and Yanagimachi, R.
Capacitation of Spermatozoa and Fertilization in Mammals.

Clermont, Y.
Spermatogenesis.

Faundes, A.
Assessment of Clinical Testing Methodology.

Fawcett, D.W.
The Structure of the Spermatozoa.
The Ultrastructure and Functions of the Sertoli Cell.

Friesen, H.G.
Prolactin.

Gemzell, C.
Induction of Ovulation.

Gibbons, I.R.
Sperm Motility.

Guillemin, R.; Davidson, J.; McCann, S.; and Sawyer, C.
Nature of Hypothalamic Hypophysiotropic Hormones Controlling the Secretion of Gonadotropins.

Hamilton, D.W.; Casillas, E.R.; Glover, T.D.; Holstein, A.F.; Jackson, H.; Johnson, A.D.; Prasad, M.R.N.; Ofner, P.; and White, I.G.
The Epididymis.

Harrison, R.A.P.
The Metabolism of Mammalian Spermatozoa.

Jensen, E.V.
Hormone Receptor Interaction in the Mechanism of Reproductive Hormone Action.

Lardner, T.J.
Bioengineering Aspects of Reproduction and Contraceptive Development.

Laurence, K.A.
Reproductive Immunology—Progress in the Last Decade.

Lieberman, S.; Gurpide, E.; Lipsett, M.; and Salhanick, H.
Steroid Hormone Secretions.

McLaren, A.
Developmental Biology.

Mann, T.
Semen: Metabolism, Antigenicity, Storage and Artificial Insemination.

Mastroianni, L. Jr.; Brackett, B.; and Hamner, C.
The Oviduct. II. Subsection on Oviductal Physiology.

Mishell, D.R., Jr.
Current Status of Injectable Contraceptive Preparations—A Review.

Morris, J.M.
The Morning-After Pill—A Report on Postcoital Contraception and Interception.

Nalbandov, A.V.; Bahr, J.M.; and Ross, G.T.
Hormonal Regulation of the Development, Maturation and Ovulation of the Ovarian Follicle.

Neaves, W.B.
Leydig Cells.

O'Malley, B.W.
Hormonal Control of Gene Expression in Reproductive Tissues.

Papkoff, H.; Ryan, R.J.; and Ward, D.N.
The Gonadotropic Hormones, LH (ICSH) and FSH: Major Advances, Current Status and Outstanding Problems.

Porter, D.G., and Finn, C.A.
The Present Status and Future Prospects of Research in the Biology of the Uterus.

Ramwell, P.W.
The Role of Prostaglandins in Reproduction.

Richart, R.M., and Darabi, K.F.
Female Sterilization.

Schwartz, N.B.; Dierschke, D.W.; McCormack, C.E.; and Waltz, P.W.
Feedback Regulation of Reproductive Cycles in Rats, Sheep, Monkeys and Humans, with Particular Attention to Computer Modeling.

Segal, S.J.
Systematic Contragestational Agents.

Setchell, B.P.
The Blood Testis Barrier.

Short, R.V., and Baird, D.T.
The Endocrinology of Ovulation and Corpus Luteum Formation, Function, and Luteolysis in Women.

Steinberger, E.; Horton, R.; and Swerdloff, R.S.
The Control of Testicular Function: Review of Investigations Conducted on Testicular Function and its Control in the Last 10-15 Years.

Tatum, H.J.
The Most Important Developments in Intrauterine Contraception Over the Past Two Decades—Past, Present and Future State of the Art.

Tietze, C.
Clinical and Epidemiological Aspects of Induced Abortion.

Wurtman, R.J.
Brain Neurotransmitters and the Hypothalamic Control of Pituitary Gonadotropin Secretion.

Yen, S.S.C.; Naftolin, F.; Lein, A.; Krieger, D.; and Utiger, R.
Hypothalamic Influences on Pituitary Function in Humans.

**Country Reports on Status and Funding of Research in the Reproductive Sciences and Contraceptive Development**

Adadevoh, B.K.; Dada, O.A.; and Olatunbosun, D.A.
East and West Africa.

Ahn, Y.O.
South Korea.

Apelo, R.A.
Philippines.

Barakat, R.M.
Iran and Turkey.

Bennett, L.
Canada.

Dusitsin, N.
Thailand.

Franchimont, P.
Belgium.

Hagenfeldt, K.
Scandinavian Countries—Denmark, Finland, Norway, and Sweden.

Hefnawi, F.
Egypt.

Hoult, J.R.
United Kingdom.

Lindner, H.R.
Israel.

Mantingh, A., and Ma, H.K.
Hong Kong.

Martini, L.
Italy.

Prasad, M.R.N.
India.

Thibault, C.
France.

van der Molen, H.J.
Netherlands.

Voigt, K.D., and Wollmer, W.
Germany.

Walters, W.A.W.
Australia and New Zealand.

**Summary Tables (Matrices) on Biomedical and Methodological Advances, Remaining Problems, and Contraceptive Possibilities**

Anderson, J.
Spermatozoan, Capacitation, and Fertilization.

Biggers, J.
Early Embryonic Development.

Brauer, J., and Naftolin, F.
CNS-Hypothalamus.

Cavazos, L., and Ofner, P.
Prostate and Seminal Vesicles.

Darney, P.
Female Accessory Organs.
Clinical Aspects of the Female.

Dym, M.
Testis.

Hamilton, D.W.
Epididymis.

Hembree, W.
Clinical Aspects of the Male.

Madhwa Raj, H.G.
Releasing Factors.

Makris, A.
Ovary.

Moyle, W.
Gonadotropins: LH, FSH and HCG.
Prolactin.

Thompson, I.
The Menstrual Cycle.

**Special Reports**

Braun, G.
Regulatory Requirements Seen from the Viewpoint of the Pharmaceutical Firms.

Diczfalusy, E.R.
Steps in the Human Reproductive Process Susceptible to Regulation and the Prospects of Developing Improved and Safe Methods for Specific Interference.

Duncan, G.W.
Survey of Contraceptive Research in Industrial and Not-for-Profit Organizations.

Fawcett, D.W.
Studies of Reproductive Biology of the Male Before 1960.
Major Advances in Our Understanding of Reproductive Biology of the Male Since 1960.

Greep, R.O.
History of Funding in Reproductive Biology and Contraceptive Development.
Major Advances and Some Remaining Gaps in Knowledge in Reproductive Biology and Contraceptive Development.

Hutt, P.B.
Regulations of Contraceptive Drugs and Devices by the Food and Drug Administration.

Kessler, A., and Standley, C.C.
Research Training in the Biomedical Aspects of Human Reproduction and Family Planning in Developing Countries.

Knobil, E., and Pullum, T.
Attitude Survey of Scientists in Reproductive Research.

Kordon, C., and Gaudin, T.
Reflections on Ethical Aspects of Contraceptive Policies and Human Testing.

Loupos, J.
The Cost of Supplies and Equipment as a Manageable Expense for Bio-Medical Research.

McLean, S.A.
Patents and Invention Financed by the Public Sector: A Case Study in Contraceptive Research. Ethics and Biomedical Research on Human Subjects: A Foundations Experience in Funding Reproductive Biology Research.

Schrogie, J.J.
Development of Contraceptive Drugs and Devices.

Segal, S.J.
Possibilities for Future Contraceptives.

Appendix B　　　　　　　　　　Human Reproduction and
　　　　　　　　　　　　　　　Prospective Means of Fertility
　　　　　　　　　　　　　　　Control

One can control a machine one does not understand, but such control is likely to be clumsy, unreliable, and potentially damaging. Accurate control can come only when that mechanism is thoroughly understood.

　The parallels with the regulation of human fertility are obvious. Contraception existed in crude and mostly unreliable forms when knowledge of reproductive physiology was almost nonexistent. Research in the first half of this century led to a sufficient understanding of the role of steroid hormones in the female reproductive system to allow the development of the pill, a contraceptive of undeniable efficacy and unparalleled social and demographic importance but one that falls short of what is required in terms of safety and the level of sophistication required of its delivery system. For the future, only painstaking further analysis of the male and female reproductive systems will reveal those points most vulnerable to interference by contraceptive techniques that will be safe, reliable, acceptable and inexpensive.

　This appendix will review some of the highlights of the modern reproductive sciences as they are currently understood and outline several points at which the chain of events necessary for reproduction could be interfered with for purposes of regulation. It is hoped this brief review will facilitate the reading of chapters 4 and 5, which describe in greater detail research developments in this field and identify crucial problems that remain to be solved.

## The Female Reproductive System

The remarkable feature of the female reproductive system is that it functions on a cyclic basis. The great triumphs of the last half-century of research in reproductive physiology center on the understanding gained of the intricate pattern of feedback relationships that exist between the three organs that maintain this cyclicity: the ovary, the pituitary gland, and a portion of the brain. Each component of this brain-pituitary-ovarian axis both secretes and responds to

This chapter by Mr. Graham Chedd summarizes the essays commissioned for the review. It is intended for the nonscientist reader.

hormones (that is, both send and receive chemical messages). The pill works by disrupting this communication. And many of the most promising avenues to new contraceptive techniques lie in finding more subtle ways of intervening in the reproductive process to sidetrack those leading to pregnancy while leaving uninterrupted the hormonal functions that contribute to normal metabolism and libido.

## The Ovary

The ovary is, first and foremost, the source of human eggs, or oocytes. Oocytes have their origins very early in embryonic development, but curiously, they do not start their life in the ovary. Studies of human female embryos just a few weeks after conception reveal the germ cells that are the precursors of oocytes migrating in the wall of the embryonic gut toward the primitive ovary (at this stage a ridge of tissue indistinguishable from the primitive testis in a male embryo). These germ cells divide repeatedly until by the fifth month of gestation there are several million settled in the tiny but by now recognizable ovaries.

There will never be as many again. As the germ cells begin their maturation into oocytes—a process that will not be completed until many years later, when they are ovulated—their numbers begin to decline in a process known as *atresia*. At birth, there are fewer than one million oocytes; by the age of seven only some three hundred thousand are left, and half of these are showing signs of degeneration. But there will be no shortage; only some three hundred fifty to four hundred ova are shed during the normal reproductive lifespan of the adult woman.

Each menstrual cycle in the adult begins with the maturation of some twenty oocytes. The ripening process takes place in *follicles*, whitish fluid-filled vesicles that bulge from the surface of the ovary. Under normal circumstances, only one follicle completes its maturation; the remainder, with their eggs, degenerate, apparently under the influence of a substance released from the prime follicle. The follicular phase of the average cycle lasts approximately fifteen days, but it may vary from about one week to more than a month. It concludes with the bursting of the prime follicle and the release of the egg to begin its journey down the oviduct and, perhaps, to a waiting sperm.

The follicle that released the egg then undergoes a remarkable change, from burst blister to a large, yellowish fleshy body protruding from the ovary. This body is known as the *corpus luteum*. It will have a crucial role to play if the egg it once contained does indeed become fertilized. If the ovum meets no successful sperm and fertilization does not take place, the corpus luteum will after a few days begin to regress. Menstruation follows, and a new batch of follicles begin to mature.

That the ovary is more than an egg-producing organ first became clear more than fifty years ago. Studies of ovaries from slaughter animals had already revealed their cyclical pattern of anatomical change from the bearing of follicles to a corpus luteum and back again. These changes had been correlated with alterations in the reproductive tract. The linking mechanism was unknown, but it was suspected to be some kind of "generative ferment" carried in the bloodstream. The concept was prophetic. In 1923 E. Allen and E.A. Doisy treated immature rats and mice with fluid removed from the ripening follicles of pigs and found that they promptly showed signs of sexual maturation. At autopsy, the uterus of each animal was found to be greatly enlarged. From then on the ovary was known not to be simply a source of eggs but an endocrine gland as well. Two estrus inducing hormones were isolated from the urine of pregnant women in the late twenties in the laboratories of Doisy and G.F. Marrian, but the hormone secreted by the ovaries was not chemically identified until 1935 by Doisy's group. More than four tons of pig ovaries provided the starting material for the isolation and characterization of this hormone. All three of these *estrogens* turned out to be related steroids that could be interconverted.

In 1932 a second class of steroids, of which *progesterone* was the most important, was identified in females by G.W. Corner and W.M. Allen. Steroids were also obtained from the adrenal glands and in males the testes were discovered to produce androgens, of which *testosterone* is the archetype. Estrogens and androgens are known respectively as "female" and "male" hormones, although each sex possesses low levels of the other's sex hormone.

The testes in males manufacture mainly androgens, but the ovaries synthesize both estrogen and progesterone. There is, however, a physiologically crucial division of labor: Follicles are the main source of estrogen; progesterone is principally the product of the corpus luteum. Accordingly, during the first half of the cycle, while the follicles are ripening, estrogen is the major ovarian steroid in the female circulation. Its levels rise to a peak just prior to ovulation. Once ovulation has occurred (or slightly before in the human female) and the corpus luteum has begun to develop, progesterone levels begin to rise—to decline precipitously at the end of the cycle in the absence of conception as the corpus luteum degenerates.

Both estrogen and progesterone have the uterus as their immediate target. The uterus provides a site for the attachment of the embryo to the mother and an environment for its development during gestation; both functions are mediated by the lining of the uterus, the *endometrium*. The uterus also provides a mechanism for the expulsion of the fetus at term by means of its smooth muscle outer coat, the *myometrium*. Under the influence of estrogen during the first, or preovulatory, phase of the menstrual cycle, the endometrium widens and contains many straight tubular glands lined with cells that store large amounts of glycogen. During the postovulatory, or luteal, phase of the cycle the endometrial glands become markedly coiled and distended with fluid that is rich in glycogen. In this state the endometrium is ready to receive and nurture a fertilized egg. In the absence of conception and due to the withdrawal of progesterone as the corpus luteum regresses, the endometrial lining of the uterus is shed during menstruation.

These "local" actions of estrogen and progesterone were the first to be discovered and are clearly crucial in the reproductive process. Estrogen is also responsible for the development of the secondary sex characteristics of females and influences sex behavior in at least some animals. Progesterone is first and foremost the hormone of pregnancy, being present in high levels in the bloodstream of women throughout gestation, in which it plays an obligatory role. But this list of functions for estrogen and progesterone leaves out the hormones' important role in regulating the sexual cycle itself, a task that makes possible their being used to inhibit ovulation, as with the pill. These actions are mediated via an inhibiting influence on the second and literally central organ in the control of the female reproductive cycle, the *pituitary gland.*

**The Pituitary**
Probably the most far reaching discovery ever made in reproductive biology was the revelation by P.E. Smith in the mid-1920s that the female and male gonads are under the control of the anterior lobe of the pituitary gland. The pituitary is a marble-sized organ located in a bony socket beneath the base of the brain and connected to it by a slender stalk. The first proof that the pituitary plays a role in the reproductive system came from experiments in which fragments of fresh tissue from the anterior pituitary produced precocious sexual maturation and an enlargement of the ovaries when implanted in immature rats and mice. This experiment was quickly bulwarked by the reverse demonstration that the removal of the pituitary gland without injury to the brain caused the total loss of function of the reproductive system in rats of both sexes.

In the early 1930s physiological experiments by F. Hisaw and H. Fevold suggested that the anterior pituitary exerted its action on the female gonad through not one hormone but two. The gonad stimulating hormone causing follicular growth became known as the *follicle stimulating hormone* (FSH); the hormone converting the follicle into a corpus luteum as *luteinizing hormone* (LH). Since then, the role of each hormone has been more precisely defined. During the first phase of the menstrual cycle FSH is the primary agent required to bring the follicles to maturity while LH stimulates them to produce estrogen. The final preovulatory enlargement of the follicle and its rupture and release of the free-floating ovum is induced by a midcycle "surge" of LH and FSH from the pituitary.

The introduction in the late sixties of the highly sensitive radioimmunoassay technique for measuring FSH and LH levels in the peripheral blood by the laboratories of E. Knobil, A. Midgley, G. Ross,

451                             Human Reproduction and Prospective
                                Means of Fertility Control


G. Abraham, R. Vande Wiele, and M. Taymor has confirmed a sharp preovulatory surge of LH and a lesser one for FSH. Both hormones appear to participate in ovulation, but LH has a far greater ovulatory potency than FSH. At other times in the cycle, the blood levels of both hormones remain low and relatively constant; the FSH profile depicts a steady, though mild, rise during the early and midfollicular phase followed by a gentle decline to a nadir just before the midcycle FSH and LH peaks. One still unexplained phenomenon revealed by the new assay technique is that LH release (and to a lesser extent that of FSH, too) is not steady but occurs in brief spurts. This episodic nature of its secretion is revealed by sharp rises in blood LH at approximately hourly intervals followed by a slow decline as the hormone decays.

One of the most intriguing—and potentially important—problems in reproductive biology concerns the formation, lifespan, and function of the corpus luteum, which in pregnancy sustains a tremendous outpouring of progesterone but which in the absence of conception dwindles and decays a few days after its formation. Before considering the corpus luteum further, two more hormones that act upon the ovaries must be introduced. The first is another hormone from the anterior pituitary, *prolactin*. Although prolactin has a common function among mammals in promoting the secretion of milk, its functional virtuosity throughout the vertebrate kingdom is amazing. It enables some species of fish to migrate between salt and fresh water; it is the hormone that brings some amphibia to the water to breed; it induces brooding behavior in birds; and in birds and mammals generally it acts to suppress the production of the other pituitary gonadotropins. Of interest here is that in some species, notably rats and mice, prolactin has a role in maintaining the function of the corpus luteum as shown by E. Astwood and by H. Evans. Whether or not is has a similar role in humans is unclear.

Another hormone having an effect on the ovary is secreted not by the pituitary but by the implanted embryo in the event of conception. This hormone, *human chorionic gonadotropin* (HCG), discovered in 1927 by S. Aschheim and B. Zondek, is present in high concentration in the blood and urine of pregnant women. Its assay, until recently by injection into immature mice, rats, rabbits, or toads but today by a simple immunological technique, forms the basis of the diagnosis of pregnancy. Possessing powerful luteinizing hormone-like activity, hCG has been used to induce ovulation in infertile women following treatment with follicle stimulating hormone.

The identity of what is known as the *luteotropic complex*—the precise combination of hormones necessary for the maintenance of the corpus luteum—differs in different species. Prolactin plays a role in rats and mice; hamsters appear to require a complex of FSH, prolactin, and estrogen; the rabbit needs estrogen alone. In women, the luteotropic agent as shown by Vande Wiele in 1970 appears to be solely LH though it remains possible that prolactin also plays a role. The task of maintaining luteal function in the event of pregnancy falls mainly upon hCG, a fact that forms the basis of one of the most promising new experimental contraceptives (see item A, "An Antipregnancy Vaccine," page 469).

Even more mysterious is the identity of the factor (or factors) responsible for ending the brief life of the corpus luteum at the end of each cycle in the absence of conception, the *luteolytic complex*. In some species, most notably the sheep, the luteolytic agent is a *prostaglandin* produced by the uterus under the influence of the ovary. There was a brief period recently when it was thought that prostaglandins, now available synthetically, might also be luteolytic in humans, providing the basis of a monthly pill that would induce the regression of the corpus luteum and consequent menstrual bleeding whether or not conception had occurred. But whatever the luteolytic agent(s) in humans, prostaglandins seem unlikely to be involved, though the search for luteolytic agents active in the human female continues (see item B, "Menstrual Regulation," page 470).

Studies since the mid-1960s have begun to unravel the structure of the gonadotropic hormones FSH, LH, and hCG. All are glycoproteins with attached sugar chains. Their amino acid sequences have been determined. Each hormone consists of two nonidentical subunits called alpha and beta of about the same size, and their most remarkable feature is that all three hormones share an identical alpha subunit. This is so regardless of the species. The hormone specificity of each molecule is provided by the beta subunit. The variability of the beta subunit has also provided the basis of the new highly sensitive radioimmunoassays by which levels of the gonadotropins in the circulation can be accurately monitored. Work is now beginning to determine the three-dimensional structure of these gonadotropins.

**Feedback Control of Gonadotropins**
FSH and LH are secreted by the anterior pituitary with the ovaries as their target. There they control the ovarian cycle: *follicular phase*, *ovulation*, and *luteal phase*. The synthesis of steroid hormones—estrogen and progesterone—by the ovaries appears to be largely under the control of LH; the precise details of this control are still under study. But although the pituitary influences the ovaries via the gonadotropins, the relationship is not unidirectional. One of the most profound discoveries in reproductive biology—and certainly the most important in practical terms—was the realization that the ovaries influence the pituitary via their own hormones, estrogen and progesterone.

That pituitary function is regulated by the ovarian hormones was first demonstrated in the early 1930s when R.K. Meyer and associates showed that estrogen suppressed the output of gonadotropins by the pituitary. The evidence was fortified by the discovery that the loss of secretory function of the ovaries, as a result of ovariectomy or menopause, was followed by a marked increase in the secretion of FSH and LH by the pituitary. That progesterone as well as estrogen has a negative feedback role became known in 1937 when A. Makepeace demonstrated that progesterone blocked postcoital ovulation in rabbits. The logical extension of this observation to the possibility of controlling fertility lay dormant for sixteen years. It was not until 1953 that the era of oral contraception was ushered in when G. Pincus and M.C. Chang showed that a number of synthetic progestational compounds are active inhibitors of ovulation in mated rabbits when given by mouth. Beginning in 1956, the most potent of these orally active compounds were tested in sexually active women of reproductive age. Their contraceptive efficacy was established beyond question. That they controlled fertility by inhibiting ovulation was inferred from an abundance of clinical evidence including the lack of a midcycle shift in basal body temperature and the absence of progesterone during the second half of the cycle (see item C, "Improvements on the Pill," page 471).

Both estrogen and progesterone, therefore, are able to suppress the output of FSH and LH by the pituitary. But that matters are not quite that simple was first hinted at as early as 1934 when W. Hohlweg showed that in some experiments estrogen increased the pituitary output of gonadotropins. A similar surprise came in the 1940s when progesterone, by now clearly shown to suppress ovulation, was found by J. Everett also to facilitate the induction of ovulation under certain circumstances. At the time of the original demonstration that estrogen could exert a positive as well as negative feedback on the pituitary, it was postulated, with what can only be termed a remarkable stroke of prescience, that the actions of the gonadal hormones on the pituitary were mediated through what was called a "sex center" in the brain. This concept was unfortunately taken lightly, disregarded altogether, or ridiculed; but thirty-five years later, history was to make amends.

### The Brain

That the brain has a direct influence on the reproductive cycle was first implied by experiments in the early thirties in which artificially changing the daylight length hastened the recrudescence of gonadal functions in seasonal breeders. Experiments of this type, clearly suggesting that the control of the gonads by the pituitary was in its turn susceptible to neural control, led to a search for neural pathways from the brain to the anterior pituitary. The search proved fruitless, yet the fact that the region of the brain immediately over the pituitary—the *hypothalamus*—was unmistakably involved in its control was confirmed time and time again by a variety of experiments in which electrical stimulation of the hypothalamus or the creation of electrolytic lesions within it altered the state of the ovary in experimental animals. Later, drugs that affect brain function were also shown to block ovulation in the rabbit if given shortly after mating. Yet the mystery remained. How, if there are no nerves connecting the hypothalamus with the pituitary, does the brain mediate its control over pituitary activity?

The answer turned out to be via a system of tiny blood vessels connecting the hypothalamus and anterior pituitary—vessels first noted by G. Popa and U. Fielding during the dissection of a cadaver in 1930. The discovery of the vessels led to speculation that they might convey neural stimuli to the pituitary by means of chemical signals, or neurohormones. It was to take many years of careful experimentation, however, before this idea gained wide acceptance in the late 1940s.

More than another decade passed before the first unequivocal demonstrations in 1960-1961 by S. McCann and by G. Harris of a substance in extracts of hypothalamic tissue that could stimulate the pituitary to release LH. In the following years hypothalamic fragments were collected from 5 million sheep and 2.5 million pigs in a herculean effort to characterize the elusive *luteinizing hormone releasing factor* (LH-RF or LRF). Success came in 1971 when, first in the pig by A. Schally and a few months later in sheep by R. Guillemin, LRF was isolated, purified, and characterized as a peptide of ten amino acids. The molecule was identical in the two species. LRF was immediately reproduced by chemical synthesis, and synthetic LRF is now available in virtually unlimited quantities and is often termed LH/FSH-RF or Gn-RF.

Both the native and synthetic LRF stimulate the secretion of FSH as well as LH. A separate FSH-RF has not been identified and it remains a moot point whether or not one exists or if LRF releases FSH as well as LH under normal physiological conditions. The synthetic peptide is already in use throughout the world to diagnose the ability of the pituitary to secrete gonadotropins in infertility and its clinical use for inducing ovulation in female patients whose infertility results from a hypothalamic deficiency is now beginning. Work is also underway on synthetic analogs of LRF as potential contraceptive agents (see item D, "Shutting Out the Brain," page 472).

The involvement of the brain in the reproductive cycle, via the hypothalamus and its chemical messenger, LRF, extends greatly the pattern of possibilities that exist for the regulation and control of reproduction. It adds a new control loop to the reproductive system. Recent research in reproductive biology has concentrated on understanding the details of this neural involvement and its integration with the pituitary-ovary axis.

The central question has concerned the nature of the feedback, mediated by estrogen and progesterone, from the ovary to pituitary. As mentioned earlier, this feedback can be both positive and negative. The conventional contraceptive pill works by negative feedback, the high levels of ovarian steroids preventing ovulation by blocking the release of hormones from the hypothalamus that control the release of FSH and LH from the pituitary. How then can even higher levels of estrogen and progesterone induce ovulation if properly timed? And where is the feedback effected, at only the hypothalamus or at both the hypothalamus and the pituitary? Are the sites of action of positive and negative feedback different?

Partial answers to questions of this sort are now beginning to emerge. One important line of research leading to these answers concerns the so-called "critical period." It has been known for some years that in the laboratory rat the LH surge that leads to ovulation occurs between 2 P.M. and 4 P.M. the day before estrus and only then. If the LH release is blocked by the administration of an anesthetic, it is delayed until the same critical two-hour period on the following day. Ovulation always takes place some ten to twelve hours after the LH surge.

Recent (1974) research by M. Takahashi and K. Yoshinaga has shown that limits of the critical period are remarkably precise. The extraordinary temporal sensitivity of this phenomenon, plus the ability of anesthesia to block it, clearly indicates a neural involvement, suggesting that the LH release is triggered by a sharp surge of LRF from the hypothalamus. Numerous other lines of evidence support the concept that the LH surge that induces ovulation is triggered by this pulse of LRF. Most clear-cut is the demonstration that acute administration of LRF will induce ovulation in all animals studied provided ripened follicles are present. The results in women are less conclusive but highly suggestive.

This leaves the obvious question: What triggers the pulse of LRF?

The role of estrogen is clearly crucial. As shown by M. Aiyer and G. Fink, when the ovaries are removed from a rat before 10 A.M. the LH surge during the critical period that same afternoon does not occur. If an injection of estrogen is given immediately after ovariectomy, the surge takes place on schedule. Other experiments suggest that the neural "clock" timing the LRF pulse in the afternoon ticks normally but that it is only effective in triggering the LRF pulse—and thus in turn the surge of LH and FSH—if it has been primed with estrogen.

Here, at least in the laboratory rat, are the beginnings of an explanation of the positive feedback effects of estrogen on ovulation. These events are clarified further by experiments that suggest the LRF preovulatory pulse is triggered not in the hypothalamus itself but just to the front of it, in the so-called preoptic area of the brain. Early indications that this is so came from an experiment in which electrodes being inserted into the hypothalamus of an anesthetized rat in an unsuccessful attempt to induce ovulation were placed accidentally in the preoptic area instead. There the electrodes stimulated ovulation. More recently, the importance of the preoptic area in triggering the LH surge has been confirmed in both rats and monkeys where severing the preoptic area from the hypothalamus surgically abolishes the acute LH discharge from the pituitary.

Of interest is the fact that in these same experimental animals the normal "baseline" secretion of FSH and LH was undisturbed. These and other experiments have led to the following explanation of the different feedback effects of estrogen on the brain-pituitary axis. Negative feedback takes place in the hypothalamus itself, inhibiting the minute rhythmic pulses of LRF that in turn trigger the tiny bursts of LH and FSH that keep the ovaries functioning throughout the cycle—a task that involves priming the follicles and ripening them preparatory to the LH surge that will trigger ovulation. This "tonic" action of the gonadotropins is also responsible for the secretion of estrogen and progesterone by the ovaries, and so is kept in balance; a fall in the level of steroids and the output of LRF and gonadotropins is increased, stimulating the ovaries further and vice versa—a rise in steroids inhibiting gonadotropins.

Superimposed upon this pattern is the *positive* feedback activity of estrogen, which has the preoptic area of the brain as its site of action. The rising levels of estrogen during the first phase of the cycle eventually trigger the preprogrammed neural clock in the preoptic area to initiate a large pulse of LRF to the pituitary, setting off the LH surge and, as a result, inducing the ovulation of a mature follicle.

Complex as this system may seem, it is by no means the end of the story. There exists other feedback loops. One, for instance, involves estrogen operating directly on the pituitary rather than the hypothalamus or preoptic area. The response to a given dose of LRF is far greater in the preovulatory period, when estrogen levels are high, than it is at any other time of the cycle in rodents, monkeys, and the human female. Apparently estrogen increases the sensitivity of the pituitary to the LRF message. Estrogen, therefore, not only triggers the message, it primes the pituitary to receive it. Other feedback loops still being explored involve the effects of gonadotropins and of releasing factors for other pituitary hormones upon LRF secretion and activity. For example, there is an undoubtedly complex story to be unraveled concerning prolactin, which appears to reduce the response of women to LRF. Research into this phenomenon might lead to an explanation of the reduced fertility of nursing mothers and perhaps open up new contraceptive possibilities.

**Pregnancy**

Although the majority of research attention has focused on the factors controlling ovulation, the fate of the egg once it has been expelled from the follicle has not been left entirely unexplored. Of late the *oviduct*, once looked upon as a conduit where sperm meets egg and the resulting zygote moves haltingly downstream to the uterus, has been the subject of rather intense investigation, with the result that it is now seen as a dynamic and complex structure. The functions of the oviduct may be broadly described as transport of gametes and provision of a suitable fluid environment for the sperm, the ovum, the fertilization process, and the newly created zygote.

The egg begins its journey at the funnel-shaped opening of the oviduct, through the contraction of its smooth muscle coat moves closer to the ovary, and efficiently picks up the ovum a few moments after its release from the follicle. After fertilization, should spermatozoa be present and successful, the resulting *zygote* is held in check for some three days while the uterus is being prepared for its reception. The mechanism by which sperm and ova are transported through the oviduct is still poorly understood, despite many studies. The oviduct is lined with *cilia*, or fine hairlike structures that in most species beat toward the uterus and are presumably involved in transport. It is also, however, becoming increasingly evident that the primary factor in the transport of both sperm and ova is the peristaltic contraction of the oviduct by means of the smooth muscle fibers that invest it.

The *oviducal fluid*, the medium in which sperm are readied for fertilization and in which the egg is fertilized and begins its development, is an extraordinary complex mixture made up from components of blood plasma and secretions from the oviduct epithelium with additional fluid contributions coming from the peritoneum, follicle, and uterus. Numerous enzymes, proteins, carbohydrates, electrolytes, and other constituents have been identified in the oviducal fluid. A detailed knowledge of fluid composition is still lacking, however, and is of fundamental importance—not only in developing suitable media for the in vitro fertilization and culture of early embryos, but also because it may provide a short step to arresting any one of the critical biological processes that occur in the oviduct. Moreover, the timing of the passage of the fertilized ovum through the oviduct is absolutely critical. If its passage could be hastened or delayed even by as much as a day, the ovum would almost certainly not find a favorable uterine environment for its continued development and implantation (see item E, "Breaking the Journey Down the Oviduct," page 473). This, of course, has only been demonstrated in laboratory animals.

Before entering the uterus, the zygote passes through the *uterotubal junction*, a gateway the sperm must earlier have passed on its way to the oviduct. In some species, the uterotubal junction is able to limit the number of sperm passing through and even monitor their quality, rejecting nonmotile sperm or sperm of the wrong species. In the human female, research interest has focused on ways that the junction may be occluded with the aid of drugs, sclerosing agents, cauterization, or the insertion of plugs or adhesives. This research has so far had limited success but an effective, simple, and safe means of closing the uterotubal junction would clearly be of major contraceptive importance (see item F, "Closing off the Tubes," page 474).

By the time the zygote enters the uterus, that organ has been prepared for its reception by the action of estrogen and progesterone from the ovaries. The embryo has by this stage undergone several cleavage divisions and has formed a hollow circular ball of cells called a *blastocyst*. The outer layer of cells of the blastocyst constitute the *trophoblast*, and it is the cells of the trophoblast that stick to the uterine wall, break through the surface epithelial layer, and burrow into the underlying stroma. As it does so, the trophoblast swallows maternal blood vessels, beginning the establishment of the *placenta*. It is this process of *implantation* that is presumably disrupted by intrauterine devices for contraception (see item G, "Preventing Implantation," page 474).

Once implantation has taken place or perhaps even earlier, the trophoblast begins the manufacture of human chorionic gonadotropin (hCG), which, as discussed earlier, prolongs the life of the corpus luteum and thus sustains the high levels of progesterone necessary for the pregnancy to proceed. In the human female the placenta itself soon starts to produce progesterone and the role of the corpus luteum becomes dispensable. Progesterone has many roles in the maintenance of pregnancy. One of the most important may be in the protection of the fetus from premature expulsion from the uterus, for progesterone in experimental rabbits has been shown to reduce the coordinated contraction of the myometrium and abolish the stimulatory effects on this contraction of the hormone oxytocin secreted by the posterior lobe of the pituitary. Labor begins, according to this hypothesis, when progesterone is withdrawn at the conclusion of gestation.

Perhaps the most remarkable recent discovery on labor initiation is that in the sheep it is determined not by the mother but by the fetus. That the fetus in any species can determine when it will be born was, to state the obvious, a departure from conventional thinking. Apparently the fetal pituitary stimulates the adrenal glands of the fetus to secrete hormones that reduce the placental synthesis of progesterone. *Labor* is initiated as a result.

Whatever the role of prostaglandins in inducing the regression of the corpus luteum in the human female (discussed earlier), it has now been firmly established that the prostaglandin content of the blood and amniotic fluid increases considerably before and during labor. Prostaglandins given by injection or infused directly into the amniotic fluid bathing the fetus have also been shown to stimulate myometrical contractions directly and have been used in midtrimester abortions and in the induction of labor at term (see item H, "Interfering with Early Embryonic Development," page 475).

The Male Reproductive System

Like its female counterpart the ovary, the *testis* functions as a nursery for germ cells. In the adult female the ovary undergoes a regular cycle, releasing one egg at a time about every twenty-eight days; but the adult testis keeps up the regular and continuous production of millions of *spermatozoa* a day throughout the male's fertile life, which may itself be twice as long as that of the female.

The testis manufactures germ cells upon a unique cellular scaffolding constructed upon the inner surface of long, thin, convoluted tubules that make up the substance of the testis. The scaffolding consists of columnar *Sertoli cells*, projecting inward and comprising part of the wall of the *seminiferous tubules*. In the interstices between the bases of the Sertoli cells lie the *spermatogonia*, the dividing cell population from which the spermatozoa are derived. As new germ cells are created by the division of spermatogonia, they climb the scaffolding provided by the Sertoli cells, squeezing upwards between the columnar cells until they emerge some sixty-four days later from the top of the scaffolding to float free in the lumen and be carried off down the tubules. During this climb the germ cells mature from primitive, relatively undifferentiated germ cells to a form clearly recognizable as a *spermatozoon*, although not quite ready to be called one.

The upward movement of germ cells between the supporting Sertoli cells poses some very peculiar problems for the control of *spermatogenesis*. It precludes the sort of physical attachment between the two cell types that other cell systems use to keep the cells in contact and in communication with each other. Indeed, a diligent search for links or communicating junctions between Sertoli cells and germ cells with even the most modern electron microscopic technique reveals no sign of such connecting structures. Apparently Sertoli cells exert their influence upon germ cell differentiation solely through the maintenance of the correct fluid environment in the spaces between them that are inhabited by the developing spermatozoa.

In this task, the Sertoli cells *do* appear to be aided by a special type of cell-to-cell contact revealed by the electron microscope. These contacts are between the Sertoli cells themselves, near their bases, and serve to seal off the lower region of the cell system, that containing the spermatogonia, from the upper portion of the scaffolding occupied by the maturing germ cells. The barrier created by these junctions constitutes one of the most effective in the body, not only trapping inside the seminiferous tubules the fluid needed for germ cell maturation, but also keeping out of the tubules agents in the bloodstream that might be damaging to the germ cells. The existence of this permeability barrier has to be taken into account when considering the design of antispermatogenic drugs, which must be able either to break up the Sertoli junctions or achieve their effect by acting on the cells outside the barrier.

Despite the effectiveness of this barrier, there clearly must exist some mechanism for permitting passage of the immature germ cells. Presumably the Sertoli junctions open and close in some manner to permit movement of the germ cells toward the lumen without breaching the permeability barrier, although the exact mechanism of the process remains a mystery—and a challenging unsolved problem of spermatogenesis.

The electron microscope also reveals open bridges between the germ cells themselves as they ascend the Sertoli scaffolding. As the spermatogonia divide, there remain links between the two resulting cells; and as these in turn divide, they also remain connected. Thus there develops a large network of interconnected cells, all of which move up through the Sertoli cells together. Networks as large as eighty conjoined cells have been seen, and they probably extend to hundreds. The existence of these interconnecting bridges is clearly important in synchronizing the complex events of germ cell development and of their movement upward past the Sertoli cells. But the cell networks also have important implications for the eventual release of the spermatozoa from the top of the Sertoli cell system. As each spermatozoon is extruded from the apex of the Sertoli cell, a tiny strand of cytoplasm connects it with a mass of residual cytoplasm left behind. This thin filament eventually breaks, freeing the spermatozoon to begin the next stage of its journey.

As in all other stages of the germ cell's differentiation within the seminiferous epithelium, this final release of the spermatozoon—along with the rest of its "class" that began the process together sixty-four days earlier—is under the control of the Sertoli cells. According to recent research these cells have emerged as playing an increasingly important and active role in spermatogenesis, certainly much more than that of a passive scaffolding or mechanical support. As this role has become apparent, though certainly not at present fully understood, the Sertoli cells have also presented an increasingly attractive target for the action of antispermatogenic agents (see item I, "Blocking Spermatogenesis," page 476).

### Androgen

Between and around the seminiferous tubules as they lie coiled in the testes are the *Leydig cells*, first noted in 1850 by F. Leydig and prophetically ascribed by French researchers B. Bouin and P. Ancel early in this century the function of producing secretions important in developing and maintaining "maleness." The Leydig cells are indeed now known to be the source of male hormones, or *androgens*, of which *testosterone* is the most important. But whereas earlier research emphasized the role of androgens carried in the blood and acting at a distance from their site of origin, principally on the male accessory glands and on the development of secondary sex characteristics, current interest focuses on the local action of androgens, acting directly on the process of spermatogenesis. The concentration of androgens in the blood and tissue fluid around the seminiferous tubules is very much higher than the levels of androgens in peripheral blood elsewhere in the body—ten times higher in some species—and it has become increasingly apparent that fertility is highly dependent upon this locally high androgen concentration.

The trick of regulating the Leydig cells' androgen production falls to the pituitary gland and to its gonadotropin, *interstitial cell stimulating hormone* (ICSH), identical to luteinizing hormone (LH) in the female. In the male, as in the female, the pituitary, acting under the direction of the hypothalamus, controls fertility through its production of luteinizing hormone (LH) and *follicle stimulating hormone* (FSH). But the names, resulting from the much more extensive research done on female reproductive endocrinology, are inappropriate for describing these functions in the male.

Although the role of LH in regulating the Leydig cells is now firmly established, the task of FSH is much less clearly defined. The original suggestion, made thirty years ago by R. O. Greep and F. L. Hisaw, was that FSH acts upon the seminiferous tubules in some way. Because in the presence of adequate levels of testosterone spermatogenesis can continue uninterrupted in rats whose pituitary glands have been removed, the importance of FSH in spermatogenesis is in doubt. On the other hand, FSH does trigger biochemical changes in the immature testes; and in very recent research it has been shown to bind to the Sertoli cells.

It is the action of FSH on the Sertoli cells that is a focus of present research. Of emerging importance is a substance known as *androgen-binding protein* (ABP), a protein produced by the Sertoli cells under the influence of FSH. Testosterone diffusing into the seminiferous tubules from the Leydig cells outside becomes bound by ABP, which may play an important role in maintaining the high concentrations of androgens necessary for spermatogenesis. The androgen-ABP complex is also carried "downstream" from the tubules to the epididymis (see next section).

Testosterone itself has recently been found capable of stimulating the synthesis by the Sertoli cells of ABP, and an unanswered question concerns the relative roles of testosterone and FSH in maintaining ABP levels. Why ABP should be necessary at all when testosterone can diffuse freely into the tubules by itself is also still unknown. Above all, perhaps, is the problem of what testosterone does to promote spermatogenesis. A *receptor protein* (distinct from ABP) that binds androgen and, presumably, carries it to the cell nucleus has recently been found in cells of the seminiferous tubules. It is not yet clear whether this receptor is in the Sertoli cells, the germ cells, or both.

In the female a feedback loop exists between LH and FSH production by the pituitary on the one hand, and progesterone and estrogen production by the ovaries on the other. In the male, a similar feedback loop between LH and the product of its target cells, testosterone, is also well established. But just as the role of FSH in spermatogenesis has long been uncertain, so too have the details of its feedback control: Its target tissue has no identified product that could feed back to the pituitary. So over forty years ago, D. R. McCullagh postulated one and named it *inhibin*. Evidence for the existence of inhibin remained scanty and indirect until 1975, when three laboratories in different parts of the world (B. P. Setchell and F. Jacks in Britain, P. Franchimont in Belgium, and D. De Kretser in Australia) obtained extracts of testis, testicular fluid, or semen with the properties of inhibin (that is, it selectively lowers FSH levels in blood while leaving LH unaffected). There is agreement among the laboratories that the substance is probably a small protein, and at the time of this writing vigorous efforts are underway to purify and characterize the active material (see item J, "Towards a Male Pill," page 477).

## From Testis to Vagina

As the newly formed spermatozoa emerge from the apexes of the Sertoli cells in the seminiferous tubules, they face a journey of well over seven meters before their eventual ejaculation. This journey begins as the spermatozoa are carried in dilute suspension to a network of thin-walled channels near the rear of the testis (the *rete testis*), and from there in a dozen or so small ducts (*ductuli efferentes*) to the beginning of the *epididymis*. This latter is a highly convoluted duct resting against the back of the testis. From the epididymis the spermatozoa pass to the *vas deferens*, a larger muscular duct that passes out of the scrotum to enter the pelvis, and terminates in a slender ejaculatory duct opening on each side of the urethra as it passes through the *prostate* just below the bladder.

On this long journey, most of which involves passage through the long epididymal duct, the spermatozoa finish their maturation. Sperm taken from the beginning of the epididymis of laboratory animals are infertile; by the time they reach its end, they have acquired fertilizing capacity and are ready to be ejaculated. There was for many years great debate as to whether spermatozoon maturation was simply a function of the length of time they spent in the epididymis or whether the specific environment provided by the epididymis was necessary for "finishing" the maturation of sperm. In recent years the electron microscope has revealed the complexities of the lining of the epididymis and has swung opinion toward the concept that the duct contributes positively to sperm maturation. However, the issue is still far from settled.

Several different cell types have been identified in the epididymis, the most numerous and most interesting of which are called the *principal cells*. These differ considerably in appearance depending upon how far along the duct they are located, but those at the beginning of the epididymis have several structural features typical of both glandular and absorptive cells. They clearly play an important role in the major compositional change that takes place in the fluid entering the epididymis from the testis. The most marked change is in volume: Some 90 percent of the fluid entering the duct is resorbed in this and adjacent segments of the duct, greatly concentrating the suspension of spermatozoa. But the biochemical makeup of the fluid is altered, too, the epididymis itself contributing several chemicals and concentrating others. How, or indeed whether, these substances contribute to the maturation of sperm during their transit, or to their maintenance during storage, are presently important gaps in our knowledge. There is no reason to assume that these compounds are the only products of the epididymis; others, of equal or perhaps

greater importance, will almost certainly be discovered in the future.
The epididymis is androgen dependent. The withdrawal of testosterone in laboratory animals through castration causes marked morphopogical and biochemical changes within the epididymis—changes that can be partially restored by the administration of testosterone. Androgens reach the epididymis from the peripheral blood, but the major contribution comes directly from the fluid flowing downstream from the seminiferous tubules, in which the androgen is bound to ABP. The initial portion of the epididymis is therefore exposed to very high levels of androgen and appears to be particularly dependent upon it (see item K, "Preventing Sperm Maturation," page 478).

Fertilization
The target of the ejaculated sperm, expelled into the upper reaches of the vagina by the rhythmic contractions of the vas deferens and bulbocavernosus muscle, is the newly released egg waiting in the oviduct. To get there, the sperm has to penetrate the cervical mucus, cross the uterus, pass through the uterotubal junction, and finally find the egg. The journey is not an easy one; tens or even hundreds of millions may have been deposited in the vagina, but only a few hundred sperm will reach the fertilization site and only one will successfully fertilize the egg. This final stage, *fertilization*, depends upon two very special design features of the sperm: the cap, or *acrosome*, molded around its head and its vigorously beating *tail*.

The acrosome is like a flat fluid-filled rubber bag fitted over the front two-thirds of the head of the spermatozoon within its outer membrane. It is filled with a complex mixture of carbohydrates and enzymes. Its task is to help the spermatozoon penetrate the coverings of the egg. These coverings consist of a transparent envelope, the *zona pellucida*, and one or more layers of adherent cells released with the egg from the ovarian follicle, the *cumulus oophorus*. In the presence of the egg, spermatozoa undergo a change called the *acrosome reaction*, in which the membrane of the acrosome fuses with the outer membrane of the spermatozoon head, releasing the enzyme-rich contents of the acrosome. These enzymes disperse the cells of the cumulus and facilitate penetration of the zona pellucida, enabling the spermatozoon to reach the surface membrane of the egg and to fuse with it.

Although the ability of a spermatozoon to swim by beating its tail is not an important factor in allowing it to reach the egg, it is vital to its fertilizing ability once it arrives. The mechanism of *sperm motility* has therefore attracted a good deal of research attention. The structural analysis of the sperm tail has been pursued almost down to the molecular level in the past decade and has provided the basis for a plausible theory of sperm motility.

One of the most intriguing puzzles in reproductive biology of recent years was posed by the discovery in 1951 by M. C. Chang that even when a spermatozoon has reached full maturity in the epididymis, it is still not capable of immediate fertilization of any egg. Further physiological changes have to take place in the female tract before the spermatozoon is able to penetrate the zona and fuse with the egg. This final preparation for fertilization is called *capacitation*.

Capacitation appears to involve changes both in the acrosome and the tail, though no morphological changes are visible in the electron microscope. As far as the acrosome is concerned, capacitation may result from a change in the sperm membrane or the removal of a component from its surface. In any case, only after this change takes place does the acrosome reaction normally occur. At the same time, there is a noticeable difference in the vigor and character of the tail movements of the sperm.

Understanding the nature of the changes involved in capacitation has been complicated by the observation that sperm already prepared for fertilization by residence in the female tract could be *decapacitated* by exposure to seminal plasma. An intensive search in the last fifteen years for an explanation of the mechanism involved has only recently begun to yield results. It seems likely that when ejaculated, spermatozoa have upon their surface components of seminal plasma that are removed by incubation in the female tract. Capacitation may consist of this removal; in decapacitation, these components are replaced by adsorption from seminal plasma.

The discovery of the need for capacitation opened up an era of research on in vitro fertilization. This has led to a much better definition of the requirements for fertilization in vivo. When capacitated spermatozoa approach the recently ovulated egg or contact the surrounding follicular cells, some unknown substance in the oviducal fluid, possibly emanating from these follicular cells, triggers the acrosome reaction. The released enzymes and the vigorously beating tail allow the sperm to penetrate as far as the cell membrane of the egg. The membranes of the sperm head and egg then fuse so that the two germ cells are enclosed by the same cell membrane. A great deal of research interest is now focused upon the region of the sperm head immediately behind the acrosome, for it is in this portion of the membrane that contact and fusion occur. This region clearly must possess special properties allowing cell-cell recognition and fusion to take place.

Sperm-egg fusion precipitates morphological and physiological changes known as *egg activation*. One of the visible manifestations of the process is the release of granules from the egg into the space between its outer membrane and the enveloping zona pellucida. These granules are probably involved in the metaphorical slamming and bolting of the door behind the successful spermatozoon that takes place upon fertilization, thus preventing other sperm from penetrating the zona pellucida. If we knew the biochemical mechanism of the block to late arriving spermatozoa, it might be possible artificially to stimulate it to prevent the entry of the first (see item L, "Preventing Fertilization," page 479).

**Item A. An Antipregnancy Vaccine**

A contraceptive that becomes activated only in the event of conception and then does so reliably and completely unobtrusively could in principle provide safe and effective protection from unwanted pregnancy. Such a possibility has been opened up by recent research on *human chorionic gonadotropin* (hCG), the hormone secreted by the conceptus and responsible for maintaining the corpus luteum in the early stages of pregnancy. hCG shares its biological and many of its chemical properties with luteinizing hormone (LH). One of the two subunits that constitute the hCG molecule, however, differs in its detailed structure from the analogous LH subunit. This allows the preparation of antibodies to the beta subunit of hCG, antibodies that are specific to the molecule and that cross react only minimally with the beta subunit of LH. The most potent and specific such antibody response is raised with a vaccine in which hCG-beta, or a fragment of the molecule, is chemically linked to a so-called *hapten*, a highly antigenic carrier molecule.

Such a vaccine could, in principle, provide the basis of an immunization program. In the event of conception, the antibody response primed by the vaccine would at once destroy the hCG-beta secreted by the conceptus, with the result that the corpus luteum would regress, progesterone levels decline, and menstruation would occur. There would be no indication that conception had ever taken place; menstruation would terminate each cycle whether or not an egg had been fertilized.

A limited number of clinical trials in women of reproductive age using a vaccine of hCG-beta bound to an inactive fragment of tetanus toxin has been encouraging, and plans are also in hand for similar tests using only fragments of hCG-beta as the antigen. Although a single inoculation with an effective vaccine could last indefinitely, there are several feasible (though still untested) means for reversing its effect. For example, the progesterone needed for a successful pregnancy could be given to a woman who wished to bear children after inoculation.

Several barriers still stand in the way of widespread clinical testing of this approach to contraception. Among the more important are considerations of the specificity of the vaccine (any anti-LH effects could disrupt the monthly cycle) and of the likelihood of its triggering autoimmune disease. But the promise of a contraceptive that could in principle be as long term, safe, and reliable as an hCG-beta vaccine is so immense that research is pressing ahead with all possible speed.

Item B. Menstrual Regulation

The key role of the corpus luteum in providing the progesterone necessary for the support of the very early stages of pregnancy makes it a very attractive target for contraceptive research. The indication from animal experiments that *prostaglandins* can cause the regression of the corpus luteum led to the testing of prostaglandins, absorbed via a vaginal or rectal suppository, for their effect in suppressing the corpus luteum in humans. The tests were unsuccessful. However, certain synthetic prostaglandins with a long half-life in the body appear to induce at least a temporary reduction in circulating progesterone and bind very firmly to receptors in the human corpus luteum. As more potent analogs are synthesized and screened for their luteolytic effects, it remains possible that orally active analogs will be developed that, taken once a month, would induce menstruation whether or not fertilization has occurred.

The search for *luteolytic agents* has extended to other substances. For example, certain orally active plant extracts, notably the Mexican plant *Montanoa tuberosa*, called *Zoapatle*, are able to suppress progesterone levels. Zoapatle taken orally also stimulates uterine contractions, which may contribute significantly to its abortifacient properties. A synthetic drug developed in India, Centchroman, is also said to interfere with luteal function. Synthetic progestins in doses large enough to suppress progesterone production by the corpus luteum but low enough not to maintain an early placentation site have also been tested, so far without notable success.

Very large doses of the synthetic estrogen *diethylstilbestrol* are used experimentally in several countries for so-called postcoital contraception. The treatment appears to be effective when given within seventy-two hours following coitus but fails to interfere with gestation once implantation has taken place. The present treatment is accompanied by a variety of unpleasant side effects, cannot be repeated often, and can be regarded only as an emergency measure. However, the approach is promising and it is at least possible that the same effect can be achieved by the use of other types of sterioids exhibiting slight or no side effects. It is also possible that high estrogen doses work not by luteolysis but by interfering with tubal transport.

**Item C. Improvements on the Pill**

Because the efficacy of the conventional contraceptive pill is high and its side effects are related to dose, present research is focused on reducing the levels of estrogen and progestin in the pill. Work in the People's Republic of China in particular has shown that the dosages of both steroids can be reduced markedly without decreasing contraceptive efficacy. Other researchers are attempting to replace synthetic estrogens with natural ones, claiming that the latter do not share all the metabolic effects of the former. Yet others are trying to diminish the burden of daily ingestion of pills by developing "weekend" pills, "once-a-month" pills, and "midcycle" pills.

A more important research trend is to develop *long-lasting sustained release preparations*, either as monthly or three-monthly injectable formulation or as biodegradable (absorbable) or easily removable implants. Two injectable compounds have been sufficiently tested to be marketed in a number of countries, but important issues regarding their safety and reversibility remain unsolved. In the present form of the implant approach, silastic tubes or rods containing the antifertility hormone are inserted (and later removed) under sterile conditions. Several progestins have been tested, but the best compound has not yet been selected. Meanwhile, clinical trials involving the use of one to six implants are underway with various compounds, and the first biodegradable implants have also been developed and placed in clinical trial.

An alternative to the implantable device is provided by intravaginal rings containing progestins. The contraceptive effectiveness and acceptability of a product based on this principle is now being studied in field trials. Three different progestins, each incorporated into silastic and molded into a ring similar in size and shape to the rim of a vaginal diaphragm, are being tested. The ring can be inserted by the woman and left in place for one month, two months, or the entire interval between spontaneous bleedings. The method is now undergoing product development with engineering and design modifications in progress to reduce cost and to standardize monthly steroid release rates.

**Item D. Shutting Out the Brain**

The identification of the brain's *releasing factor* for luteinizing hormone led to immediate speculation that analogs of the molecule may be useful in fertility control. The idea was to synthesize a slightly modified version of the molecule that would compete with the natural releasing factor and reduce or eliminate its effectiveness, thus preventing the surge of LH from the pituitary that triggers ovulation. Natural LRF works by recognizing and binding to receptor sites in the pituitary; an analog of the molecule that also binds to the receptors but does so without triggering LH release would block the access of the body's own LRF to the receptors.

With this in mind, large numbers of LRF analogs have been synthesized and the relationship between the structure of the molecule and its function is being studied in detail. Analogs that will compete effectively with natural LRF in test tube systems of cultured pituitary cells have been synthesized. Other LRF analogs have a longer half-life in the body than the natural molecule that acts locally and breaks down rapidly. Such longer lived versions of the molecule will also be necessary in any pharmaceutical application. So far no single revision of the releasing factor has emerged as promising enough for clinical trials. But the elegance of the concept of a once-a-month pill that will prevent ovulation by blocking the initial event in the sequence that triggers it has sufficient appeal for research in this area to continue actively.

### Item E. Breaking the Journey Down the Oviduct

The journey of the fertilized ovum down the oviduct normally takes some two or three days. The chances of a successful implantation in the uterus at the end of the journey could be markedly diminished by interfering with the timing of its arrival. Unfortunately, the factors that govern the speed of the egg's passage are still very poorly understood, though they are likely to include the motility of the oviduct and its secretions. Tubal motility is influenced in vitro by various prostaglandins and several pituitary and ovarian hormones. In animals, relatively minor adjustments in the estrogen-progesterone balance can modify tubal transport enough to prevent implantation, but there is little evidence that the same is true in humans.

There is as yet no immediate prospect of developing methods of fertility control based on interference with tubal transport. This may well be due to gaps in existing knowledge. For example, the contraceptive action of large estrogen doses immediately after coitus may work less through blocking ovulation than by interfering with tubal transport, but real information on this topic is still very scanty.

### Item F. Closing Off the Tubes

Female *sterilization* remains an important method of fertility control; some 20 million women are sterilized each year. Present research is aimed at replacing major surgery with "office" procedures and the development of reversible procedures. Among the former is a technique for cauterizing the opening of the oviduct from the uterus with an electric current, viewing the operation—performed transcervically—with the aid of a fiber-optic endoscope. Other approaches involve chemicals, such as quinacrine or even a powerful tissue adhesive, to close off the uterotubal junction. In all cases, however, several years of research remain before the safety and efficacy of these techniques can be established unequivocally.

### Item G. Preventing Implantation

Intrauterine devices are used by approximately 15 million women, a number limited by the failure of present devices in rare instances (their efficacy is slightly lower than that of the Pill) and by side effects such as pain, bleeding, and expulsion of the IUD. Although there is still some research on the development of conventional insert devices, there are already so many of them that few can ever be tested in a large enough group of women. And meanwhile, research attention has swung toward *medicated* IUDs, devices that release microgram quantities per day of copper, progesterone, or other substances.

The copper-carrying devices are the most advanced in testing and have been available in many countries since 1973. One of the devices, the Copper T, is being introduced in India. Several versions are under test; TCu 200, the one studied longest, can be left in place for at least two years and probably no longer than four. This time limitation is being overcome with newer models, which may also have a higher contraceptive effectiveness than TCu 200. The Copper 7, now on sale in the United States, has been designed and tested for replacement every two years, making it less appealing for use in developing countries.

The progesterone releasing IUD, also in the form of a T-shaped plastic device, is similarly limited in its lifespan. Present technology permits progesterone release for no more than a year. The contraceptive efficacy of this device has not yet been fully established. Various nonsteroidal antifertility agents are also being tested in IUDs, but research in this area is still at a very early stage.

Another important line of development concerns devices that can be implanted immediately following the birth of a baby (or, more precisely, the delivery of the placenta). This is the only opportunity to provide women with a contraceptive in many countries. The development of such devices will require, however, a better knowledge of the variability and systematic change in the size of the human uterus after delivery.

In general, the factors controlling implantation are still poorly understood. A better knowledge of the hormonal and biochemical requirements might well yield its rewards in the form, for instance, of improved hormonal methods of interfering with implantation without the need for an IUD at all. Meanwhile, perhaps the most intellectually fascinating problem of implantation concerns the toleration by the mother's body of what is, in effect, a foreign organ graft, the conceptus. Why is the fetus not rejected as would be a heart or a kidney? A better understanding of the underlying immunological phenomena might again provide a promising lead for new methods of fertility control in the somewhat more distant future.

Item H. Interfering with Early Embryonic Development

Intervening in the reproductive process after implantation of the embryo is a far less desirable method of fertility regulation than contraception. However, more than half of the world's population lives now in countries with legalized *abortion* for broadly interpreted medical and social-medical indications. It follows, therefore, that there is a need for improved methods for interrupting gestation. Natural and synthetic *prostaglandins* have been tested for this purpose. The advantages of the present generation are outweighed by the disadvantages and further progress will depend upon the synthesis of prostaglandin analogs that are more potent when acting on the myometrium than on other smooth muscle. Meanwhile, for midtrimester terminations of pregnancy, prostaglandins are safer than intraamniotic saline or hysterotomy. No other chemical abortifacient is currently under overt investigation although the need for safe, self-administered methods for the interruption of the very early stages of gestation is clear. At present, *suction curretage* still seems to be the method of choice in many countries for the termination of early pregnancy (up to about twelve weeks).

### Item I. Blocking Spermatogenesis

Although spermatogenesis is a natural target for pharmacological intervention, it is also a very elusive one. Even today, very little is known of the biochemical details of the development of spermatozoa from spermatogonia (although the technology that should allow such knowledge to be garnered has lately become available). Those compounds that have been found to interfere with sperm production have been mainly drugs first explored for their effectiveness in blocking cell division in general, in the context of cancer chemotherapy. But these drugs, which mimic the effects of ionizing radiation, are often cumulative in their effects, highly toxic, and, perhaps worst of all, potentially mutagenic. Their principal target cells are the spermatogonia and spermatocytes, which they either destroy or prevent from dividing.

These radiomimetic drugs are good examples of what an effective, safe, antispermatogenic compound should not be. By acting on the spermatogonia, they threaten the basic renewing germ cell population and so pose the risk of inducing permanent sterility. Furthermore, their mutagenic potential suggests the possibility that spermatozoa surviving the drug may be genetically damaged, raising the specter of congenital abnormality in the resulting fetus. The "ideal" antispermatogenic agent, in contrast, should leave the spermatogonia unharmed and should be free of mutagenic properties.

Somewhat closer to this ideal is a group of heterocyclic compounds that inhibit spermatogenesis with no evidence of mutagenesis and with complete restoration of fertility on withdrawal. Several have proven too toxic in animal studies to receive clinical trials; but others, especially compounds known as nitroimidazoles, are probably worthy of further study. One heterocyclic compound, a bis(dichloroacetyl)diamine, was brought to clinical trial in human volunteers with very promising result until the discovery that the drug is incompatible with alcohol, having an Antibuse-like effect. Its effect is to cause the final stages of spermatozoon formation to go awry, resulting in misshapen sperm that are retained within the Sertoli cells, which, from these and other studies, are emerging as potentially sensitive to a variety of chemical and physical agents. The new awareness of the importance and responsiveness of Sertoli cells suggests that drugs may be found that will indirectly inhibit spermatogenesis or prevent sperm release, avoiding risk to the spermatogonia or genetic damage to the spermatozoa themselves.

Item J. Toward a Male Pill

The male analog of the female pill—an orally active drug that will suppress germ cell production by interfering with its hormonal regulation—has in recent years become a realistic goal. Nearest to success is a steroid combination that suppresses LH levels and thus diminishes testosterone production by the Leydig cells, lowering the concentration of testosterone in the seminiferous tubules below that necessary for sperm production. Further off is the still untested possibility of a nonsteroid drug that would bring about the same result by interfering with the production by Sertoli cells of the androgen-binding protein (ABP).

Local testosterone levels in the tubules can be effectively lowered by increasing testosterone in the peripheral blood and taking advantage of its feedback control mechanism to shut down LH production by the pituitary. However, the high dosage of testosterone needed causes changes in lipoprotein metabolism and blood cell formation. A recently studied synthetic analog of testosterone, Danazol, appears to act directly on Leydig cells to suppress testosterone production; and preliminary trials have been carried out in which small doses of testosterone in combination with Danazol were administered in the belief that they might act synergistically. The results were promising and suggest that further exploration of this approach may yield a safe and effective contraceptive.

Estrogens and/or progestogens also lower LH levels in the male but are accompanied by breast enlargement and marked loss of libido. Among the encouraging developments in male contraceptive research are current studies with combinations of estrogens or progestagens with testosterone in the hope of suppressing LH levels without these undesirable effects on libido. Promising results have been obtained both with silastic implants of progestogen and testosterone and with oral preparations of a synthetic testosterone with estrogen. In both cases, sperm counts dropped to infertile levels without obvious side effects and with a return to normal fertility upon cessation of the treatment.

The prospects for male contraception through the suppression of FSH rather than LH are even more exciting but, at present, less tangible. Interfering with Sertoli cell synthesis of ABP by lowering FSH theoretically might affect local testosterone levels in the tubules while leaving the concentration of hormone in peripheral blood unchanged. Hopes for this possibility rest with the current progress being made in the isolation and characterization of inhibin, the once hypothetical feedback product of the seminiferous tubules.

### Item K. Preventing Sperm Maturation

Spermatozoa first acquire the capacity to fertilize during their slow passage through the epididymis, so this would seem the ideal site of action for a fertility control agent. The discovery that the epididymis requires an unusually high concentration of testosterone suggested that an antitestosterone might interfere with its physiological integrity. One such agent, cyproterone acetate, has been successfully tested in silastic implants in rodents; and because the drug is already in clinical use in some countries with minimal side effects, clinical trials of its antifertility effects are now underway. Preliminary results show that cyproterone acetate does induce infertility, but significant depression of sperm counts, as well as sperm motility, suggest that its effects are not limited to the epididymis.

Another agent, alpha-chlorohydrin, is also thought to exert its antifertility effects in the epididymis and has lately been found to affect sperm directly. The compound may be too toxic for use in humans, but an analog has recently been found that, it is hoped, will prove sufficiently free of toxicity for human trials.

Enthusiasm for the development of a posttesticular fertility control agent continues to run high, and the rapid progress of recent years in understanding epididymal function makes it likely that the approach will ultimately succeed.

Item L. Preventing Fertilization

Because spermatozoa are the cells actually transferred between the sexes, they are the logical target for contraception, as the various mechanical and chemical devices aimed at keeping the sperm from reaching the egg testify. Research is now identifying other vulnerable steps between ejaculation and fertilization that may lend themselves to pharmacological or immunological interruption.

The discovery of the phenomenon of capacitation and of the existence within the seminal plasma of substances that can decapacitate sperm has led to much speculation that the identification of these substances will provide an excellent means of fertility control. Similarly, attempts are being made to purify the enzymes released by the acrosome reaction that are needed for the successful penetration of the egg. Some evidence now exists that these enzymes could be prepared. From such research may emerge a successful immunological approach to contraception.

Once a sperm has penetrated an egg, the *zona reaction* (the release of granules between the outer membrane of the egg and the zona pellucida) prevents other sperm from gaining entry. Although the nature of these granules is unknown, various treatments have been shown to mimic their action. Trypsin-like enzymes, for instance, will render the zona impenetrable; and nonenzyme plant substances known as lectins have also been reported to block fertilization. It seems likely that a great variety of agents that alter the molecular configuration of the zona pellucida may prove equally effective in preventing sperm penetration. However, whether these substances are safe as well as effective and whether they can be delivered to the site of fertilization in a high enough concentration remain to be seen.

Appendix C                    Country Funding Data

The following tables have been compiled from data submitted by the funding agencies and the national and geographical consultants commissioned by this review of the reproductive sciences and contraceptive development. Expenditures for reproductive research supported by agency and by type of research were requested. The funds shown for each country are only those available from agencies located in that specific country (with the exception of South Korea) for the purpose of research conducted internally unless specified as donated to foreign funding agencies or research projects. In an effort to minimize definitional difficulties, the research subjects considered relevant to the study were given to each funding agency and consultant.

Expenditures were categorized as follows: *Fundamental studies in the reproductive sciences* includes grants and contracts supporting basic research; core support to research centers, conferences, and symposia; and information dissemination. *Training* is for a variety of mechanisms to train researchers in both the fundamental and clinical aspects of reproductive research. *Contraceptive development* is defined as all studies concerning agents being clinically tested for their fertility regulating effect. Pharmaceutical company funds and core support to clinical research centers were categorically assigned to this research area. *Studies on safety* of current fertility control methods includes research on agents presently available to the public.

In many instances, especially in the developing world, this information was extremely difficult to compile and therefore data were not available for all years and all categories. These data possibly do not represent all the research funds in the field, but every effort was made to obtain all available information on documentable support. Where it was known that money had been spent but exact totals were not available, "NA" appears. A dash (—) indicates that it was not known whether or not any funds had been expended; the notation (e) after an amount means that it is estimated.

Many consultants reported the amounts in their national currencies and all reported in the current value. In order to compile these data, money amounts were converted to U.S. dollars using the exchange rate of the relevant year as specified in the *Statistical Yearbook* of the United Nations (1967; 1974). For 1973-1974 exchange rates, *Information Please Almanac* was used (Golenpaul Associates 1974).

Exchange rates for 1975 were not available at the time the tables were compiled; these funds were converted using 1974 rates.

The *Statistical Yearbook* also served as the source for converting current dollar values to constant dollars. The annual Consumer Price Index was used for each country involved (1970=100) (U.N. 1974; Pardo 1975).

**Definition of the Reproductive Sciences and Contraceptive Development**

**Laboratory Research: Hypothalamus, Pituitary, CNS**
Hypothalamus-median eminence
  Neurotransmitters
  Releasing and inhibiting factors
Hypothalamic influences on pituitary function
Gonadotropins
  FSH and LH
  Prolactin
Pituitary-gonadol interrelationships
Induction of ovarian development and ovulation
Feedback regulation of reproductive cycles in rodents and primates
Immune effects

**Ovarian Function (Endocrine, Gametogenic)**
Steroidogenesis
Steroid hormone secretions
Follicle growth, maturation and atresia
Ovulation, CL formation, function, and luteolysis
Prostaglandins
Mechanism of hormone action
Induction of ovarian development and ovulation
Immune effects

**Testicular Function (Endocrine, Gametogenic)**
Spermatogenesis
Sertoli cells
Leydig cells
Sperm
  Metabolism
  Motility
  Morphology
Blood-testis barrier
Mechanism of hormone action
Pituitary-gonadal interrelationships
Prostaglandins
Immune effects
Steroidogenesis

**Male Accessory**
Epididymis
Semen
Prostaglandins

**Female Genital Tract**
Uterus
Oviduct
Gamete transport
Mechanism of hormone action
Prostaglandins

**Fertilization Process**
Fertilization and capacitation

**Early Embryonic Development**
Developmental biology
Molecular biology
Implantation
Chorionic gonadotropin

**Contraceptives: Evaluation of Safety of Current Contraceptives and Current Leads**

Chemical
Female
  Oral methods
  Injectable methods
  Implants
  Postcoital pill
Male
  Infertility
  Hormonal therapy
  Oral methods
  Injectable methods
  Implants

Mechanical
Female
  Abortion

  Clinical-epidemological
  Experimental
  Sterilization
  Bioengineering
Male
  Sterilization
  Bioengineering

**Exchange rate sources**

1965
United Nations, Department of Economic and Social Affairs Statistical Office. 1967. *Statistical Yearbook 1966*, 18th ed., p. 578. New York: United Nations.
1966-1972
_____. 1974. 25th ed., p. 599.
1973
Golenpaul Associates. 1974. *Information Please Almanac Atlas and Yearbook 1974*, 28th ed., p. 112. New York: Simon and Schuster.

**Consumer Price Index**

1965-1973
*Statistical Yearbook 1973*. 1974. 25th ed., New York: United Nations, p. 450.
1974
Pardo, M. (United Nations Statistical Department). 1975. Personal communication.

Table C.1 Africa—Internal Expenditures: Reproductive Research and Contraceptive Development (by Purpose and Proportion of Governmental Medical Research Funds, 1965, 1969-1974, 1975 [est.], in Current and Constant U.S. Dollars [1970=100])

|  | 1965 | 1969 | 1970 |
|---|---|---|---|
| **Fund. St./Training** | | | |
| Government (estimate) | | | |
| Subtotal | | | |
| Current $ | — | — | — |
| Constant '70 $[a] | — | — | — |
| **Contraceptive Development and Safety** | — | — | — |
| Donations to foreign | | | |
| funding agencies or | | | |
| research projects | — | — | — |
| Total | | | |
| Current $ | — | — | — |
| Constant '70 $[a] | — | — | — |
| A. Government total | | | |
| Current $ | NA | NA | NA |
| Constant $ | | | |
| B. Government medical research | | | |
| budget—Current $ | NA | NA | NA |
| Constant $ | | | |
| % A of B | | | |

[a] Nigerian CPI used because most of funds estimated are from that country.

## Country Funding Data

| 1971 | 1972 | 1973 | 1974 | 1975 (est.) |
|---|---|---|---|---|
| 717,000 | 717,000 | 717,000 | 717,000 | — |
| 628,947 | 612,820 | 592,561 | 504,930 | — |
| — | — | — | — | — |
| — | — | — | — | — |
| 717,000 | 717,000 | 717,000 | 717,000 | — |
| 628,947 | 612,820 | 592,561 | 504,930 | |
| NA | NA | NA | NA | NA |
| NA | NA | NA | NA | NA |

Table C.2 Australia—Internal Expenditures: Reproductive Research and Contraceptive Development (by Purpose and Proportion of Governmental Medical Research Funds, 1965, 1969-1974, 1975 [est.], in Current and Constant U.S. Dollars [1970=100])

|  | 1965 | 1969 | 1970 |
|---|---|---|---|
| Fund. St./Training |  |  |  |
| Government |  |  |  |
| NH & MRC[a] |  | 53,622 | 73,894 |
| ARGC[b] |  | 63,739 | 76,957 |
| Dept. of Primary Industry | 403,530 | 303,889 | 426,510 |
| University research grants |  | 51,124 | 46,908 |
| Miscellaneous |  | 22,474 | 72,953 |
| Philanthropy |  |  |  |
| Endocrinology Foundation |  | 16,769 | 22,284 |
| Bushell Trust |  |  | 5,236 |
| Other |  | 5,142 | 3,454 |
| Industry |  |  |  |
| Aus. Wool Corporation |  | 49,803 | 58,558 |
| Aus. Pig Industry Research Council |  |  |  |
| Ind. & private individual donations |  | 8,160 | 6,016 |
| Aus. meat research |  |  |  |
| Subtotal |  |  |  |
| Current $ | 403,530 | 574,722 | 792,770 |
| Constant '70 $ | 469,221 | 598,669 | 792,770 |
| Contraceptive Development and Safety | — | — | — |
| Donations to foreign funding agencies or research projects | — | — | — |
| Total |  |  |  |
| Current $ | 403,530 | 574,722 | 792,770 |
| Constant '70 $ | 469,221 | 598,669 | 792,770 |
| A. Government total |  |  |  |
| Current $ | — | 53,622 | 73,894 |
| Constant $ | — | 55,856 | 73,894 |
| B. Government medical research budget[c]—Current $ |  | 1,492,107 | NA |
| Constant $ |  | 1,554,278 |  |
| % A of B |  | 3.59 |  |

[a]NH & MRC—National Health and Medical Research Council
[b]ARGC—Australian Research Grants Committee
[c]NH & MRC only

| 1971 | 1972 | 1973 | 1974 | 1975 [est.) |
|---|---|---|---|---|
| 120,184 | 106,763 | 285,188 | 984,060(e) | 984,060(e) |
| 80,372 | 161,809 | | | |
| 525,940 | 552,513 | 654,688 | | |
| 66,480 | 94,031 | 102,634 | | |
| 45,144 | 15,300 | 15,864 | | |
| 23,820 | 51,000 | 99,150 | 127,800(e) | 127,800(e) |
| 5,597 | 5,992 | 6,657 | | |
| 26,598 | 3,503 | 28,753 | | |
| 125,369 | 117,699 | 138,769 | 166,140(e) | 166,140(e) |
| 56,984 | 13,687 | 28,456 | | |
| 15,444 | 12,558 | 12,606 | | |
| | 21,440 | 10,536 | | |
| 1,091,932 | 1,156,295 | 1,383,301 | 1,278,000(e) | 1,278,000(e) |
| 1,030,125 | 1,032,406 | 1,124,634 | 900,000(e) | 900,000(e) |
| — | — | — | — | — |
| — | — | — | — | — |
| 1,091,932 | 1,156,295 | 1,383,301 | 1,278,000(e) | 1,278,000(e) |
| 1,030,125 | 1,032,406 | 1,124,634 | 900,000(e) | 900,000(e) |
| 120,184 | 106,763 | 285,188 | 984,060(e) | 984,060(e) |
| 113,381 | 95,324 | 231,860 | 863,210(e) | 863,210(e) |
| 2,328,193 | NA | 3,518,225 | NA | NA |
| 2,196,408 | | 2,860,345 | | |
| 5.16 | | 8.10 | | |

Table C.3 Belgium—Internal Expenditures: Reproductive Research and Contraceptive Development (by Purpose and Proportion of Governmental Medical Research Funds, 1965, 1969-1974, 1975 [est.], in Current and Constant U.S. Dollars [1970=100])

|  | 1965 | 1969 | 1970 |
|---|---|---|---|
| **Fund. St./Training** | | | |
| Government | | | |
| IRSCIA[a] | — | 22,500 | 44,500 |
| University | 23,345 | — | 41,659 |
| Private | 7,955 | — | 112,500 |
| Industry | — | — | 114 |
| Subtotal | | | |
| Current $ | 31,300 | 22,500 | 198,773 |
| Constant '70 $ | 37,262 | 23,437 | 198,773 |
| **Contraceptive Development and Safety** | | | |
| Industry | — | 10,000 | 15,000 |
| Subtotal | | | |
| Current $ | — | 10,000 | 15,000 |
| Constant '70 $ | — | 10,416 | 15,000 |
| Donations to foreign funding agencies or research projects | — | — | — |
| **Total** | | | |
| Current $ | 31,300 | 32,500 | 213,773 |
| Constant '70 $ | 37,262 | 33,854 | 213,773 |

Note: The following additional data was obtained too late to include in the tables in the text of this report.

| | | | |
|---|---|---|---|
| **Fund. St./Training** | | | |
| Government | | | |
| FNRS/FRSM[b] | 90,587 | 171,984 | 126,188 |
| **Revised Total** | | | |
| Current $ | 121,887 | 204,484 | 339,961 |
| Constant '70 $ | 145,103 | 213,004 | 339,961 |
| A. Government total | | | |
| Current $ | 113,932 | 194,484 | 212,347 |
| Constant $ | 135,633 | 202,588 | 212,347 |
| B. Government medical research | | | |
| budget—Current $ | NA | NA | 2,147,500 |
| Constant $ | NA | NA | 2,147,500 |
| % A of B | — | — | 9.9 |

[a]IRSCIA—Institute for Scientific Research in Industry and Agriculture.
[b]FNRS/FRSM—Fonds Nationale de Recherche Scientifique.

| 1971 | 1972 | 1973 | 1974 | 1975 (est.) |
|---|---|---|---|---|
| 82,000 | 82,000 | 13,438 | | |
| 151,259 | 167,448 | 109,083 | 109,083 | 27,083 |
| 163,734 | 142,477 | 238,496 | 246,117 | 242,053 |
| 105 | 216 | 176,520 | 179,499 | 90,675 |
| | | 10,945 | 18,988 | 65,493 |
| 397,098 | 392,141 | 548,482 | 553,687 | 425,304 |
| 381,825 | 356,491 | 464,815 | 416,306 | 319,777 |
| 20,000 | 25,000 | 60,000 | 70,000 | 85,000 |
| 20,000 | 25,000 | 60,000 | 70,000 | 85,000 |
| 19,230 | 22,727 | 50,847 | 52,631 | 63,909 |
| — | — | — | — | — |
| 417,098 | 417,141 | 608,482 | 623,687 | 510,304 |
| 401,056 | 379,219 | 515,663 | 468,937 | 383,687 |
| 165,181 | 217,590 | 166,457 | 357,067 | 274,879 |
| 582,279 | 634,731 | 774,939 | 981,294 | 785,183 |
| 559,883 | 577,028 | 656,727 | 737,815 | 590,363 |
| 398,440 | 467,038 | 527,474 | 712,817 | 544,015 |
| 383,115 | 424,580 | 535,945 | 535,945 | 409,034 |
| 3,000,000 | 4,030,000 | 4,918,100 | 6,017,375 | NA |
| 2,844,615 | 3,663,636 | 4,167,881 | 4,524,342 | NA |
| 13.3 | 11.6 | 10.7 | 11.8 | — |

Table C.4 Britain—Internal Expenditures: Reproductive Research and Contraceptive Development (by Purpose and Proportion of Governmental Medical Research Funds, 1965, 1969-1974, 1975 [est.], in Current and Constant U.S. Dollars [1970=100])

|  | 1965 | 1969 | 1970 |
|---|---|---|---|
| **Fund. St./Training** | | | |
| Government | | | |
| MRC[a] (estimate) | 929,160 | 1,198,561 | 1,377,033 |
| ARC[b] (estimate) | | 1,511,271 | 1,708,612 |
| SRC[c] | | | |
| Philanthropy | | | |
| Wellcome Trust | 123,084 | 189,040 | 216,017 |
| Nuffield Foundation | | 7,022 | 150,186 |
| Subtotal | | | |
| Current $ | 1,052,244 | 2,905,894 | 3,451,848 |
| Constant '70 $ | 1,315,305 | 3,091,376 | 3,451,848 |
| **Contraceptive Development and Safety** | | | |
| Industry | 41,700 | 69,000 | 107,100 |
| Subtotal | | | |
| Current $ | 41,700 | 69,000 | 107,100 |
| Constant '70 $ | 52,125 | 73,404 | 107,100 |
| Donations to foreign funding agencies or research projects | — | — | — |
| **Total** | | | |
| Current $ | 1,093,944 | 2,974,894 | 3,558,948 |
| Constant '70 $ | 1,367,430 | 3,164,780 | 3,558,948 |
| A. Government total | | | |
| Current $ | 929,160 | 2,709,832 | 3,085,645 |
| Constant $ | 1,161,450 | 2,882,800 | 3,085,645 |
| B. Government medical research | | | |
| budget[d]—Current $ | 75,630,252 | 88,729,016 | 105,263,150 |
| Constant $ | 94,537,815 | 94,392,570 | 105,263,150 |
| % A of B | 1.22 | 3.05 | 2.93 |

[a]MRC—Medical Research Council
[b]ARC—Agriculture Research Council
[c]SRC—Science Research Council
[d]Estimates from the Office of Health Economics (incl. MRC, UGC, and DHSS)

# Country Funding Data

| 1971 | 1972 | 1973 | 1974 | 1975 (est.) |
|---|---|---|---|---|
| 1,738,363 | 1,764,788 | 2,324,267 | 2,343,256 | 2,220,000(e) |
| 2,130,690 | 2,042,254 | 2,048,000(e) | 2,220,000(e) | 2,220,000(e) |
|  |  | 108,911 |  |  |
| 190,355 | 185,200 | 111,673 | 51,698 |  |
| 7,022 | 7,022 |  |  |  |
| 4,066,430 | 3,999,264 | 4,592,851(e) | 4,614,954(e) | 4,440,000(e) |
| 3,730,669 | 3,418,174 | 3,588,157(e) | 3,118,212(e) | 3,000,000(e) |
| 153,800 | 120,900 | NA | NA | NA |
| 153,800 | 120,900 |  |  |  |
| 141,100 | 103,333 |  |  |  |
| — | — | — | — | — |
| 4,220,230 | 4,120,164 | 4,592,851(e) | 4,614,954(e) | 4,440,000(e) |
| 3,871,770 | 3,521,507 | 3,588,157(e) | 3,118,212(e) | 3,000,000(e) |
| 3,869,053 | 3,807,042 | 2,433,178(e) | 2,343,256(e) | 4,440,000(e) |
| 3,549,589 | 3,253,882 | 1,900,920(e) | 1,583,281(e) | 3,000,000(e) |
| 127,877,230 | 129,107,980 | 163,366,330 | NA | NA |
| 117,318,559 | 110,348,700 | 127,629,945 |  |  |
| 3.02 | 2.94 | 2.8 |  |  |

Table C.5 Canada—Internal Expenditures: Reproductive Research and Contraceptive Development (by Purpose and Proportion of Governmental Medical Research Funds, 1965, 1969-1974, 1975 [est.], in Current and Constant U.S. Dollars [1970=100])

|  | 1965 | 1969 | 1970 |
|---|---|---|---|
| **Fund. St./Training** |  |  |  |
| Government |  |  |  |
| MRC[a] | 192,149 | 473,819 | 676,112 |
| NRC[b] |  |  | 3,959 |
| Ontario Ministry of Health |  |  |  |
| University budgets (estimates) | 191,475 | 472,825 | 699,666 |
| Philanthropy |  |  |  |
| Banting Research Foundation |  | 446 | 19,796 |
| Subtotal |  |  |  |
| Current $ | 383,624 | 947,090 | 1,399,533 |
| Constant '70 $ | 462,198 | 976,381 | 1,399,533 |
| **Contraceptive Development and Safety** |  |  |  |
| Government |  |  |  |
| IDRC | 0 | 0 | 0 |
| Private | — | — | — |
| Industry | 170,000 | 368,500 | 663,260 |
| Subtotal |  |  |  |
| Current $ | 170,000 | 368,500 | 663,260 |
| Constant '70 $ | 204,819 | 379,896 | 663,260 |
| Donations to foreign funding agencies or research projects |  |  |  |
| CIDA[c] |  |  |  |
| IDRC (estimate) |  |  |  |
| Subtotal |  |  |  |
| Current $ |  |  |  |
| Constant '70 $ |  |  |  |
| **Total** |  |  |  |
| Current $ | 553,624 | 1,315,590 | 2,062,793 |
| Constant '70 $ | 667,017 | 1,356,278 | 2,062,793 |
| A. Government total[e] |  |  |  |
| Current $ | 192,149 | 473,819 | 680,071 |
| Constant $ | 231,504 | 488,473 | 680,071 |
| B. Government medical research |  |  |  |
| budget[e]—Current $ | 8,564,815 | 28,870,093 | 100,489,230 |
| Constant $ | 10,319,054 | 29,762,982 | 100,489,230 |
| % A of B | 2.24 | 1.64 | 0.68 |

[a]MRC—Medical Research Council
[b]NRC—National Research Council

## Country Funding Data

| 1971 | 1972 | 1973 | 1974 | 1975 (est.) |
|---|---|---|---|---|
| 786,121 | 812,959 | 945,965 | 1,032,000(e) | 1,032,000(e) |
| 3,991 | 3,767 | 6,030 | NA | |
| | | 10,032 | NA | |
| 815,826 | 847,073 | 991,112 | NA | |
| 27,440 | 30,836 | 29,034 | 32,250(e) | 32,250(e) |
| 1,633,378 | 1,694,635 | 1,982,173 | 1,064,250(e) | 1,064,250(e) |
| 1,585,804 | 1,569,106 | 1,708,770 | 825,000(e) | 825,000(e) |
| 0 | — | 37,211 | 37,743 | — |
| — | — | — | — | — |
| 652,000 | 695,110 | 338,160 | 377,000 | 288,000 |
| 652,000 | 695,110 | 375,371 | 414,743 | 288,000 |
| 633,009 | 643,620 | 323,595 | 321,506 | 223,255 |
| | | 1,006,036 | 1,275,510 | 1,161,000(e) |
| 49,921 | 42,407 | 56,876 | 890,032 | 850,000 |
| 49,921 | 42,407 | 1,062,912 | 2,165,542 | 2,011,000 |
| 48,468 | 39,267 | 916,303 | 1,678,714 | 1,558,914 |
| 2,335,299 | 2,432,152 | 3,420,456 | 3,644,535 | 3,363,250 |
| 2,267,281 | 2,251,993 | 2,948,668 | 2,825,220 | 2,607,170 |
| 840,033 | 859,133 | 1,046,082 | NA | NA |
| 815,566 | 795,493 | 901,794 | NA | NA |
| 107,166,663 | 110,527,300 | 120,223,340 | NA | NA |
| 104,045,303 | 102,340,111 | 103,640,810 | NA | NA |
| 0.77 | 0.77 | 0.87 | | |

[c]CIDA—Canadian International Development Agency
[d]IDRC—International Development Research Centre
[e]Government totals include IDRC, MRC and NRC.

Table C.6 Denmark—Internal Expenditures: Reproductive Research and Contraceptive Development (by Purpose and Proportion of Governmental Medical Research Funds, 1965, 1969-1974, 1975 [est.], in Current and Constant U.S. Dollars [1970=100])

|  | 1965 | 1969 | 1970 |
|---|---|---|---|
| Fund. St./Training | | | |
| Government | | | |
| MRC[a] | — | 16,150 | 48,240 |
| Philanthropy | — | — | — |
| Subtotal | | | |
| Current $ | — | 16,150 | 48,240 |
| Constant '70 $ | | 17,181 | 48,240 |
| Contraceptive Development and Safety | — | — | — |
| Donations to foreign funding agencies or research projects DANIDA[b] | | | |
| Subtotal | | | |
| Current $ | | | |
| Constant '70 $ | | | |
| Total | | | |
| Current $ | — | 16,150 | 48,240 |
| Constant '70 $ | | 17,181 | 48,240 |
| A. Government total[c] | NA | | |
| Current $ | | 16,150 | 48,240 |
| Constant $ | | 17,181 | 48,240 |
| B. Government medical research budget[c]—Current $ | NA | 234,310 | 1,122,230 |
| Constant $ | | 249,265 | 1,122,230 |
| % A of B | | 6.89 | 4.30 |

[a]MRC—Medical Research Council
[b]DANIDA—Danish International Development Agency
[c]MRC only

| 1971 | 1972 | 1973 | 1974 | 1975 (est.) |
|---|---|---|---|---|
| 79,660 | 88,750 | 113,480 | 71,550 | 71,000(e) |
| — | — | — | — | — |
| 79,660 | 88,750 | 113,480 | 71,550 | 71,000(e) |
| 75,151 | 78,540 | 92,260 | 50,387 | 50,000(e) |
| — | — | — | — | — |
|  | 200,000 | 364,000 | 500,000 | 600,000 |
|  | 200,000 | 364,000 | 500,000 | 600,000 |
|  | 176,991 | 295,934 | 352,112 | 422,535 |
| 79,660 | 288,750 | 477,480 | 571,550 | 671,000(e) |
| 75,151 | 255,531 | 388,195 | 402,500 | 472,535(e) |
| 79,660 | 88,750 | 113,480 | 71,550 | 71,000(e) |
| 75,151 | 78,540 | 92,260 | 50,387 | 50,000(e) |
| 1,492,350 | 1,938,690 | 2,415,460 | 2,810,090 | NA |
| 1,407,877 | 1,715,654 | 1,963,788 | 1,985,978 |  |
| 5.33 | 4.57 | 4.69 | 2.53 |  |

Table C.7 Egypt—Internal Expenditures: Reproductive Research and Contraceptive Development (by Purpose and Proportion of Governmental Medical Research Funds, 1965, 1969-1974, 1975 [est.], in Current and Constant U.S. Dollars [1970=100])

|  | 1965 | 1969 | 1970 |
|---|---|---|---|
| Fund. St./Training | — | — | — |
| Contraceptive Development and Safety Government | — | 185,000 | 92,000 |
| Subtotal Current $ | — | 185,000 | 92,000 |
| Constant '70 $ |  | 192,708 | 92,000 |
| Donations to foreign funding agencies or research projects | — | — | — |
| Total Current $ | — | 185,000 | 92,000 |
| Constant '70 $ |  | 192,708 | 92,000 |
| A. Government total Current $ Constant $ | NA | NA | NA |
| B. Government medical research budget—Current $ Constant $ % A of B | NA | NA | NA |

Country Funding Data

| 1971 | 1972 | 1973 | 1974 | 1975 (est.) |
|---|---|---|---|---|
| — | — | — | — | — |
| 107,000 | 116,000 | 134,000 | 157,000 | 45,000 |
| 107,000 | 116,000 | 134,000 | 157,000 | 45,000 |
| 103,883 | 110,476 | 121,818 | 128,689 | 36,885 |
| — | — | — | — | — |
| 107,000 | 116,000 | 134,000 | 157,000 | 45,000 |
| 103,883 | 110,476 | 121,818 | 128,689 | 36,885 |
| NA | NA | NA | NA | NA |
| NA | NA | NA | NA | NA |

Table C.8 Finland—Internal Expenditures: Reproductive Research and Contraceptive Development (by Purpose and Proportion of Governmental Medical Research Funds, 1965, 1969-1974, 1975 [est.], in Current and Constant U.S. Dollars [1970=100])

|  | 1965 | 1969 | 1970 |
|---|---|---|---|
| **Fund. St./Training** | | | |
| Government | | | |
| MRC[a] | 8,830 | 11,430 | 21,530 |
| Philanthropy | | | |
| J. Stiftelse | 15,530 | 15,950 | 19,860 |
| Subtotal | | | |
| Current $ | 24,360 | 27,380 | 41,390 |
| Constant '70 $ | 26,769 | 28,227 | 41,390 |
| **Contraceptive Development and Safety** | — | — | — |
| Donations to foreign funding agencies or research projects | | | |
| FIN-AID[b] | | | |
| Subtotal | | | |
| Current $ | — | — | — |
| Constant '70 $ | | | |
| **Total** | | | |
| Current $ | 24,360 | 27,380 | 41,390 |
| Constant '70 $ | 26,769 | 28,227 | 41,390 |
| A. Government total | | | |
| Current $ | 8,830 | 11,430 | 21,530 |
| Constant $ | 9,703 | 11,783 | 21,530 |
| B. Government medical research | | | |
| budget—Current $ | 185,710 | 204,050 | 238,100 |
| Constant $ | 204,076 | 210,360 | 238,100 |
| % A of B | 4.75 | 5.60 | 9.04 |

[a]MRC—Medical Research Council
[b]FIN-AID—Finish Agency for International Development

| 1971 | 1972 | 1973 | 1974 | 1975 (est.) |
|---|---|---|---|---|
| 47,370 | 38,800 | 40,050 | 38,450 | 37,500(e) |
| 17,460 | 15,850 | 22,540 | 19,730 | 19,730(e) |
| 64,830 | 54,650 | 62,590 | 58,180 | 57,230(e) |
| 61,160 | 47,939 | 48,898 | 38,787 | 38,153(e) |
| — | — | — | — | — |
|  |  |  | 81,000 | 100,000 |
| — | — | — | 81,000 | 100,000 |
|  |  |  | 54,000 | 66,666 |
| 64,830 | 54,650 | 62,590 | 139,180 | 157,230(e) |
| 61,160 | 47,939 | 48,898 | 92,787 | 104,820(e) |
| 47,370 | 38,800 | 40,050 | 38,450 | 37,500(e) |
| 44,688 | 34,035 | 31,289 | 25,633 | 25,000(e) |
| 270,730 | 307,320 | 333,330 | 570,860 | NA |
| 255,405 | 269,578 | 260,414 | 380,573 |  |
| 17.49 | 12.62 | 12.01 | 6.73 |  |

Table C.9 France—Internal Expenditures: Reproductive Research and Contraceptive Development (by Purpose and Proportion of Governmental Medical Research Funds, 1965, 1969-1974, 1975 [est.], in Current and Constant U.S. Dollars [1970=100])

|  | 1965 | 1969 | 1970 |
|---|---|---|---|
| **Fund. St./Training** | | | |
| Government | | | |
| INSERM[a] | d | d | 398,550 |
| CNRS[b] | 214,898 | 434,532 | 591,304 |
| Universities | 960,408[d] | 1,532,734[d] | 1,487,862 |
| INRA[c] | 1,112,245 | 1,703,237 | 1,712,138 |
| Research & Industrial | | | |
| Development | | | 110,326 |
| University (training | | | |
| & conferences) | 130,816 | 266,548 | 207,066 |
| Philanthropy | — | — | — |
| Subtotal | | | |
| Current $ | 2,418,367 | 3,937,051 | 4,507,246 |
| Constant '70 $ | 2,985,638 | 4,144,264 | 4,507,246 |
| **Contraceptive Development and Safety** | | | |
| Industry | | | |
| Subtotal | | | |
| Current $ | — | — | — |
| Constant '70 $ | | | |
| Donations to foreign funding agencies or research projects | — | — | — |
| **Total** | | | |
| Current $ | 2,418,367 | 3,937,051 | 4,507,246 |
| Constant '70 $ | 2,985,638 | 4,144,264 | 4,507,246 |
| A. Government total | | | |
| Current $ | 2,418,367 | 3,937,051 | 4,507,246 |
| Constant $ | 2,985,638 | 4,144,264 | 4,507,246 |
| B. Government medical research budget[e] | | | |
| Current $ | 215,140,000 | 302,390,000 | 317,190,000 |
| Constant $ | 265,604,938 | 318,305,263 | 317,190,000 |
| % A of B | 1.12 | 1.30 | 1.42 |

[a]INSERM—National Institute of Health
[b]CNRS—National Center for Scientific Research
[c]INRA—National Institute of Agriculatural Research
[d]Combined University and INSERM
[e]Ministries of Research and Industrial Development (Research Institutes-research funds only), Agriculture (INRA), Education (University budgets est., and CNRS), and Health (INSERM).

| 1971 | 1972 | 1973 | 1974 | 1975 (est.) |
|---|---|---|---|---|
| 617,433 | 818,129 | 1,121,145 | 1,029,350 | |
| 799,233 | 1,238,012 | 1,446,256 | 1,398,323 | |
| 1,955,939 | 2,160,624 | 2,092,058 | 2,836,478 | |
| 2,071,073 | 2,526,706 | 3,244,493 | 3,245,283 | |
| 380,268 | 414,425 | 562,996 | 509,434 | |
| 200,958 | 340,180 | 246,256 | 293,501 | |
| — | — | — | — | — |
| 6,024,904 | 7,498,076 | 8,713,204 | 9,312,369 | 11,533,482(e) |
| 5,683,872 | 6,694,711 | 7,261,003 | 6,797,349 | 8,418,600(e) |
| | | | 1,433,962 | |
| — | — | — | 1,433,962 | — |
| | | | 1,046,687 | |
| — | — | — | — | — |
| 6,024,904 | 7,498,076 | 8,713,204 | 10,746,331 | 11,533,482(e) |
| 5,683,872 | 6,694,711 | 7,261,003 | 7,844,037 | 8,418,600(e) |
| 6,024,904 | 7,498,076 | 8,713,204 | 9,312,369 | 11,533,482(e) |
| 5,683,872 | 6,694,711 | 7,261,003 | 6,797,349 | 8,418,600(e) |
| 380,780,000 | 450,730,000 | 594,000,000 | 625,969,998 | NA |
| 359,226,415 | 402,437,500 | 495,000,000 | 456,912,408 | |
| 1.58 | 1.66 | 1.46 | 1.37 | |

Table C.10 Germany—Internal Expenditures: Reproductive Research and Contraceptive Development (by Purpose and Proportion of Governmental Medical Research Funds, 1969-1974, 1975 [est.], in Current and Constant U.S. Dollars [1970=100])

|  | 1965 | 1969 | 1970 |
|---|---|---|---|
| Fund. St./Training |  |  |  |
| Government |  |  |  |
| DFG[a] | — | 399,002 | 650,407 |
| Philanthropy | — | — | — |
| Subtotal |  |  |  |
| Current $ |  | 399,002 | 650,407 |
| Constant '70 $ |  | 411,342 | 650,407 |
| Contraceptive Development and Safety |  |  |  |
| Government |  |  |  |
| Ministry of Research and Technology |  |  |  |
| Industry |  |  |  |
| Subtotal |  |  |  |
| Current $ |  |  |  |
| Constant '70 $ |  |  |  |
| Donations to foreign funding agencies or research projects | — | — | — |
| Total |  |  |  |
| Current $ | — | 399,002 | 650,407 |
| Constant '70 $ |  | 411,342 | 650,407 |
| A. Government total |  |  |  |
| Current $ | — | 399,022 | 650,407 |
| Constant $ |  | 411,342 | 650,407 |
| B. Government medical research |  |  |  |
| budget[b]—Current $ | — | 14,905,149 | 21,917,808 |
| Constant $ |  | 15,366,132 | 21,917,808 |
| % A of B |  | 2.67 | 2.97 |

[a]DFG—Deutsche Forschungsgemeinschaft
[b]DFG Life Sciences Division Budget

| 1971 | 1972 | 1973 | 1974 | 1975 (est.) |
|---|---|---|---|---|
| 1,162,080 | 1,406,250 | 2,042,253 | 2,159,000(e) | 2,159,000(e) |
| — | — | — | — | — |
| 1,162,080 | 1,406,250 | 2,042,253 | 2,159,000(e) | 2,159,000(e) |
| 1,106,742 | 1,266,892 | 1,716,178 | 1,700,000(e) | 1,700,000(e) |
| | | | 320,755 | 293,103 |
| | | 10,000 | 250,000 | 250,000 |
| | | 10,000 | 570,755 | 543,103 |
| | | 8,403 | 449,413 | 427,640 |
| — | — | — | — | — |
| 1,162,080 | 1,406,250 | 2,052,253 | 2,729,755(e) | 2,702,103(e) |
| 1,106,742 | 1,266,892 | 1,724,582 | 2,149,413(e) | 2,127,640(e) |
| 1,162,080 | 1,406,250 | 2,052,253 | 2,479,755(e) | 2,452,103(e) |
| 1,106,742 | 1,266,892 | 1,716,178 | 1,952,562(e) | 1,930,789(e) |
| 33,639,143 | 37,500,000 | 59,859,154 | NA | NA |
| 32,037,279 | 33,783,783 | 50,301,810 | | |
| 3.45 | 3.75 | 3.41 | | |

Table C.11 Hong Kong—Internal Expenditures: Reproductive Research and Contraceptive Development (by Purpose and Proportion of Governmental Medical Research Funds, 1965, 1969-1973, 1975 [est.], in Current and Constant U.S. Dollars [1970=100])

|  | 1965 | 1969 | 1970 |
|---|---|---|---|
| Fund. St./Training | — | — | — |
| Contraceptive Development and Safety Government (estimate) | — | 33,000 | 33,000 |
| Subtotal Current $ | — | 33,000 | 33,000 |
| Constant '70 $ |  | 35,483 | 33,000 |
| Donations to foreign funding agencies or research projects | — | — | — |
| Total Current $ | — | 33,000 | 33,000 |
| Constant '70 $ |  | 35,483 | 33,000 |
| A. Government total Current $ Constant $ | NA | NA | NA |
| B. Government medical research budget—Current $ Constant $ | NA | NA | NA |
| % A of B |  |  |  |

| 1971 | 1972 | 1973 | 1974 | 1975 (est.) |
|---|---|---|---|---|
| — | — | — | — | — |
| 33,000 | 33,000 | 33,000 | | |
| 33,000 | 33,000 | 33,000 | — | — |
| 32,039 | 30,275 | 25,581 | | |
| — | — | — | — | — |
| 33,000 | 33,000 | 33,000 | — | — |
| 32,039 | 30,275 | 25,581 | | |
| NA | NA | NA | NA | NA |
| NA | NA | NA | NA | NA |

Table C.12 India—Internal Expenditures: Reproductive Research and Contraceptive Development (by Purpose and Proportion of Governmental Medical Research Funds, 1965, 1969-1974, 1975 [est.], in Current and Constant U.S. Dollars [1970=100])

|  | 1965 | 1969 | 1970 |
|---|---|---|---|
| **Fund. St./Training** |  |  |  |
| Government |  |  |  |
| ICMR[a] |  |  |  |
| Rep. physiology unit | 29,525 | 92,418 | 141,129 |
| Reproductive biology course (training) |  | 3,333 | 3,333 |
| Meetings, seminars |  | 11,333 | 11,333 |
| Miscellaneous |  | 22,667 | 13,913 |
| Research grants | 23,725 | 92,080 | 186,598 |
| CSIR[b] |  | 25,000 | 25,000 |
| Ministry of Health & Family Planning |  | 5,333 | 5,333 |
| Other gov't. funds | 116,663 | 492,734 | 653,828 |
| Subtotal |  |  |  |
| Current $ | 169,913 | 744,898 | 1,040,467 |
| Constant '70 $ | 217,837 | 784,103 | 1,040,467 |
| **Contraceptive Development and Safety** |  |  |  |
| Government |  |  |  |
| ICMR |  |  |  |
| Contraceptive testing, university (IRR)[c] | 30,086 | 33,100 |  |
| Clinical trials |  |  | 2,197 |
| Central Drug Research Institute | 6,555 | 9,973 | 13,793 |
| Subtotal |  |  |  |
| Current $ | 36,641 | 43,073 | 15,990 |
| Constant '70 $ | 46,975 | 45,340 | 15,990 |
| Donations to foreign funding agencies or research projects | — | — | — |
| Total |  |  |  |
| Current $ | 206,554 | 787,971 | 1,056,457 |
| Constant '70 $ | 264,813 | 829,443 | 1,056,457 |
| A. Government total |  |  |  |
| Current $ | 206,554 | 787,971 | 1,056,457 |
| Constant $ | 264,183 | 829,443 | 1,056,457 |
| B. Government medical research |  |  |  |
| budget—Current $ | 1,176,400 | 4,062,000 | 4,648,000 |
| Constant $ | 1,508,205 | 4,275,789 | 4,648,000 |
| % A of B | 17.55 | 19.4 | 22.7 |

[a]ICMR—Indian Council of Medical Research
[b]CSIR—Council of Scientific and Industrial Research
[c]IRR—Institute for Reproduction Research

Country Funding Data

| 1971 | 1972 | 1973 | 1974 | 1975 (est.) |
|---|---|---|---|---|
| 395,216 | 240,000 | 160,000 | 186,666 | |
| 5,333 | 3,333 | | 3,333 | |
| 11,333 | 9,333 | 5,333 | | |
| 20,820 | 22,635 | 22,667 | 22,667 | |
| 216,533 | 191,706 | 167,472 | 11,581 | |
| 25,000 | 25,000 | 25,000 | 25,000 | |
| 5,333 | 5,333 | | | |
| 84,764 | 243,912 | 389,195 | | |
| | | | | |
| 764,332 | 741,252 | 769,667 | 249,247 | — |
| 742,069 | 673,864 | 601,302 | 151,058 | |
| | | | | |
| 47,333 | 55,413 | 55,333 | 57,333 | |
| 14,423 | 14,852 | 16,188 | 21,092 | |
| | | | | |
| 61,756 | 70,265 | 71,521 | 78,425 | — |
| 59,957 | 63,877 | 55,875 | 47,530 | |
| | | | | |
| — | — | — | — | — |
| | | | | |
| 826,088 | 811,517 | 841,188 | 327,672 | |
| 802,027 | 737,741 | 657,178 | 198,589 | — |
| | | | | |
| 826,088 | 811,517 | 841,188 | 327,672 | — |
| 802,027 | 737,741 | 657,178 | 19,589 | |
| | | | | |
| 4,471,000 | 4,849,000 | 5,211,000 | NA | — |
| 4,340,776 | 4,408,181 | 4,071,093 | | |
| 18.5 | 16.7 | 16.1 | | |

Table C.13 Iran—Internal Expenditures: Reproductive Research and Contraceptive Development (by Purpose and Proportion of Governmental Medical Research Funds, 1965, 1969-1974, 1975 [est.], in Current and Constant U.S. Dollars [1970=100])

|  | 1965 | 1969 | 1970 |
|---|---|---|---|
| Fund. St./Training | — | — | — |
| Contraceptive Development and Safety | | | |
| Government | | | |
| Ministry of Health | | | |
| University | | | |
| Ministry of Science | | | |
| Subtotal | | | |
| Current $ | — | — | — |
| Constant '70 $ | — | — | — |
| Donations to foreign funding agencies or research projects | — | — | — |
| Total | | | |
| Current $ | — | — | — |
| Constant '70 $ | — | — | — |
| A. Government total | NA | NA | NA |
| Current $ | | | |
| Constant $ | | | |
| B. Government medical research budget—Current $ | NA | NA | NA |
| Constant $ | | | |
| % A of B | | | |

# Country Funding Data

| 1971 | 1972 | 1973 | 1974 | 1975 (est.) |
|------|------|------|------|-------------|
| — | — | — | — | — |
|   |   | 500,000 | 175,000 |   |
|   |   |   | 75,000 |   |
|   |   |   | 15,000 |   |
| — | — | 500,000 | 265,000 | — |
| — | — | 409,836 | 190,647 | — |
| — | — | — | — | — |
| — | — | 500,000 | 265,000 | — |
| — | — | 409,836 | 190,647 | — |
| NA | NA | NA | NA | NA |
| NA | NA | NA | NA | NA |

Table C.14 Israel—Internal Expenditures: Reproductive Research and Contraceptive Development (by Purpose and Proportion of Governmental Medical Research Funds, 1965, 1969-1974, 1975 [est.], in Current and Constant U.S. Dollars [1970=100])

|  | 1965 | 1969 | 1970 |
|---|---|---|---|
| **Fund. St./Training** |  |  |  |
| Government |  |  |  |
| Academy of Science |  | 5,000 |  |
| Nat'l. Council for R&D |  |  |  |
| National Lottery |  |  |  |
| Ministry of Agriculture | 5,000 | 5,000 | 5,000 |
| Ministry of Health | 5,000 | 15,000 | 20,000 |
| University Grants Comm. (Ministry of Finance, Education & Jewish Agency) | 20,000 | 40,000 | 58,000 |
| Philanthropy |  |  |  |
| Miscellaneous | 2,000 | 3,000 | — |
| Industry |  |  |  |
| Elgin Hydronautics |  | 5,000 | 5,000 |
| Subtotal |  |  |  |
| Current $ | 32,000 | 73,000 | 88,000 |
| Constant '70 $ | 39,024 | 77,660 | 88,000 |
| **Contraceptive Development and Safety** | — | — | — |
| Donations to foreign funding agencies or research projects | — | — | — |
| Donations from outside government |  |  |  |
| World Bank |  |  |  |
| Private (Misc.) | — | — | 24,000 |
| Industry |  |  |  |
| Sandoz | 30,000 |  |  |
| Schering |  |  |  |
| IBM |  |  |  |
| Subtotal |  |  |  |
| Current $ | 30,000 | — | 24,000 |
| Constant '70 $ | 36,585 |  | 24,000 |
| **Total** |  |  |  |
| Current $ | 62,000 | 73,000 | 112,000 |
| Constant '70 $ | 75,609 | 77,660 | 112,000 |

| 1971 | 1972 | 1973 | 1974 | 1975 (est.) |
|---|---|---|---|---|
|  | 15,000 | 27,000 | 7,000 | 7,000 |
|  | 14,000 | 17,000 | 5,000 | 5,000 |
| 25,000 | 25,000 | 28,000 | 20,000 | 15,000 |
| 25,000 | 30,000 | 52,000 | 55,000 | 53,000 |
| 73,000 | 94,000 | 114,000 | 140,000 | 118,000 |
| — | 6,000 | 6,000 | 5,000 | 5,000 |
| 5,000 |  |  |  |  |
| 128,000 | 184,000 | 244,000 | 232,000 | 203,000 |
| 114,286 | 146,032 | 160,526 | 109,434 | 95,754 |
| — | — | — | — | — |
| — | — | — | — | — |
|  |  | 10,000 |  |  |
| 4,000 | 3,000 | 3,000 | — | — |
|  |  | 10,000 | 10,000 | 5,000 |
|  |  | 26,000 | 12,000 |  |
| 4,000 | 3,000 | 49,000 | 22,000 | 5,000 |
| 3,571 | 2,380 | 32,236 | 10,377 | 2,358 |
| 132,000 | 187,000 | 293,000 | 254,000 | 208,000 |
| 117,857 | 148,412 | 192,763 | 119,811 | 98,113 |

Table C.15 Italy—Internal Expenditures: Reproductive Research and Contraceptive Development (by Purpose and Proportion of Governmental Medical Research Funds, 1965, 1969-1974, 1975 [est.], in Current and Constant U.S. Dollars [1970=100])

|  | 1965 | 1969 | 1970 |
|---|---|---|---|
| **Fund. St./Training** | | | |
| Government | | | |
| CNR[a] | 11,622 | 128,289 | 202,785 |
| University (estimate) | — | 1,145,000 | 1,145,000 |
| Philanthropy | — | — | — |
| Subtotal | | | |
| Current $ | 11,622 | 1,273,289 | 1,347,785 |
| Constant '70 $ | 13,513 | 1,340,304 | 1,347,785 |
| **Contraceptive Development and Safety** | | | |
| Industry | 119,700 | 194,400 | 367,100 |
| Subtotal | | | |
| Current $ | 119,700 | 194,400 | 367,100 |
| Constant '70 $ | 139,186 | 204,631 | 367,100 |
| Donations to foreign funding agencies or research projects | — | — | — |
| **Total** | | | |
| Current $ | 131,322 | 1,467,689 | 1,714,885 |
| Constant '70 $ | 152,700 | 1,544,935 | 1,714,885 |
| A. Government total | NA | NA | NA |
| Current $ | | | |
| Constant $ | | | |
| B. Government medical research budget—Current $ | NA | NA | NA |
| Constant $ | | | |
| % A of B | | | |

[a]CNR—Consiglio Nazionale delle Ricerche

## Country Funding Data

| 1971 | 1972 | 1973 | 1974 | 1975 (est.) |
|---|---|---|---|---|
| 274,999 | 309,889 | 284,112 | 190,573 | — |
| 1,145,000 | 1,145,000 | 1,145,000 | 1,145,000 | — |
| — | — | — | — | |
| 1,419,999 | 1,454,889 | 1,429,112 | 1,335,573 | 1,314,000(e) |
| 1,352,380 | 1,310,710 | 1,161,879 | 914,776 | 900,000(e) |
| 421,500 | 427,000 | 367,900 | 459,200 | 222,180 |
| 421,500 | 427,000 | 367,900 | 459,200 | 222,180 |
| 401,428 | 384,685 | 299,105 | 314,520 | 152,178 |
| — | — | — | — | — |
| 1,841,499 | 1,881,889 | 1,797,012 | 1,794,773 | 1,536,180(e) |
| 1,753,808 | 1,695,395 | 1,460,985 | 1,229,296 | 1,052,178(e) |
| NA | NA | NA | NA | NA |
| NA | NA | NA | NA | NA |

514                            Country Funding Data

Table C.16 Japan—Internal Expenditures: Reproductive Research and
Contraceptive Development (by Purpose and Proportion of Governmental
Medical Research Funds, 1965, 1969-1974, 1975 [est.], in Current and
Constant U.S. Dollars [1970=100])

|  | 1965 | 1969 | 1970 |
|---|---|---|---|
| **Fund. St./Training** | | | |
| Government | | | |
| University | | | |
| Grants in Aid | | | |
| Philanthropy | | | |
| Subtotal | | | |
| Current $ | | | |
| Constant '70 $ | | | |
| **Contraceptive Development and Safety** | | | |
| Industry | | | |
| Subtotal | | | |
| Current $ | | | |
| Constant '70 $ | | | |
| Donations to foreign funding agencies or research projects | | | |
| **Total** | | | |
| Current $ | | | |
| Constant '70 $ | | | |
| A. Government total | NA | NA | NA |
| Current $ | | | |
| Constant $ | | | |
| B. Government medical research budget—Current $ | NA | NA | NA |
| Constant $ | | | |
| % A of B | | | |

Source: This information was compiled from responses to a survey conducted
by Richard Mahoney, Program Officer of the Ford Foundation.

| 1971 | 1972 | 1973 | 1974 | 1975 (est.) |
|---|---|---|---|---|
| | 217,168 | | | |
| | 196,444 | | | |
| | — | | | |
| | 413,612 | | | |
| | 372,623 | | | |
| | 928,245 | | | |
| | 928,245 | | | |
| | 836,256 | | | |
| | — | | | |
| | 1,341,857 | | | |
| | 1,208,880 | | | |
| NA | NA | NA | NA | NA |
| NA | NA | NA | NA | NA |

Table C.17 Netherlands—Internal Expenditures: Reproductive Research and Contraceptive Development (by Purpose and Proportion of Governmental Medical Research Funds, 1965, 1969-1974, 1975 [est.], in Current and Constant U.S. Dollars [1970=100])

|  | 1965 | 1969 | 1970 |
|---|---|---|---|
| Fund. St./Training |  |  |  |
| Government |  |  |  |
| TNO[a] |  |  |  |
| ZWO[b]-FUNGO[c] |  |  |  |
| University budget (est) |  |  |  |
| Philanthropy | — | — | — |
| Subtotal |  |  |  |
| Current $ |  |  |  |
| Constant '70 $ |  |  |  |
| Contraceptive Development and Safety |  |  |  |
| Government | — | — | — |
| Industry | — | 60,000 | 200,000 |
| Subtotal |  |  |  |
| Current $ |  | 60,000 | 200,000 |
| Constant '70 $ |  | 62,500 | 200,000 |
| Donations to foreign funding agencies or research projects | — | — | — |
| Total |  |  |  |
| Current $ | — | 60,000 | 200,000 |
| Constant '70 $ |  | 62,500 | 200,000 |
| A. Government total | NA | NA | NA |
| Current $ |  |  |  |
| Constant $ |  |  |  |
| B. Government medical research budget[d]—Current $ | NA | NA | NA |
| Constant $ |  |  |  |
| % A of B |  |  |  |

[a]TNO—Netherlands Organization for Applied Scientific Research
[b]ZWO—Netherlands Organization for the Advancement of Pure Research
[c]FUNGO—Foundation for Fundamental Medical Research
[d]Total expenditure on scientific research in medical faculties of the universities, on life sciences via ZWO and on health research via TNO.

Country Funding Data

| 1971 | 1972 | 1973 | 1974 | 1975 (est.) |
|---|---|---|---|---|
| 5,808 | 6,230 | 7,334 | 8,370 | |
| | | 121,222 | 162,592 | |
| 1,666,667 | 2,000,000 | 2,333,333 | 2,666,667 | |
| — | — | — | — | — |
| 1,672,475 | 2,006,230 | 2,461,889 | 2,837,629 | 2,837,629(e) |
| 1,548,587 | 1,729,508 | 1,969,511 | 2,071,262 | 2,071,262 |
| — | — | — | — | — |
| 350,000 | 400,000 | 450,000 | 450,000 | 500,000 |
| 350,000 | 400,000 | 450,000 | 450,000 | 500,000 |
| 324,074 | 344,827 | 360,000 | 328,467 | 364,963 |
| — | — | — | — | — |
| 2,022,475 | 2,406,230 | 2,911,889 | 3,287,629 | 3,337,629 |
| 1,872,662 | 2,074,336 | 2,329,511 | 2,399,729 | 2,436,225 |
| 1,672,475 | 2,006,230 | 2,461,889 | 2,837,629 | NA |
| 1,548,587 | 1,729,508 | 1,969,511 | 2,071,262 | |
| 39,859,000[d] | 51,300,000 | NA | NA | NA |
| 36,906,481 | 44,224,137 | | | |
| 4.19 | 3.91 | | | |

Table C.18 New Zealand—Internal Expenditures: Reproductive Research and Contraceptive Development (by Purpose and Proportion of Governmental Medical Research Funds, 1965, 1969-1974, 1975 [est.], in Current and Constant U.S. Dollars [1970=100])

|  | 1965 | 1969 | 1970 |
|---|---|---|---|
| Fund. St./Training | | | |
| Government | | | |
| MRC[a] | 13,422 | 42,481 | 55,710 |
| Philanthropy | — | — | — |
| Subtotal | | | |
| Current $ | 13,422 | 42,481 | 55,710 |
| Constant '70 $ | 16,990 | 45,193 | 55,710 |
| Contraceptive Development and Safety | — | — | — |
| Donations to foreign funding agencies or research projects | — | — | — |
| Total | | | |
| Current $ | 13,422 | 42,481 | 55,710 |
| Constant '70 $ | 16,990 | 45,193 | 55,710 |
| A. Government total | NA | NA | NA |
| Current $ | | | |
| Constant $ | | | |
| B. Government medical research budget—Current $ | NA | NA | NA |
| Constant $ | | | |
| % A of B | | | |

[a]MRC—Medical Research Council

| 1971 | 1972 | 1973 | 1974 | 1975 (est.) |
|---|---|---|---|---|
| 85,755 | 89,251 | 96,317 | — | — |
| — | — | — | — | — |
| 85,755 | 89,251 | 96,317 | 106,500(e) | 106,500(e) |
| 77,959 | 75,636 | 75,248 | 75,000(e) | 75,000(e) |
| — | — | — | — | — |
| — | — | — | — | — |
| 85,755 | 89,251 | 96,317 | 106,500(e) | 106,500(e) |
| 77,959 | 75,636 | 75,248 | 75,000(e) | 75,000(e) |
| NA | NA | NA | NA | NA |
| NA | NA | NA | NA | NA |

Table C.19 Norway—Internal Expenditures: Reproductive Research and Contraceptive Development (by Purpose and Proportion of Governmental Medical Research Funds, 1965, 1969-1974, 1975 [est.], in Current and Constant U.S. Dollars [1970=100])

|  | 1965 | 1969 | 1970 |
|---|---|---|---|
| **Fund. St./Training** | | | |
| Government | | | |
| MRC[a] | — | 26,570 | 29,120 |
| Agriculture Research Council | — | — | — |
| Philanthropy | — | — | — |
| Subtotal | | | |
| Current $ | — | 26,570 | 29,120 |
| Constant '70 $ | | 29,722 | 29,120 |
| **Contraceptive Development and Safety** | — | — | — |
| Donations to foreign funding agencies or research projects | | | |
| NOR-AID[b] | — | — | — |
| Subtotal | | | |
| Current $ | | | |
| Constant '70 $ | | | |
| **Total** | | | |
| Current $ | — | 26,570 | 29,120 |
| Constant '70 $ | | 29,722 | 29,120 |
| A. Government total[c] | NA | | |
| Current $ | | 26,570 | 29,120 |
| Constant $ | | 29,722 | 29,120 |
| B. Government medical research budget[c]—Current $ | NA | 784,300 | 756,300 |
| Constant $ | | 871,444 | 756,300 |
| % A of B | | 3.41 | 3.85 |

[a]MRC—Medical Research Council
[b]NOR-AID—Norwegian Agency for International Development
[c]MRC only

Country Funding Data

| 1971 | 1972 | 1973 | 1974 | 1975 (est.) |
|---|---|---|---|---|
| 47,940 | 52,660 | 63,710 | 83,240 | 80,400(e) |
| — | 197,290 | 250,000 | 310,380 | 268,000(e) |
| — | — | — | — | — |
| 47,940 | 249,950 | 313,710 | 393,620 | 348,400(e) |
| 45,226 | 219,254 | 257,139 | 293,746 | 260,000(e) |
| — | — | — | — | — |
| — | 454,000 | 666,000 | 1,190,000 | 1,500,000(WHO) |
| | 454,000 | 666,000 | 1,190,000 | 1,500,000 |
| | 398,245 | 545,901 | 888,059 | 1,119,402 |
| 47,940 | 703,950 | 979,710 | 1,583,620 | 1,848,400(e) |
| 45,226 | 617,500 | 803,040 | 1,181,805 | 1,379,402(e) |
| 47,940 | 52,660 | 63,710 | 83,240 | 80,400(e) |
| 45,226 | 46,192 | 52,221 | 62,119 | 60,000 |
| 923,990 | 1,067,670 | 1,521,740 | 1,712,200 | NA |
| 871,683 | 936,552 | 1,247,327 | 1,277,761 | |
| 5.18 | 4.93 | 4.18 | 4.86 | |

Table C.20 Philippines—Internal Expenditures: Reproductive Research and Contraceptive Development (by Purpose and Proportion of Governmental Medical Research Funds, 1965, 1969-1974, 1975 [est.], in Current and Constant U.S. Dollars [1970=100])

|  | 1965 | 1969 | 1970 |
|---|---|---|---|
| Fund. St./Training | — | — | — |
| Contraceptive Development and Safety | | | |
| Government | | | |
| University Santo Tomas | | | |
| Committee on Population | | | |
| Subtotal | | | |
| Current $ | — | — | — |
| Constant '70 $ | | | |
| Donations to foreign funding agencies or research projects | — | — | — |
| Total | | | |
| Current $ | — | — | — |
| Constant '70 $ | | | |
| A. Government total | NA | NA | NA |
| Current $ | | | |
| Constant $ | | | |
| B. Government medical research budget—Current $ | NA | NA | NA |
| Constant $ | | | |
| % A of B | | | |

| 1971 | 1972 | 1973 | 1974 | 1975 (est.) |
|------|------|------|------|-------------|
| —    | —    | —    | —    | —           |
|      |      | 3,687 | 3,687 |           |
|      |      | 20,839 | 35,068 |         |
| —    | —    | 24,526 | 38,755 | —       |
|      |      | 16,030 | 18,026 |         |
| —    | —    | —    | —    | —           |
| —    | —    | 24,526 | 38,755 | —       |
|      |      | 16,030 | 18,026 |         |
| NA   | NA   | NA   | NA   | NA          |
| NA   | NA   | NA   | NA   | NA          |

Table C.21 South Korea—Internal Expenditures: Reproductive Research and Contraceptive Development (by Purpose and Proportion of Governmental Medical Research Funds, 1965, 1969-1974, 1975 [est.], in Current and Constant U.S. Dollars [1970=100])

|  | 1965 | 1969 | 1970 |
|---|---|---|---|
| Fund. St./Training | — | — | — |
| Contraceptive Development and Safety |  |  |  |
| Government |  | 3,500 | 2,500 |
| University | 1,500 | 2,000 | 1,500 |
| Philanthropy |  | 5,000 |  |
| Subtotal |  |  |  |
| Current $ | 1,500 | 10,500 | 4,000 |
| Constant '70 $ | 2,679 | 12,209 | 4,000 |
| Donations to foreign funding agencies or research projects | — | — | — |
| Donations from Outside |  |  |  |
| Wyeth |  |  | 4,000 |
| CMB[a] |  |  |  |
| UNFPA[b] |  |  |  |
| Subtotal |  |  |  |
| Current $ | — | — | 4,000 |
| Constant '70 $ |  |  | 4,000 |
| Total |  |  |  |
| Current $ | 1,500 | 10,500 | 8,000 |
| Constant '70 $ | 2,679 | 12,209 | 8,000 |
| A. Government total |  |  |  |
| Current $ | 1,500 | 5,500 | 4,000 |
| Constant $ | 2,679 | 6,395 | 4,000 |
| B. Government medical research |  |  |  |
| budget—Current $ | NA | 339,180 | 1,088,250 |
| Constant $ |  | 394,395 | 1,088,250 |
| % A of B |  | 1.62 | 0.36 |

[a]CMB—China Medical Board
[b]UNFPA—United Nations Fund for Population Activities

| 1971 | 1972 | 1973 | 1974 | 1975 (est.) |
|---|---|---|---|---|
| — | — | — | — | — |
| 4,000 | 6,000 | 19,000 | 6,250 | |
| 2,000 | 2,000 | 2,000 | 3,000 | |
| 5,000 | | 11,000 | | |
| 11,000 | 8,000 | 32,000 | 9,250 | |
| 9,649 | 6,299 | 24,427 | 5,674 | — |
| — | — | — | — | — |
| 4,000 | | | | |
| | | 1,500 | | |
| | | | 110,500 | |
| 4,000 | — | 1,500 | 110,500 | — |
| 3,508 | | 1,145 | 67,791 | |
| 15,000 | 8,000 | 33,500 | 119,750 | |
| 13,157 | 6,299 | 25,572 | 73,466 | |
| 6,000 | 8,000 | 21,000 | 9,250 | |
| 5,263 | 6,299 | 16,030 | 5,688 | |
| 1,351,850 | 1,814,350 | 1,705,500 | 1,545,000 | — |
| 1,185,833 | 1,428,587 | 1,027,410 | 947,852 | |
| 0.44 | 0.44 | 1.56 | 0.60 | |

Table C.22 Sweden—Internal Expenditures: Reproductive Research and Contraceptive Development (by Purpose and Proportion of Governmental Medical Research Funds, 1965, 1969-1974, 1975 (est.), in Current and Constant U.S. Dollars [1970=100]

|  | 1965 | 1969 | 1970 |
|---|---|---|---|
| **Fund. St./Training** | | | |
| Government | | | |
| MRC[a] | — | 298,000 | 368,000 |
| SIDA[b] | — | — | 210,000 |
| Philanthropy | — | — | — |
| Subtotal | | | |
| Current $ | — | 298,000 | 578,000 |
| Constant '70 $ | | 320,430 | 578,000 |
| **Contraceptive Development and Safety** | | | |
| Industry | 184,000 | 452,800 | 311,000 |
| Subtotal | | | |
| Current $ | 184,000 | 452,800 | 311,000 |
| Constant '70 $ | 230,000 | 486,881 | 311,000 |
| Donations to foreign funding agencies or research projects | | | |
| SIDA | — | — | 50,000 |
| Subtotal | | | |
| Current $ | — | — | 50,000 |
| Constant '70 $ | | | 50,000 |
| **Total** | | | |
| Current $ | 184,000 | 750,800 | 939,000 |
| Constant '70 $ | 230,000 | 807,312 | 939,000 |
| A. Government total[c] | | | |
| Current $ | NA | 298,000 | 368,000 |
| Constant $ | | 320,430 | 368,000 |
| B. Government medical research | | | |
| budget[c]—Current $ | NA | 5,240,000 | 6,830,000 |
| Constant $ | | 5,634,408 | 6,830,000 |
| % A of B | | 5.68 | 5.39 |

[a]MRC—Medical Research Council
[b]SIDA—Swedish International Development Authority
[c]MRC only

| 1971 | 1972 | 1973 | 1974 | 1975 (est.) |
|---|---|---|---|---|
| 398,000 | 426,000 | 430,000 | 422,000 | 493,000 |
| 120,000 | 40,000 | — | — | — |
| — | — | — | — | — |
| 518,000 | 466,000 | 430,000 | 422,000 | 493,000 |
| 484,112 | 408,771 | 352,459 | 314,925 | 367,910 |
| 706,600 | 859,200 | 674,200 | 554,000 | — |
| 706,600 | 859,200 | 674,200 | 554,000 | — |
| 660,373 | 753,684 | 552,622 | 413,432 | |
| — | 3,000,000 | 4,000,000 | 5,000,000 | 5,000,000 |
| — | 3,000,000 | 4,000,000 | 5,000,000 | 5,000,000 |
| — | 2,631,578 | 3,278,688 | 3,731,343 | 3,731,343 |
| 1,224,600 | 4,325,200 | 5,104,200 | 5,976,000 | 5,493,000 |
| 1,144,486 | 3,794,035 | 4,183,770 | 4,459,701 | 4,099,253 |
| 398,000 | 426,000 | 430,000 | 422,000 | 493,000 |
| 371,962 | 373,684 | 352,459 | 314,925 | 367,910 |
| 8,230,000 | 9,050,000 | 10,510,000 | 11,550,000 | 13,740,000 |
| 7,691,588 | 7,938,596 | 8,614,754 | 8,619,402 | 10,253,731 |
| 4.83 | 4.70 | 4.09 | 3.65 | 3.58 |

Table C.23 Thailand—Internal Expenditures: Reproductive Research and Development (by Purpose and Proportion of Governmental Medical Research Funds, 1965, 1969-1974, 1975 [est.], in Current and Constant U.S. Dollars [1970=100])

|  | 1965 | 1969 | 1970 |
|---|---|---|---|
| Fund. St./Training | — | — | — |
| Contraceptive Development and Safety | | | |
| Government | | | |
| NRC[a] | | 446 | 125 |
| Subtotal | | | |
| Current $ | — | 446 | 125 |
| Constant '70 $ | | 451 | 125 |
| Donations to foreign funding agencies or research projects | — | — | — |
| Total | | | |
| Current $ | — | 446 | 125 |
| Constant '70 $ | | 451 | 125 |
| A. Government total | NA | | |
| Current $ | | 446 | 125 |
| Constant $ | | 451 | 125 |
| B. Government medical research budget[b]—Current $ | NA | 110,960 | 93,099 |
| Constant $ | | 112,080 | 93,099 |
| % A of B | | 0.4 | 0.1 |

[a]NRC—National Research Council
[b]National Research Council (NRC) budget

| 1971 | 1972 | 1973 | 1974 | 1975 (est.) |
|---|---|---|---|---|
| — | — | — | — | — |
|  | 3,185 | 1,810 |  |  |
| — | 3,185 | 1,810 | — | — |
|  | 3,005 | 1,534 |  |  |
| — | — | — | — | — |
| — | 3,185 | 1,810 | — | — |
|  | 3,005 | 1,534 |  |  |
| — | 3,185 | 1,810 | — | — |
| — | 3,005 | 1,534 |  |  |
| 102,130 | 102,618 | 125,000 | — | — |
| 100,127 | 96,809 | 105,932 |  |  |
| 0.0 | 3.1 | 1.4 |  |  |

Table C.24 Turkey—Internal Expenditures: Reproductive Research and Contraceptive Development (by Purpose and Proportion of Governmental Medical Research Funds, 1965, 1969-1974, 1975 [est.], in Current and Constant U.S. Dollars [1970=100])

|  | 1965 | 1969 | 1970 |
|---|---|---|---|
| **Fund. St./Training** | — | — | — |
| **Contraceptive Development and Safety** Government | | | |
| Subtotal Current $ Constant '70 $ | — | — | — |
| Donations to foreign funding agencies or research projects | — | — | — |
| Total Current $ Constant '70 $ | — | — | — |
| A. Government total Current $ Constant $ | NA | NA | NA |
| B. Government medical research budget—Current $ Constant $ % A of B | NA | NA | NA |

| 1971 | 1972 | 1973 | 1974 | 1975 (est.) |
|---|---|---|---|---|
| — | — | — | — | — |
|  |  | 37,000 | 75,000 | 50,000 |
| — | — | 37,000 | 75,000 | 50,000 |
|  |  | 24,183 | 41,209 | 27,472 |
| — | — | — | — | — |
| — | — | 37,000 | 75,000 | 50,000 |
|  |  | 24,183 | 41,209 | 27,472 |
| NA | NA | NA | NA | NA |
| NA | NA | NA | NA | NA |

Table C.25 United States—Internal Expenditures: Reproductive Research and Contraceptive Development (by Agency and Proportion of Governmental Medical Research Funds, 1965, 1969-1974, 1975 [est.], in Current and Constant U.S. Dollars [1970=100])

|  | 1965 | 1969 | 1970 |
| --- | --- | --- | --- |
| Government |  |  |  |
| CPR[a] (NICHD)[b,c] | 4,496,290 | 11,448,169 | 14,388,145 |
| NIH[d] (excluding NICHD) | 1,731,878 | 3,367,869 | 4,820,641 |
| NIMH[e] | 237,041 | 559,727 | 658,726 |
| NSF[f] | 531,037 | 482,377 | 213,700 |
| FDA[g] | 130,000 | 673,990 | 1,006,500 |
| AEC[h] | 629,800 | 1,016,600 | 994,000 |
| AID[i] | 0 | 906,213 | 2,905,063 |
| Miscellaneous | 1,496,579 | 277,000 | 320,000 |
| Subtotal |  |  |  |
| Current $ | 9,252,625 | 18,731,945 | 25,306,775 |
| 1970 Constant $ | 11,422,993 | 19,927,601 | 25,306,775 |
| Philanthropy |  |  |  |
| Ford Foundation | 4,871,730 | 7,507,980 | 10,568,447 |
| Rockefeller Foundation | 58,630 | 532,895 | 1,687,285 |
| Scaife | 1,450,000 | 2,000,000 | 2,000,000 |
| Population Council[j] | — | — | 2,111,257 |
| Miscellaneous | 223,887 | 354,762 | 429,496 |
| Subtotal |  |  |  |
| Current $ | 6,604,247 | 10,395,637 | 16,796,485 |
| 1970 Constant $ | 8,153,391 | 11,059,188 | 16,796,485 |
| Industry | 10,071,000 | 12,409,000 | 12,906,000 |
| Subtotal |  |  |  |
| Current $ | 10,071,000 | 12,409,000 | 12,906,000 |
| 1970 Constant $ | 12,433,333 | 13,201,063 | 12,906,000 |
| Total |  |  |  |
| Current $ | 25,927,872 | 41,536,582 | 55,009,260 |
| 1970 Constant $ | 32,009,717 | 44,187,852 | 55,009,260 |
| A. Government total |  |  |  |
| Current $ | 9,252,625 | 18,731,945 | 25,306,775 |
| Constant $ | 11,422,993 | 19,927,601 | 25,306,775 |
| B. Government medical research |  |  |  |
| budget—Current $ | 1,229,580,000 | 1,764,150,000 | 1,743,000,000 |
| Constant $ | 1,518,000,000 | 1,857,000,000 | 1,743,000,000 |
| % A of B | .8 | 1.1 | 1.5 |

Note: For breakdown into research categories see Agency Funding Data in Appendix D.

[a]CPR—Center for Population Research

[b]NICHD—National Institute of Child Health and Human Development

[c]CPR (NICHD) differs from Table D.3, Appendix D for 1970-1974 as

# Country Funding Data

| 1971 | 1972 | 1973 | 1974 | 1975 (est.) |
|---|---|---|---|---|
| 20,678,832 | 29,656,076 | 34,912,539 | 31,304,183 | 36,000,000 |
| 4,549,320 | 5,145,613 | 5,831,301 | 7,376,933 | 4,000,000 |
| 876,871 | 1,056,021 | 1,244,457 | 1,423,747 | 1,200,000 |
| 797,600 | 851,500 | 771,900 | 461,300 | 400,000 |
| 1,374,554 | 1,428,744 | 859,354 | 640,265 | — |
| 922,900 | 988,700 | 714,900 | 771,981 | — |
| 3,864,396 | 7,583,418 | 6,484,911 | 6,255,967 | 3,400,000 |
| 251,000 | — | — | — | — |
| 33,315,473 | 46,710,072 | 50,819,362 | 48,234,376 | 45,900,000 |
| 32,034,108 | 43,250,066 | 44,578,387 | 37,979,823 | 36,141,732 |
| 17,806,875 | 12,192,543 | 11,106,360 | 9,530,961 | 6,200,000 |
| 3,174,011 | 3,562,154 | 2,485,603 | 4,150,212 | 3,000,000 |
| 2,250,000 | 2,250,000 | 2,250,000 | 2,250,000 | 250,000 |
| 45,825 | — | — | — | — |
| 211,356 | 221,080 | 301,600 | 149,320 | — |
| 23,488,067 | 18,225,777 | 16,143,563 | 16,080,493 | 9,450,000 |
| 22,584,679 | 16,875,719 | 14,161,020 | 12,661,805 | 7,440,944 |
| 14,859,000 | 14,789,000 | 15,107,000 | 14,789,000 | 14,637,000 |
| 14,859,000 | 14,789,000 | 15,107,000 | 14,789,000 | 14,637,000 |
| 14,287,500 | 13,693,518 | 13,251,754 | 11,644,881 | 11,525,196 |
| 71,662,540 | 79,724,849 | 82,069,925 | 79,103,869 | 69,987,000 |
| 68,906,287 | 73,819,303 | 71,991,161 | 62,286,509 | 55,083,000 |
| 33,315,473 | 46,710,072 | 50,819,362 | 48,234,376 | |
| 32,034,108 | 43,250,066 | 44,578,387 | 37,979,823 | NA |
| 2,011,600,000 | 2,225,880,000 | 2,311,920,000 | 2,886,710,000 | |
| 1,880,000,000 | 2,061,000,000 | 2,028,000,000 | 2,273,000,000 | NA |
| 1.7 | 2.1 | 2.2 | 1.7 | |

funds received from AID in those years have not been included in the above amounts to eliminate double counting.

[d]NIH—National Institutes of Health
[e]NIMH—National Institute of Mental Health
[f]NSF—National Science Foundation
Notes continue next page

Table C.25 (continued)

gFDA—Federal Drug Administration
hAEC—Atomic Energy Commission
iAID—Agency for International Development
jTo eliminate double counting, expenditures by Population Council are not shown with the exception of those in excess of the funds received from the Ford Foundation, Rockefeller Foundation, Scaife Family Charitable Trusts, Agency for International Development and the Center for Population Research.

Appendix D                    Agency Funding Data

The following tables were compiled from data received from the respective agencies. When available, grant abstracts were used to code each grant in the various categories of interest in accord with the definitions presented in appendix C. Grant titles served the purpose for research sponsored by CPR or other institutes of NIH (excluding NICHD). The tables were then drawn up by computer.

The amounts in the tables are actual yearly expenditures for reproductive research. The only major exceptions are multiple year grants made by the Ford Foundation and AID. In these two cases estimates of the yearly payments were calculated by dividing the total appropriations by the scheduled number of years of the grant or contract. The category "Other Agency" (meaning funds given to one of the other funding agencies listed here) was used to signal funds expended by another agency. In compiling the tables in chapter 11, these funds were not added into the totals as they were accounted for in the expenditures of the recipient agency.

For conversion to constant 1970 dollars weighted Consumer Price Index (CPI) numbers for the fifteen other industrialized nations and nine developing nations were used. The CPI source for each country was the *Statistical Yearbook*. Variances in the total constant dollar values reflected in the tables and in those of chapter 11 are due to the use of different CPI rates. In the following tables, CPI conversion values for the nations where recipients were located were used. Constant dollars in chapter 11 utilized the CPI of the nations that were the sources of funds.

Table D.1 Expenditures for the Reproductive Sciences and Contraceptive Development, by Funding Agency and Location of Research, in Current and Constant U.S. Dollars (1970=100) Based on CPI Values for Nations in which Research was Conducted, 1965, 1969-1974

| Agency for International Development (AID) | 1965 | 1969 | 1970 |
|---|---|---|---|
| **United States** | | | |
| Fundamental studies | — | — | 94,000 |
| Contraceptive development | — | 94,356 | 1,254,689 |
| Monitoring safety | — | — | 304,333 |
| Training | — | — | — |
| Other agency | — | 600,000 | 998,250 |
| Subtotal | — | 694,356 | 2,651,272 |
| Constant 1970 $ | — | 738,676 | 2,651,272 |
| **Developing Nations** | | | |
| Fundamental studies | — | — | 41,934 |
| Contraceptive development | — | — | — |
| Monitoring safety | — | 211,857 | 211,857 |
| Subtotal | — | 211,857 | 253,791 |
| Constant 1970 $ | — | 225,379 | 253,791 |
| Total | — | 906,213 | 2,905,063 |
| Constant 1970 $ | — | 964,055 | 2,905,063 |

Note: Multiple year grants and contracts pf AID were divided by their scheduled number of years to give estimated yearly expenditures.

Table D.2 Expenditures for the Reproductive Sciences and Contraceptive Development, by Funding Agency and Location of Research, in Current and Constant U.S. Dollars (1970=100) Based on CPI Values for Nations in which Research was Conducted, 1965, 1969-1974

| Atomic Energy Commission (AEC) | 1965 | 1969 | 1970 |
|---|---|---|---|
| **United States** | | | |
| Fundamental studies | 618,000 | 1,006,200 | 994,000 |
| Subtotal | 618,100 | 1,006,200 | 994,000 |
| Constant 1970 $ | 763,086 | 1,070,425 | 994,000 |
| **Other Industrial Nations** | | | |
| Fundamental studies | 11,700 | 10,400 | — |
| Subtotal | 11,700 | 10,400 | — |
| Constant 1970 $ | 14,268 | 10,947 | — |
| Total | 629,800 | 1,016,600 | 994,000 |
| Constant 1970 $ | 777,354 | 1,081,372 | 994,000 |

| 1971 | 1972 | 1973 | 1974 |
|---|---|---|---|
| 60,000 | — | — | — |
| 2,248,022 | 3,346,644 | 4,509,420 | 3,980,099 |
| 304,333 | 304,333 | — | — |
| — | 444,333 | 444,333 | 444,333 |
| 998,250 | 998,250 | 998,250 | — |
| 3,610,605 | 5,093,560 | 5,952,003 | 4,424,432 |
| 3,471,735 | 4,716,259 | 5,221,055 | 3,483,804 |
| | | | |
| 41,934 | 41,934 | 41,934 | — |
| — | 273,667 | 279,117 | 481,464 |
| 211,857 | 2,174,257 | 211,857 | 1,286,450 |
| 253,791 | 2,489,858 | 532,908 | 1,831,535 |
| 232,835 | 2,110,048 | 413,106 | 1,159,198 |
| | | | |
| 3,864,396 | 7,583,418 | 6,484,911 | 6,255,967 |
| 3,704,570 | 6,826,307 | 5,634,161 | 4,643,002 |

| 1971 | 1972 | 1973 | 1974 |
|---|---|---|---|
| 910,300 | 976,200 | 702,400 | 771,981 |
| 910,300 | 976,200 | 702,400 | 771,981 |
| 875,288 | 903,888 | 616,140 | 607,859 |
| | | | |
| 12,600 | 12,500 | 12,500 | — |
| 12,600 | 12,500 | 12,500 | — |
| 11,775 | 11,061 | 10,000 | — |
| | | | |
| 922,900 | 988,700 | 714,900 | 771,981 |
| 887,063 | 914,949 | 626,140 | 607,859 |

Table D.3 Expenditures for the Reproductive Sciences and Contraceptive Development, by Funding Agency and Location of Research, in Current and Constant U.S. Dollars (1970=100) Based on CPI Values for Nations in which Research was Conducted, 1965, 1969-1974

| Center for Population Research (CPR)— National Institute of Child Health and Human Development (NICHD) | 1965 | 1969 | 1970 |
|---|---|---|---|
| **United States** | | | |
| Fundamental studies | 367,976 | 8,102,379 | 11,299,470 |
| Contraceptive development | — | 1,215,993 | 178,520 |
| Monitoring safety | 59,000 | 401,281 | 1,588,408 |
| Training | 653,932 | 1,551,891 | 1,640,384 |
| Other agency | — | — | — |
| Subtotal | 4,392,608 | 11,271,544 | 14,706,782 |
| Constant 1970 $ | 5,422,972 | 11,991,004 | 14,706,782 |
| **Other Industrialized Nations** | | | |
| Fundamental studies | 23,400 | 88,144 | 10,316 |
| Contraceptive development | — | — | — |
| Monitoring safety | — | 42,704 | 42,720 |
| Training | 76,832 | 45,777 | 26,577 |
| Subtotal | 100,232 | 176,625 | 79,613 |
| Constant 1970 $ | 122,234 | 185,921 | 79,613 |
| **Developing Nations** | | | |
| Fundamental studies | 3,450 | — | — |
| Subtotal | 3,450 | — | — |
| Constant 1970 $ | 4,207 | — | — |
| **Total** | 4,496,290 | 11,448,169 | 14,786,395 |
| Constant 1970 $ | 5,549,413 | 12,176,925 | 14,786,395 |

# Agency Funding Data

| 1971 | 1972 | 1973 | 1974 |
|---|---|---|---|
| 13,480,440 | 20,090,976 | 26,521,162 | 20,262,437 |
| 2,243,125 | 3,625,727 | 2,340,257 | 3,513,407 |
| 3,134,320 | 4,439,645 | 4,978,362 | 5,106,614 |
| 1,580,208 | 2,322,660 | 865,691 | 1,665,218 |
| — | — | 522,165 | 463,908 |
| 20,488,093 | 29,479,008 | 35,227,637 | 31,011,584 |
| 19,700,089 | 27,295,378 | 28,901,436 | 24,418,570 |
| 220,857 | 277,890 | 74,652 | 292,232 |
| — | 110,772 | — | 60,699 |
| 299,507 | 172,000 | — | 337,918 |
| — | 14,656 | 8,500 | — |
| 520,364 | 575,318 | 83,152 | 690,849 |
| 486,304 | 509,130 | 66,521 | 479,756 |
| 68,625 | — | — | — |
| 68,625 | — | — | — |
| 62,958 | — | — | — |
| 21,077,082 | 30,054,326 | 35,310,789 | 31,702,433 |
| 20,249,351 | 27,804,508 | 30,967,956 | 24,962,546 |

Agency Funding Data

Table D.4 Expenditures for the Reproductive Sciences and Contraceptive Development, by Funding Agency and Location of Research, in Current and Constant U.S. Dollars (1970=100) Based on CPI Values for Nations in which Research was Conducted, 1965, 1969-1974

| Food and Drug Administration (FDA) | 1965 | 1969 | 1970 |
|---|---|---|---|
| **United States** | | | |
| Fundamental studies | 40,000 | 60,000 | 60,000 |
| Contraceptive development | 45,000 | 70,000 | 50,000 |
| Monitoring safety | 45,000 | 543,990 | 731,398 |
| Subtotal | 130,000 | 673,990 | 841,398 |
| Constant 1970 $ | 160,493 | 717,010 | 841,398 |
| **Other Industrialized Nations** | | | |
| Monitoring safety | — | — | 165,112 |
| Subtotal | — | — | 165,112 |
| Constant 1970 $ | — | — | 165,112 |
| Total | 130,000 | 673,990 | 1,006,500 |
| Constant 1970 $ | 160,493 | 717,010 | 1,006,500 |

Table D.5 Expenditures for the Reproductive Sciences and Contraceptive Development, by Funding Agency and Location of Research, in Current and Constant U.S. Dollars (1970=100) Based on CPI Values for Nations in which Research was Conducted, 1965, 1969-1974

| National Institutes of Health (NIH) (excluding NICHD) | 1965 | 1969 | 1970 |
|---|---|---|---|
| **United States** | | | |
| Fundamental studies | 1,503,728 | 3,116,269 | 4,546,153 |
| Contraceptive development | 22,761 | 35,574 | — |
| Monitoring safety | 144 | 194,890 | 245,843 |
| Training | 89,018 | — | — |
| Subtotal | 1,615,651 | 3,346,733 | 4,792,001 |
| Constant 1970 $ | 1,994,630 | 3,560,354 | 4,792,001 |
| **Other Industrialized Nations** | | | |
| Fundamental studies | 53,561 | — | 12,240 |
| Subtotal | 53,561 | — | 12,240 |
| Constant 1970 $ | 65,318 | — | 12,240 |
| **Developing Nations** | | | |
| Fundamental studies | 62,666 | 21,136 | 16,400 |
| Subtotal | 62,666 | 21,136 | 16,400 |
| Constant 1970 $ | 76,421 | 22,485 | 16,400 |
| Total | 1,731,878 | 3,367,869 | 4,820,641 |
| Constant 1970 $ | 3,809,466 | 3,582,839 | 4,820,641 |

| 1971 | 1972 | 1973 | 1974 |
|---|---|---|---|
| 60,000 | 80,000 | 80,000 | 80,000 |
| 50,000 | 50,000 | 30,000 | 30,000 |
| 980,517 | 949,780 | 400,720 | 243,890 |
| 1,090,517 | 1,079,780 | 510,720 | 353,890 |
| 1,048,574 | 999,796 | 448,000 | 278,653 |
| 284,037 | 348,964 | 348,634 | 286,375 |
| 284,037 | 348,964 | 348,634 | 286,375 |
| 265,455 | 308,817 | 278,907 | 198,871 |
| 1,374,554 | 1,428,744 | 859,354 | 640,265 |
| 1,314,029 | 1,308,613 | 726,907 | 477,524 |

| 1971 | 1972 | 1973 | 1974 |
|---|---|---|---|
| 3,641,296 | 4,115,545 | 4,862,539 | 5,050,998 |
| 630,598 | 653,083 | 642,780 | 844,602 |
| 277,426 | 376,985 | 325,982 | 1,481,332 |
| — | — | — | — |
| 4,549,320 | 5,145,613 | 5,831,301 | 7,376,933 |
| 4,374,346 | 4,764,456 | 5,115,176 | 5,806,608 |
| — | — | — | — |
| — | — | — | — |
| — | — | — | — |
| — | — | — | — |
| — | — | — | — |
| — | — | — | — |
| 4,549,320 | 5,145,613 | 5,831,301 | 7,376,933 |
| 4,374,346 | 4,764,456 | 5,115,176 | 5,808,608 |

Table D.6 Expenditures for the Reproductive Sciences and Contraceptive Development, by Funding Agency and Location of Research, in Current and Constant U.S. Dollars (1970=100) Based on CPI Values for Nations in which Research was Conducted, 1965, 1969-1974

| National Institute of Mental Health (NIMH) | 1965 | 1969 | 1970 |
|---|---|---|---|
| **United States** | | | |
| Fundamental studies | 228,401 | 549,602 | 658,726 |
| Subtotal | 228,401 | 549,602 | 658,726 |
| Constant 1970 $ | 281,976 | 584,682 | 658,726 |
| **Other Industrialized Nations** | | | |
| Fundamental studies | 8,640 | 10,125 | — |
| Subtotal | 8,640 | 10,125 | — |
| Constant 1970 $ | 10,536 | 10,657 | — |
| Total | 237,041 | 559,727 | 658,726 |
| Constant 1970 $ | 292,512 | 595,339 | 658,726 |

Table D.7 Expenditures for the Reproductive Sciences and Contraceptive Development, by Funding Agency and Location of Research, in Current and Constant U.S. Dollars (1970=100) Based on CPI Values for Nations in which Research was Conducted, 1965, 1969-1974

| National Science Foundation (NSF) | 1965 | 1969 | 1970 |
|---|---|---|---|
| **United States** | | | |
| Fundamental studies | 531,037 | 482,377 | 213,700 |
| Subtotal | 531,037 | 482,377 | 213,700 |
| Constant 1970 $ | 655,601 | 513,167 | 213,700 |
| Total | 531,037 | 482,377 | 213,700 |
| Constant 1970 $ | 655,601 | 513,167 | 213,700 |

| 1971 | 1972 | 1973 | 1974 |
|---|---|---|---|
| 876,871 | 1,056,021 | 1,244,457 | 1,423,747 |
| 876,871 | 1,056,021 | 1,244,457 | 1,423,747 |
| 843,145 | 977,797 | 1,091,628 | 1,121,060 |
| — | — | — | — |
| — | — | — | — |
| — | — | — | — |
| 876,871 | 1,056,021 | 1,244,457 | 1,423,747 |
| 843,145 | 977,797 | 1,091,628 | 1,121,060 |

| 1971 | 1972 | 1973 | 1974 |
|---|---|---|---|
| 797,600 | 851,500 | 771,900 | 461,300 |
| 797,600 | 851,500 | 771,900 | 461,300 |
| 766,923 | 788,425 | 677,105 | 363,228 |
| 797,600 | 851,500 | 771,900 | 461,300 |
| 766,923 | 788,425 | 677,105 | 363,228 |

Table D.8 Expenditures for the Reproductive Sciences and Contraceptive Development, by Funding Agency and Location of Research, in Current and Constant U.S. Dollars (1970=100) Based on CPI Values for Nations in which Research was Conducted, 1965, 1969-1974

| United Nations Fund for Population Activities[a] (UNFPA) | 1965 | 1969 | 1970 |
|---|---|---|---|
| Other Industrialized Nations | | | |
| Other agency | — | — | — |
| Subtotal | — | — | — |
| Constant 1970 $ | — | — | — |
| Total | — | — | — |
| Constant 1970 $ | — | — | — |

[a]Funds given to WHO

Table D.9 Expenditures for the Reproductive Sciences and Contraceptive Development, by Funding Agency and Location of Research, in Current and Constant U.S. Dollars (1970=100) Based on CPI Values for Nations in which Research was Conducted, 1965, 1969-1974

| World Health Organization (WHO) | 1965 | 1969 | 1970 |
|---|---|---|---|
| **United States** | | | |
| Fundamental studies | 12,700 | 2,000 | 4,000 |
| Contraceptive development | — | — | — |
| Monitoring safety | — | — | — |
| Subtotal | 12,700 | 2,000 | 4,000 |
| Constant 1970 $ | 15,679 | 2,127 | 4,000 |
| **Other Industrialized Nations** | | | |
| Fundamental studies | 32,630 | 28,925 | 21,000 |
| Contraceptive development | — | 6,000 | — |
| Monitoring safety | — | — | 5,600 |
| Training | — | — | — |
| Subtotal | 32,630 | 34,925 | 26,600 |
| Constant 1970 $ | 39,792 | 36,763 | 26,600 |
| **Developing Nations** | | | |
| Fundamental studies | 10,000 | 5,300 | 11,250 |
| Contraceptive development | — | — | — |
| Monitoring safety | — | — | 14,000 |
| Training | — | — | — |
| Subtotal | 10,000 | 5,300 | 25,250 |
| Constant 1970 $ | 12,195 | 5,638 | 25,250 |
| Total | 55,330 | 42,225 | 55,850 |
| Constant 1970 $ | 67,666 | 44,528 | 55,850 |

| 1971 | 1972 | 1973 | 1974 |
|---|---|---|---|
| 206,935 | 539,742 | 1,407,943 | 337,208 |
| 206,935 | 539,742 | 1,407,943 | 337,208 |
| 193,397 | 477,647 | 1,126,354 | 234,172 |
| 206,935 | 539,742 | 1,407,943 | 337,208 |
| 193,397 | 477,647 | 1,126,354 | 234,172 |

| 1971 | 1972 | 1973 | 1974 |
|---|---|---|---|
| — | 63,200 | 489,700 | 705,000 |
| — | 232,200 | 695,205 | 1,057,780 |
| — | — | 23,700 | 53,100 |
| — | 295,400 | 1,208,605 | 1,815,880 |
| — | 273,518 | 1,060,188 | 1,429,826 |
| 59,390 | 318,200 | 669,950 | 1,147,950 |
| 6,000 | 291,700 | 539,610 | 926,250 |
| — | 70,390 | 102,800 | 218,850 |
| — | 454,000 | 566,000 | 670,000 |
| 65,390 | 1,134,290 | 1,878,360 | 2,963,050 |
| 61,112 | 1,003,796 | 1,502,688 | 2,057,673 |
| 55,564 | 865,406 | 581,490 | 416,025 |
| 147,954 | 626,652 | 904,430 | 1,360,450 |
| 17,000 | 93,875 | 256,390 | 305,210 |
| — | 581,500 | 700,000 | 815,000 |
| 220,518 | 2,167,433 | 2,442,310 | 2,896,685 |
| 202,310 | 1,836,807 | 1,893,262 | 1,964,774 |
| 285,908 | 3,597,123 | 5,529,275 | 7,685,615 |
| 263,422 | 3,114,121 | 4,456,138 | 5,452,273 |

Table D.10 Expenditures for the Reproductive Sciences and Contraceptive Development, by Funding Agency and Location of Research, in Current and Constant U.S. Dollars (1970=100) Based on CPI Values for Nations in which Research was Conducted, 1965, 1969-1974

| Ford Foundation 1965 | 1965 | 1969 | 1970 |
|---|---|---|---|
| **United States** | | | |
| Fundamental studies | 2,960,491 | 1,708,365 | 3,414,983 |
| Contraceptive development | — | 365,856 | 476,999 |
| Monitoring safety | — | 118,675 | 159,597 |
| Training | 792,130 | 1,701,214 | 1,664,783 |
| Other agency | — | 1,478,750 | 1,750,000 |
| Subtotal | 3,752,621 | 5,372,860 | 7,466,362 |
| Constant 1970 $ | 4,632,865 | 5,715,808 | 7,466,362 |
| **Other Industrialized Nations** | | | |
| Fundamental studies | 296,000 | 1,260,886 | 1,510,898 |
| Contraceptive development | 132,050 | 71,731 | 92,898 |
| Monitoring safety | 109,000 | 106,583 | 118,333 |
| Training | — | 83,100 | 46,800 |
| Other agency | — | — | — |
| Subtotal | 537,050 | 1,522,300 | 1,768,929 |
| Constant 1970 $ | 654,939 | 1,602,421 | 1,768,929 |
| **Developing Nations** | | | |
| Fundamental studies | 336,059 | 207,015 | 142,158 |
| Contraceptive development | 5,000 | — | 15,000 |
| Monitoring safety | 50,000 | 268,612 | 925,158 |
| Training | 191,000 | 137,193 | 250,840 |
| Subtotal | 582,059 | 612,820 | 1,333,156 |
| Constant 1970 $ | 709,828 | 651,936 | 1,333,156 |
| **Total** | 4,871,730 | 7,507,980 | 10,568,447 |
| Constant 1970 $ | 5,997,632 | 7,970,165 | 10,568,447 |

Note: Multiple year grants of the Ford Foundation were divided by their scheduled number of years to give estimated yearly expenditures.

| 1971 | 1972 | 1973 | 1974 |
|---|---|---|---|
| 6,427,718 | 3,198,729 | 2,451,246 | 2,986,766 |
| 759,930 | 673,453 | 1,377,482 | 747,909 |
| 373,111 | 170,611 | 170,611 | 105,313 |
| 1,544,031 | 1,703,938 | 1,901,596 | 1,412,085 |
| 5,568,400 | 3,132,250 | 2,645,995 | 1,698,070 |
| 14,673,190 | 8,878,981 | 8,546,930 | 6,950,143 |
| 14,108,836 | 8,221,278 | 7,497,307 | 5,472,577 |
| | | | |
| 1,531,647 | 1,400,116 | 934,636 | 1,420,666 |
| 224,815 | 205,696 | 207,151 | 238,174 |
| 109,815 | 163,228 | 107,363 | 77,438 |
| 220,275 | 224,975 | 225,772 | 234,872 |
| — | 600,000 | 400,000 | — |
| 2,086,552 | 2,594,015 | 1,874,922 | 1,971,150 |
| 1,950,048 | 2,295,588 | 1,499,937 | 1,368,854 |
| | | | |
| 364,201 | 358,854 | 313,053 | 195,283 |
| 6,698 | 30,000 | 16,500 | — |
| 350,903 | 121,888 | 149,700 | 289,005 |
| 325,331 | 208,805 | 205,250 | 125,380 |
| 1,047,133 | 719,547 | 684,503 | 609,668 |
| 960,672 | 609,785 | 468,610 | 385,865 |
| 17,806,875 | 12,192,543 | 11,106,355 | 9,530,961 |
| 17,019,556 | 11,126,651 | 9,465,854 | 7,227,296 |

Table D.11 Expenditures for the Reproductive Sciences and Contraceptive Development, by Funding Agency and Location of Research, in Current and Constant U.S. Dollars (1970=100) Based on CPI Values for Nations in which Research was Conducted, 1965, 1969-1974

| Population Council | 1965 | 1969 | 1970 |
|---|---|---|---|
| **United States** | | | |
| Fundamental studies | 540,794 | 2,177,878 | 4,538,573 |
| Contraceptive development | 118,784 | 127,289 | 483,485 |
| Monitoring safety | 356,131 | 211,910 | 114,400 |
| Training | 89,413 | 191,294 | 192,882 |
| Subtotal | 1,105,172 | 2,708,371 | 5,329,340 |
| Constant 1970 $ | 1,364,409 | 2,881,245 | 5,329,340 |
| **Other Industrialized Nations** | | | |
| Fundamental studies | 152,473 | 244,400 | 591,209 |
| Contraceptive development | 47,000 | 65,325 | 31,760 |
| Monitoring safety | 22,581 | 5,000 | — |
| Training | 2,169 | 2,708 | 4,170 |
| Subtotal | 224,223 | 317,433 | 627,139 |
| Constant 1970 $ | 273,442 | 334,140 | 627,139 |
| **Developing Nations** | | | |
| Fundamental studies | 87,321 | 166,008 | 279,070 |
| Contraceptive development | — | 220,571 | 138,245 |
| Monitoring safety | 106,028 | 37,827 | 19,810 |
| Training | — | — | 67,653 |
| Subtotal | 193,359 | 424,406 | 504,778 |
| Constant 1970 $ | 235,803 | 451,495 | 504,778 |
| **Total** | 1,522,754 | 3,450,210 | 6,461,257 |
| Constant 1970 $ | 1,873,654 | 3,666,880 | 6,461,257 |

| 1971 | 1972 | 1973 | 1974 |
|---|---|---|---|
| 6,049,337 | 3,254,096 | 3,243,713 | 3,413,705 |
| 1,148,324 | 1,517,779 | 474,104 | 1,065,457 |
| 113,550 | 112,027 | 58,109 | 61,400 |
| 188,088 | 123,780 | 180,514 | 141,257 |
| 7,499,299 | 5,007,632 | 3,956,440 | 4,681,819 |
| 7,210,864 | 4,636,742 | 3,470,561 | 3,686,471 |
| 456,212 | 504,644 | 325,166 | 237,343 |
| 151,100 | 119,184 | 135,890 | 193,209 |
| — | — | — | — |
| — | 4,000 | — | — |
| 608,012 | 627,828 | 461,056 | 430,552 |
| 568,235 | 555,600 | 368,844 | 298,994 |
| 215,845 | 215,029 | 200,673 | 262,698 |
| 355,329 | 276,065 | 179,536 | 126,561 |
| 21,600 | 15,635 | 38,056 | — |
| 14,140 | 15,518 | 59,357 | 74,992 |
| 606,914 | 522,257 | 477,672 | 464,251 |
| 556,801 | 442,590 | 370,288 | 293,829 |
| 8,714,225 | 6,157,767 | 4,895,168 | 5,576,622 |
| 8,335,900 | 5,634,932 | 4,209,693 | 4,279,194 |

Table D.12 Expenditures for the Reproductive Sciences and Contraceptive Development, by Funding Agency and Location of Research, in Current and Constant U.S. Dollars (1970=100) Based on CPI Values for Nations in which Research was Conducted, 1965, 1969-1974

| Rockefeller Foundation | 1965 | 1969 | 1970 |
|---|---|---|---|
| **United States** | | | |
| Fundamental studies | — | 526,050 | 1,297,853 |
| Contraceptive development | 25,000 | — | — |
| Training | — | — | 387,522 |
| Other agency | — | — | — |
| Subtotal | 25,000 | 526,050 | 1,685,385 |
| Constant 1970 $ | 30,864 | 559,627 | 1,685,385 |
| **Other Industrialized Nations** | | | |
| Fundamental studies | — | — | — |
| Training | — | — | — |
| Subtotal | — | — | — |
| Constant 1970 $ | — | — | — |
| **Developing Nations** | | | |
| Fundamental studies | 30,000 | 5,320 | — |
| Training | 3,630 | 1,525 | 1,900 |
| Subtotal | 33,630 | 6,845 | 1,900 |
| Constant 1970 $ | 41,012 | 7,281 | 1,900 |
| **Total** | 58,630 | 532,895 | 1,687,285 |
| Constant 1970 $ | 71,876 | 566,908 | 1,687,285 |

Table D.13 Expenditures for the Reproductive Sciences and Contraceptive Development, by Funding Agency and Location of Research, in Current and Constant U.S. Dollars (1970=100) Based on CPI Values for Nations in which Research was Conducted, 1965, 1969-1974

| Scaife Family Charitable Trusts | 1965 | 1969 | 1970 |
|---|---|---|---|
| **United States** | | | |
| Fundamental studies | — | — | — |
| Other agency | 1,450,000 | 2,000,000 | 2,000,000 |
| Subtotal | 1,450,000 | 2,000,000 | 2,000,000 |
| Constant 1970 $ | 1,790,123 | 2,127,659 | 2,000,000 |
| **Total** | 1,450,000 | 2,000,000 | 2,000,000 |
| Constant 1970 $ | 1,790,123 | 2,127,659 | 2,000,000 |

| 1971 | 1972 | 1973 | 1974 |
|---|---|---|---|
| 2,375,368 | 2,594,534 | 2,086,769 | 2,666,737 |
| 10,500 | — | — | — |
| 281,233 | 366,728 | 297,757 | 419,410 |
| 500,000 | 500,000 | — | 1,000,000 |
| 3,167,101 | 3,461,262 | 2,384,526 | 4,086,147 |
| 3,045,289 | 3,203,946 | 2,091,689 | 3,217,438 |
| — | 41,725 | 78,765 | 28,569 |
| 2,000 | — | — | — |
| 2,000 | 41,725 | 78,765 | 28,569 |
| 1,869 | 36,924 | 63,012 | 19,839 |
| 2,660 | 59,167 | 22,312 | 35,496 |
| 2,250 | — | — | — |
| 4,910 | 59,167 | 22,312 | 35,496 |
| 4,504 | 50,141 | 17,296 | 22,465 |
| 3,174,011 | 3,562,154 | 2,485,603 | 4,150,212 |
| 3,051,662 | 3,291,011 | 2,171,997 | 3,259,742 |

| 1971 | 1972 | 1973 | 1974 |
|---|---|---|---|
| 250,000 | 250,000 | 250,000 | 250,000 |
| 2,000,000 | 2,000,000 | 2,000,000 | 2,000,000 |
| 2,250,000 | 2,250,000 | 2,250,000 | 2,250,000 |
| 2,163,461 | 2,083,333 | 1,973,684 | 1,771,653 |
| 2,250,000 | 2,250,000 | 2,250,000 | 2,250,000 |
| 2,163,461 | 2,083,333 | 1,973,684 | 1,771,653 |

Appendix E	Estimates of Funding Needs
for Adequate Exploitation
of Existing Knowledge

Projected funding levels for each of the four branches of the field—fundamental studies, contraceptive development, safety studies, and training—were determined inductively from the review's assessments of the current state of knowledge and significant remaining questions that could be answered with additional investigation. Although the approach to estimating future needs varied somewhat in dealing with each branch, all shared a common origin in the information gathered in the essays that served as the starting point for estimating the studies, manpower, and facilities required. For each of these factors, current average dollar costs were used and summed. The result was an estimate of the funds required to support a program in 1976 that would adequately exploit the field's existing knowledge base (the program would pursue vigorously the outstanding fundamental research questions that appear ready for investigation, the developmental studies necessary to test both advanced and less advanced approaches to new contraceptives, the range of safety issues related to current contraceptive methods, and the efforts required as a result to strengthen the field's professional capacity). These 1976 estimates were then inflated in chapter 12 to account for anticipated inflation between 1976 and 1980. The estimating procedures and assumptions for each of the branches are outlined in the following sections.

**Fundamental Studies**

The estimation of future needs in fundamental research began with the forty-one essays (see volume II) prepared by experts in several nations. In addition to reviewing advances in each subspecialty, each author was asked to delineate major gaps in knowledge. In aggregate, nearly three hundred such topics were listed, but there was some overlap between the different authors. We estimated that the essays included 230 unique topics, which are listed by author in summary E.1 (located at the end of this appendix).

The list is varied, ranging from detailed subjects that conceivably could be investigated with a concentrated effort in a few years to much broader questions that will need more extensive exploration by a greater number of researchers. Examination of the list by members of the steering committee led to the working assumption that although time and effort would obviously vary for each topic, an *average* of five "research units" working on each topic over a period of five years would be needed. The "research unit" concept employed in this estimate is a team headed by a senior investigator (Ph.D. or M.D.) with a research assistant, other supporting staff, equipment, supplies, facilities, and services. We reiterate that this estimate embodies an average: Some of the topics could be resolved with less time and effort; others might require more. In lieu of attempting the virtually impossible task of preparing detailed work schedules for investigation of each topic and guessing how quickly each study would yield definitive results, it seemed preferable to employ an averaging concept that would capture the likely range of variation in the research required for these topics.

We then attempted to estimate the current annual costs of supporting a single "research unit" of the type specified in both industrialized and developing nations. Salaries and other costs are of course higher in industrialized nations, but more professional staff are involved in developing-nation laboratories by custom and of necessity because fewer research service facilities are available. When the current average costs were assembled, there was surprisingly little or no cost difference in the two settings; a research unit of the type specified appears to require funding of about $120,000 a year whether it is located in an industrialized or developing nation.

The annual cost per research unit in an industrialized nation was estimated as follows:

| Items | Annual Cost |
|---|---|
| Ph.D. or M.D. group leader | $ 25,000 |
| Research assistant | 15,000 |
| Chemical technician, full time | 13,000 |
| Assistant technician, ½ time | 7,000 |
| Fringe benefits @ 20% | 12,000 |
| Space and services | 30,000 |
| Equipment purchase and maintenance | 6,000 |
| Expendable supplies and animals | 10,000 |
| Publications, travel, misc. | 2,000 |
| | $120,000 |

554	Estimates of Funding Needs
for Adequate Exploitation
of Existing Knowledge

Table E.1 Estimated Annual Costs of Fundamental Studies in the Reproductive Sciences

| | |
|---|---:|
| Inventory of identified research topics 230 @ annual cost of $120,000 per research unit; 5 units per topic (230 X 120,000 X 5) | $138,000,000 |
| Continuing support of ongoing basic research exclusive of listed topics | 24,000,000 |
| Support of Symposia, Conferences, Publications | 3,000,000 |
| Total | $165,000,000 |

These unit costs were then multiplied by the number of topics to be pursued (230) and the average number of research units for each topic (5) to arrive at the sum of $138 million. Because the 230 topics do not encompass all current investigations, we added an estimate of $24 million to cover continuation of ongoing research not covered in the inventory. Symposia, conferences, and publications were estimated at $3 million annually. Table E.1 summarizes these estimates for fundamental research.

Contraceptive Development

Estimation of the costs of contraceptive development studies required first that current leads be classified into those in more advanced stages of development and those less advanced. For the more advanced leads, costs were estimated with considerable specificity for each of the steps required, beginning with studies of pharmacology and mechanism of action and ending up with field studies and production start-up costs. For the less advanced leads, less detailed estimates could be prepared. The estimates are incorporated in the tables accompanying this section.

Table E.2a shows the cost breakdown for development of current advanced leads over a five-year period. Each step is specified in the column headings. Because the advanced leads vary both in the amount of work that has already been done and in the nature of the activities still required, the estimates for each lead also vary, based on the judgment of close observers of the worldwide contraceptive research effort. In some cases (such as biodegradable implant) the testing of a second version of the same basic method is costed at lower figures than the first, on the assumption that some of the results of the studies on the first version would not need to be replicated. Other assumptions in the estimating procedure include the following:

1. Expenditures for pharmacology and mechanism of action include studies in animals to determine effects on specific physiological and biochemical parameters and to determine the biokinetics of drug metabolism. The cost estimates are based on an annual cost of $120,000 per research unit, as with fundamental studies.

2. The cost of dosage form development includes designing of the dosage form with the desired drug release rate, determining stability of the dosage form, manufacturing materials for initial study, and developing and applying control procedures. The research unit cost basis is the same as for the previous item.

3. Toxicity testing in animals includes standard subacute and chronic toxicity studies and special investigations of teratology, mutagenicity, immunological responses, and metabolic effects, depending upon the method. The standard subacute and chronic toxicity studies include three-month, one-year, and two-year studies in three species. Taken as a unit, the cost of these investigations is estimated at $775,000. The five-year estimates also include two years of "lifetime" studies in two species estimated at $100,000 per year. The costs of completing the lifetime studies are not included because they could not be completed within a single five-year period.

In making these estimates, consideration is taken of the fact that some modes of administration represent a simple extrapolation of methods already well studied and so should not require a full battery of toxicity tests.

4. The clinical pharmacology (phase I) budgets include estimates of routine probing for effects on general health and organ systems, studies of drug metabolism in humans, and special tests (such as effects on clotting time, diabetogenic effects, and liver function tests) where appropriate. The immunological approach includes titrations of antibody response and special studies of possible interaction with organs and hormones. The budgets for the biodegradable materials

556  Estimating of Funding Needs
for Adequate Exploitation
of Existing Knowledge

Table E.2a Breakdown of Five-Year Cost Requirements for Advanced Lead Development, 1976-1980 (U.S. dollars in 000s)

|  | Pharma-cology Mechanism of Action | Dosage Form Development | Toxicology Assessment— Animals |
|---|---|---|---|
| **Immunological Approaches** |  |  |  |
| β-hCG vaccine | 800 | 1,400 | 2,600 |
| Improved hCG vaccine | 1,200 | 2,600 | 2,000 |
| Other antigens | 2,000 | 2,600 | 2,800 |
| **Luteolytic Approaches** |  |  |  |
| Nonsteroidal estrogen | 600 | 600 | 1,800 |
| Plant product | 1,200 | 2,000 | 1,800 |
| Prostaglandin analog (luteolysis) | 600 | 800 | 1,200 |
| Natural estrogen | 400 | 400 | 800 |
| **Sustained Release Forms** |  |  |  |
| Sustained release progestin implants | 400 | 600 | 1,800 |
| Second progestin implants | 400 | 400 | 1,800 |
| Biodegradable implant | 800 | 1,600 | 2,800 |
| Second biodegradable implant | 600 | 1,000 | 1,600 |
| Injectable biodegradable carrier | 800 | 1,600 | 2,600 |
| Second injectable steroid | 600 | 1,000 | 1,600 |
| Vaginal ring | 100 | 600 | 800 |
| Second steroid in vaginal ring |  | 400 | 800 |
| **Postpartum IUD** |  | 100 |  |
| **Nonsurgical Abortion** |  |  |  |
| Protaglandin |  | 400 | 600 |
| Plant product | 1,200 | 1,000 | 1,600 |
| **Suppression of Spermatogenesis** (Synthetic steroid) | 1,400 | 400 | 2,600 |
| Second steroid spermatogenesis | 1,000 | 400 | 2,000 |
| **Total—1976-1980** |  |  |  |

Estimating of Funding Needs
for Adequate Exploitation
of Existing Knowledge

| Clinical Pharmacology Phase I | Clinical Effect and Side Effects Phase II and III | Continuing Evaluation of Phases I, II, III Subjects | Field Studies | Scale-Up for Production | Management Costs | 5-Year Total |
|---|---|---|---|---|---|---|
| 1,400 | 2,400 | 4,000 | 6,400 | 2,000 | 3,150 | 24,150 |
| 1,600 | 2,800 | 4,000 | 6,400 | 2,400 | 3,450 | 26,450 |
| 1,600 | 3,000 | 4,000 | 6,400 | 2,400 | 3,720 | 28,520 |
| 800 | 1,200 | 1,400 | 3,400 | 1,000 | 1,620 | 12,420 |
| 900 | 1,200 | 1,400 | 3,400 | 1,200 | 1,960 | 15,060 |
| 900 | 900 | 1,000 | 3,500 | 800 | 1,460 | 11,160 |
| 800 | 1,200 | 900 | 3,300 | 300 | 1,220 | 9,320 |
| 600 | 1,200 | 1,800 | 2,000 | 800 | 1,530 | 10,730 |
| 600 | 1,200 | 1,800 | 3,000 | 300 | 1,420 | 10,920 |
| 1,400 | 1,500 | 3,000 | 3,000 | 1,000 | 2,260 | 17,360 |
| 1,000 | 1,200 | 2,000 | 3,000 | 600 | 1,650 | 12,650 |
| 1,400 | 1,400 | 3,000 | 3,000 | 1,000 | 2,220 | 17,020 |
| 1,000 | 1,200 | 2,000 | 3,000 | 600 | 1,650 | 12,650 |
| 600 | 800 | 600 | 2,000 | 400 | 880 | 6,780 |
| 600 | 800 | 600 | 2,000 | 300 | 820 | 6,320 |
|  | 800 | 600 | 2,000 |  | 530 | 4,030 |
| 600 | 800 | 300 | 2,000 |  | 710 | 5,410 |
| 1,000 | 1,000 | 800 | 2,500 | 1,200 | 1,550 | 11,850 |
| 1,600 | 2,400 | 4,000 | 4,000 | 400 | 2,520 | 19,320 |
| 1,000 | 2,400 | 4,000 | 4,000 | 400 | 2,280 | 17,480 |
|  |  |  |  |  |  | 279,600 |

are enlarged to take into account the probable necessity of studying the metabolism of the carrier.

5. Clinical studies (phases II and III) are expected to involve four thousand to ten thousand patients for each method and extend from one to three years, depending on the regimen. Both here and in the clinical pharmacology studies, the estimate encompasses the possibility that more than one dosage variant may reach the point of initial trial.

6. Continuing evaluation of phase I, II, and III subjects covers the continued study of patients for long-term effects beyond the normal duration of studies of effectiveness and safety and after the product is in general use.

7. Field studies include large-scale studies of acceptability, effectiveness, and side effects of the method under field conditions. Included in these estimates are the costs of the dosage form used in the studies; expenditures to support the necessary clinical and paramedical personnel; expenditures to prepare instructional materials and train personnel; and the costs of record keeping, follow-up, and analysis. The estimate contemplates such studies of each lead in four to six countries, with costs estimated at $500,000 to $1 million per country.

8. Plant and labor costs incident to going from the laboratory to pilot scale production are included in the scale-up for production estimates.

9. Experience has shown that some modifications of dosage form and schedule of administration result in the need to repeat parts of the studies. The estimates of the costs of development of the leads therefore allow for this replication of the earlier phases of the development process (pharmacology and mechanism of action, dosage form development, toxicology assessment in animals, and clinical pharmacology) by multiplying initial estimates for these steps by 1.5.

In addition to these studies, safety studies (prospective cohort and case control studies) of the following leads would be expected to begin within the first five-year interval. The cost estimates for these studies are presented in table E.2b.[1]

[1] See notes to table E.3 for the costing formulae used in safety studies.

Table E.2b  Estimated Costs of Long-Term Safety Evaluation of Advanced Leads, 1976-1980 (U.S. Dollars)

| | |
|---|---:|
| Prostaglandin analog for luteolysis | 9,360,000 |
| Natural estrogen | 11,250,000 |
| Sustained release implants | 8,420,000 |
| Steroid in vaginal ring | 8,420,000 |
| Postpartum IUD | 2,400,000 |
| Prostaglandin—nonsurgical abortion | 9,600,000 |
| Total | 49,450,000 |

A number of leads are in less advanced stages of development and thus cannot be specified as definitively as those encompassed in table E.2a. The following research leads are expected to enter the development process during the next five years. To facilitate this and to begin to carry out initial steps of the process would require support as estimated in table E.2c.

These estimates of the 1976-1980 costs of an adequate contraceptive development program are assembled and annualized in table E.2d.

## Safety Studies of Current Fertility Control Methods

A number of considerations were taken into account in preparing the estimates of the cost of safety studies needed on methods of fertility control currently in use. A basic issue involved a question of research strategy as to the relative emphasis to be placed on prospective and retrospective studies. In prospective studies, a large group of patients is followed for long periods of time and comparisons are made between users of the method(s) under study and controls who use other methods or no method. In retrospective case-control studies, subjects with the condition being investigated are identified. A matched control not showing the condition is selected for each such subject. The history of the cases and controls are examined to compare the proportion of each group using the fertility control method under investigation. Prospective studies are costlier but yield infor-

Table E.2c Estimated Costs of Development of Less Advanced Leads, 1976-1980 (U.S. Dollars)

| | |
|---|---|
| Releasing hormone antagonists | 21,850,000 |
| Testicular gonadotropin regulator | 24,150,000 |
| Receptor regulation | 17,250,000 |
| Sperm antigens | 29,900,000 |
| Ovulation detection | 9,200,000 |
| Prolongation of lactation | 17,250,000 |
| Intravaginal or intracervical spermicides | 13,500,000 |
| Nonsteroid blocker of sperm viability | 27,600,000 |
| Improved female sterilization techniques | 15,200,000 |
| Total | 175,900,000 |

Table E.2d Summary of 1976-1980 Estimates and Annual Costs of Contraceptive Development Studies (U.S. Dollars)

| | 1976-1980 Estimates | Annual Estimates |
|---|---|---|
| Advanced lead development | 279,600,000 | 55,920,000 |
| Long-term safety evaluation of advanced leads | 49,450,000 | 9,890,000 |
| Development of less advanced leads | 175,900,000 | 35,180,000 |
| Total | 504,950,000 | 100,990,000 |

mation on incidence rates not possible to obtain directly through case-control studies. Retrospective case-control studies, which are less expensive, have been the most frequent type of safety evaluation conducted thus far on fertility control methods.

The estimates presented here reflect the judgment that the time has come to initiate several large prospective studies of major current methods. The judgment is based on the fact that concern over safety has continued to increase as the results of small-scale and selective retrospective studies have been published. Issues have been raised by these studies that will remain unresolved and subject to dispute until they are explored by carefully designed and executed prospective studies.

There certainly can be—and undoubtedly are—differences of opin-

## Estimates of Funding Needs for Adequate Exploitation of Existing Knowledge

ion among researchers and research administrators regarding an appropriate strategy. Some would choose to continue to rely principally on retrospective studies and to initiate few or no prospective studies, either on cost-effectiveness grounds (not only are they less expensive but they require less time and can be replicated more readily) or perhaps because they doubt that sufficient funds will ever be made available to support the prospective studies required. In addition, some consider it unrealistic to expect that prospective studies can be done meaningfully in poorer nations with inadequate medical record systems or even that they can be properly carried out in many industrialized nations. They would advocate that such studies be limited to a few countries in which conditions are propitious.

These considerations cannot be dismissed as inconsequential. A strategy emphasizing any particular methodology might result in different estimates, either higher or lower. We believe that the ultimate judgment on these questions can best be made in the concrete circumstances of assessing the appropriateness of alternative research designs to elucidate specific issues and in the face of the resources actually made available to support the range of studies required. We present these estimates as indicating the order of magnitude of the resources required to provide better answers than are currently available regarding the safety of existing methods. More detailed planning and consultation in implementing such a program could well lead to an altered distribution of emphasis than is conceptualized here. We are convinced, however, that these estimates represent useful approximations of the sums required for an adequate program to evaluate the safety of existing fertility control methods.

The estimates presented in table E.3 are based on these assumptions:

1. The laboratory investigations of phase IV pharmacology studies are estimated on the basis of research unit costs similar to those for fundamental studies.

2. Considering the incidence of such morbidity as thromboembolic disease, stroke, and myocardial infarction, we estimate that fifty thousand treated patients and fifty thousand controls would be needed in prospective studies for meaningful evaluation. Based on the

Estimates of Funding Needs
for Adequate Exploitation
of Existing Knowledge

Table E.3 Estimated Costs of Studies on Safety of Fertility Control
Methods Currently in Use, 1976-1980 (U.S. Dollars)

|  | Estimated Cost of Project[a] | Multiplier for Number Similar Projects Required[a] | Total |
|---|---|---|---|
| **Orals** |  |  |  |
| Phase IV pharmacology |  |  |  |
| Effect on milk volume and quality | 500,000 | 6 | 3,000,000 |
| Passage of steroid into milk | 250,000 | 2 | 500,000 |
| Immune suppression | 500,000 | 5 | 2,500,000 |
| Hematologic effects | 500,000 | 6 | 3,000,000 |
| Metabolic effects | 400,000 | 10 | 4,000,000 |
| Prospective cohort studies |  |  |  |
| Studies to evaluate incidence of thromboembolic disease, hypertension, gall bladder disease, stroke, diabetes, and similar events in users | 20,000,000 | 2 | 40,000,000 |
| Extended studies to evaluate incidence of events occurring only after long exposure—carcinogenesis, rate of return to fertility, etc. | 20,000,000 | 1 | 20,000,000 |
| Teratogenesis in offspring (Hospital-based studies) | 3,000,000 | 2 | 6,000,000 |
| Functional development; sexual maturation of offspring | 4,000,000 | 2 | 8,000,000 |
| Case-Control Studies |  |  |  |
| Thromboembolic disease, stroke carcinogenesis, etc.; comprehensive studies | 6,000,000 | 4 | 24,000,000 |
| Single disease entities | 1,000,000 | 12 | 12,000,000 |
| Offspring |  |  |  |
| Teratology, etc. | 1,000,000 | 3 | 3,000,000 |
| Functional development, maturation | 1,000,000 | 4 | 4,000,000 |
| Subtotal |  |  | 130,000,000 |
| **Injectables** |  |  |  |
| Phase IV pharmacology |  |  |  |
| Effect on milk volume and quality | 500,000 | 3 | 1,500,000 |
| Passage of steroid into milk | 250,000 | 2 | 500,000 |
| Metabolic effect | 400,000 | 4 | 1,600,000 |
| Hematologic effect | 500,000 | 3 | 1,500,000 |
| Prospective cohort studies |  |  |  |
| To evaluate incidence of thromboembolic disease, reduced adrenal function, diabetes, |  |  |  |

Estimates of Funding Needs
for Adequate Exploitation
of Existing Knowledge

Table E.3 (continued)

|  | Estimated Cost of Project[a] | Multiplier for Number Similar Projects Required[b] | Total |
|---|---|---|---|
| carcinogenesis, and other possible side effects | 20,000,000 | 1 | 20,000,000 |
| Teratogenesis in offspring (Hospital-based studies) | 3,000,000 | 1 | 3,000,000 |
| Functional development; sexual maturation of offspring | 4,000,000 | 1 | 4,000,000 |
| **Case-control studies** |  |  |  |
| Thromboembolic disease, carcinogenesis, other possible side effects—comprehensive | 6,000,000 | 2 | 12,000,000 |
| —single disease entities | 1,000,000 | 4 | 4,000,000 |
| Offspring |  |  |  |
| Teratology, etc. | 1,000,000 | 2 | 2,000,000 |
| Functional development, maturation | 1,000,000 | 2 | 2,000,000 |
| **Subtotal** |  |  | 52,100,000 |
| **Postcoital Compounds** |  |  |  |
| Phase IV pharmacology |  |  |  |
| Metabolic effect | 500,000 | 3 | 1,500,000 |
| Hematologic effect | 500,000 | 2 | 1,000,000 |
| **Prospective cohort studies** |  |  |  |
| To evaluate effects of repeated use on carcinogenesis, thromboembolic disease, etc. | 20,000,000 | 1 | 20,000,000 |
| Teratology where unsuccessful (Hospital-based studies) | 3,000,000 | 1 | 3,000,000 |
| **Case-control studies** |  |  |  |
| Carcinogenesis and other possible side effects—comprehensive | 6,000,000 | 2 | 12,000,000 |
| —single disease entities | 1,000,000 | 8 | 8,000,000 |
| Offspring |  |  |  |
| Teratology | 1,000,000 | 2 | 2,000,000 |
| **Subtotal** |  |  | 47,500,000 |
| **IUDs** |  |  |  |
| Phase IV pharmacology/pathology |  |  |  |
| Effects on endometrial histology of long-term exposure | 250,000 | 5 | 1,250,000 |
| Effects on tubal patency | 250,000 | 4 | 1,000,000 |
| Effects on endometrial enzymes and proteins | 250,000 | 2 | 500,000 |
| **Prospective cohort studies** |  |  |  |
| Pelvic inflammatory disease, |  |  |  |

564  Estimates of Funding Needs
for Adequate Exploitation
of Existing Knowledge

Table E.3 (continued)

|  | Estimated Cost of Project[a] | Multiplier for Number Similar Projects Required[b] | Total |
|---|---|---|---|
| bleeding patterns, anemia, perforations, ectopic pregnancies (septic abortions) | 10,000,000[c] | 2 | 20,000,000 |
| Extended to studies to evaluate incidence of events occurring only after prolonged exposure—carcinogenesis, rate of return to fertility, et cetera | 10,000,000[c] | 1 | 10,000,000[c] |
| Teratogenesis in offspring (Hospital-based studies) | 3,000,000 | 2 | 6,000,000 |
| Case-control studies | | | |
| Ectopics, carcinogenesis, pelvic inflammatory disease | 1,000,000 | 6 | 6,000,000 |
| Teratogenecity | 1,000,000 | 3 | 3,000,000 |
| Subtotal | | | 62,050,000 |
| **Surgical Sterilization—Women** | | | |
| Comparative morbidity of different methods | 2,000,000 | 3 | 6,000,000 |
| Case-control studies | | | |
| Selected health factors | 1,000,000 | 6 | 6,000,000 |
| Subtotal | | | 47,750,000 |
| **Surgical Sterilization—Men** | | | |
| Phase IV pharmacology | | | |
| Immunological studies | 500,000 | 8 | 4,000,000 |
| Hormone studies | 500,000 | 5 | 2,500,000 |
| Prospective cohort studies | | | |
| Immunologic disease, general health | 10,000,000 | 1 | 10,000,000 |
| Case-control studies | | | |
| Immunologic diseases | 1,000,000 | 4 | 4,000,000 |
| Reversibility | 500,000 | 4 | 2,000,000 |
| Subtotal | | | 22,500,000 |
| **Abortion** | | | |
| Immediate morbidity-comparison of different techniques including endometrial aspiration | 1,500,000 | 3 | 4,500,000 |
| Prospective cohort studies | | | |
| Ectopic rate, infertility, prematurity, psychological effects | 10,000,000 | 2 | 20,000,000 |

Estimates of Funding Needs
for Adequate Exploitation
of Existing Knowledge

Table E.3 (continued)

|  | Estimated Cost of Project[a] | Multiplier for Number Similar Projects Required[b] | Total |
|---|---|---|---|
| **Case-control studies** |  |  |  |
| Ectopic rate, infertility, prematurity, psychological effects | 1,000,000 | 6 | 6,000,000 |
| Subtotal |  |  | 30,500,000 |
| **Spermicides** |  |  |  |
| **Clinical pharmacology** |  |  |  |
| Absorption and metabolism | 200,000 | 4 | 800,000 |
| **Prospective cohort studies** |  |  |  |
| Teratology (Hospital-based studies) | 3,000,000 | 2 | 6,000,000 |
| **Case-control studies** |  |  |  |
| Venereal disease, ectopic pregnancy, vaginal inflammation, carcinogenicity | 1,000,000 | 8 | 8,000,000 |
| Teratology | 1,000,000 | 3 | 3,000,000 |
| Subtotal |  |  | 17,800,000 |
| **Total 1976-1980** |  |  | 360,150,000 |
| **Average Annual Cost** |  |  | 72,030,000 |

[a]Costs are estimated over a five-year interval. Some projects will be completed in less time and some prospective studies will not have been completed in the interval.

[b]The multiplier takes into account the need to examine safety issues in populations with different diets, genetic makeup, and general health. It also takes some account of multiple products within the same product line.

[c]Account is taken of the fact that other prospective studies (as, for example, studies on orals) will gather information on IUD users as members of the control group.

experience of the Contraceptive Evaluation Branch of NICHD's Center for Population Research, we estimate that such a study would cost $5 million per year and would continue, on average, for four years. Because the control patients would also be using contraceptive methods, studies of several methods would be actually truncated into a few large programs studying several methods concurrently. For example, because IUDs and orals could be studied concurrently, only fifty thousand additional IUD patients and controls would be needed.

3. The prospective studies of teratogenesis would evaluate the incidence of abnormal offspring among about one hundred thousand mothers entering the hospital to give birth. The particular fertility control method used, if any, would be identified at the time of entry. An estimated cost of $30 per infant has been used.

4. The estimates of the costs of prospective studies of functional development and sexual maturation are based on following through puberty children born of mothers using the method at the time of conception and children born of mothers not using the method. The cost estimates are predicated on twenty thousand subjects examined three times during the five-year period of this cost estimation at $50 per examination.

5. Case-control studies involve identification of subjects with particular symptoms or disease entity and matching each such subject with a control not showing the condition. The percentage of each group using the contraceptive method is then identified. Two types of case-control studies are budgeted. One type, budgeted at $200 per record, is comprehensive and involves simultaneously processing patient records for a series of symptoms and diseases. The second involves focusing on a single symptom or disease and is estimated at $40 per record.

6. Evaluation of morbidity of surgical sterilization of women and abortion is based on an estimated $10 per record, plus $500,000 for analysis. Phase IV pharmacology, prospective cohort, and case-control studies for these methods are based on the cost formulae noted previously.

## Training to Strengthen Professional Capacity

The estimates of the costs of training programs and other mechanisms to strengthen professional capacity in the field are based on current costs and moderate projected increases in the number of research centers and investigators at different levels. The training programs, awards, and core support grants would include personnel and facilities in the United States, other industrialized nations, and the developing nations. These estimates are presented in table E.4.

## Projected vs. Actual Funding Levels

Table E.5 assembles the annual estimates from each of the previous

Table E.4 Estimate of the Annual Cost of Training to Strengthen Professional Capacity (U.S. Dollars)

| Programs | |
|---|---:|
| Research training programs 25 @ $250,000<br>United States (12)<br>Other industrialized countries (8)<br>Developing countries (5) | 6,250,000 |
| Postdoctoral fellowships 250 @ $10,000<br>5% per annum replacement of 2,500<br>5% growth per annum | 2,500,000 |
| Career development awards 50 @ $20,000<br>2% increase in junior positions per year | 1,000,000 |
| Career scientist awards 50 @ $35,000<br>2% increase in senior positions per year | 1,750,000 |
| Institutional facilities support 40 @ $250,000<br>Population research centers<br>Core facility support programs<br>Research and training centers<br>Clinical research centers | 10,000,000 |
| Specialized courses<br>Workshops<br>Technical training in methodology<br>Reproductive biology courses | 500,000 |
| Scientist exchange programs | 500,000 |
| Total | 22,500,000 |

Table E.5 Actual Funding of Reproductive Sciences Related to Contraception vs. Projected Needs for Program to Exploit Adequately Existing Knowledge (current dollars)

|  | Available 1974 | Projected 1976 |
|---|---|---|
| Fundamental studies in the reproductive sciences | 67,100,000 | 165,000,000[a] |
| Contraceptive development | 34,400,000 | 100,990,000 |
| Studies on safety of methods currently in use | 10,900,000 | 72,030,000 |
| Training to strengthen professional capacity | 6,300,000 | 22,500,000[a] |
| Total | 118,700,000 | 360,520,000 |

[a] "Institutional Facilities Support" (table E.4), included in the 1976 estimate under training, was formerly classified under fundamental studies. This change was made for a number of reasons, the most important being that funds for core support are more appropriately categorized with other programs to strengthen the field's professional capacity. As a result, however, the 1974 figures for both fundamental studies and training, derived from actual grant and contract lists, are not strictly comparable to the proposed amounts required for 1976.

tables and compares the projected 1976 programs with the funding levels actually available in 1974. The estimated funding needs for fundamental studies, contraceptive development, and training range from two and a half to three and a half times the amounts available in 1974; and the estimates for safety studies would be almost seven times greater, reflecting the emphasis placed in this review on the urgent need to expand this branch of the field rapidly.

**Summary E.1**

Fundamental Studies—Reproductive Sciences

The following is a list of the gaps in our knowledge of the reproductive sciences related to fertility control as cited by scientists active in the field. They range from the very detailed to broad conceptual hiatuses. There is some overlap from author to author. However, an attempt has been made to retain the essence of what each author(s) felt were the major hiatuses in his or her area of specialization. This list is adapted from the scientific essays requested by the review published in volume II, *Frontiers in Reproduction and Fertility Control*.

**Gaps in knowledge of the reproductive sciences**
Papkoff, H., R. S. Ryan, and D. N. Ward: The Gonadotropic Hormones, LH(ICSH) and FSH
1. Chemical and physical structure of gonadotropins
    a. conformational structures needed for biological activity
    b. carbohydrate moieties—location and amounts of each sugar; role in biological activity
2. Chemical nature of circulating gonadotropins
    a. purify, isolate, and characterize
3. Metabolism and excretion of gonadotropins
    a. tissues capable of inactivating and metabolizing gonadotropins
    b. mechanism of inactivation
4. Biosynthesis of the gonadotropins
5. Nature and properties of gonadotropin receptors
6. Role of prostaglandins in relationship to action of gonadotropins at the target level
7. Role of protein kinases in gonadotropin action
8. Classification of cytoplasmic receptors for gonadotropins

Bahl, O. P.: Chemistry and Biology of Human Chorionic Gonadotropin and its Subunits
1. Fine details of the carbohydrate structure
2. Role of the carbohydrate structure in biological and immunological properties and in the subunit interaction and conformation of hCG.
3. Elucidation of the molecular steps involved in the stimulation of protein synthesis by target cells
4. Isolation and characterization of the hCG-LH receptor
5. Biosynthesis of hCG and its regulation
6. Use of carboxy-terminal fragments in the detection and control of fertility

7. Study of disease-state hCG (hydatidiform mole and choriocarcinoma)

Friesen, H. G. L.: Prolactin (PRL)
1. Biochemical mechanisms involved in mediating the actions of PRL in animals (including man)
2. Interrelationship between PRL and gonadotropin secretion
3. Receptors for PRL in human
4. Minimal sequence required for biological activity
5. Neuroendocrine control of PRL secretion

Lieberman, S., E. Gurpide, M. Lipsett, and H. Salhanick: Steroid Hormone Secretion
1. Agents that disturb the timing or amount of, or inhibit effectively the production of, ovarian hormones
2. Control of steroidogenesis
   a. inhibitors to binding of gonadotropins to ovarian receptors
   b. block placed at the multienzyme system level that synthesizes progesterone
   c. accelerate degradative processes
   d. specific inhibitors to cytochromes that oxidize or hydroxylate steroids
3. Characteristics of transport system of androgen into cells of seminiferous tubules
4. Biochemistry of androgen and estrogen action on cells of seminiferous tubules
5. Study of increasing selective aromatizing capacity of testis
6. Selective manipulation of FSH in male

Bahr, J. M., G. T. Ross, and A. B. Nalbandov: Hormonal Regulation of Development, Maturation, and Ovulation of the Ovarian Follicle
1. Coordination of gametogenesis and steroidogenesis
2. Basis for selection of a cohort of follicles destined to begin maturation, causes of atresia, selection in primates of the one follicle that matures and ovulates
3. Changes in microcirculation during follicular maturation, ovulation, and CL development

4. Effect of gonadotropins on vascular bed of ovary
5. Cause of the ovulatory spurt (increase in growth rate)
6. Cause of induction of increasing numbers of receptor sites
7. Binding sites for sex steroids in the ovary
8. Appropriate mix of gonadotropins needed to cause ovulation
9. Role of androgens in female
10. Hormonal stimuli involved in steroid synthesis

Baird, D. T., and R. V. Short: Endocrinology of Ovulation and Corpus Luteum Formation, Function, and Luteolysis in Women
1. Dynamics of follicular growth and atresia in human ovary
2. Mechanism of follicular rupture
3. Luteotrophic control of human CL
4. Mechanism of human luteal regression and its prevention in early pregnancy
5. Biochemistry of human endometrium and the mechanism of menstruation
6. Merits and disadvantages of cyclical as opposed to continuous ovarian suppression as a form of contraception
7. Medical control of menstruation on the morbidity due to disorders of menstruation

Gemzell, C.: Induction of Ovulation
1. Initiation of the first step of development of the primordial follicle
2. Timing of dependence of primordial follicle on gonadotropic stimulation
3. Basis for selection of follicular maturation
4. Mechanism that keeps follicles in storage from degenerating
5. Mechanism behind the great difference in sensitivity to gonadotropins between ovaries
6. Studies of anovulatory women and women with abnormal ovarian function

Guillemin, R., S. McCann, C. Sawyer, and J. Davidson: Nature of Hypothalamic Hypophysiotropic Hormones Controlling the Secretion of Gonadotropins

1. Clarification of the existence of FSH-RF
2. Cellular elements involved in the synthesis of LRF
3. Biosynthetic pathways of LRF
4. Physiological significance of serotonin and melatonin in inhibiting gonadotropin release

Wurtman, R. J.: Brain Neurotransmitters and the Hypothalamic Control of Pituitary
1. Development of agents acting selectively on brain synapses that accelerate or suppress the secretion of specific gonadotropic hormones
2. Chemical identification of neurotransmitter(s) involved in gonadotropin production and release
3. Role of nonhypothalamic releasing factor polypeptides in the brain

Yen, S. S. C., F. Naftolin, A. Lein, D. Krieger, and R. Utiger: Hypothalamic Influences on Pituitary Function in Humans
1. Study of neural mediators expanded so that specific and practical blocking or depleting agents might be found
2. Biosynthetic pathway of LRF
3. Storage mechanism of LRF
4. Study of LRF and analogs to elucidate action of LRF
5. Biosynthesis and storage of gonadotropic hormones

Blandau, R. J., R. M. Brenner, J. L. Boling, and S. H. Broderson: The Oviduct Transport Mechanisms; Mastroianni, L. Jr., B. Brackett, and C. Hamner: The Oviduct: Oviductal Physiology
1. Physiological role of the oviduct in sperm capacitation
2. Physiological role of the secretory products of the oviduct
3. Role of the oviduct in continued embryonal survival
4. Effect of recently ovulated ovum on oviduct function
5. Pharmacological influences on oviduct function
6. Mechanism of transport of agents from the blood to the oviduct fluid
7. Effect of oviduct fluid on gamete and embryo transport

Estimates of Funding Needs
for Adequate Exploitation
of Existing Knowledge


McLaren, A.: Embryogenesis
1. Development of an embryotoxic or embryostatic agent or substance to act on the embryo without affecting the mother
2. Interaction between embryo and mother (re: inhibitory substance produced by mother, embryonic stimulus)
3. Comparative studies on embryos of different mammalian species
4. More studies on human embryos during preimplantation period to ascertain details of biochemistry and general pattern of development
5. Development of an organ culture in which pieces of human uterus could be maintained (induce follicular growth and full oocyte maturation)
6. Studies of the causal mechanisms underlying developmental events
7. Delivery system possibilities if embryo to be target for contraception (more information needed on medicated IUD release rates)

Tietze, C.: Clinical and Epidemiological Aspects of Induced Abortion
1. Most suitable means of analgesia and/or anesthesia
2. Methods of dilating the cervical canal less traumatic than the customary metal instruments
3. Study of long-term health sequelae associated with multiple abortion
4. Optimal timing for endometrial aspiration as an early abortion procedure.

Segal, S. J.: Systemic Contragestational Agents
 1. Screening of natural plant products used for abortion
 2. Analysis of anthropological literature for claimed folkloric abortifacients
 3. Synthesis and study of prostaglandin analogs for abortifacient activity
 4. Study of latent effect of hCG on human corpus luteum
 5. Clearance rate of hCG following abortion, kinetic study
 6. Screening of estrogens and antiestrogens for postcoital antifertility activity
 7. Development of animal models applicable to early abortifacient activity in women

8. Screening of analogs of compounds known to have antiprogestational activity
9. Isolation and study of naturally occurring luteolytic agents
10. Analysis of proteolytic enzyme involved in shedding of the zona pellucida

Tatum, H. J.: The Most Important Developments in Intrauterine Contraception over the Past Two Decades
1. Mechanism by which copper exerts its contraceptive effect
2. Mechanism of action by which IUDs effect contraception
3. Investigation of the functional significance of the morphological changes and alterations in enzymatic systems in the uterine epithelium brought about by the presence of an IUD
4. Study of the uterine sensitivity to endogenous ovarian hormones in the presence of an IUD
5. Research on development of an acceptable and efficient IUD for insertion postpartum
6. Mechanisms involved in bleeding associated with IUDs
7. Development of IUDs that result in less bleeding

Morris, J. M.: The Morning After Pill
1. Metabolic requirements of the blastocyst for preimplantation survival
2. Dependence of the blastocyst on gonadotropin for preventing immunologic rejection
3. Initiation of secretion of gonadotropins by the fertilized ovum
4. Clarification of the role of histamine in decidualization
5. Relationship of blastokinin or uterotropin to carbonic anhydrase or other similar enzymes
6. Role of carbonic anhydrase inhibitors
7. Role of prostaglandins in interception of fertilized ovum
8. Effect of estrogen on Mullerian duct development and on the fetal vagina
9. Structural alterations in the steroid molecule or other estrogen or progestogen analogs that might yield an effective coital or postcoital pill

Mishell, D. R.: Current Status of Injectable Contraceptive Preparation: A Review
1. Clarification of cause of prolongation of inhibition of ovulation by both DMPA and norethindrone enanthate
2. Clarification of harmful effect to human female due to prolongation of low levels of circulating estrogens
3. Appropriateness of the beagle dog as a model for testing safety of contraceptive agents
4. Predictability of resumption of ovulation and fertility in women
5. Clarification of irreversible effects (sterility) of long-acting steroids in women

Ramwell, P.: The Role of Prostaglandins (PG) in Reproduction
1. Study of PG's role in hypothalamic releasing factor release and regulation of adeno-hypophyseal hormone secretion
2. Regulation of ovarian PG synthesis and actions of these agents
3. Role of PGs in spermiogenesis and sperm maturation and transport
4. Study of PGs in seminal fluid—effects on sperm transport in both male and female reproductive tracts
5. Possible teratogenic effects of PGs and PG synthetase inhibitors
6. Possible side effects of infants whose mothers were treated with PGs for labor induction

O'Malley, B. W.: Hormonal Control of Gene Expression in Reproductive Tissues
1. Chromatin structure and function in relation to hormone action
2. Biochemistry of hormone receptor complex interaction with chromatin
3. Define precise biochemical sequence of events that occur following binding of the hormone receptor complex to chromatins and the appearance of the first new mRNA molecules
4. Biochemistry of receptor—amount of cellular receptor, its activation
5. Study of sex steroid competitors that bind covalently to the receptor and are themselves inactive
6. Means of destroying steroid hormone receptor

Jensen, E. V., K. J. Catt, J. Gorski, and H. G. Williams-Ashman: Hormone Receptor—Interaction in the Mechanism of Reproductive Hormone Action

*Steroid Hormones*

1. Control of biosynthesis and intracellular levels of the receptor protein in a target cell
2. Molecular basis of steroid induced transformation and translocation of the receptor protein
3. Precise nature of the acceptor site in the chromatin
4. Detailed biochemical mechanisms by which the transformed complex modulates RNA synthesis in target cell nuclei
5. Nature of the restriction on genome function acquired during differentiation that results in the phenomenon of hormone dependency in certain cells
6. Precise intranuclear localization of steroid and receptor under physiological conditions
7. Mechanism by which the hormone leaves the nucleus
8. Isolation of native and hormone-transformed modification of the receptor proteins in amounts sufficient for determination of composition and structure
9. Antibodies to receptor proteins useful in isolating the native aggregating forms as well as indicating the tissue and species specificity of receptor structure
10. Correlation of receptor proteins with reproductive processes of ovulation, ovum transport, spermatogenesis, capacitation, fertilization, implantation, and feedback control mechanisms

*Gonadotropic Hormones*

11. Thermodynamic aspects of hormone-receptor interaction
12. Control of gonadotropin receptor induction
13. Exact nature of the hormone receptor interaction that leads to activation of gonadal target cells
14. Factors that modulate gonadotropin binding in vivo and the way in which the activation signal is transmitted
15. Nature of the signal that initiates steroidogenesis
16. Clarification of cAMP involvement and the steps following protein kinase activation in gonadotropin hormone action

17. Receptors for long-term effects (cell growth and division) and acute effects (cAMP and steroid hormone formation) need to be distinguished

Porter, D. G., and C. A. Finn: The Present Status and Future Prospects of Research in the Biology of the Uterus
1. Factors involved in inhibiting human myometrium during pregnancy
2. Factors involved in inducing the state of refractoriness to implantation that pertains through much of the cycle
3. Attachment of fertilized ovum to endometrial epithelium
4. Study of antibodies to decidual tissue
5. Structure, composition, and mode of secretion of the mucus by the cervix
6. Mechanism of parturition

Clermont, Y.: Spermatogenesis
1. Mechanisms that control differentiation of stem cells into either new stem cells or differentiating spermatogonia
2. Molecular processes that lead to the synapsis of homologous chromosomes and crossing-over in meiosis
3. Controlling molecular mechanisms involved in the structural modification of the nucleus or cytoplasmic organelles during spermiogenesis
4. Influence of hormones (both gonadotropins and androgens) on spermatogenesis
5. Relationship between Sertoli cells and germ cells

Fawcett, D.W.: The Ultrastructure and Functions of the Sertoli Cells
1. Extent of production of androgens in Sertoli cells and dependence of spermatogenesis on steroids or their metabolites produced by Sertoli cells
2. Physiological role of androgen-binding protein in the Sertoli cell
3. Process of spermiation and how it could be controlled
4. Inhibin and agent involved in local feedback between tubules and Leydig cells need to be isolated, characterized, and their site of origin determined

5. Mechanism controlling the occluding junctions between Sertoli cells and determination if barrier is essential for fluid secretion and germ cell differentiation

Neaves, W. B.: Leydig Cells
1. Physiological relationship between Leydig cells and seminiferous epithelium
2. Role of Leydig cells in gametogenesis
3. Explanation of spermatogenic inhibition following suppression of Leydig cell function
4. Effects of various contraceptive methods (such as vasectomy) on Leydig cell function
5. Ability of exogenous steroids to substitute for normal Leydig cell function
6. An understanding of relationship between fine structure of Leydig cells and synthesis of steroids

Setchell, B. P.: The Blood-Testis Barrier
1. Differences in permeability between seminiferous tubules and rete-testis
2. Fluid movement within the testis
3. Method for localizing and measuring the concentration of naturally occurring compounds in the germinal epithelium
4. Permeability of testicular capillaries
5. Changes in function of the blood-testis barrier and its control
6. Information on composition of fluid inside tubular system

Fawcett, D. W.: The Structure of the Spermatozoan
1. Effect of structural and maturational defects on fertilizing capacity of sperm
2. Biochemical mechanisms of chromatin condensation, its biological significance and its rapid reversal to form the male pronucleus
3. Factors involved in triggering acrosomal reaction
4. Relative importance of zona lysis and mechanical penetration by a motile sperm
5. Nature of cortical granules and mechanism by which they block polyspermy

6. Characterization of acrosomal hydrolases (possibility of antigenic properties)
7. Chemical basis for specificity of postacrosomal region and molecular mechanisms involved in gamete attachment and fusion
8. Detachment of sperm tail
9. More characterization of components of the tail
10. Study of chemical nature and function of fibrous components peculiar to sperm tail
11. Isolation and characterization of the outer fibers and fibrous sheath typical to sperm tails and their effect upon pattern of motility
12. More studies regarding sperm membranes

Harrison, R. A. P.: The Metabolism of Mammalian Spermatozoa
1. Specific roles for sperm enzymes in egg penetration
2. Substrate utilization by human spermatozoa
3. Biochemistry of sperm motility, especially enzymology of energy transduction
4. Biology of sperm investments—the sperm coating material, the plasma membrane in inner and outer acrosomal membranes, the postacrosomal cap, and the nuclear membrane

Mann, T.: Semen: Metabolism, Antigenicity, Storage, and Artificial Insemination
1. Influence of seminal plasma in stored semen on sperm aging
2. Reasons for return of fertility to sperm of some infertile males by suspending them in seminal plasma from fertile males
3. Substances or spermicides that are able to pass into semen after ingestion or parenteral administration
4. Effect of spermicides on semen metabolism
5. Incidence of sperm autoimmunity in vasectomized men and in certain infertile but not vasectomized patients
6. More effective methods for utilizing purified and properly identified sperm antigens as potential antifertility agents
7. Nature of changes in sperm antigens in the female tract and the influence these changes exert on the fertilizing potential of sperm
8. Influence of sperm antigens on the sex ratio of progeny

9. Reasons for variation in sperm from different men in fertilizing ability after storage
10. Thawed spermatozoa and the distinction between their fertilizing ability and motility
11. Risks to the conceptus from using stored semen in artificial insemination

Chang, M. C., C. R. Austin, J. M. Bedford, B. G. Brackett, R. H. F. Hunter, and R. Yanigamachi: Capacitation of Spermatozoa and Fertilization in Mammals
1. Control of capacitation, further characterization of the chemical nature of the sperm cell surface and the changes occurring in this surface in the epididymis and female tract
2. Definition of the functional changes in the sperm resulting in its ability to undergo the acrosome reaction and penetrate the egg
3. Penetrability of eggs in various stages of maturation
4. Measurement of fertilizing ability of sperm

Gibbons, I. R.: Sperm Motility
1. Transformation of the chemical energy released by hydrolysis of ATP into the mechanical energy of sliding tubules
2. Coordination mechanisms responsible for producing flagellar motion of sperm tail
3. Nature of major protein components of mammalian sperm flagella
4. Function and chemistry of the supplementary extraaxonemal structures in mammaliam sperm flagella
5. Mechanisms responsible for changing the form of the flagellar beat as a response to differing physiological conditions

Laurence, K. A.: Reproductive Immunology Progress in the Last Decade
 1. Antibodies to steroid receptors
 2. Antibodies to impair sperm motility or fertilizing capability
 3. Antibody enhancing factors for human use
 4. Specific antigens of the female reproductive tract
 5. Specific antigens of the male reproductive tract

6. Tissue localized antibody production in the reproductive tract
7. Problems of overcoming immunological tolerance
8. Antibodies to enzymes in steroidogenesis
9. Passive transfer of antibodies to reproductive system antigens
10. Immunization as a basis for predetermination of sex of offspring
11. Combined immunization for reproductive and health benefits
12. Micro-organisms and infertility

Schwartz, N. B., D. W. Dierschke, C. E. McCormack, and P. B. Waltz: Feedback Regulation of Reproductive Cycles in Rats, Sheep, Monkeys, and Humans, with Particular Attention to Computor Modeling
1. Importance in primates of the "internal" type of negative feedback (between pituitary and hypothalamus) of gonadotropins in regulating gonadotropin secretion
2. Mechanism of the pulsatile release of the LH
3. Role of progesterone triggering the preovulatory surge of gonadotropins
4. Role of plasma binding proteins in the feedback relationships
5. Relationship of neural tissue to the preovulatory surge and tonic levels of GnTH
6. Mechanism by which progesterone exerts its facilitative effect on LH secretion
7. Responsiveness of the anterior pituitary (output of both LH and FSH) to injected LRF during pregnancy
8. Cause for the lactation induced delay of the return of cyclic ovulation after parturition
9. Regulation of the differential release of LH and FSH
10. Mechanism by which the steroids exert a differential negative feedback on the secretion of gonadotropins
11. Mechanism by which estrogen exerts its local positive feedback in the ovary facilitating follicular growth

Steinberger, E., R. S. Swerdloff, and R. Horton: The Control of Testicular Function
1. Relationship between pulsatile patterns of gonadotropins and gonadal steroid secretions

2. Metabolic clearance rates and production rates of FSH and LH
3. Relative roles of testosterone and estradiol on inhibition of LH
4. Physiologic dosage of steroids that regulate the gonadotropins
5. Identification, isolation, and purification of "inhibin"
6. Physiologic mechanisms resulting in the increase in blood LH and FSH observed in children
7. Influence of pheromones in humans
8. Influence of light, smell, and nutrition on reproductive functions of both sexes
9. Details of enzymology in steroidogenesis have not been worked out entirely
10. Intracellular transport of steroidogenic intermediates and secretion of the steroid
11. Physiologic role of 17 $\alpha$-hydroxyprogesterone and estradiol in males
12. Control of the Leydig cell by factors other than LH (such as FSH, substances in plasma, PGs)
13. Direct role of LH (ICSH) on growth and function of the seminiferous tubules
14. Mode of action of the 5 $\alpha$-reduced metabolites and C21 steroids on maintenance of spermatogenesis
15. Physiologic role of androgen-binding protein (ABP) and the cytoplasmic androgen receptor molecule in mediation of hormonal effects in the testis

Hamilton, D. W., E. R. Cusillas, A. F. Holstein, H. Jackson, A. D. Johnson, M. R. N. Prasad, P. Ofner, and I. G. White.: Epididymis
  1. Understanding of the role of androgen-binding protein in the epididymis
  2. Understanding of the role of steroids in sperm maturation. Do sterioids that are formed or transported by the epididymis interact with sperm and do they control metabolic activity of sperm
  3. Specificity of the role of the epididymis in the maturation process of spermatozoa
  4. Role of the initial segment in sperm maturation
  5. Function of the lymphocytelike cells, especially with respect to immunological consequences of vasectomy

6. Detailed knowledge of the mechanisms of transport and accumulation of small molecular weight substances by the epididymis
7. An understanding of the dynamic aspects (secretion and absorption) of control of the luminal environment by the epididymal epithelium
8. Interactions between epididymal sperm from various segments and their environment under physiological conditions
9. Role of gonadotropic hormones and other hormones (such as thyroxine) in control of epididymal function
10. Biochemical and physiological processes involved in sperm transport in the human male ducts
11. Role of sperm antigens in fertility
12. Role of the epithelium in affecting sperm motility patterns
13. Role of intraluminal and systemic androgens in control of epididymis function
14. Role of epididymis in sperm removal

Cavazos, L. F., and P. Ofner: The Mammalian Accessory Sex Gland—A Morphological and Functional Analysis
1. Additional information of the direct role of the accessory sex glands in the fertilization process
2. Role of zinc in physiology of accessory sex glands
3. What is the high polyamine content in prostatic secretion
4. Possibility of blocking enzymes responsible for metabolic transformation of testosterone
5. Need to understand fully the role of each metabolite of testosterone (and other circulating androgens) in accessory sex gland function
6. Androgen metabolism and inhibition of 5 $\alpha$-androstanedoil formation: Is there a particular metabolite that is responsible for sex gland secretion
7. Search for agents that are competitive for cytoplasmic receptor of testosterone and/or 5 $\alpha$-dihydrotestosterone
8. Basis for the differential effects of estrogens on the accessory sex glands
9. Comparative effects of estrogen and steroidal and nonsteroidal antiandrogens on accessory sex glands

10. Limited information on role of apical plasma membrane and glycocalyx of accessory sex glands . . . function of glycocalyx in fertility and sterility—possibility to induce infertility by obliterating the mucopolysaccharide layer with a chemical agent

11. Role of prolactin in synergizing androgen action on accessory reproductive organs

12. Endrogenous factors other than androgen that influence/stimulate accessory sex gland secretions

# List of Tables

1.1 Table of Accounts of Human Reproduction, ca. 1975 (p. 40)
3.1 Criteria of Personal and Social Acceptability of Contraceptive Methods (p. 69)
3.2 Characteristics of Contraceptives Related to Life-Cycle Stage (p. 71)
3.3 Acceptability and Continuity of Use of Current Contraceptive Methods (p. 74)
7.1 Published Research Papers on Reproduction and Contraceptive Development, by Geographic Area of Authors, 1970-1975 (p. 301)
9.1 Principal FDA Requirements for Testing and Approval of New Drugs and Contraceptives (p. 353)
10.1 Private Agency Reproductive Research Expenditures in the U.S., 1922-1945 (p. 371)
10.2 Private Agency Reproductive Research Expenditures in the U.S., 1946-1960 (p. 378)
10.3 Reproductive Research Expenditures in the U.S., 1961-1965 (p. 383)
11.1 Total Expenditures for Research in the Reproductive Sciences and Contraceptive Development by Country of Origin, 1965, 1969-1974 (p. 398)
11.2 Total Expenditures and Percentage Distribution for Research in the Reproductive Sciences and Contraceptive Development by Country of Origin and Sector, 1965, 1969-1974 (p. 402)
11.3 Expenditures in the Reproductive Sciences and Contraceptive Development of Major Funding Agencies in Industrialized Nations, by Sector, 1965, 1969-1974 (p. 406)
11.4 Percent Distribution of Expenditures in all Reporting Countries for the Reproductive Sciences and Contraceptive Development, by Purpose, 1965, 1969-1974 (p. 408)
11.5 Percent Distribution of Expenditures in the Reproductive Sciences and Contraceptive Development by Research Area, by Purpose and Source of Funds, 1965, 1969-1974 (p. 410)
11.6 Percent Distribution of Expenditures for Research in the Reproductive Sciences and Contraceptive Development, by Geographic Area in which Research Is Conducted, 1965, 1969-1974 (p. 414)
11.7 Relative Growth of Expenditures for Research in the Reproductive Sciences and Contraceptive Development in Institutions Located in the United States, Other Industrialized Countries, and Developing Countries, 1965, 1969-1974 (p. 414)
11.8 Expenditures in the Reproductive Sciences and Contraceptive Development Per Woman of Reproductive Age, 1965, 1969-1973 (p. 416)
11.9 Government Expenditures for Medical Research and Research in the Reproductive Sciences and Contraceptive Development, U.S. and Selected Countries, 1965, 1969-1974 (p. 420)

11.10 National Planning Association Measures of U.S. Government Outlays for Specific Categories of Medical Research, 1974 (p. 422)
11.11 Estimated 1975 Reproductive Research Expenditures Compared to 1973 in U.S. and Fifteen Other Industrialized Nations, by Sector (p. 424)
12.1 Percent of Total NIH Research Expenditures Devoted to Cancer, Heart and Lung, and Reproductive Sciences and Contraceptive Development, FY 1972-1975 (p. 428)
12.2 Percent of Council Approved Grants Awarded by NCI, NHLI, and CPR, FY 1969, 1971, 1973, and 1974 (p. 430)
12.3 Projected 1980 Funding Levels for Reproductive Research and Contraceptive Development Required to Match 1974 Expenditures (p. 432)
12.4 1974 Expenditures for Reproductive Research Compared to Projected 1976 Program Needs (p. 435)
12.5 Three Levels of Projected Funding for the Reproductive Sciences and Contraceptive Development, 1974-1980 (p. 437)
C.1 Africa—Internal Expenditures: Reproductive Research and Contraceptive Development (by Purpose and Proportion of Governmental Medical Research Funds, 1965, 1969-1974, 1975 [est.], in Current and Constant U.S. Dollars [1970=100]) (p. 484)
C.2 Australia—Internal Expenditures: Reproductive Research and Contraceptive Development (by Purpose and Proportion of Governmental Medical Research Funds, 1965, 1969-1974, 1975 [est.], in Current and Constant U.S. Dollars [1970=100]) (p. 486)
C.3 Belgium—Internal Expenditures: Reproductive Research and Contraceptive Development (by Purpose and Proportion of Governmental Medical Research Funds, 1965, 1969-1974, 1975 [est.], in Current and Constant U.S. Dollars [1970=100]) (p. 488)
C.4 Britain—Internal Expenditures: Reproductive Research and Contraceptive Development (by Purpose and Proportion of Governmental Medical Research Funds, 1965, 1969-1974, 1975 [est.], in Current and Constant U.S. Dollars [1970=100]) (p. 490)
C.5 Canada—Internal Expenditures: Reproductive Research and Contraceptive Development (by Purpose and Proportion of Governmental Medical Research Funds, 1965, 1969-1974, 1975 [est.], in Current and Constant U.S. Dollars [1970=100]) (p. 492)
C.6 Denmark—Internal Expenditures: Reproductive Research and Contraceptive Development (by Purpose and Proportion of Governmental Medical Research Funds, 1965, 1969-1974, 1975

## List of Tables

C.7 Egypt—Internal Expenditures: Reproductive Research and Contraceptive Development (by Purpose and Proportion of Governmental Medical Research Funds, 1965, 1969-1974, 1975 [est.], in Current and Constant U.S. Dollars [1970=100]) (p. 494)

C.7 Egypt—Internal Expenditures: Reproductive Research and Contraceptive Development (by Purpose and Proportion of Governmental Medical Research Funds, 1965, 1969-1974, 1975 [est.], in Current and Constant U.S. Dollars [1970=100]) (p. 496)

C.8 Finland—Internal Expenditures: Reproductive Research and Contraceptive Development (by Purpose and Proportion of Governmental Medical Research Funds, 1965, 1969-1974, 1975 [est.], in Current and Constant U.S. Dollars [1970=100]) (p. 498)

C.9 France—Internal Expenditures: Reproductive Research and Contraceptive Development (by Purpose and Proportion of Governmental Medical Research Funds, 1965, 1969-1974, 1975 [est.], in Current and Constant U.S. Dollars [1970=100]) (p. 500)

C.10 Germany—Internal Expenditures: Reproductive Research and Contraceptive Development (by Purpose and Proportion of Governmental Medical Research Funds, 1965, 1969-1974, 1975 [est.], in Current and Constant U.S. Dollars [1970=100]) (p. 502)

C.11 Hong Kong—Internal Expenditures: Reproductive Research and Contraceptive Development (by Purpose and Proportion of Governmental Medical Research Funds, 1965, 1969-1974, 1975 [est.], in Current and Constant U.S. Dollars [1970=100]) (p. 504)

C.12 India—Internal Expenditures: Reproductive Research and Contraceptive Development (by Purpose and Proportion of Governmental Medical Research Funds, 1965, 1969-1974, 1975 [est.], in Current and Constant U.S. Dollars [1970=100]) (p. 506)

C.13 Iran—Internal Expenditures: Reproductive Research and Contraceptive Development (by Purpose and Proportion of Governmental Medical Research Funds, 1965, 1969-1974, 1975 [est.], in Current and Constant U.S. Dollars [1970=100]) (p. 508)

C.14 Israel—Internal Expenditures: Reproductive Research and Contraceptive Development (by Purpose and Proportion of Governmental Medical Research Funds, 1965, 1969-1974, 1975 [est.], in Current and Constant U.S. Dollars [1970=100]) (p. 510)

C.15 Italy—Internal Expenditures: Reproductive Research and Contraceptive Development (by Purpose and Proportion of Governmental Medical Research Funds, 1965, 1969-1974, 1975 [est.], in Current and Constant U.S. Dollars [1970=100]) (p. 512)

C.16 Japan—Internal Expenditures: Reproductive Research and Contraceptive Development (by Purpose and Proportion of Governmental Medical Research Funds, 1965, 1969-1974, 1975 [est.], in Current and Constant U.S. Dollars [1970=100]) (p. 514)

C.17 Netherlands—Internal Expenditures: Reproductive Research and Contraceptive Development (by Purpose and Proportion of Governmental Medical Research Funds, 1965, 1969-1974, 1975 [est.], in Current and Constant U.S. Dollars [1970=100]) (p. 516)

C.18 New Zealand—Internal Expenditures: Reproductive Research and Contraceptive Development (by Purpose and Proportion of Governmental Medical Research Funds, 1965, 1969-1974, 1975 [est.], in Current and Constant U.S. Dollars [1970=100]) (p. 518)

C.19 Norway—Internal Expenditures: Reproductive Research and Contraceptive Development (by Purpose and Proportion of Governmental Medical Research Funds, 1965, 1969-1974, 1975 [est.], in Current and Constant U.S. Dollars [1970=100]) (p. 520)

C.20 Philippines—Internal Expenditures: Reproductive Research and Contraceptive Development (by Purpose and Proportion of Governmental Medical Research Funds, 1965, 1969-1974, 1975 [est.], in Current and Constant U.S. Dollars [1970=100]) (p. 522)

C.21 South Korea—Internal Expenditures: Reproductive Research and Contraceptive Development (by Purpose and Proportion of Governmental Medical Research Funds, 1965, 1969-1974, 1975 [est.], in Current and Constant U.S. Dollars [1970=100]) (p. 524)

C.22 Sweden—Internal Expenditures: Reproductive Research and Contraceptive Development (by Purpose and Proportion of Governmental Medical Research Funds, 1965, 1969-1974, 1975 [est.], in Current and Constant U.S. Dollars [1970=100]) (p. 526)

C.23 Thailand—Internal Expenditures: Reproductive Research and Contraceptive Development (by Purpose and Proportion of Governmental Medical Research Funds, 1965, 1969-1974, 1975 [est.], in Current and Constant U.S. Dollars [1970=100]) (p. 528)

C.24 Turkey—Internal Expenditures: Reproductive Research and Contraceptive Development (by Purpose and Proportion of Governmental Medical Research Funds, 1965, 1969-1974, 1975 [est.], in Current and Constant U.S. Dollars [1970=100]) (p. 530)

C.25 United States—Internal Expenditures: Reproductive Research and Contraceptive Development (by Agency and Proportion of Governmental Medical Research

# List of Tables

Funds, 1965, 1969-1974, 1975 [est.], in Current and Constant U.S. Dollars [1970=100]) (p. 532)

D.1 Agency for International Development: Expenditures for the Reproductive Sciences and Contraceptive Development, by Funding Agency and Location of Research, in Current and Constant U.S. Dollars (1970=100] Based on CPI Values for Nations in which Research was Conducted, 1965, 1969-1974 (p. 536)

D.2 Atomic Energy Commission: Expenditures for the Reproductive Sciences and Contraceptive Development, by Funding Agency and Location of Research, in Current and Constant U.S. Dollars (1970=100) Based on CPI Values for Nations in which Research was Conducted, 1965, 1969-1974 (p. 536)

D.3 Center for Population Research—National Institute of Child Health and Human Development: Expenditures for the Reproductive Sciences and Contraceptive Development, by Funding Agency and Location of Research, in Current and Constant U.S. Dollars (1970=100) Based on CPI Values for Nations in which Research was Conducted, 1965, 1969-1974 (p. 538)

D.4 Food and Drug Administration: Expenditures for the Reproductive Sciences and Contraceptive Development, by Funding Agency and Location of Research, in Current and Constant U.S. Dollars (1970=100) Based on CPI Values for Nations in which Research was Conducted, 1965, 1969-1974 (p. 540)

D.5 National Institutes of Health: Expenditures for the Reproductive Sciences and Contraceptive Development, by Funding Agency and Location of Research, in Current and Constant U.S. Dollars (1970=100) Based on CPI Values for Nations in which Research was Conducted, 1965, 1969-1974 (p. 540)

D.6 National Institute of Mental Health: Expenditures for the Reproductive Sciences and Contraceptive Development, by Funding Agency and Location of Research, in Current and Constant U.S. Dollars (1970=100) Based on CPI Values for Nations in which Research was Conducted, 1965, 1969-1974 (p. 542)

D.7 National Science Foundation: Expenditures for the Reproductive Sciences and Contraceptive Development, by Funding Agency and Location of Research, in Current and Constant U.S. Dollars (1970=100) Based on CPI Values for Nations in which Research was Conducted, 1965, 1969-1974 (p. 542)

D.8 United Nations Fund for Population Activities: Expenditures for the Reproductive Sciences and Contraceptive Development, by Funding Agency and Location of Research, in Current and Constant U.S. Dollars

(1970=100) Based on CPI Values for Nations in which Research was Conducted, 1965, 1969-1974 (p. 544)

D.9 World Health Organization: Expenditures for the Reproductive Sciences and Contraceptive Development, by Funding Agency and Location of Research, in Current and Constant U.S. Dollars (1970=100) Based on CPI Values for Nations in which Research was Conducted, 1965, 1969-1974 (p. 544)

D.10 Ford Foundation: Expenditures for the Reproductive Sciences and Contraceptive Development, by Funding Agency and Location of Research, in Current and Constant U.S. Dollars (1970=100) Based on CPI Values for Nations in which Research was Conducted, 1965, 1969-1974 (p. 546)

D.11 Population Council: Expenditures for the Reproductive Sciences and Contraceptive Development, by Funding Agency and Location of Research, in Current and Constant U.S. Dollars (1970=100) Based on CPI Values for Nations in which Research was Conducted, 1965, 1969-1974 (p. 548)

D.12 Rockefeller Foundation: Expenditures for the Reproductive Sciences and Contraceptive Development, by Funding Agency and Location of Research, in Current and Constant U.S. Dollars (1970=100) Based on CPI Values for Nations in which Research was Conducted, 1965, 1969-1974 (p. 550)

D.13 Scaife Family Charitable Trust: Expenditures for the Reproductive Sciences and Contraceptive Development, by Funding Agency and Location of Research, in Current and Constant U.S. Dollars (1970=100) Based on CPI Values for Nations in which Research was Conducted, in 1965, 1969-1974 (p. 550)

E.1 Estimated Annual Costs of Fundamental Studies in the Reproductive Sciences (p. 554)

E.2a Breakdown of Five-Year Cost Requirements for Advanced Lead Development, 1976-1980 (U.S. dollars in 000s) (p. 556)

E.2b Estimated Costs of Long-Term Safety Evaluation of Advanced Leads, 1976-1980 (p. 559)

E.2c Estimated Costs of Development of Less Advanced Leads, 1976-1980 (p. 560)

E.2d Summary of 1976-1980 Estimates and Annual Costs of Contraceptive Development Studies (p. 560)

E.3 Estimated Costs of Studies on Safety of Fertility Control Methods Currently in Use, 1976-1980 (p. 562)

E.4 Estimate of the Annual Cost of Training to Strengthen Professional Capacity (p. 567)

E.5 Actual Funding of Reproductive Sciences Related to Contraception vs. Projected Needs for Program to Exploit Adequately Existing Knowledge (current dollars) (p. 568)

# List of Figures

S.1 Total worldwide expenditures for the reproductive sciences and contraceptive development (p. 18)
S.2 Percent of worldwide expenditures for reproductive sciences and contraceptive development by geographic distribution: source of funds and location of research activity (p. 19)
S.3 Total and percent of worldwide expenditures for reproductive sciences and contraceptive development by sector: government, philanthropy, and pharmaceutical firms (p. 20)
S.4 Total and percent of worldwide expenditures for reproductive sciences and contraceptive development by purpose: fundamental reproductive research, contraceptive development, evaluation of safety of contraceptives, and training to strengthen professional capacity (p. 22)
S.5 Three levels of projected funding for the reproductive sciences and contraceptive development (p. 24)
1.1a Index of relative incidence of selected adverse outcomes of pregnancy, by maternal age (p. 44)
1.1b Index of relative incidence of selected adverse outcomes of pregnancy, by maternal age (p. 45)
4.1 A schematic representation of follicular growth, maturation, and ovulation; corpus luteum formation; follicular atresia; and luteolysis (p. 82)
4.2 A schematic representation of the neuroendocrine mechanism controlling the gonad stimulating functions of the anterior pituitary (p. 92)
4.3 The sequence of ovarian events and the pituitary and ovarian hormonal actions involved in the control of the uterine endometrium during a fertile cycle (p. 98)
4.4 The amino acid sequence of luteinizing hormone releasing hormone (LRF) (p. 104)
4.5 Showing elevation of plasma FSH and LH induced by a single injection of LRF (also termed GnRH) (p. 105)
4.6 Profiles of the plasma concentration of LH, FSH, prolactin, progesterone, and estradiol-17$\beta$ during the estrous cycle of the rat (p. 110)
4.7 Profiles of the plasma concentration of FSH and LH during the normal menstrual cycle of the human female (p. 111)
4.8 Showing the pulsatile secretion of LH by four ovariectomized rhesus monkeys (p. 112)
4.9 Showing the pulsatile secretion of FSH and LH in the human female on day 7 during a normal menstrual cycle and after menopause when the plasma level of gonadotropin is elevated (p. 113)
4.10 Profiles of the mean plasma concentration of FSH, LH, estradiol, estrone, and progesterone during the menstrual cycle in the rhesus monkey (p. 114)

4.11 Illustrating the increased responsiveness to LRF at midcycle in the human female (p. 117)
4.12 A schematic representation of the interrelationships between the brain, pituitary, and ovary (p. 119)
4.13 Illustrating both the positive and negative feedback action of estrogen on LH secretion in the female rhesus monkey (p. 121)
4.14 A model of the profiles of FSH, LH, estradiol and progesterone during the human menstrual cycle (p. 126)
4.15 Showing schematically ovarian follicular development, ovulation, corpus luteum formation, and processes involved as the ovum passes through the oviduct to become embedded in the wall of the uterus (p. 133)
5.1 Cutaway diagram of the architecture of the human testis and its excurrent duct system (p. 166)
5.2 Seminiferous tubules and interstitial tissue as seen in histological sections (p. 168)
5.3 The steps of spermatogenesis in the rat (p. 170)
5.4 Composite drawing of the excurrent ducts and accessory glands of the male reproductive tract (p. 174)
5.5 Diagram presenting the typical location and the components of the junctional complex between Sertoli cells (p. 181)
5.6 Drawing illustrating the manner in which the occluding junctions between Sertoli cells divide the seminiferous epithelium into a basal compartment occupied by the spermatogonia and preleptotene spermatocytes and an adluminal compartment containing more advanced stages of the germ cell population (p. 182)
5.7 Schematic representation of the syncytial nature of the mammalian germ cells (p. 187)
5.8 Diagram of the stages of sperm release (p. 190)
5.9 Diagram of the six recognizable cell associations or stages of the cycle of the human seminiferous epithelium (p. 193)
5.10 Schematic representation of the arrangement of the blood and lymph vessels and cellular and extracellular components of an intertubular space in the human testis (p. 206)
5.11 Scheme of the successive steps in the synthesis of testosterone from acetate, with indication of the cytoplasmic organelle of the Leydig cell in which each step is believed to take place (p. 210)
5.12 Schematic representation of the endocrine regulation of spermatogenesis (p. 216)
5.13 Drawing of a generalized mammalian spermatozoon with the cell membrane removed to show the arrangement of the underlying structural components (p. 233)
5.14 The barriers around the recently ovulated egg that the capacitated spermatozoa must traverse to reach the perivitelline space and achieve activation and fertilization of the ovum (p. 234)

5.15 Successive stages of the acrosome reaction (p. 236)
5.16 Schematic representation of the current interpretation of the organization of the axoneme of cilia and sperm flagella (p. 238)
5.17 The stages in fusion of the spermatozoon with the egg as seen in electron micrographs (p. 245)

**Contents of Companion Volume**
*Frontiers in Reproduction and Fertility Control: A Review of the Reproductive Sciences and Fertility Control*

1. The Gonadotropic Hormones, LH (ICSH) and FSH: Major Advances, Current Status, and Outstanding Problems
   Harold Papkoff, Robert J. Ryan, and Darrell N. Ward
2. The Chemistry and Biology of Human Chorionic Gonadotropin and Its Subunits
   Om P. Bahl
3. Prolactin
   Henry G. Freisen
4. Steroid Hormone Secretions
   Seymour Lieberman, Erlio Gurpide, Mortimer Lipsett, and Hilton Salhanick
5. Hormonal Regulation of the Development, Maturation, and Ovulation of the Ovarian Follicle
   J. M. Bahr, G. T. Ross, and A. V. Nalbandov
6. The Endocrinology of Ovulation and Corpus Luteum Formation, Function, and Luteolysis in Women
   D. T. Baird and R. V. Short
7. Induction of Ovulation
   Carl Gemzell
8. Feedback Regulation of Reproductive Cycles in Rats, Sheep, Monkeys, and Humans, with Particular Attention to Computer Modeling
   Neena B. Schwartz, Donald J. Dierschke, Charles E. McCormack, and Paul W. Waltz
9. Gonadotropin Secretion
   Roger Guillemin, S. M. McCann, C. H. Sawyer, and J. M. Davidson
10. Brain Neurotransmitters and the Hypothalamic Control of Pituitary Gonadotropin Secretion
    Richard J. Wurtman
11. Hypothalamic Influences on Pituitary Function in Humans
    S. S. C. Yen, F. Naftolin, A. Lein, D. Krieger, and R. Utiger
12. Reproductive Immunology—Progress in the Last Decade
    K. A. Laurence
13. The Oviduct. I. Transport Mechanisms
    R. J. Blandau, R. M. Brenner, J. L. Boling, and S. H. Broderson
14. The Oviduct. II. Oviductal Physiology
    L. Mastroianni, Jr., B. Brackett, and C. Hamner
15. Present Status and Future Prospects of Research on the Biology of the Uterus
    D. G. Porter and C. A. Finn
16. Embryogenesis
    Ann McLaren
17. Clinical and Epidemiological Aspects of Induced Abortion
    Christopher Tietze
18. Systemic Contragestational Agents
    Sheldon J. Segal
19. Female Sterilization
    Ralph M. Richart and Katherine F. Darabi
20. Intrauterine Contraception
    Howard J. Tatum
21. The Morning-After Pill: A Report on Postcoital Contraception and Interception
    John McLean Morris
22. Injectable Contraceptive Preparations
    Daniel R. Mishell, Jr.
23. Bioengineering Aspects of Reproduction and Contraceptive Development
    Thomas J. Lardner

24. The Role of Prostaglandins in Reproduction
 Peter W. Ramwell
25. Hormonal Control of Gene Expression in Reproductive Tissues
 Bert W. O'Malley
26. Hormone-Receptor Interaction in the Mechanism of Reproductive Hormone Action
 E. V. Jensen, K. J. Catt, J. Gorski, and H. G. Williams-Ashman
27. The Control of Testicular Function
 Emil Steinberger, Ronald S. Swerdloff, and Richard Horton
28. Spermatogenesis
 Yves Clermont
29. The Ultrastructure and Functions of the Sertoli Cell
 Don W. Fawcett
30. Leydig Cells
 William B. Neaves
31. The Blood-Testis Barrier
 B. P. Setchell
32. The Structure of the Spermatozoon
 Don W. Fawcett
33. The Metabolism of Mammalian Spermatozoa
 R. A. P. Harrison
34. The Mammalian Accessory Sex Glands: A Morphological and Functional Analysis
 L. F. Cavazos
35. The Epididymis
 David W. Hamilton
36. Semen: Metabolism, Antigenicity, Storage, and Artificial Insemination
 Thaddeus Mann
37. Capacitation of Spermatozoa and Fertilization in Mammals
 M. C. Chang, C. R. Austin, J. M. Bedford, B. G. Brackett, R. H. F. Hunter, and R. Yanagimachi
38. Sperm Motility
 I. R. Gibbons
39. Regulation of Male Fertility
 C. Alvin Paulsen
40. Assessment of Clinical Testing Methodology
 Aníbal Faúndes

# Name Index

Abdul-Karim, R., cit., 134
Aberle, S. D., cit., 368
Abraham, G. E., cit., 118
Abramowicz, M., cit., 40
Adadevoh, B. K., cit., 310, 311
Adams, C. E., cit., 232, 235
Afzelius, B. A., cit., 240
Ahmad, N., cit., 213
Aiyer, M. A., cit., 123
Albritton, E., 380
Allen, E., 368; cit., 81, 99, 167
Allen, J. M., cit., 222
Allen W. M., cit., 81, 97
Allison, A. C., cit., 235
Amos, L., cit., 239
Ancel, P., cit., 171, 207
Anderson, E., cit., 196, 198
Anderson, L. L., cit., 136
Anderson, O. F., cit., 248
Aoki, A., cit., 183, 185, 208, 209, 211
Aonuma, S., cit., 243
Appelgren, L. E., cit., 227, 231
Archibald, D., cit., 129
Arimura, A., cit., 118
Armstrong, D. T., cit., 125, 207
Aschheim, S., cit., 85, 86, 171
Ashitaka, Y., cit., 124
Atkinson, L. E., cit., 63, 118
Auletta, F. J., cit., 137
Austin, C. R., cit., 232, 242

Baccetti, B., cit., 240
Badawy, S., cit., 130
Bagshaw, K. D., cit., 108
Bahl, O. P., cit., 100
Baillie, A. H., cit., 208, 226
Baker, C., cit., 83
Baker, V. F., cit., 230, 231
Banks, J., cit., 176
Barber, B., cit., 348
Barnes, A., cit., 431
Barraclough, C. A., cit., 122
Barrini, A. G., cit., 240
Barros, C., cit., 235, 244
Barry, M., cit., 165
Bartsch, G., cit., 219
Bates, R. W., cit., 88
Bavister, B. D., cit., 246
Becker, O., cit., 175
Bedford, J. M., cit., 223, 224, 225, 235, 237
Behrman, H. R., cit., 129

Bellisario, R., cit., 100
Bellvé, A. R., cit., 196, 198, 203
Benda, C., cit., 165, 169
Beneden, E. van, cit., 167
Bennett, G., cit., 226, 330, 331
Benoit, M. J., cit., 175, 221, 228
Berelson, B., cit., 39, 55, 57, 69
Bergstrom, S., cit., 177
Berson, S.A., cit., 108
Berstein, M. H., cit., 232
Bewley, T. A., cit., 100
Beyler, A. L., cit., 201
Bhatnagar, Y. M., cit., 196
Black, T. R. L., cit., 63
Blackshaw, A. W., cit., 222
Blake, C. A., cit., 109
Blandau, R. J., 131, 223, 224
Blaquier, J. A., cit., 215, 229
Bogdanove, E. M., cit., 102
Bone, M., cit., 39, 54
Bongaarts, J., cit., 59
Borrie, W. D., cit., 54
Bouin, B., cit., 171, 207
Bowen, R. H., cit., 169
Brackett, B. G., cit., 242, 243
Branca, A., cit., 173
Branton, D., cit., 180
Braun, G., cit., 348, 355
Brenner, R. M., cit., 132
Bressler, R. S., cit., 204
Briggs, M., cit., 219, 354
Broder, D., cit., 380
Brokelmann, J., cit., 178
Brown, P. D. C., cit., 228
Brown-Woodman, H. D. C., cit., 231
Bruce, W. R., cit., 202
Buckley, J., cit., 232, 235, 237
Bunge, R. G., cit., 200
Burgos, M. H., cit., 178, 189, 196, 197
Burgus, R., cit., 103
Bustos-Obregon, E., cit., 192, 205
Butcher, R. L., cit., 115, 118
Butenandt, A., cit., 81, 83, 171

Caldwell, B. V., cit., 129, 137
Callandra, R. S., cit., 229
Calvin, H. I., cit., 224, 225
Campbell, G. T., cit., 102
Campbell, J. A., cit., 137
Canfield, R. E., cit., 100
Carlsen, R. B., cit., 100
Castro, A. E., cit., 214

# Name Index

Catchpole, H. R., cit., 88
Catt, K. J., cit., 108
Cavazos, L., cit., 228
Cerceo, E., cit., 84
Chaikoff, I. L. cit., 209
Chang, M. C., cit., 93, 228, 242, 243, 246, 309, 310, 376
Channing, C. P., cit., 87, 124, 128
Chari, S., cit., 218
Chemes, H. E., cit., 212
Chin, W., cit., 217
Chow, B. F., cit., 87, 212
Christensen, A. K., cit., 202, 205, 207, 208
Clegg, R. C., cit., 232
Cleland, K. W., cit., 175
Clemens, J. A., cit., 106
Clermont, Y., cit., 169, 171, 183, 191, 192, 204, 232
Coale, A., cit., 49, 50
Cohen, E., cit., 231
Cohen, M. P., cit., 223
Cole, H. H., cit., 86
Collins, W. E., cit., 115, 118
Colton, F. B., 376
Connell, E., cit., 60, 61, 64
Cooper, G. W., cit., 224, 225
Cooper, T. G., cit., 207, 228
Coppola, J. A., cit., 230, 231
Corner, G. W., cit., 97, 368
Corner, G. W., Jr., cit., 99
Coulston, F., cit., 201
Counce, S. J., cit., 195
Coutinho, E. M., cit., 219
Crabo, B., cit., 221, 222, 227, 231
Cuyler, W. K., cit., 213

Dada, O. A., cit., 310
Danzo, B. J., cit., 221
Daughaday, W. H., cit., 108, 115
Dauzier, L., cit., 246
Davies, D. V., cit., 176
Davies, J., cit., 221, 228
Dawson, R. M. C., cit., 175, 224, 226, 228
De Graf, R., 165
de Kretser, D. M., cit., 196, 205, 209
Desjardin, C., cit., 209, 213
Dickinson, R. L., 369
Diczfalusy, E., cit., 354
Dierschke, D. J., cit., 109

Djerassi, C., 93, 376; cit., 309, 326, 330, 332, 333, 348, 355, 377
Doisy, E. A., cit., 81
Domm, L. V., cit., 171
Donovan, B. T., cit., 90
Dorrington, J. H., cit., 124
Douglas, C. D., cit., 396, 423
Dragoja, B. M., cit., 232
Drobeck, H. P., cit., 201
Dumond, D. E., cit., 68
Duncan, G., 394
Duncan, G. W., cit., 327, 331, 379
Dunckley, G., cit., 209
Dusitsin, N., cit., 309
Dykshorn, S. W., cit., 88
Dym, M., cit., 180, 183, 186, 192, 213

Ebner, V. von, cit., 165, 167
Edelman, G. M., cit., 237
Edstrom, K., cit., 40
Eik-Nes, K. B., cit., 209, 212, 213, 226
Einer-Jensen, N., cit., 183
El-Fouly, M., cit., 128
Elkington, J. S. H., cit., 215
Engelberg, W., cit., 108
Engle, E. T., cit., 89, 171
Erickson, G. F., cit., 87
Erickson, R. R., cit., 137
Ericsson, R. J., cit., 230, 231
Erlanger, B. F., cit., 108
Euler, U. S. von, cit., 177
Evans, H. M., cit., 87, 89
Everett, J. W., cit., 91, 95, 120
Ewing, L., cit., 213

Faiman, C., cit., 108
Fawcett, D. W., cit., 178, 180, 183, 184, 185, 186, 189, 192, 194, 196, 197, 198, 201, 203, 204, 205, 208, 211, 222, 223, 224, 226, 228, 229, 232, 237, 241
Felizet, G., cit., 173
Fels, E., cit., 81
Feng, T. D., 310
Ferguson, M. M., cit., 208, 226
Ferin, M., cit., 131
Fevold, H. L., cit., 86, 171, 172, 212
Fielding, U., cit., 93
Filshie, G. M., cit., 129
Fink, G., cit., 123
Finkle, J. L., cit., 55

Finn, C. A., cit., 138
Fischer, M. I., cit., 176
Flageolle, B. Y., cit., 201
Fleischer, B., cit., 226
Fleischer, S., cit., 226
Flickinger, C. J., cit., 180
Flores, F., cit., 122
Flores, M. N., cit., 201, 205
Folman, Y., cit., 212
Foote, R. H., cit., 191
Forbes, A. D., cit., 129
Franchimont, P., cit., 217, 218
Frankel, A. I., cit., 226
Franklin, L., cit., 244
Fray, C. S., cit., 224
Freedman, R., cit., 57, 69
Freeman, M. E., cit., 115
French, F. S., cit., 214, 228, 229
Frick, J., cit., 219
Friedman, H. M., cit., 103
Friedman, M. H., cit., 89
Friend, D. S., cit., 237
Friesen, H., cit., 106, 108, 115
Frisch, R. E., cit., 70
Fritz, I. B., cit., 214, 227
Fugino, M., cit., 103, 106
Fugo, N. W., cit., 115, 118
Furer, R., cit., 202

Gaddum-Rosse, P., cit., 224
Gallagher, T. F., cit., 171
Garcia, C. R., cit., 94
Gardner, W. U., cit., 99
Gatenby, J. B., cit., 169
Gaud, W. S., cit., 384
Gaudin, T., cit., 348
Gemzell, C., 130
Gerson, T., cit., 209
Gibbons, I. R., cit., 239, 240
Gibian, H., cit., 371
Giesel, L. O., cit., 169
Gillim, S. W., cit., 208
Gilula, N. B., cit., 183, 185
Glover, J. J., cit., 108
Glover, T. D., cit., 222, 228
Go, V. L. W., cit., 188
Goding, J. R., cit., 120
Goldberg, E., cit., 248
Goldblum, J. F., cit., 374
Gondos, B., cit., 186
Goodfriend, L., cit., 108
Gorski, R. A., cit., 122

Gould, T. C., cit., 176
Graff, M., cit., 373
Green, J. D., cit., 93
Greenwald, G. S., cit., 87
Greenwood, F. C., cit., 108
Greep, R. O., cit., 81, 87, 90, 129, 130, 171, 209, 212
Gunn, S. A., cit., 176
Gurin, S., cit., 84
Gutman, E. B., cit., 176
Guyda, H., cit., 106, 108, 115
Gwatkin, R. B. L., cit., 248

Haddad, A., cit., 226
Hafez, E. S. E., 131; cit., 135
Hagelstein, M. T., cit., 218
Hagenas, L., cit., 214
Halasz, B., cit., 122
Halban, J., cit., 81
Hall, P. F., cit., 208, 209, 211, 213
Haltmeyer, G. C., cit., 212, 213
Hamilton, D. W., cit., 222, 226, 228, 241
Hammond, J., cit., 81
Hammond, J., Jr., cit., 87
Hanisch, G., cit., 171
Hanley-Bowdoin, L., cit., 196
Hansson, V., cit., 214, 215, 228, 229
Harkavy, O., cit., 377, 385, 394
Harris, G. W., cit., 90, 91, 93, 103
Harris, M. E., cit., 207, 226, 228
Harrison, R. G., cit., 173
Hart, D. M., cit., 208
Hart, G. H., cit., 86
Hartman, C., 369
Hartree, E. F., cit., 232, 235
Heape, W., cit., 81
Heard, R. D. H., cit., 83
Hefnawi, F., cit., 312
Heidger, P. M., cit., 180, 204, 205
Heller, C. G., cit., 201, 218, 219
Herbert, V., cit., 108
Hertwig, O., cit., 165
Hertz, R., cit., 373
Hill, R., cit., 373
Hilliard, J., cit., 129
Himes, N. E., 369; cit., 68
Hisaw, F. L., cit., 86, 96, 99, 171, 172
Hisaw, F. L., Jr., cit., 96
Hoagland, H., cit., 376
Hodgen, G. D., cit., 217
Hoffer, A. P., cit., 222, 223, 224, 228, 229

Hohlweg, W., cit., 90
Hollenberg, R. D., cit., 223
Hollinshead, W. H., cit., 91
Holstein, A. F., cit., 205, 221, 222
Horan, A. H., cit., 223, 225
Horowitz, B., cit., 375
Hosi, T., cit., 246
Hotchkiss, J., cit., 118
Hoult, J. R., cit., 388
Hovatta, O., cit., 205
Huckins, C., cit., 186, 188, 192, 194
Hudson, B., cit., 226, 228
Hugenholz, A. D., cit., 188
Huggins, C., cit., 176
Hughes, W., cit., 171
Humphrey, G. F., cit., 176
Hunter, W. L., cit., 108
Hutchinson, C. F., cit., 248
Hutt, P. B., cit., 348, 354
Hwang, P., cit., 106, 108, 115

Igarashi, M., cit., 348, 351
Inano, H., cit., 227
Ingersoll, F. W., cit., 94
Ishihama, A., cit., 381
Ito, S., cit., 186
Iwamatsu, T., cit., 246

Jacks, F., cit., 218
Jackson, H., cit., 200
Jacobs, L. S., cit., 106, 115
Jacobsohn, D., cit., 90
Jaffe, F. S., cit., 385
Jaffe, R. B., cit., 118
Johansson, E. D. B., cit., 137, 219
Johnson, L., 383; cit., 384
Johnson, M. H., cit., 184
Johnson, S. G., cit., 217
Jolly, W. A., cit., 81
Jones, A. L., cit., 208, 211, 226, 228
Jones, E. F., cit., 64
Jones, I. C., cit., 172
Jukn, M., cit., 171
Junkman, K., cit., 90

Kamberi, I. A., cit., 106
Kammerman, C., cit., 124
Karim, S. M., cit., 129
Kasius, R. V., cit., 375
Kastin, A. J., cit., 103, 118
Kato, J., cit., 116
Kennedy, J. F., 380

Kerr, J. B., cit., 205
Kessler, A., cit., 304
Ketzen, L., cit., 214
Kierzenbaum, A. L., cit., 195
King, L. S., cit., 93
Kinsey, A., 369
Kirton, K. T., cit., 129, 230
Klug, A., cit., 239
Knauer, E., cit., 81
Knobil, E., cit., 109, 115, 118, 120, 137, 315
Koide, S. S., cit., 124
Kolliker, R. A. von, cit., 165
Kopec, B., cit., 214
Kordon, C., cit., 348
Kormano, M., cit., 180, 204, 222
Kruegar, D. M., cit., 217
Krutscher, W., cit., 176
Kuznets, S., cit., 49

Ladman, A. J., cit., 221
Lam, D. M. K., cit., 202
Landace, R. L., cit., 177
Laporte, P., cit., 226
Lasagna, L., cit., 355
Laughead, R., cit., 177
Laurence, K. A., cit., 130
La Valette, St. G. von, cit., 165, 167
Leak, L. V., cit., 180, 204, 205
Leblond, C. P., cit., 169, 171, 183, 191, 226, 232
Lee, C. M., cit., 64, 124
Lee, P. A., cit., 218
Lee, V. K., cit., 217, 218
Leeuwenhoek, A. van, 165, 231
Lehman, F., cit., 137
Lehmeyer, J. E., cit., 106
Leonard, A., cit., 422
Leonard, S. L., cit., 86, 172
Lewit, S., cit., 64, 382
Leydig, F., 171
Leyendecker, G., cit., 217
Li, C. H., cit., 100, 102, 103
Liang, H., 310
Lieberman, S., cit., 84, 108
Liggins, G. C., cit., 137
Liu, W. K., cit., 100
Loeb, L., cit., 87, 99
Loewit, K. K., cit., 130
Long, C. N. H., cit., 88
Long, I. R., cit., 204
Long, J. A., cit., 208, 211

Loriaux, D. L., cit., 218
Lubicz-Nawrocki, C. M., cit., 228
Lynn, W. S., cit., 84

McArthur, J. W., cit., 70, 94
McCann, S. M., cit., 103
McCormick, Mrs. S., 376
MacCorquodale, D. W., cit., 81
McCracken, J. A., cit., 129, 137
McCullagh, D. R., cit., 172, 213, 217
MacDonald, G. J., cit., 129, 130, 209
McGadey, J., cit., 226
McGee, L. C., cit., 171, 221
Machino, A., cit., 227
McLean, S. A., cit., 344, 348
McLean, W. S., cit., 215
MacLeod, J., cit., 218
MacLeod, R. M., cit., 106
Madhwa Raj, H. G., cit., 130
Maier, J., cit., 394
Makepeace, A. W., 376; cit., 89
Malacara, J. M., cit., 118
Malthus, T., 147
Malven, P. V., cit., 89
Mancini, R. E., cit., 214
Maneely, R. B., cit., 228
Mann, T., cit., 176, 177, 226, 227
Markee, J. E., cit., 91
Marquies, N. R., cit., 227
Marrian, G. F., cit., 81
Marsh, J. M., cit., 84, 128
Marshall, F. H. A., cit., 81, 91
Mason, N. R., cit., 84, 202, 205, 207, 221
Massa, E. M., cit., 208
Mastroianni, L., Jr., cit., 134
Matron, L. J., cit., 201
Matsuo, H., cit., 103
Mauldin, W. P., cit., 56
Mawson, C. A., cit., 176
Mazur, R. S., cit., 137
Means, A. R., cit., 213
Meistrich, M. L., cit., 202
Meites, J., cit., 106
Melo, J. F., cit., 219
Merkow, L., cit., 209
Messier, B., cit., 171
Metz, C. B., cit., 247
Meves, F., cit., 165, 169
Meyer, R. K., cit., 89
Michaele, I., cit., 237

Midgley, A. R., Jr., cit., 108, 118, 124, 125
Miller, R. G., cit., 202
Millette, C. F., cit., 237
Mills, S. C., cit., 224
Milunsky, A., cit., 40
Minlon, H., 309
Mishell, D. R., cit., 118
Miyamoto, H., cit., 246
Moens, P. B., cit., 188
Moghissi, K. S., cit., 138
Monesi, V., cit., 195
Monita, J., cit., 221
Monroe, S. E., cit., 118
Moon, Y. S., cit., 124
Moor, H., cit., 180
Moore, C. R., cit., 89, 171, 172, 173, 217, 221
Morris, M. D., cit., 209
Morris, N. M., cit., 43
Morton, D. B., cit., 235
Moses, M., cit., 194, 195
Moudgal, N. R., cit., 130, 209
Moyle, W. R., cit., 207, 209
Murad, F., cit., 209
Murota, S., cit., 211
Murphy, B. E. P., cit., 108
Murphy, H. D., cit., 214
Murstein, M. J., cit., 59

Naftolin, F., cit., 122
Nagano, T., cit., 183, 202
Nalbandov, A., 128
Neaves, W. B., cit., 180, 205, 209
Neill, J. D., cit., 115, 116, 137
Nekola, M. V., cit., 85
Nelson, W. O., cit., 172, 200, 201
Neri, R., cit., 185
Neumann, H. C., cit., 213, 230
Neutra, M., cit., 226
Nevo, A. C., cit., 237
Niall, H. D., cit., 103
Nicander, L., cit., 175, 180, 222
Nicoll, C., cit., 88
Nicolson, G. L., cit., 248
Niemi, M., cit., 204, 222
Nishikawa, Y., cit., 221
Niswender, G. D., cit., 108
Nixon, R., cit., 385
Noda, Y., cit., 244
Nolin, J. M., cit., 102
Nortman, D., cit., 40, 43, 54, 55, 56

Nygren, K. G., cit., 219

Oakberg, E. F., cit., 186, 188
Odell, W. D., cit., 108
Oikawa, T., cit., 248
Okumura, K., cit., 202
Olatunbosun, D. A., cit., 310
Oliphant, G., cit., 242, 243
Olson, G., cit., 240, 241
Omran, A. R., cit., 43
Oppenheimer, W., cit., 381
Orebin-Crist, M. C., cit., 221, 223, 225, 228
Orr, A. H., cit., 108
Ortovant, R., cit., 169, 191
Oura, C., cit., 244
Ozawa, H., cit., 226

Pallini, V., cit., 240
Papkoff, H., cit., 100
Parker, M. L., cit., 108
Parkes, A., 380; cit., 379
Parlow, A. F., cit., 100
Parsons, U., cit., 177
Patanelli, D. J., cit., 200, 201
Pattee, C. J., cit., 108
Paul, R., cit., 231
Paul, W. E., cit., 108
Paulos, A., cit., 224
Paulsen, C. A., cit., 205, 219
Paulsen, D. F., cit., 195
Pearlman, W. H., cit., 84
Peckham, W. P., cit., 102
Pederson, H., cit., 240
Penardi, R., cit., 129
Perey, B., cit., 183
Perkin, G., cit., 300, 306
Peyre, A., cit., 226
Pfaff, D. W., cit., 116
Pfeiffer, C. A., cit., 90
Phariss, B. B., cit., 129, 137
Phillips, D. M., cit., 189, 198, 223, 224, 232
Phillips, R., cit., 202
Pierce, J. G., cit., 100
Pincus, G., 376, 377; cit., 79, 93, 94
Piotrow, P. T., cit., 64
Ploen, L., cit., 196, 202, 214
Podesta, E. J., cit., 212
Pohley, F. M., cit., 94
Pokel, J. D., cit., 209
Polakoski, K. L., cit., 232, 247

Popa, G. T., cit., 93
Porter, D. G., cit., 138
Potts, G. O., cit., 230
Prasad, M. R. N., cit., 226, 230, 308, 389
Presl, J., cit., 124
Price, D., cit., 89, 172, 217
Price, J. M., cit., 240
Pullum, T., cit., 315

Raj, M., cit., 213
Rajalakshmi, M., cit., 226, 230
Rajaniemi, H., cit., 124
Rao, J. M., cit., 218
Raud, H. R., cit., 108
Rayford, P. L., cit., 108
Rebbe, H., cit., 240
Reddy, P. R. K., cit., 218
Regaud, C., cit., 167, 175
Reichert, L. E., cit., 124
Reichlin, S., cit., 118
Reid, B. L., cit., 175
Ritzen, E. M., cit., 214, 228, 229
Rivarola, M. A., cit., 212
Robson, J. M., cit., 87
Rock, J., cit., 94
Rockefeller, J. D., III, 377
Rolshoven, E., cit., 167
Romrell, L. J., cit., 196, 198, 203
Roosen-Runge, E. C., cit., 169
Rose, B., cit., 108
Rosen, S., cit., 223
Ross, G. T., cit., 108, 115, 118
Ross, J., cit., 74
Ross, M. H., cit., 204
Rowe, A. J., cit., 239
Rowlands, I. W., cit., 175, 226
Rumery, R. E., cit., 223, 224
Ruschig, H., cit., 81
Russo, J., cit., 189
Ruzicka, L., cit., 171
Ryan, K. J., cit., 83, 84, 87
Ryan, R. J., cit., 108, 124
Ryan, W. T., cit., 230
Ryder, N. B., cit., 43, 54, 57, 58

Sacerdote, F. L., cit., 189
Sairam, M. R., cit., 100
Salamon, S., cit., 231
Saldarini, R. J., cit., 231
Samisoni, J. I., cit., 222
Samuels, L. T., cit., 84

## Name Index

Sanborn, B. M., cit., 214, 215
Sanger, M., 376
Satir, P., cit., 239
Saunders, L., cit., 422
Savard, K., cit., 84, 128
Sawyer, C. H., cit., 91, 95, 109, 129
Saxena, B. B., cit., 217
Schaar, C., cit., 106
Schally, A. V., cit., 103, 109, 131
Scherins, R. J., cit., 217, 218, 219
Schindler, H., cit., 237
Schmidt, J., cit., 83
Schneider, I., cit., 217
Schomberg, D. W., cit., 137
Schumacher, G. F. B., cit., 232
Schumadier, C. F. B., cit., 232, 247
Schwartz, N. B., cit., 115, 116, 118
Schwenk, E., cit., 83
Scott, T. W., cit., 224, 226
Segal, S. J., cit., 304, 307, 382
Sehon, A. H., cit., 108
Seiguer, A. C., cit., 214
Sertoli, E., cit., 165
Setchell, B. P., cit., 180, 183, 184, 185, 207, 218, 222, 226, 228
Seyler, L. E., Jr., cit., 118
Shafer, K. D., cit., 382
Shah, U., cit., 134
Shapiro, S., cit., 40
Sherins, R. J., cit., 218, 219
Shikita, M., cit., 211
Shoglund, R. D., cit., 219
Shome, B., cit., 100
Short, R. V., cit., 83, 87
Shortland, F., cit., 209
Siegel, E., cit., 43
Simpson, M. E., cit., 89
Singh, S. P., cit., 230
Sivin, I., cit., 55, 56, 62, 64
Slater, J. J., cit., 222
Slaunwhite, W. R., cit., 84
Slautterback, D. L., cit., 186
Slotta, K. H., cit., 81
Smith, D. M., cit., 85
Smith, M., cit., 63
Smith, M. S., cit., 115, 116
Smith, O. W., cit., 84
Smith, P. E., cit., 85, 86, 171
Solari, A. J., cit., 195
Solomon, S., cit., 84
Southam, A. L., cit., 382
Spallanzani, L., 165

Speroff, L., cit., 137
Sproul, E. E., cit., 176
Srivastava, P. N., cit., 232, 235
Stambaugh, R., cit., 232, 235, 237
Standley, C. C., cit., 304
Staple, E., cit., 84
Steelman, S. L., cit., 94
Stefanini, M., cit., 244
Steinberger, A., cit., 214
Steinberger, E., cit., 203, 213, 214, 215
Stephens, R. E., cit., 239
Stevens, V. C., cit., 131
Stratton, L. G., cit., 213
Strauch, B. S., cit., 209
Stricker, P., cit., 88
Stubblefield, E., cit., 194
Stumpf, W. E., cit., 116
Summers, K. E., cit., 239
Suzuki, S., cit., 183
Swain, D. C., cit., 372
Sweeney, C. M., cit., 100
Swierstra, E. E., cit., 191

Takahashi, M., cit., 123, 131
Taleisnik, S., cit., 103
Talwar, G. P., cit., 131
Tamaoki, B., cit., 211, 227
Tanner, J. M., cit., 70
Tatum, H. J., cit., 62
Teichman, R. J., cit., 232
Teitelbaum, M. S., cit., 43
Thayer, S. A., cit., 81
Thibault, C., cit., 246, 388
Thomas, M., cit., 84
Tichenor, P., cit., 228
Tietze, C., cit., 40, 59, 64, 370, 382
Tillson, S. A., cit., 115, 137
Tilney, L. G., cit., 239
Toren, D., cit., 209
Toyama, Y., cit., 204
Toyoda, Y., cit., 246
Tregear, G. W., cit., 108
Tres, L., cit., 195
Tsai, C. C., cit., 120
Tsong, Y. Y., cit., 124
Tuck, R. R., cit., 183

Usui, N., cit., 235
Utiger, R., cit., 108

Vaitukaitis, J., 213

# Name Index

Vale, W., cit., 106
van der Molen, J. H., cit., 208
Van Dyke, H. B., cit., 87, 212
Van Hall, E. V., cit., 102
Van Thiel, D. H., cit., 217
Vande Wiele, R., cit., 84, 128
Vanha-Pertulla, T., cit., 124
Vaughn, M., cit., 209
Veler, C. D., cit., 81
Verney, E. B., cit., 91
Vernon, R. G., cit., 214
Vitale-Calpe, R., cit., 180
Voglmayr, J. K., cit., 180, 183, 207, 224
Voigt, K. D., cit., 388
von Baer, 165
von Berswordt-Wallrabe, R., cit., 213

Wade, N., cit., 430
Waida, Y., cit., 221
Waites, G. M. H., cit., 180, 183, 185, 207, 222, 226, 228
Walsh, E. L., cit., 172, 213
Ward, D. N., cit., 100
Wardell, W., cit., 355
Warner, F. D., cit., 239
Warwick, D., cit., 360
Weinman, D. E., cit., 243
Weinstein, G. L., cit., 89
Weinstock, E., cit., 58
Werkin, H., cit., 83
Werthessen, N. T., cit., 83
Westoff, C. F., cit., 43, 54, 64
White, A., cit., 88
White, I. G., cit., 224, 226, 228, 231
Whitelaw, R. G., cit., 99
Wilde, C. E., cit., 108
Williams, R. O., cit., 231
Williams, W. L., cit., 232, 243
Williams-Ashman, H. G., cit., 176
Wilson, J. D., cit., 212
Wiltbank, G. D., cit., 137
Winterberger, S., cit., 246
Wintersteiner, O., cit., 149
Wishik, S. M., cit., 385
Wislocki, G. B., cit., 93, 228
Wolbergs, H., cit., 176
Wollmer, W., cit., 388
Woolley, D. M., cit., 240
Woolley, H. B., cit., 431
Worcester, J., cit., 94
Wotiz, H. H., cit., 84

Wray, J. D., cit., 43
Wright, N. H., cit., 62
Wright, P. A., cit., 137
Wyngarden, L. J., cit., 129

Yallow, R. S., cit., 108
Yamaji, T., cit., 120
Yanagimachi, R., cit., 235, 244, 246, 248
Yen, S. S. C., cit., 109, 116, 118, 120
Yokoyama, M., cit., 246
Yoshinaga, K., cit., 131
Young, W. C., cit., 175, 221, 223, 225
Youngdale, G. A., cit., 230

Zambone, L., cit., 244
Zaneveld, L. J. D., cit., 232, 237, 247
Zarate, A., cit., 118
Zeleznik, A. J., cit., 124
Zemjanis, R., cit., 186
Zuckerman, S., cit., 1, 379

# Subject Index

Abortifacients, chemical, 280-281
Abortion(s)
  improvements in surgical techniques for, 6
  induced, 53-54, 56, 64
  legal, ratio to live births of, 58-59
  number of (1975), 39
  prostaglandins used for, 280-281
  U.S. Supreme Court's invalidation of state laws restricting, 58
  vacuum aspiration technique for, 278
ABP. *See* Androgen-binding protein
Acceptability
  and continuity of use of contraceptive methods, 73-75
  criteria of contraceptive methods, 68-70
  safety and, of pill and IUDs, 60-65
Accessory glands, of male reproductive tract, 175-177
Acetate, 83
  conversion to cholesterol, 84
Acrosin
  acrosome reaction and, 235-237
  shown to be sperm specific, 247
Acrosomal cap, 232
  origin of, 196-197
Acrosome, and acrosome reaction, 232-237, 248
Activation
  and capacitation, sperm, 242-243
  egg, 244
Adenosine monophosphate (AMP), 209, 213, 214, 215
Adluminal compartment, of seminiferous epithelium, 180, 183, 184, 186
Adrenocortical glucocorticoids, impact of administration of, on delivery, 137-138
Adrenocorticotropin hormone (ACTH), impact of administration of, on delivery, 137-138
Africa, subsaharan, research in, 310-311
Agricultural Experiment Stations, 374
Agricultural Research Council (Great Britain), 303
Albritton report, 380
All India Institute of Medical Sciences, 283, 304, 307, 308
All-Union Scientific Institute of Obstetrics and Gynaecology (Moscow), 303

Alphachlorohydrin
  concentration of, in epididymis, 227
  toxicity of, 231
American Cancer Society
  and Committee on Growth, 375
  scholar awards of, 318
AMP. *See* Adenosine monophosphate
Androgen(s)
  aromatization of, to estrogens, 84
  conversion of pregnenolone and progesterone to, 84
  crystallization of, from extracts of urine and testis, 171
  found in males and females, 83
  ultrastructure of Leydig cells and biosynthesis of, 207-212
Androgen-binding protein (ABP), 220
  and cytoplasmic androgen receptor, 214-215
  presence of, in epididymis, 228-229
Androstenedione, 83
Androsterone, synthesis of, 171
Animal fertility, improvement of, 252
Antiprogestins, luteolysis and, 143-144
Antisera
  to gonadotropins, 130-131
  to releasing factors, 131
Argentina, research in, 13
Atresia, follicle, 85, 143
Atropine, inhibition of ovulation by, 93
Australia
  contraceptive patterns in, 54
  governmental expenditures on reproductive sciences and contraceptive development in, 404
  regulation of fertility control drugs in, 351
  research in, 303, 412-413
Austria, and ICCR, 334
Axonemes
  of mammalian sperm tail, 237-242
  subfibers A and B of, 239

Bangladesh, contraceptive patterns in, 55
Basal compartment, of seminiferous epithelium, 180, 184
Battelle Memorial Institute, 333, 394
Belgium
  governmental expenditures for repro-

Belgium (*continued*)
 ductive sciences and contraceptive development in, 404
 research in, 303
"Big science," emergence of, 367
Biodegradable coated steroid preparation, placement of, into cervix, 284-285
Births, number of (1975), 39-41
Blood-testis barrier, 180-186, 249
Bracelets, contraceptive, 286
Brazil
 and ICCR, 334
 immunization of women in, with chorionic gonadotropin, 283
 research in, 13, 304, 306
Bulbourethral glands, 176
Bulgaria, abortion ratios in, 59

Canada
 contributions to WHO Expanded Programme of Research from, 336, 387, 400
 distribution of progesterone-releasing IUDs in, 280
 expenditures by, for contraceptive research and development, 329, 330, 404, 419
 regulation of fertility control drugs in, 351
 research in, 303, 412-413
 training efforts in, 313
Capacitation
 changes in sperm surface and, 248
 in vitro fertilization and, 246
 sperm activation and, 242-243
Carnitine
 decrease in, resulting from castration, 228
 presence of, in epididymis, 227
Castration, studies of morphological effects on epididymis, 228
Cell membranes, emergence of new information on, 294
Cells, principal and basal, 175
Centchroman, 146
 suppression of corpus luteum function by, 144, 283, 284
Center for Population Research. See under U.S. National Institute of Child Health and Human Development

Central Drug Research Institute (India), 284
Central Family Planning Institute (India), 308
Cervix
 chemical composition of mucus of, 138
 function of mucus of, 99-100, 146-147
 intracervical devices, 286-287
 placement of biodegradable coated steroid preparation into, 284-285
Chemical abortifacients, 280-281
Chile
 immunization of women in, with chorionic gonadotropin, 283
 research in, 13, 304, 306
China, People's Republic of
 contraceptive patterns in, 56
 development of vacuum aspiration technique for abortion in, 278
 research in, 309-310, 395
Chinese Academy of Sciences
 Institute of Organic Chemistry of, 309, 310
 Institute of Physiology of, 310
Chlormadinone acetate, suppression of plasma progesterone levels by, 129
α-Chlorohydrin, doses of, producing reversible infertility, 230-231
Chlorpromazine, prolactin secretion and, 106
Cholesterol, 83
 acetate conversion to, 84
 side chain cleavage to pregnenolone and progesterone, 84, 209
 synthesis of, 208-209
Chorionic gonadotropin. See human chorionic gonadotropin (hCG)
Chromatin, process of condensation of, 197, 224-225, 244-246
Chulalongkorn University (Thailand), 308-309
CIBA (Switzerland), expenditures of, on endocrine research, 371
Ciliogenesis, 132-134
Classification of fertility control methods, 53, 56
Colombia, research in, 306
Columbia University, 375
Commission on Biomedical Research, Presidential, 430
Committee on Growth, 375

Committee on Human Reproduction, 374
Committee on Population and Family Planning, 385
Committee on Research in Endocrinology, 370
Committee for Research in Problems of Sex, 368-369, 370, 371, 375
Communicating junctions, 179-180
Conceptions, number of (1975), 39
Condom, 73
 acceptability and continuity of use of, 74
 changes in appeal of latex, 278
Conference on Mechanisms Concerned with Conception (1959), 379
Connecting piece
 fixation of outer dense fibers in mammalian sperm tails by, 240
 joining condensed nucleus to outer dense fibers of flagellum, 198
Continuity of use of contraceptive methods, acceptability and, 73-75
Contraceptive(s). *See also* Intrauterine devices
 advances in development of, 4-13
 characteristics of, related to reproductive life cycle, 70-73
 conventional, 53, 56
 development, expenditures for, 405-412
 hormonal, 6, 53, 284-287
 injectable, 284-287
 immunological approaches, 142, 282-283
 implants, 6, 285
 limitations of, 56-60
 male, pharmacologic, 289-291
 need for broad array of, 3-4
 oral
  acceptability and continuity of use of, 74
  advances in development of, 4-5, 6
  emergence of, 3, 148, 278
  introduction of, to industrial and developing nations, 54-56
  limitations of, 57
  pharmaceutical industry's research on, 327
  safety and acceptability of, 60-65
  source of funds for development of, 376

 postcoital, 128-129, 278, 281
 safety and acceptability of, 60-65
Copper T
 development of, 279-280
 ICCR's work on, 279, 334-335
Cornell University, 375
Corpus luteum
 exertion of regressive influence on, by uterus, 136-137
 formation and demise of, 127-129
 formation, life span, and function of, 87-88
 influence of uterus on life and function of, 99
 pharmacologic suppression of, 283-284
 as primary source of progesterone, 83
 protrusion of, from surface of ovary, 81
Council of Scientific and Industrial Research (India), 308
CPR (Center for Population Research). *See under* U.S. National Institute of Child Health and Human Development
Critical period
 discovery of, in rats, 95
 refining limits of, 123
Cumulus cells, 232, 235
Cyproterone acetate
 altering of epididymal function by, 230
 impairment of spermatogenesis by, 230, 290
Cytokinesis, incomplete, in spermatogenesis, 186, 187, 192
Cytoplasmic droplet, 224, 225
Cytoplasmic receptor protein (CR), 215

Danazol, suppression of spermatogenesis by, 218-219
Decapacitation, sperm, 243
Decapeptide, studies aimed at use of, to regulate ovulation, 292
Declaration of Helsinki, 359
Dehydroepiandrosterone, presence of, in epididymis, 226
Délégation générale à la Recherche Scientifique et Technique (DGRST; France), 388

Denmark
  contributions to WHO Expanded
    Programme of Research from, 336,
    387, 400
  governmental expenditures for reproductive sciences and contraceptive
    development in, 404
  reproductive research in, 303
Depo-Provera, withholding of approval
  of, in United States, 351
Desmosomes, 179
Deutsche Forschungsgemeinschaft
  (DFG; Germany), 303, 388
Developing nations. *See also* Industrialized nations
  contraceptive patterns in, 54-56
  limits of contraceptive technology in,
    56-60
  research in reproductive sciences and
    contraceptive development in, 303-306
    ethical considerations of, 359-361
    expenditures for, 397-400, 412-413
    funding of, 389
    recommendations for participation
      in, 30-32
Diamines, inhibition of spermatogenesis by, 200, 201-202
Diaphragm, 73
  acceptability and continuity of use
    of, 74
Dibenamine, inhibition of ovulation
  by, 93
Diethylstilbestrol
  experimental use of, for postcoital
    contraception, 128
  synthesis of compounds related to,
    284
Dihydrotestosterone (DHT)
  activity of, 212, 229
  formation of, 212
  metabolizing testosterone to, 229
Dinitropyrroles, inhibition of spermatogenesis by, 200
Dissemination of results of reproductive research, recommendations for,
  35-36
DNA, 195, 197, 198
Dominican Republic, immunization of
  women in, with chorionic gonadotropin, 283
Dopamine, prolactin secretion and, 106

Down's Syndrome, 43
Drugs
  antispermatogenic, effect on germ
    cells of, 199-203
  consequences of regulatory requirements for, 352-355
  ethical considerations for testing new,
    356
  interruption of spermatogenesis by
    radiomimetic, 200
  for male contraception, 289-291
  regulation of fertility control, 348-352
Ductuli efferentes, 173, 175, 220
  results of ligation of, 229
  ultrastructural studies of, 221-223
Ductus epididymidis, 173-175
  ultrastructural studies of, 221-223
Dynein, 239, 240

Egg activation, 244
Egypt, research in, 311-312, 389
Ejaculatory duct, 175
Electron microscope, effect of introduction of, 178, 231-232, 250
Endocrine Study Section (initially
  Endocrinology and Metabolism
  Study Section), 373
Endometrium, 146
  control of, by estrogen and progesterone, 96-99
Epididymis, 173, 177
  absorption and secretion in, 175
  function(s) of, 220-221
    biochemical, 225-227
    control of, as approach to contraception, 229-231
    regulation of, 228-229
  maturational changes of spermatozoa
    in, 223-225, 242
  ultrastructural studies of, 222
Epithelium, seminiferous
  compartments of, 180, 183, 184
  dynamics of cell movement within,
    188-189
Ergocornine, prolactin secretion and,
  106
Ergocryptine, prolactin secretion and,
  106
17-$\beta$ Estradiol, obtaining of, 81
Estriol, isolation of, 81
Estrogen(s)
  action of, on the uterus, 96-99

Estrogen(s) (*continued*)
  administration of, for postcoital contraception, 128-129, 278, 281
  aromatization of androgens to, 84
  changes in levels of, during sexual cycle, 115-116
  feedback effects of, on pituitary secretion of gonadotropins, 118-120
  found in males and females, 83
  inhibition of ovulation by, 94
  male contraception and, 219, 289
  secretion of pituitary gonadotropins and, 89-90
  stimulation of preoptic area by, 122
Estrone, isolation of, 81
Ethical considerations, in conducting biomedical research, 31, 356-361, 362
Ethinyl estradiol, experimental use of, for postcoital contraception, 128
Excurrent ducts, of the testis, 173-175
  structure and function of, 220-231
Expanded Programme of Research, Development, and Research Training in Human Reproduction. *See under* World Health Organization
Expenditures (on research in reproductive sciences and contraceptive development). *See also* Financing
  compared with all medical research, 415-422
  estimates of future, 433-436
  by United States, industrialized nations, and developing nations, 397-400, 412-413
Experimentation, ethical considerations of, 356-361, 362

Family planning programs
  in India, 60, 279
  national initiation of, 3
Family Planning Services, 385, 404
"Fertility drugs," multiple births in women treated with, 130
Fertility regulation, 2-4
  new opportunities in, 291-295
  recommendations for achievement of, improved, 25-30
Fertilization
  contraceptive strategies directed against spermatozoon and, 247-248
  spermatozoon and, 231-248

  zona penetration and, 243-246
Fetus, initiation of delivery mechanism rests with, 137-138
Fibrous sheath, in mammalian spermatozoon, 241
Financing (of research on regulation of fertility), 16-25. *See also* Expenditures
  early 1960s, 379-383
  future, 23-25, 27-28
  governmental involvement in (1965-1975), 383-389
  past and present, 16-23
  post - World War II period (1940s and 1950s), 372-379
  pre - World War II period (1920s and 1930s), 368-371
Finland
  contributions to WHO Expanded Programme of Research from, 336, 387, 400
  governmental expenditures for reproductive sciences and contraceptive development in, 404
  and ICCR, 334
  immunization of women in, with chorionic gonadotropin, 283
  reproductive research in, 303
Flagellum, connecting piece joining condensed nucleus to outer dense fibers of, 198
Food and Drug Amendments (1962), 352. *See also* U.S. Food and Drug Administration
Follicle stimulating hormone (FSH), 85
  changes in serum levels of, during sexual cycle, 109, 115
  and discovery of pituitary and placental gonad stimulating hormones, 86-87
  effects of, on male reproductive tract, 172
  effects of, on seminiferous tubules, 212-214
  formation and demise of corpus luteum and, 127-128
  influence on secretory function of gonads of, 100
  introduction of improved bioassay for, 94-95
  LRF and, 103

Follicle stimulating hormone (FSH) (*continued*)
  need for further studies on structure, binding, receptors, and mechanism of action of, 138-141
  primary structure of, 100-102
  receptors, analysis of location of, 293
  release, negative feedback control of, 217-218, 220
  RIA procedures for, 108
  role of, in maintaining levels of androgen-binding protein, 215
Food for Freedom Act (1966), 384
Ford Foundation, 1, 284
  contributions to WHO Expanded Programme of Research from, 336, 387
  funding of Biomedical Division of Population Council by, 377
  and male contraception, 291
  and patent agreements, 344
  principal source of support for reproductive research from, 377
  and progesterone-releasing IUDs, 280
  program on population of, 381
  reduction in grants from, 401, 404
  and research in developing nations, 305, 389
  support of, for ICCR, 334
  support of, for Latin American Program of Research in Human Reproduction, 307
  support of, for research in Egypt, 311, 312
  training and fellowship awards from, 313
France, reproductive research in, 13, 303, 388
FSH. *See* Follicle stimulating hormone
Fund for Population Activities. *See under* United Nations
Fundamental studies in reproductive sciences, expenditures for, 405-412
Funding strategies, need for contraceptive research and, 319-321

Galactosyltransferase, 226
Gap junctions, 179, 183
General Service Foundation, support for ICCR from, 334
Geographical distribution of responsibility for research, 30-32
Germ cells, 249
  contraception by agents acting upon multiplication and differentiation of, 199-203
  relationship of, to Sertoli cells, 178-186
  syncytial nature of, 186-189
Germany, West
  regulation of fertility control drugs in, 351
  reproductive research in, 13, 303, 388
  governmental expenditures for, 419
Glycerophosphorylcholine
  decrease in, resulting from castration, 228
  synthesis of, by epididymal epithelium, 225-226, 227
Glycoproteins, chemistry of gonadotropic, 292-293
Golgi apparatus, 196
  function of, 211
Gonad stimulating hormones, discovery of pituitary and placental, 85-89
Gonadotropin(s)
  advances in knowledge of chemistry of, 8-9
  antisera to, 130-131
  elucidation of structure of, 12, 100-102
  factors influencing secretion of pituitary, 89-95
  feedback effects of estrogen on pituitary secretion of, 118-120
  need for further study of structure, binding, receptors, and mechanism of action of, 138-141
  production and release of, controlled by luteinizing hormone releasing factor, 7-8
  receptor proteins, and mechanism of hormone action, ovarian binding of, 124-125
  receptors, analysis of location of, 293
  release of, in the male, 172
  suppression of, as contraceptive strategy, 218-220
  three principal human, structural analysis of, 292-293
Gonorrhea, number of adolescents having (1973), 42

Governmental funding for reproductive research level of, 400-404
Great Britain
contributions to WHO Expanded Programme of Research from, 336, 387
distribution of contraceptives in, 63
distribution of progesterone-releasing IUDs in, 280
governmental expenditures for reproductive sciences and contraceptive development in, 404, 419
popularity of pill in, 54
regulation of fertility control drugs in, 351
research in, 13, 303, 388
tests on male oral contraceptives in, 219

Heterocyclic compounds, inhibition of spermatogenesis by, 200-201
Hormonal contraceptives. See Contraceptives, oral; Steroids
Hormones. See also Androgen(s); Estrogen(s); Follicle stimulating hormone; Human chorionic gonadotropin; Luteinizing hormone; Progesterone; Prolactin; Testosterone
interaction of, with genetic material of cell nuclei, 9-10
ovarian, 81-83
  feedback of, 116-120
  steroid, biosynthesis of, 83-84
  secretion rates and circulating blood levels of, 10-11
Human chorionic gonadotropin (hCG)
influence on secretory function of gonads of, 100
possibility of developing vaccine against pregnancy using $\beta$ subunit of, 142, 282-283, 335
primary structure of, 101-102
receptors, analysis of location of, 293
RIA procedures for, 108
testing of vaccine against, 6
Human Embryology and Development Study Section, 373
Hungary, reproductive research in, 303
Hyaluronidase, acrosomal, 235, 247
Hygienic Laboratory, 372
Hypophyseal portal system, and neurovascular mechanism regulating pituitary functions, 93
Hypophysectomy
regression of Leydig cells and seminiferous epithelium after, 171
studies of morphological effects on epididymus of, 228
Hypophysis, anterior, male and female reproductive systems under functional control of, 85-86
Hypothalamus
bisexuality of, 122-123
role of, in induction of ovulation, 118
Hysterectomy, persistence of corpora lutea in guinea pig following, 99

ICCR (International Committee for Contraception Research). See under Population Council
IDRC. See International Development Research Centre of Canada
Immunoendocrinology, reproductive, 130-131
Immunological approach to fertility control, 142, 282-283
Implants, development of subdermal contraceptive, 6, 285
India
contraceptive patterns in, 55
expenditures by, for contraceptive research and development, 329, 419
family planning program in, 60, 279
and ICCR, 334
immunization of women with chorionic gonadotropin in, 282-284
introduction of Copper T in, 279
policy on human research in, 356
research in, 304, 307-308, 389
study of Centchroman in, 284
Indian Council for Medical Research, 14
evaluation program of, for new IUDs, 287
governmental funding of, 389
research and training at, 308
Industrialized nations. See also Developing nations
contraceptive patterns in, 54-56
limits of contraceptive technology in, 56-60
research in reproductive sciences and contraceptive technology in, 302-303

Industrialized Nations (*continued*)
  ethical considerations of, 359-360
  expenditures for, 397-400, 412-413
  funding of, 387-389
Inflation, impact of, on reproductive research, 431-433
Inhibin
  current efforts to isolate and characterize, 217-218, 220
  unsuccessful efforts to isolate, 172, 217
Injectable contraceptives, 284-287
Insemination
  artificial, 374
  with frozen sperm, 370
Institute of Materia Medica (Peking), 310
Institutional and professional capacity, recommendations for strengthening, 32-35
Interagency Committee on Population Research, 393-394
International Committee for Contraception Research. *See under* Population Council
International Development Research Centre (IDRC) of Canada, 1
  contributions to WHO Expanded Programme of Research from, 336, 387
  on patent policy and research grants, 344-345
  support of, for ICCR, 334
  support of, for Latin American Program of Research in Human Reproduction, 307
International Planned Parenthood Federation (IPPF). *See also* Planned Parenthood Federation of America
  Central Medical Committee of, 62
  survey by, of world needs in family planning, 65
International Postpartum Program, 55
  and discontinuation rates for IUD and pill users, 64
Interstitial cell stimulating hormone (ICSH), 86. *See also* Luteinizing hormone (LH)
Intracervical devices, 286-287
Intrauterine devices (IUDs), 53. *See also* Copper T
  acceptability and continuity of use of, 74-75
  copper-releasing, 279-280
  emergence of, 3, 148, 278
  improvement in, 4-5, 6, 287-288
  introduction of, to industrial and developing nations, 54-56
  limitations of, 57
  medicated, pharmaceutical industry's development of, 327
  progesterone-releasing, 280
  revival of interest in, 381-382
  safety and acceptability of, 60-65
Iran, research in, 311, 389
Israel
  governmental expenditures on reproductive sciences and contraceptive development in, 404
  published papers in, 301
  research in, 303, 412-413
  training efforts in, 313
Italy
  government expenditures for reproductive sciences and contraceptive development in, 404
  research in, 303
IUDs. *See* Intrauterine devices

Jamaica, contraceptive patterns in, 55
Japan
  published papers in, 301
  research in, 13, 303, 412-413
  training efforts in, 313

Lactic dehydrogenase-X, as sperm-specific isoenzyme, 248
Lalor Foundation, fellowships awarded by, 313
Lamina propria, of seminiferous tubules, 203-205
Latin America, widespread reproductive research in, 306-307
Latin American Program of Research in Human Reproduction (PLAMIRH), 307
Leydig cells
  localization of LH receptors to, 293
  regression of, after hypophysectomy, 171
  relationship of, to seminiferous tubules, 205-207
  testosterone from, 12, 172

Leydig cells (*continued*)
  ultrastructure of, and biosynthesis of androgens, 207-212
LH. *See* Luteinizing hormone
Long-acting forms of contraceptive steroids, 6, 284-287
LRF. *See* Luteinizing hormone releasing factor
Luteinizing hormone (LH)
  changes in serum levels of, during sexual cycle, 109, 115
  and discovery of pituitary and placental gonad stimulating hormones, 86-87, 88
  effects of, on male reproductive tract, 172
  formation and demise of corpus luteum and, 127-128
  heightened output of, at midcycle, 94
  inducing ovulation by raising blood level of, 7-8
  influence on secretory function of gonads of, 100
  and LRF, 103
  need for further studies on structure, binding, receptors, and mechanism of action of, 138-141
  primary structure of, 101-102
  receptors, analysis of location of, 293
  RIA procedures for, 108
  role of, in control of Leydig cell function, 212-213
  role of preoptic area in triggering ovulatory discharge of, 120-122
  stimulus to ovum maturation due to, 85
Luteinizing hormone releasing factor (LRF)
  feedback action of ovarian hormones and, 116-120
  hypothalamic, isolation and structure of, 103-106
  inhibition of induced ovulation by immunization with antisera to, 131
  isolation and synthesis of, 7-8
  response to, in women who are breast feeding, 107
Luteolysis
  and antiprogestins, 143-144
  pharmacologic approaches to, 283-284

McGill University, 375

Malaysia, research in, 308
Male
  contraception, pharmacologic, 289-291
  need for more study of reproductive process in, 29-30
  new methods of fertility regulation for, 6-7
  testing of contraceptive injection for, 279
Manchette, composition of, 197
Markle (John and Mary R.) Foundation
  scholar awards of, 318
  support of, for Committee on Research in Endocrinology, 370
Mayo Foundation, 303
Medical Institute of the Social Security Service of Mexico, 284
Medical Research Council (MRC; Great Britain), 303, 388
Medroxyprogesterone, suppression of plasma progesterone levels by, 129
Meiosis
  and synthetic activities during spermatogenesis, 194-196
  defined, 167
Melatonin, possible role of, in regulating human fertility, 292
Mestranol, inhibition of ovulation by, 94
Mexico
  contributions to WHO Expanded Programme of Research from, 336, 387
  distribution of progesterone-releasing IUDs in, 280
  funding of reproductive sciences in, 395
  manufacture of Copper Ts in, 280
  policy on human research in, 356, 358-359
  research in, 13, 304, 306, 395
Middle East, research in, 311-312
Milbank Memorial Fund, 375
Miscarriages, spontaneous, and stillbirths, number of (1975), 39
Mitochondrial sheath, in mammalian spermatozoon, 241
*Montanoa tuberosa*, (zoapatle) extracts of, used to suppress progesterone levels, 144, 283-284

Myoid cells, 204
Myometrium, 146
  control of, by estrogen and progesterone, 96

National Academy of Sciences (NAS)
  Division of Medical Sciences of, 369
  National Research Council (NRC) of, 368, 369, 370, 374, 375
National Cancer Act (1971), 428
National Cancer Institute, 372, 430
National Commission for the Protection of Human Subjects of Biomedical and Behavioral Research, 356
National Committee on Maternal Health (NCMH), 369-370, 374
National Fertility Study (1970), 57-58, 64
National Heart and Lung Institute, 430
National Institute of Arthritis and Metabolic Diseases (NIAMD), 373
National Institute of Child Health and Human Development, National Institutes of Health. *See* U.S. National Institute . . .
National Planning Association (NPA), 419
National Research Council. *See under* National Academy of Sciences
National Research Council of Thailand, 308
National Science Foundation (NSF)
  decline in allocations from, 404
  establishment of, 373
Nepal, contraceptive patterns in, 55
Netherlands
  governmental expenditures in reproductive sciences and contraceptive development in, 404
  research in, 303
Neural influences on reproductive functions, 90-93
Neural mechanism regulating release of prolactin, 106-107
Neuroendocrinology of reproduction, 141-142
Neurovascular mechanism regulating pituitary function, hypophyseal portal system and, 93
New Jersey Agricultural Experiment Station, 374

*New York Times*, 362
Nitrofurans, suppression of spermatogenesis by, 200
Nitroimidazoles, suppression of spermatogenesis by, 200-201
Norethindrone
  inhibition of spermatogenesis by, 219
  suppression of plasma progesterone levels by, 129
Norethindrone enanthate, development of, 329
Norethisterone
  inhibition of ovulation by, 93
  synthesis of, 376
Norethynodrel
  inhibition of ovulation by, 93-94
  suppression of plasma progesterone levels by, 129
  synthesis of, 376
Norgestrel, suppression of plasma progesterone levels by, 129
Norgestrienone, inhibition of spermatogenesis by, 219
Northwestern University, research on fertility control at, 341
Norway, contributions to WHO Expanded Programme of Research from, 336, 387, 400
Nuffield Foundation, fellowships awarded by, 313
Nuremberg Code, 359

Occluding junctions, 180, 183, 184, 249
Oliver Byrd Fund, fellowships awarded by, 313
Oocytes, maturation of, 84-85, 143
Ooplasm, rapid transformation of sperm nucleus in, 244-246
Oregon Regional Primate Research Center, 303
Organization for Economic Cooperation and Development (OECD), Development Assistance Committee of, 21, 422
Ortho Parmaceutical Corporation, 354
Outer dense fibers
  of flagellum, 198
  in mammalian sperm tails, 240-242
Ovarian binding of gonadotropins, receptor proteins, and mechanism of hormone action, 124-125

Ovarian follicles, as primary source of estrogen, 83
Ovarian hormones, 81-83
 feedback action of, 116-120
Ovarian steroid hormones, biosynthesis of, 83-84
Ovarian stroma, function of, in reproductive cycle, 129-130
Ovary
 biosynthesis of ovarian steroid hormones, 83-84
 and luteolysis and antiprogestins, 143-144
 maturation of oocytes in, 84-85, 143
 ovarian hormones, 81-83
Oviduct
 ciliogenesis in, 132-134
 fluid in, 134-135
 need for further study of, 144-145
 transport of ova and sperm through, 131-132
 uterotubal junction and, 135-136
Ovulation
 in human female, 125-127
 inhibition of, by orally active progestational steroids, 93-94
 role of hypothalamus in induction of, 118
 use of decapeptide to regulate, 292

Pakistan, contraceptive patterns in, 55
Parturition, mechanism initiating, 137-138
Patent policy, 343-346
Pathfinder Fund, 384
 International IUD Program of, 341-342
Peer review, 357, 359, 360
 committees, 358-359
Perphenazine, prolactin secretion and, 106
Peru, research in, 304, 306
Pharmaceutical industry
 burden assumed by, 362-363
 collaboration with, 343-346
 expectations of, on contraceptive development, 330-333
 expenditures for contraceptive research and development by, 327-328, 379, 394, 397, 400-404
 role of, in contraceptive development, 13-14, 15, 326-330

Philanthropic funding for reproductive research, level of, 400-404. *See also* Ford Foundation; Population Council; Rockefeller Foundation; Scaife Family Charitable Trusts
Philippines, research in, 308, 389
Pill. *See* Contraceptives, oral
Pineal gland, role of, in regulating pituitary-ovarian function, 292
Pituitary
 gonadotropins, factors influencing secretion of, 89-95
 and placental gonad stimulating hormones, discovery of, 85-89
 role of, in spermatogenic function of testis, 171-172, 177
Planned Parenthood Federation of America (PPFA), 374-375, 376. *See also* International Planned Parenthood Federation
 and Conference on Mechanisms Concerned with Conception, 379
Poland, reproductive research in, 303
Polyspermy, block to, 244, 248
Population
 growth rates of world, 46-48
 impact of fertility decline on growth of, 48-50
 impact of growth of, on study of reproductive sciences related to regulation of fertility, 393
Population Council, 55, 333
 Biomedical Division of, 377
 and Conference on Mechanisms Concerned with Conception, 379
 contraceptive development programs supported by, 332
 and Copper T, 279
 early studies by, of IUDs, 327
 establishment of (1952), 377
 funds for reproductive research from, 378
 International Committee for Contraception Research (ICCR) of, 14
 and Copper T, 279, 288, 334, 335
 immunization of women with chorionic gonadotropin by, 283
 on male contraception, 291
 program of, described, 334-335
 testing of long-acting forms of contraceptive steroids by, 284-286

Population Council (*continued*)
and National Committee on Maternal Health, 370
and patent agreements, 344
and Population Research Centers, 302
and progesterone-releasing IUDs, 280
and research in developing nations, 305
and revival of interest in intrauterine contraception, 381-382
training and fellowship awards from, 313
Population Research Act (1970), 385-386, 404
Population Research Centers, 302
Portal system, hypophyseal, and neurovascular mechanism regulating pituitary functions, 93
Postacrosomal sheath, 244
Postcoital contraception, administration of estrogens in high doses for, 128-129, 278, 281
Pregnancy
  impact of, on unmarried adolescent women, 41-42
  outcomes, adverse, 2, 41-46
  possibility of developing vaccine against, 142, 279, 282-283
Pregnant mare serum gonadotropin (PMSG)
  influence on secretory function of gonads of, 100
  injections of, to synchronize fertility periods in sheep, 374
Pregnenolone, 83
  conversion of, to progesterone and to androgen, 84
  resulting from cholesterol side chain cleavage, 84, 209-211
Premarital sexual activity, variations in, from culture to culture, 70-72
Preoptic area, role of, in triggering ovulatory discharge of LH, 120-122
Primate reproductive cycle, phases of, 86
PRL. *See* Prolactin
Progestational steroids, orally active, inhibition of ovulation by, 93-94
Progesterone, 83
  action of, on the uterus, 96-99
  changes in levels of, during sexual cycle, 115-116

cholesterol side chain cleavage to, 84, 211
conversion of, to androgen, 84
conversion of pregnenolone to, 84
determination of structure of, 81-83
extracts of *Montanoa tuberosa* (Zoapatle) used to suppress levels of, 144, 283-284
impact on premature delivery of administration of, 138
isolation of, 81
-releasing IUDs, 280
secretion of pituitary gonadotropins and, 89
sources of, 83
stimulation of preoptic area by, 122
stimulatory and inhibitory role of, in reproductive physiology, 95
suppression of plasma levels of, by doses of contraceptive progestins, 129
Progestins
  for male contraception, 289-290
  new synthesis and testing of, 284
  suppression of plasma progesterone levels by, 129
Progestogens, suppression of spermatogenesis by, 219
Program for Applied Research on Fertility Regulation (PARFR), 341
Prolactin (PRL), 12
  discovery and function of, 88-89, 102-103
  need for further study of, 138, 141
  neural mechanism regulating release of, 106-107
  RIA procedures for, 108
  role of, in regulation of luteal function, 87, 100
Prolactin inhibiting factor (PIF), 106, 107
Prolactin releasing factor (PRF), 106-107
Prostaglandin $F_{2\alpha}$
  15-methyl, for use as intramuscular injection for abortion, 281, 283, 284
  identification of uterine luteolytic factor as, in sheep, 129, 137
Prostaglandins
  research by pharmaceutical firms on, 327

Prostaglandins (*continued*)
  synthetic, used for midtrimester abortions, 5-6, 280-281
Prostate, 175-176
Public Health Service Act of 1944, 372
Public sector agencies
  collaboration of, with industry, 14, 343-346
  programs of, in contraceptive development, 13-14, 326, 333-334
  recommendations for, to protect patentable inventions, 33
  relationships among, and with private industry, 342-343
Published research papers, on reproduction and contraceptive development, 301-302
Puerto Rico, contraceptive patterns in, 56

Radioimmunoassay (RIA), 109
  development of, 107-108, 115
Radiomimetic drugs, interruption of spermatogenesis by, 200
Rats, discovery of "critical period" in, 95
Receptors, gonadotropin, structural features and location of, 293-294
  steroidal, 293-294
Regulation
  and consequences of regulatory requirements, 352-355, 361-362
  of fertility control agents, 348-352
Reproduction
  immunology of, 142, 282-283
  neuroendocrinology of, 141-142
Reproduction Research Information Service, 301
Reproductive hormones, changes in levels of, 109-115
Reproductive immunoendocrinology, 130-131
Reproductive life cycle
  characteristics of contraceptives related to, 70-73
  four stages of, 68-70
Research
  in developing nations, 303-306
  and funding strategies, need for contraceptive, 319-321
  groups, numbers of, in United States and other industrial nations, 13
  in industrialized nations, 302-303
  institutions, basic, 300-302
  in Latin America, 306-307
  in Middle East, 311-312
  organizations, nonprofit, 333
  in People's Republic of China, 309-310
  in South Asia, 307-308
  in Southeast Asia, 308-309
  stability of support for, 322-323
  in subsaharan Africa, 310-311
  support of basic and applied, 321-322
Reserpine, prolactin secretion and, 106
Rete testis, 173, 180
Rhythm method
  acceptability and continuity of use of, 74
  limitations of, 148
RIA. *See* Radioimmunoassay
RNA, 195, 213
Rockefeller Foundation, 1
  decline in allocations from, 404
  fellowships awarded by, 313, 318
  funding of Biomedical Division of Population Council by, 377
  percentage of funds for reproductive research from, 378
  programs on population at, 381
  support for Committee for Research in Problems of Sex from, 368, 375
  support for ICCR from, 334
Rockefeller University (formerly Rockefeller Institute), 377, 381

Safety
  and acceptability, of pill and IUDs, 60-65
  contraceptive, issue of, 347-348
  of current fertility control methods, expenditures for, 405-412
  recommendation for higher priority, 26, 29
Salk Institute for Biological Studies, 303
Scaife Family Charitable Trusts
  decline in allocations from, 404
  fellowships awarded by, 313
  funding of Biomedical Division of Population Council by, 377

Schering AG (Germany)
  development of norethindrone enanthate by, 329
  expenditures of, on endocrine research, 371
  on regulatory requirements, 354, 355
Scientific community, involvement of international, 412-413
Scientists, reproductive, views of, 315-319
  on contraceptive research and funding strategies, 319-321
  on stability of support for research, 322-323
  on support of basic and applied research, 321-322
  on training, 323-324
Scrotum, thermoregulatory function of, 173, 177
Searle (G. D.) Company, 376, 377
Seminal vesicles, 175, 176
Seminiferous epithelium, 167
  cycle of, defined, 169-171
  organization of, 177-186
  regression of, after hypophysectomy, 171
Seminiferous tubules
  biochemistry of, 212-220
  boundary tissue of, 203-205
  effects of FSH on, 212-214
  nineteenth century studies of, 165
  organization of interstitial tissue of, 205-207
  spermatogenic function of, 171
Sertoli cells
  and blood-testis barrier, 249
  effect of diamines on, 202
  and FSH receptors, 293
  function of, in reproduction, 11-12, 165-167
  and mechanism of sperm release, 188-189
  relationship of germ cells to, 178-186
*Sex and Internal Secretions*, 369
Sex steroids
  measurement of, in plasma based on competitive protein binding, 108
  secretion of pituitary gonadotropins and, 89-90
Sialic acids
  decrease in, resulting from castration, 228
  identified in epididymis, 226, 227
Side chain cleavage, 84
Singapore
  contraceptive patterns in, 55, 56
  research in, 308
Skin, absorption of steroids through, for contraception, 286
Social Hygiene, Inc., 368
Social Security Act of 1935, 372
Society of Andrology, 30
South Asia, research in, 307-308
South Korea
  contraceptive patterns in, 56
  research in, 308, 389
Southeast Asia, research in, 308-309
Sperm release, mechanism of, 188-189, 249
Spermatids, 167
  bridges connecting groups of, 186, 189
  transformation into spermatozoa of, 199
Spermatocytes, 189
  bridges connecting groups of, 186-188
  exclusion of, as target for drug action, 199
Spermatogenesis
  and contraception by agents acting upon multiplication and differentiation of germ cells, 199-203
  control of, 212-220
  cyclic nature of, 169, 177
  defined, 167
  meiosis and synthetic activities during, 194-196
  and morphogenesis of spermatozoa, 196-199
  process of, 11-12
  stages of, 169-171, 177, 183, 249
  and stem cell renewal and spermatogenic cycle, 191-194
  syncytial nature of germ cells and mechanism of sperm release, 186-189
  temperature sensitivity of, 173, 177
Spermatogonia
  bridges joining chains of, 186-188
  exclusion of, as target for drug action, 199
  and stem cell renewal, 191-194
  two types of, 167, 169

Spermatozoa
  activation and capacitation of, 242-243
  development of, 165-169, 177
  and fertilization, 231-248
    contraceptive strategies directed against, 247-248
  maturational changes of, in epididymis, 223-225, 229-230
  and mechanism of sperm release, 188-189
  morphogenesis of, 196-199
  motor apparatus of, 237-242
  penetration of zona pellucida by, 243-246
  two centuries of studies of, 165
Spermiogenesis
  defined, 167
  documentation of morphogenetic events of, 167-169
  and morphogenesis of spermatozoa, 196, 197
  and separation of spermatozoa from germ cell syncytia, 189
Stanford Research Institute, 333
Stem cell renewal, 191-194
Sterilization
  acceptability and continuity of, 74, 75
  female, 54, 64, 288-289
  improvements in surgical techniques for, 6
  male, 64, 291
  voluntary, 53, 54, 56
Steroids
  contraceptive, long-acting forms of, 6, 284-287
  synthetic, in oral contraceptives, 5
Stroma, ovarian, function of, in reproductive cycle, 129-130
Subdermal implant, investigations on, 6, 285
Survey, conducted to determine levels of funding for reproductive sciences and contraceptive development, 394-397
Sweden
  contributions from, for WHO Expanded Programme of Research, 336, 387, 400
  and ICCR, 334
  immunization of women in, with chorionic gonadotropin, 283
  regulation of fertility control drugs in, 351
  reproductive research in, 13, 303
Synaptonemal complex, 194-195
Syntex Corporation, 330, 355, 376-377
Syphilis, number of adolescents having (1973), 42

Taiwan, contraceptive patterns in, 56
Testicle, account of structure of, 165
Testis
  barrier, blood-, 180-186, 249
  endocrine functions of, 171, 173
  influences on functions of, other than those mediated by hormones, 172-173
  new concepts of function of, 11-12
  role of pituitary in spermatogenic function of, 171-172
  separation of components of, 202-203
  structure and function of peritubular and interstitial tissue of, 203-212
Testosterone, 83
  administration of excess, resulting in infertility, 218-220
  biosynthesis of, 208-212, 249
  inhibition of sperm production by, 289, 290
  presence of, in epididymis, 226-227
  regulation of epididymal function by use of, 228-229
  role of, in maintaining levels of androgen-binding protein, 215
  secreted from Leydig cells, 12, 172, 207
  synthesis of, 171
Thailand
  contraceptive patterns in, 56
  research in, 208-209, 389
Thalidomide tragedy, 352
Thiophenes, suppression of spermatogenesis by, 200
Thyroid stimulating hormone releasing factor (TRF), 106-107
Training
  expenditures for, 405-412
  process of, to develop research skills, 312-315
  views of reproductive scientists on, 323-324

Trinidad, contraceptive patterns in, 55
Tubal ligation, simplification of procedures for, 278
Tubuli recti, 173
Tubulin, as principal protein of axoneme of sperm tail, 239, 241-242
Turkey, research in, 311, 389

Union of Soviet Socialist Republics, reproductive research in, 303, 395
United Nations
Fund for Population Activities (UNFPA) of, 387
contributions of government agencies to, 401
contributions for WHO Expanded Programme of Research from, 336, 387
purchase of Copper Ts by, 279
recommendations for its support of reproductive research, 28, 31
support of research in developing countries by, 389
support of research in Egypt by, 312
policy change in, due to concern over population growth, 16
Population Conference of (1965), 382
on population research, 386-387
resolution of, "Population Growth and Economic Development," 382
United States
basic research institutions in, 300-301
expenditures for research in reproductive sciences and contraceptive development by, 397-400, 412-413
regulation of fertility control drugs in, 351-352
U.S. Agency for International Development (AID)
contraceptive development programs supported by, 332
governmental funding of, 384
International Fertility Research Program (IFRP) of, 287, 342
levels of funding from, 401-404
Office of Population of
Research Division of, 14, 334, 340-342
Office of the War on Hunger of, 384
Population Information Program of, 342

Program for Applied Research on Fertility Regulation (PARFR), 341
U.S. Department of Agriculture
support of reproductive research by, 373-374
work sponsored by, in animal husbandry, 370
U.S. Department of Health, Education and Welfare, 432
policy of, on protection of human subjects, 356, 358, 359
programs of, in research and delivery of services, 384, 385
reaction of, to Albritton report, 380
U.S. Food and Drug Administration (FDA)
Advisory Committee on Obstetrics and Gynecology of, 386
approval of postcoital estrogens by, 281
examination of IUD infection rates by, 62
expenditures by, 404
new drug applications approved by, 335
regulatory requirements of, 352-354
restrictions on oral contraceptives by, 61
use-patent protection after product approval by, 332, 333
U.S. National Institute of Child Health and Human Development (NICHD)
allocation of $1 million by Congress to (1967), 386
Center for Population Research (CPR) of, 14, 385
awards for research grants and training activities by, 396
Contraceptive Development Branch of, 291, 332, 334, 337-340
contraceptive development funds of, used for studies of the male, 30
Contraceptive Evaluation Branch of, 339
and development of norethindrone enanthate, 329
establishment of, 313, 385
expenditures for reproductive sciences and contraceptive development by, 428-430
levels of funding from, 401-404, 423

U.S. National Institute of Child Health and Human Development (*continued*)
  program of support for Population Research Centers by, 302-303
  testing of long-acting forms of contraceptive steroids by, 284, 286
  testing of new varieties of IUDs by, 287
  establishment of, 380-381
  patent policy of, 346
U.S. National Institutes of Health (NIH)
  appropriations for, 386
  expansion of (1940s), 372
  expenditures for reproductive sciences and contraceptive development by, 428-430
  funds for heart and lung and cancer research from, 23, 428-429
  and inflation, 431
  levels of funding from, 401-404
  and National Science Foundation, 373
  patent policy of, 345-346
  and peer review committees, 358
  policy change in, due to concern over population growth, 16
  Senior Research Fellowships of, 318
  study by, of funds available for research in reproductive sciences and contraceptive development, 393
  survey of research on reproduction by (1961), 380
U.S. Public Health Service, Center for Disease Control of, 62
U.S. Supreme Court, invalidation of state laws restricting abortion by, 58
University budgets, effect of changes in, on funds available for research, 396-397
University of California, 375
University of Chicago, 375
University of Delhi Department of Zoology, 308
University Grants Commission (India), 308
University of Ibadan (Nigeria), 311
University of Lund (Sweden), 375
University of Rochester, 375
Uruguay, research in, 306
Uterotubal junction, 135-136
  need for further studies on, 146

Uterus
  actions of estrogen and progesterone on, 96-99
  exertion of regressive influence on corpus luteum by, 136-137
  period of sensitivity in, contrasted with refractory state, 146
  production of prostaglandins by, 129

Vacuum aspiration technique, for first trimester abortion, 278
Vagina
  as route for administration of contraceptive agents, 147
  testing on suppository for, to induce abortion, 279
Vaginal rings, development and testing of new, 285-286
Vas deferens, 175
  development of methods of reversibly occluding, 291
Vasectomy, efforts to develop methods for reversibility of, 291

Wales, contraceptive patterns in, 54
Worcester Foundation for Experimental Biology, 327, 333, 376, 377
World Bank, recommendations for its support of reproductive research, 28, 31
World Health Organization, 284
  contributions of government agencies to, 401
  on Eastern European reproductive research institutes, 303
  Expanded Programme of Research, Development, and Research Training in Human Reproduction, 14, 283, 307, 332
  contributions from industrialized nations to, 336, 387, 400
  and progesterone-releasing IUDs, 280
  program of, described, 334, 335-337, 387
  on research in developing nations, 304, 389
  research training grants awarded by, 313
  studies by, on effects of contraception on Thai population, 309
  task force on the male of, 30

World Health Organization (*continued*)
  testing of long-acting forms of contraceptive steroids by, 284, 286
  testing of new varieties of IUDs by, 287-288
  Unit in Human Reproduction of, 382
  work of, on male contraception, 291
  expenditures by, 409-412
  Maternal and Child Health Unit of, 382
  patent policy of, 345
  policy change in, due to concern over population growth, 16
  recommendations for additional funding of, 32
  Scientific Group on the Biology of Human Reproduction of, 382
World Population Plan of Action, provisions of, 25

XY body, 195

Yugoslavia
  abortion ratios in, 59
  reproductive research in, 303

Zoapatle, 146. *See also Montanoa tuberosa*
  extracts of, used to suppress progesterone levels, 144, 283-284
Zona pellucida, 235
  enclosing ovulated egg, 232
  penetration of, and fertilization, 243-246
  prevention of sperm penetration of, 247, 248
Zonula occludens, 179
Zygote, biochemical relationship between oviduct function and, 145